The Main Functional Groups *(continued)*

Structure	Class of Compound	Specific Example	Name	Use
C. Containing nitrogen				
$-NH_2$	primary amine	$CH_3CH_2NH_2$	ethylamine	intermediate for dyes, medicinals
$-NHR$	secondary amine	$(CH_3CH_2)_2NH$	diethylamine	pharmaceuticals
$-NR_2$	tertiary amine	$(CH_3)_3N$	trimethylamine	insect attractant
$-C\equiv N$	nitrile	$CH_2=CH-C\equiv N$	acrylonitrile	orlon manufacture
D. Containing oxygen and nitrogen				
$-\overset{+}{N}\overset{O}{\underset{O^-}{}}$	nitro compounds	CH_3NO_2	nitromethane	rocket fuel
$-\overset{O}{\overset{\|}{C}}-NH_2$	primary amide	$\overset{O}{\overset{\|}{HCNH_2}}$	formamide	softener for paper
E. Containing halogen				
$-X$	alkyl or aryl halide	CH_3Cl	methyl chloride	refrigerant, local anesthetic
$-\overset{O}{\overset{\|}{C}}-X$	acid (acyl) halide	$\overset{O}{\overset{\|}{CH_3CCl}}$	acetyl chloride	acetylating agent
F. Containing sulfur				
$-SH$	thiol	CH_3CH_2SH	ethanethiol	odorant to detect gas leaks
$-S-$	thioether	$(CH_2=CHCH_2)_2S$	allyl sulfide	odor of garlic
$-\overset{O}{\underset{O}{\overset{\|}{\underset{\|}{S}}}}-OH$	sulfonic acid	$CH_3-\langle\text{ring}\rangle-SO_3H$	*para*-toluenesul-fonic acid	strong organic acid

ELEVENTH EDITION

Supplementary Text based on

Organic Chemistry
A SHORT COURSE

Harold Hart
Michigan State University

Leslie E. Craine
Central Connecticut State University

David J. Hart
The Ohio State University

Houghton Mifflin Company
Boston New York

Vice President and Publisher: Charles Hartford
Executive Editor: Richard Stratton
Editorial Associate: Marisa Papile
Project Editor: Tracy Williams
Editorial Assistant: Mollie Young
Senior Production/Design Coordinator: Sarah Ambrose
Senior Manufacturing Coordinator: Florence Cadran
Senior Marketing Manager: Katherine Greig

Custom Publishing Editor: Dee Renfrow
Custom Publishing Production Manager: Kathleen McCourt
Project Coordinator: Anisha Sandhu

Cover Design: Joel Gendron

ISBN: 0-618-54274-4
N-03965

3 4 5 6 7 8 9 – CM – 06 05

Houghton Mifflin
Custom Publishing

222 Berkeley Street • Boston, MA 02116

Address all correspondence and order information to the above address.

CONTENTS

3   Alkenes and Alkynes   71

4 Aromatic Compounds   116

8

Ethers and Epoxides   231

9

Aldehydes and Ketones   250

10 Carboxylic Acids and Their Derivatives 283

11   Amines and Related Nitrogen Compounds   321

The following chapters have been deleted at
the Instructor's Request: 6, 12, 13, 14, 15, 18.

PREFACE

Purpose

This year marks the fiftieth anniversary of the publication of this text. Although the content and appearance of the book have changed over time, our purpose in writing it remains constant: to present in a clear and engaging manner a brief introduction to modern organic chemistry.

This book was written for students who, for the most part, will not major in chemistry, but whose main interest—agriculture, biology, human or veterinary medicine, pharmacy, nursing, medical technology, health sciences, engineering, nutrition, forestry, and more—requires some knowledge of organic chemistry. To encourage these students to enjoy the subject as we do, we have made a special effort to illustrate the practical applications of organic chemistry to everyday life and to biological processes. The success of this approach is demonstrated by the widespread use of this text by hundreds of thousands of students in the United States and worldwide, via numerous translations.

The text is designed for a one-semester introductory course, but it is readily adapted to other formats. It is often used in a one- or two-quarter course. In some countries (France and Japan, for example) it serves as an introductory text for chemistry majors, followed by a longer and more detailed full-year text. It has even been used in the United States for a one-year science majors course (with suitable supplementation by the instructor). And in a number of high schools, it is used as the text for a second-year course, following the usual introductory general chemistry.

New in the Eleventh Edition

The entire text was carefully revised to sharpen the writing and clarify difficult sections. In addition to many small changes, users of the previous edition will notice the following more substantial changes: (1) Introductions to a number of chapters have been revised to provide a better connection between the chemistry and our world. (2) Artwork has been revised on a number of occasions to help clarify difficult concepts. (3) Format changes have been adopted that will make it easier for students to distinguish text from examples and problems. (4) *A Closer Look At* sections have been added throughout the text. This new idea presents students with questions and Web activities that will provide more in-depth coverage of selected topics. These sections are denoted by the icon 🖾. (5) The ◉ icon has been introduced to direct students to a student CD or to the Web for animations that should help clarify structural and mechanistic questions that frequently arise.

Six new *A Word About* sections have been added in this edition, and three former sections of this type have been deleted. We hope that students and teachers alike will enjoy the following timely and interesting topics: S_N2 Reactions in Nature: Biological Methylations; Halogenated Organic Compounds from the Sea; Water Treatment and the Chemistry of Enols and Enolates; Green Chemistry and Ibuprofen: A Case Study; Polyacetylene and Conducting Polymers; and The Human Genome.

We are very conscious of the need to keep the book to a manageable size for the one-semester course. Wherever possible, some old material has been deleted to make room for the new material that has been added. Users will find that this edition is nearly identical in length to the previous one.

Organization

The organization is fairly classical, with some exceptions. After an introductory chapter on bonding, isomerism, and an overview of the subject (Chapter 1), the next three chapters treat in sequence saturated, unsaturated, and aromatic hydrocarbons. The concept of reaction mechanism is presented early, and examples are included in virtually all subsequent chapters. Stereoisomerism is also introduced early, briefly in Chapters 2 and 3, and then given separate attention in a full chapter (Chapter 5). Halogen compounds are used in Chapter 6 as a vehicle for introducing aliphatic substitution and elimination mechanisms and dynamic stereochemistry.

Chapters 7 through 10 take up oxygen functionality in order of increasing oxidation state of carbon (alcohols and phenols, ethers, aldehydes and ketones, acids and their derivatives). Brief mention of sulfur analogs is made in these chapters. Chapter 11 deals with amines.

Chapters 2 through 11 treat all of the main functional groups and constitute the heart of the course. Chapter 12 then takes up spectroscopy, with an emphasis on NMR and applications to structure determination. It handles the student's question, How do you know that those molecules really have the structures you say they have?

Next come two chapters on topics not always treated in introductory texts but especially important in practical organic chemistry—Chapter 13 on heterocyclic compounds and Chapter 14 on polymers. The book ends with four chapters on biologically important substances—lipids, carbohydrates, amino acids and proteins, and nucleic acids.

A Word About Essays

Although relevant applications of organic chemistry are stressed throughout the text, short sections under the general rubric *A Word About* emphasize applications to other branches of science and to human life. These sections, which have been a popular feature, appear at appropriate places within the text rather than as isolated essays. Set in a distinctive format, they stand out from the text so that instructors can easily require these sections or not, as desired. There are thirty-eight of these essays, six new in this edition.

A Closer Look At Web-based Activities

Many resources on topics relevant to organic chemistry have become readily available through the World Wide Web. *A Closer Look At* features a guided tour of selected topics in organic chemistry through directed activities based on selected web sites. Instructors may assign these activities as desired, using them as a basis for class discussion or as a springboard for projects. There are eight of these activities in this edition.

Examples and Problems

Problem solving is essential to learning organic chemistry. **Examples** (worked-out problems) appear at appropriate places within each chapter to help students develop these skills. These examples and their solutions are clearly marked. Unsolved **problems** that provide immediate learning reinforcement are included within each chapter and are supplemented with an abundance of end-of-chapter problems. The combined number of examples and problems is 819, or an average of more than 45 per chapter.

Student Ancillaries

Study Guide and Solutions Manual   Written by the text authors, this guide contains chapter summaries and learning objectives, reaction summaries, mechanism summaries, answers to all text problems, and sample test questions. Users can customize this ancillary to include the Study Guide alone, the Solutions Manual alone, a Partial Solutions Manual (solutions to only odd-numbered problems), or any combination thereof.

Laboratory Manual   Written by the text authors, this manual contains 30 experiments that have been time-tested with thousands of students. A substantial number of the preparative experiments contain procedures on both the **macro-** and **microscale,** thus adding considerable flexibility for the instructor and the opportunity for both types of laboratory experience for the student. Hazardous chemicals on the OSHA list have been avoided, care has been taken to minimize contact with solvents, and many caution notes and waste disposal instructions are included. The experiments, capable of being completed in a two- or three-hour lab period, are a good mix of techniques, preparations, tests, and applications.

Student CD-ROM

Featuring material developed by the text authors, this CD contains an interactive package of 3D chemical structures and animations. The structures can be rotated and measurements of bond angles and bond lengths can be obtained. Animations of reaction mechanisms and other concepts can be played forward and backward or frame by frame.

Instructor Ancillaries

Instructor's Resource Manual with Test Bank   This ancillary contains a transition guide, chapter summaries, answers to the Study Guide test questions, and a 730-item multiple-choice test bank.

Computerized Testing   These disks present the Test Bank questions in a computerized testing program by ESA Test. Instructors can produce chapter tests, midterms, and final exams easily and with excellent graphics capability. The instructor can also edit existing questions and add new ones as desired, or preview questions on screen and add them to the test with a single keystroke. The testing program is available in both Microsoft Windows and Macintosh formats.

Instructor's Resource Manual for the Laboratory Manual   Written by the text authors, this manual contains comments on each experiment in the Laboratory Manual as well as answers to the questions.

Transparencies   This package contains a set of 100 full-color transparencies depicting figures and tables from the eleventh edition.

Instructor CD-ROM and PPM   Drawings and tables from the text are included on this CD, along with a simple-to-use classroom presentation program. The disc is specifically designed to allow instructors to facilitate active learning and to enhance multimedia classroom presentations. See your Houghton Mifflin sales representatives for additional information.

Acknowledgments

We would like to thank the following reviewers for diligently contributing their insights to the eleventh edition:

> Edwin Jahngen, University of Massachusetts, Lowell; John Tanaka, University of Connecticut; Brian Yates, University of Tasmania; Mark Warkentin, University of Western Ontario; Richard Laursen, Boston University; Benjamin Gung, Miami University; Stephen Bergmeier, Ohio University; James Barborak, University of North Carolina, Greensboro; Gregory Friestad, University of Vermont; Edward Parish, Auburn State University; Julie Smist, Springfield College; Thomas Hays, Texas A&M University, Kingsville; Robert Evans, Hanover College.

We have incorporated many of their recommendations, and the book is much improved as a consequence.

One pleasure of authorship is receiving letters from students who have benefited from the book, and from their teachers. We thank all who have written to us, from all parts of the world, since the last edition; we have incorporated many of the suggestions in this revision. We will be happy to hear from users and nonusers, faculty and students, with suggestions for further improvements.

HAROLD HART
EMERITUS PROFESSOR OF CHEMISTRY
MICHIGAN STATE UNIVERSITY
EAST LANSING, MI 48824

LESLIE E. CRAINE
DEPARTMENT OF CHEMISTRY
CENTRAL CONNECTICUT STATE UNIVERSITY
1615 STANLEY STREET
NEW BRITAIN, CT 06050

DAVID J. HART
DEPARTMENT OF CHEMISTRY
THE OHIO STATE UNIVERSITY
100 WEST 18TH AVENUE
COLUMBUS, OH 43210

11

AMINES AND RELATED NITROGEN COMPOUNDS

In this chapter we will discuss the last of the major families of simple organic compounds—the amines—relatives of ammonia that abound in nature and play an important role in many modern technologies. Examples of important amines include the painkiller morphine, found in poppy seeds, and putrescine, one of several polyamines responsible for the unpleasant odor of decaying flesh. A diamine that is largely the creation of man is 1,6-diaminohexane, used in the synthesis of nylon. Amine derivatives, known as quaternary ammonium salts, also touch our daily lives in the form of synthetic detergents. Several neurotoxins also belong to this family of compounds. They are toxic because they interfere with the key role that acetylcholine, also a quaternary ammonium salt, plays in the transmission of nerve impulses.

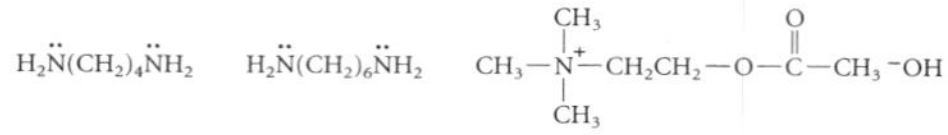

$$H_2\ddot{N}(CH_2)_4\ddot{N}H_2 \qquad H_2\ddot{N}(CH_2)_6\ddot{N}H_2 \qquad CH_3-\overset{\overset{CH_3}{|}}{\underset{\underset{CH_3}{|}}{N}}{}^+-CH_2CH_2-O-\overset{\overset{O}{\parallel}}{C}-CH_3\ {}^-OH$$

putrescine   1,6-diaminohexane   acetylcholine

In this chapter we will first describe the structure, preparation, chemical properties, and uses of some simple amines. Later in the chapter, we will discuss a few natural and synthetic amines with important biological properties.

11.1 Classification and Structure of Amines

The relation between ammonia and amines is illustrated by the following structures:

H—N̈—H   R—N̈—H   R—N̈—R   R—N̈—R
    | | | |
    H   H   H   R

ammonia   primary amine   secondary amine   tertiary amine

▲ The painkiller morphine is obtained from opium, the dried sap of the unripe seed of the poppy *Papaver somniferum.*

321

Each **Chapter Opener Photo** highlights an application of organic chemistry to everyday life and includes a molecular model of the appropriate substance.

The **Chapter Outline** gives students an overview of the topics to come.

Alkanes are **saturated hydrocarbons,** containing only carbon–carbon single bonds. **Cycloalkanes** contain rings. **Unsaturated hydrocarbons** contain carbon–carbon double or triple bonds. **Aromatic hydrocarbons** are cyclic compounds structurally related to benzene.

...formulas of the first ten unbranched alkanes

...of	Molecular formula	Structural formula	Number of structural isomers
	CH_4	CH_4	1
	C_2H_6	CH_3CH_3	1
	C_3H_8	$CH_3CH_2CH_3$	1
	C_4H_{10}	$CH_3CH_2CH_2CH_3$	2
	C_5H_{12}	$CH_3(CH_2)_3CH_3$	3
	C_6H_{14}	$CH_3(CH_2)_4CH_3$	5
	C_7H_{16}	$CH_3(CH_2)_5CH_3$	9
octane	8	C_8H_{18}	$CH_3(CH_2)_6CH_3$	18
nonane	9	C_9H_{20}	$CH_3(CH_2)_7CH_3$	35
decane	10	$C_{10}H_{22}$	$CH_3(CH_2)_8CH_3$	75

All alkanes fit the general molecular formula C_nH_{2n+2}, where n is the number of carbon atoms. Alkanes with carbon chains that are unbranched (Table 2.1) are called **normal alkanes.** Each member of this series differs from the next higher and the next lower member by a —CH_2— group (called a **methylene group**). A series of compounds in which the members are built up in a regular, repetitive way like this is called a **homologous series.** Members of such a series have similar chemical and physical properties, which change gradually as carbon atoms are added to the chain.

Unbranched alkanes are called **normal alkanes,** or *n*-alkanes.

Compounds of a **homologous series** differ by a regular unit of structure and share similar properties.

> **EXAMPLE 2.1**
>
> What is the molecular formula of an alkane with six carbon atoms?
>
> *Solution* If $n = 6$, then $2n + 2 = 14$. The formula is C_6H_{14}.
>
> **PROBLEM 2.1** What is the molecular formula of an alkane with 14 carbon atoms?
>
> **PROBLEM 2.2** Which of the following are alkanes?
>
> a. C_7H_{18} b. C_7H_{16} c. C_8H_{16} d. $C_{27}H_{56}$

2.2 Nomenclature of Organic Compounds

In the early days of organic chemistry, each new compound was given a name that was usually based on its source or use. Examples (Figures 1.12 and 1.13) include limonene (from lemons), α-pinene (from pine trees), coumarin (from the tonka bean, known to South American natives as *cumaru*), and penicillin (from the mold that produces it, *Penicillium notatum*). Even today, this method of naming can be used to give a short, simple name to a molecule with a complex structure. For example, cubane (p. 3) was named after its shape.

It became clear many years ago, however, that one could not rely only on common or trivial names and that a systematic method for naming compounds was needed. Ideally, the rules of the system should result in a unique name for each

Following many concepts within the chapter are **Examples** that walk students through the thought processes involved in problem-solving, carefully outlining all the steps involved. These are followed by **Problems** to reinforce the information just presented.

A Word About boxes show students how chemistry is relevant to their everyday lives. Topics relate to students' future careers in the health and environmental fields. ▶

A WORD ABOUT ...

Isomers, Possible and Impossible

Table 2.1 shows that there are 75 structural isomers of the alkane $C_{10}H_{22}$. How many such isomers do you think there might be if we double the number of carbons ($C_{20}H_{42}$)? The answer is 366,319! And if we double the number of carbons again ($C_{40}H_{82}$)? Exactly 62,481,801,147,341. Of course, no one sits down with pencil and paper or molecular models and determines these numbers by constructing all the possibilities; it could take a lifetime. Complex mathematical formulas have been developed to compute these numbers.

Although we can write some isomers' formulas on paper, they are structurally impossible and cannot be syn-

thesized. Consider, for example, the series of alkanes obtained by replacing the hydrogens of methane with methyl groups and then repeating that process on the product indefinitely. You can see from the drawings below, even though they are only two dimensional, that in this way we build up molecules with a central core of carbon atoms and a surface of hydrogen atoms.

$$CH_4 \longrightarrow C(CH_3)_4 \longrightarrow$$
$$C[C(CH_3)_3]_4 \longrightarrow C[C[C(CH_3)_3]_3]_4$$
$$CH_4 \longrightarrow C_5H_{12} \longrightarrow C_{17}H_{36} \longrightarrow C_{53}H_{108}$$

In three dimensions, the molecules are nearly spherical. Of these compounds, only the first two are known (methane and 2,2-dimethylpropane). The $C_{17}H_{36}$ hydrocarbon (tetra-t-butylmethane or, more accurately, 3,3-di-t-butyl-2,2,4,4-tetramethylpentane) has not yet been synthesized, and if it ever is, it will be an exceptionally strained molecule. The reason is simply that there is not enough room for all the methyl groups on the surface of the molecule.

The first thing to look at in a pair of isomers is their bonding patterns (or atom connectivities). If the bonding patterns are *different*, the compounds are **structural (or constitutional) isomers.** But if the bonding patterns are the *same*, the compounds are stereoisomers. Examples of structural isomers are ethanol and methoxymethane (pages 21–22) or the three isomeric pentanes (page 22). Examples of stereoisomers are the staggered and eclipsed forms of ethane (page 55) or the *cis* and *trans* isomers of 1,2-dimethylcyclopentane (page 59).

If compounds are stereoisomers, we can make a further distinction as to isomer type. If *single-bond rotation* easily interconverts the two stereoisomers (as with staggered and eclipsed ethane), we call them conformers. If the two stereoisomers can be interconverted only by breaking and remaking bonds (as with *cis*- and *trans*-1,2-dimethylcyclopentane), we call them **configurational isomers.**[*]

Configurational isomers (such as *cis–trans* isomers) are stereoisomers that can only be interconverted by breaking and remaking bonds.

PROBLEM 2.17 Classify each of the following isomer pairs according to the scheme in Figure 2.7.

a. 1-iodopropane and 2-iodopropane
b. *cis*- and *trans*-1,2-dimethylcyclohexane
c. chair and boat forms of cyclohexane

[*]Remember that conformers are different conformations of the same molecule, whereas configurational isomers are different molecules. Geometric isomers (*cis–trans* isomers) are one type of configurational isomer. As we will see in Chapter 3, geometric isomerism also occurs in alkenes. Also, we will see other types of configurational isomers in Chapter 5.

PROBLEM 2.6 Name the following compounds by the IUPAC system:

a. CH_3CHFCH_3 b. $(CH_3)_3CCH_2CHClCH_3$

PROBLEM 2.7 Write the structure for 3,3-dimethylpentane.

PROBLEM 2.8 Explain why 1,3-dichlorobutane is a correct IUPAC name, but 1,3-dimethylbutane is *not* a correct IUPAC name.

2.6   Sources of Alkanes

The two most important natural sources of alkanes are **petroleum** and **natural gas.** Petroleum is a complex liquid mixture of organic compounds, many of which are alkanes or cycloalkanes. For more details about how petroleum is refined to obtain gasoline, fuel oil, and other useful substances, read "A Word About Petroleum, Gasoline, and Octane Number" on pages 106–107.

Natural gas, often found associated with petroleum deposits, consists mainly of methane (about 80%) and ethane (5 to 10%), with lesser amounts of some higher alkanes. Propane is the major constituent of liquefied petroleum gas (LPG), a domestic fuel used mainly in rural areas and mobile homes. Butane is the gas of choice in some areas. Natural gas is becoming an energy source that can compete with and possibly surpass oil. In the United States, there are about a million miles of natural gas pipelines distributing this energy source to all parts of the country. Natural gas is also distributed worldwide via huge tankers. To conserve space, the gas is liquefied ($-160°C$), because 1 cubic meter (m^3) of liquefied gas is equivalent to about 600 m^3 of gas at atmospheric pressure. Large tankers can carry more than 100,000 m^3 of liquefied gas.

Petroleum and **natural gas** are the two most important natural sources of alkanes.

Marginal definitions highlight important terms for easy review. ◀

A CLOSER LOOK AT ...

Natural Gas

Log on to http://www.college.hmco.com and follow the links to the student text web site. Use the Web Link section for Chapter 2 to answer the following questions and find more information on the topics below.

U.S. Department of Energy—Fossil Fuels

1. What are the physical properties of natural gas? Why does natural gas used for home heating (or from the Bunsen burner in your chemistry lab) have an odor?
2. How is natural gas produced in nature? What keeps it from escaping to the surface of the earth? How is it obtained from its natural source?
3. How many miles (or kilometers) of pipeline for transporting natural gas currently exist in the United States?

What percent of U.S. energy needs are supplied by natural gas? What are the major geographic locations of natural gas in the United States and Canada?

World Map of Petroleum Basins

1. Where are the greatest potential sources of natural gas located?
2. Using the map, go to your region of the world to explore where potential natural gas resources are located. Report on your findings.
3. In the region you have selected, view the natural gas resources data in tabular form. How many trillion cubic feet of natural gas can potentially be found in your country according to the table you are viewing?

A Closer Look At boxes take text material beyond the pages and allow students to explore concepts using the Internet. ◀

REACTION SUMMARY

1. Reactions of Alkanes and Cycloalkanes

a. Combustion (Sec. 2.12a)

$$C_nH_{2n+2} + \left(\frac{3n+1}{2}\right)O_2 \longrightarrow nCO_2 + (n+1)H_2O$$

b. Halogenation (Sec. 2.12b)

$$R-H + X_2 \xrightarrow[\text{or light}]{\text{heat}} R-X + H-X \quad (X = Cl, Br)$$

MECHANISM SUMMARY

1. Halogenation (Sec. 2.13)

initiation $:\ddot{X}-\ddot{X}: \longrightarrow 2 :\ddot{X}\cdot$

termination $2 :\ddot{X}\cdot \longrightarrow :\ddot{X}-\ddot{X}:$

$2 \ R\cdot \longrightarrow R-R$

$R\cdot + :\ddot{X}\cdot \longrightarrow R-\ddot{X}:$

propagation $R-H + :\ddot{X}\cdot \longrightarrow R\cdot + H-\ddot{X}:$

$R\cdot + :\ddot{X}-\ddot{X}: \longrightarrow R-\ddot{X}: + :\ddot{X}\cdot$

ADDITIONAL PROBLEMS

Alkane Nomenclature and Structural Formulas

2.26 Write structural formulas for the following compounds:

a. 3-methylpentane
b. 2,2-dimethylbutane
c. 4-ethyl-2,2-dimethylhexane
d. 2-bromo-3-methylpentane
e. 1,1-dichlorocyclopropane
f. 2-iodopropane
g. 1,1,3-trimethylcyclohexane
h. 1,1,3,3-tetrachloropropane

2.27 Write expanded formulas for the following compounds, and name them using the IUPAC system:

a. $CH_3(CH_2)_2CH_3$
b. $(CH_3)_2CHCH_2CH_2CH_3$
c. $(CH_3)_3CCH_2CH_2CH_3$
d. $(CH_2)_4$
e. $CH_3CH_2CHFCH_3$
f. $CH_3CCl_2CBr_3$
g. i-PrCl
h. MeBr
i. CH_2ClCH_2Cl
j. $(CH_3CH_2)_2CHCH(CH_3)CH_2CH_3$

2.28 Give both common and IUPAC names for the following compounds:

a. CH_3I
b. CH_3CH_2Br
c. CH_2Cl_2
d. CHI_3
e. $(CH_3)_2CHBr$
f. CH_3CH_2
g. $(CH_3)_3CBr$

2.29 Write a structure for each of the compounds listed. Explain why the name given here is incorrect in each case.

a. 1-methylbutane
b. 2,3-dibromopropane
c. 2-ethylbutane
d. 4-chloro-3-methylbutane
e. 1,3-dimethylcyclopropane
f. 1,1,3-trimethylpentane

3.36 Explain why the following names are incorrect, and give a correct name in each case:

a. 3-pentene
b. 3-butyne
c. 2-methylcyclohexene
d. 2-ethyl-1-propene
e. 3-methyl-1,3-butadiene
f. 1-methyl-2-butene
g. 3-pentyne-1-ene
h. 3-buten-1-yne

3.37

a. What are the usual lengths for the single (sp^3-sp^3), double (sp^2-sp^2), and triple ($sp-sp$) carbon–carbon bonds?

 b. The *single* bond in each of the following compounds has the length shown. Suggest a possible explanation for the observed shortening.

$$CH_2=CH-CH=CH_2 \qquad CH_2=CH-C\equiv CH \qquad HC\equiv C-C\equiv CH$$
$$1.47\text{ Å} \qquad\qquad 1.43\text{ Å} \qquad\qquad 1.37\text{ Å}$$

3.38 Which of the following compounds can exist as *cis–trans* isomers? If such isomerism is possible, draw the structures in a way that clearly illustrates the geometry.

a. 2-pentene
b. 1-hexene
c. 1-chloropropene
d. 3-bromopropene
e. 1,3,5-hexatriene
f. 1,2-dichlorocyclodecene

3.39 The mold metabolite and antibiotic *mycomycin* has the formula

$$HC\equiv C-C\equiv C-CH=C=CH-CH=CH-CH=CH-CH_2-\overset{\overset{\displaystyle O}{\parallel}}{C}-OH$$

Number the carbon chain, starting with the carbonyl carbon.

a. Which multiple bonds are conjugated?
b. Which multiple bonds are cumulated?
c. Which multiple bonds are isolated?

Electrophilic Addition to Alkenes

3.40 Write the structural formula and name of the product when each of the following reacts with 1 mole of bromine:

a. 2-butene
b. vinyl chloride
c. 1,4-cyclohexadiene
d. 1,3-cyclohexadiene
e. 2,3-dimethyl-2-butene

3.41 What reagent will react by addition to what unsaturated hydrocarbon to form each of the following compounds?

a. $CH_3CHClCHClCH_3$
b. $(CH_3)_2CHOSO_3H$
c. $(CH_3)_3COH$
d. 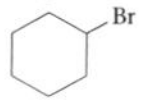 [cyclohexane with Br]
e. $CH_3CH=CHCH_2Br$
f. $CH_3CCl_2CCl_2CH_3$

g. 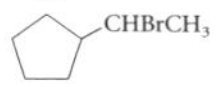 [cyclopentane with $CHBrCH_3$]

3.42 Which of the following reagents are electrophiles? Which are nucleophiles?

a. HCl
b. H_3O^+
c. Br^-
d. $AlCl_3$
e. HO^-

3.43 Water can act as an electrophile or as a nucleophile. Explain.

$\mathbf{c_9}$ = concept connections

Marginal annotations:

A **Reaction Summary** and a **Mechanism Summary** are located at the conclusion of applicable chapters to help students review key concepts of the chapter.

Additional Problems are organized according to the major topics covered in the chapter.

A special **concept connection** icon is used to designate problems that involve the synthesis of several concepts or the use of concepts introduced in a previous chapter.

TO THE STUDENT

I n this introduction we will tell you briefly about organic chemistry and why it is important in a technological society. We will also explain how this course is organized and give you a few hints that may help you to study more effectively.

What Is Organic Chemistry About?

The term *organic* suggests that this branch of chemistry has something to do with *organisms,* or living things. Originally, organic chemistry did deal only with substances obtained from living matter. Years ago, chemists spent much of their time extracting, purifying, and analyzing substances from animals and plants. They were motivated by a natural curiosity about living matter and also by the desire to obtain from nature ingredients for medicines, dyes, and other useful products.

It gradually became clear that most compounds in plants and animals differ in several respects from those that occur in nonliving matter, such as minerals. In particular, most compounds in living matter are made up of the same few elements: **carbon, hydrogen, oxygen, nitrogen,** and sometimes sulfur, phosphorus, and a few others. Carbon is virtually always present. This fact led to our present definition: **Organic chemistry** is the chemistry of carbon compounds. This definition broadens the scope of the subject to include not only compounds from nature but also synthetic compounds—compounds invented by organic chemists and prepared in their laboratories.

▲ Natural and synthetic organic compounds are everywhere in the environment and in our material culture.

1

Organic chemistry is the chemistry of carbon compounds.

Synthetic Organic Compounds

Scientists used to think that compounds that occurred in living matter were different from other substances and that they contained some sort of intangible **vital force** that imbued them with life. This idea discouraged chemists from trying to make organic compounds in the laboratory. But in 1828 the German chemist Friedrich Wöhler, then 28 years old, accidentally prepared **urea,** a well-known constituent of urine, by heating the inorganic (or mineral) substance ammonium cyanate. He was quite excited about this result, and in a letter to his former teacher, the Swedish chemist J. J. Berzelius, he wrote, "I can make urea without the necessity of a kidney, or even of an animal, whether man or dog." This experiment and others like it gradually discredited the vital-force theory and opened the way for modern synthetic organic chemistry.

Synthesis consists of piecing together small simple molecules to make larger, more complex molecules.

    Synthesis usually consists of piecing together small, relatively simple molecules to make larger, more complex ones. To make a molecule that contains many atoms from molecules that contain fewer atoms, one must know how to link atoms to each other—that is, how to make and break chemical bonds. Wöhler's preparation of urea was accidental, but synthesis is much more effective if it is carried out in a controlled and rational way, so that, when all the atoms are assembled, they will be connected to one another in the correct manner to give the desired product.

    Chemical bonds are made or broken during chemical reactions. In this course, you will learn about quite a few reactions that can be used to make new bonds and that are therefore useful in synthesis.

Why Synthesis?

At present, the number of organic compounds that have been synthesized in research laboratories is far greater than the number isolated from nature. Why is it important to know how to synthesize molecules? There are several reasons. For one, it might be important to synthesize a natural product in the laboratory in order to make the substance more widely available at lower cost than it would be if the compound had to be extracted from its natural source. Some examples of compounds first isolated from nature but now produced synthetically for commercial use are vitamins, amino acids, the dye indigo, and the moth-repellent camphor. Although the term *synthetic* is sometimes frowned on as implying something artificial or unnatural, these synthetic natural products are in fact identical to the same compounds extracted from nature.

    Another reason for synthesis is to create new substances that may have new and useful properties. Synthetic fibers such as nylon and Orlon, for example, have properties that make them superior for some uses to natural fibers such as silk, cotton, and hemp. Most drugs used in medicine are synthetic (including aspirin, ether, Novocain, and ibuprofen). The list of synthetic products that we take for granted is long indeed—plastics, detergents, insecticides, and oral contraceptives are just a few. All of these are compounds of carbon; all are organic compounds.

    Finally, organic chemists sometimes synthesize new compounds to test chemical theories—and sometimes they synthesize compounds just for the fun of it. Certain geometric structures, for example, are aesthetically pleasing, and it can be a challenge to make a molecule in which the carbon atoms are arranged in some regular way. One example is the hydrocarbon cubane, C_8H_8. First synthesized in 1964, its molecules have eight carbons at the corners of a cube, each carbon with one hydrogen and three other carbons connected to it. Cubane is more than just aesthetically pleasing. The bond angles in cubane are distorted from normal because of its geometry. Studying the chemistry of cubane therefore gives chemists information about how the distortion of carbon–carbon and carbon–hydrogen bonds affects their chemical behavior. Although initially of only theoretical interest, the special properties of cubane may eventually lead to its practical use in medicine and in explosives.

cubane, C_8H_8
mp 130–131°C
P. E. Eaton (U. of Chicago), 1964

Organic Chemistry in Everyday Life

Organic chemistry touches our daily lives. We are made of and surrounded by organic compounds. Almost all the reactions in living matter involve organic compounds, and it is impossible to understand life, at least from the physical point of view, without knowing some organic chemistry. The major constituents of living matter—proteins, carbohydrates, lipids (fats), nucleic acids (DNA, RNA), cell membranes, enzymes, hormones—are organic, and later in the book, we will describe their chemical structures. These structures are quite complex. To understand them, we will first have to discuss simpler molecules.

Other organic substances include the gasoline, oil, and tires for our cars, the clothing we wear, the wood of our furniture and the paper of our books, the medicines we take, plastic containers, camera film, perfume, carpeting, and fabrics. Name it, and the chances are good that it is organic. Daily, in the paper or on television, we encounter references to polyethylene, epoxys, Styrofoam, nicotine, polyunsaturated fats, and cholesterol. All of these terms refer to organic substances; we will study them and many more in this book.

In short, organic chemistry is more than just a branch of science for the professional chemist or for the student preparing to become a physician, dentist, veterinarian, pharmacist, nurse, or agriculturist. It is part of our technological culture.

Organization

Organic chemistry is a vast subject. Some molecules and reactions are simple; others are quite complex. We will proceed from the simple to the complex by beginning with a chapter on bonding, with special emphasis on bonds to carbon. Next, we have three chapters on organic compounds that contain only two elements, carbon and hydrogen (called hydrocarbons). The second of these chapters (Chapter 3) contains an introduction to organic reaction mechanisms and a discussion of reaction equilibria and rates. These are followed by a chapter that deals with the three dimensionality of organic compounds. Next we add other elements to the carbon and hydrogen framework, halogens in Chapter 6, oxygen and sulfur in Chapters 7 through 10, and nitrogen in Chapter 11. At that point, we will have completed an introduction to all the main classes of organic compounds.

Spectroscopy is a valuable tool for determining organic structures—that is, the details of how atoms and groups are arranged in organic molecules. We take up this topic in Chapter 12. Next comes a chapter on heterocyclic compounds, many of which are important in medicine and in natural products. It is followed by a chapter on polymers, which highlights one of the most important industrial uses of organic chemistry. The last four chapters deal with the organic chemistry of four major classes of biologically important molecules: the lipids, carbohydrates, proteins, and nucleic acids. Because the structures of these molecules of nature are rather complex, we leave them for last. But with the background knowledge of simpler molecules that you will have acquired by then, these compounds and their chemistry will be clearer and more understandable.

To help you organize and review new material, we have placed a *Reaction Summary* and a *Mechanism Summary* at the end of each chapter in which new reactions and new reaction mechanisms are introduced.

A Word About

In each chapter after the first, you will find special sections under the general heading *A Word About*. These are short, self-contained articles that expand on the main subject of the chapter. They may deal with intellectual curiosities (the first one, on impossible organic structures); industrial applications (petroleum, gasoline, and octane number in Chapter 3 or industrial alcohols in Chapter 7); organic chemistry in biology or medicine (polycyclic aromatic hydrocarbons and cancer in Chapter 4 or morphine and other nitrogen-containing drugs in Chapter 13); or just fun topics (sweetness and sweeteners in Chapter 16). They provide a convenient break at various points in each chapter, and we hope that you will enjoy them.

A Closer Look At

Throughout the text, there are a number of sections entitled, *A Closer Look At*. These sections feature a variety of web-based activities on selected topics of organic chemistry. *A Closer Look At* activities provide a wonderful basis for classroom discussion, as well as offer a springboard for potential projects.

The Importance of Problem Solving

One key to success in studying organic chemistry is problem solving. Each chapter in this book contains a large number of facts that must be digested. Also, the subject matter builds continuously, so that to understand each new topic, it is essential to have the preceding information clear in your mind and available for recall. To learn all this material, careful study of the text is necessary, but it is *not sufficient*. Practical knowledge of how to use the facts is required, and such skill can be obtained only through the solving of problems.

This book contains several types of problems. Some, called *Examples,* contain a *Solution,* so that you can see how to work such problems. Throughout a chapter, examples are usually followed by similar *Problems,* designed to reinforce your learning immediately by allowing you to be sure that you understand the new material just presented. At the end of each chapter, *Additional Problems* enable you to practice your problem-solving skills. The problems are grouped by topics. In general, problems that simply test your knowledge come first and more challenging problems follow. Problems that require you to make connections between new concepts and concepts introduced in previous chapters are often marked with a special icon: ↻ .

Try to work as many problems as you can. If you have trouble, seek help from your instructor or from the study guide that accompanies this text. The study guide provides answers to the problems and explains how to solve them. Problem solving is time-consuming, but will pay off in an understanding of the subject.

And now let us begin.

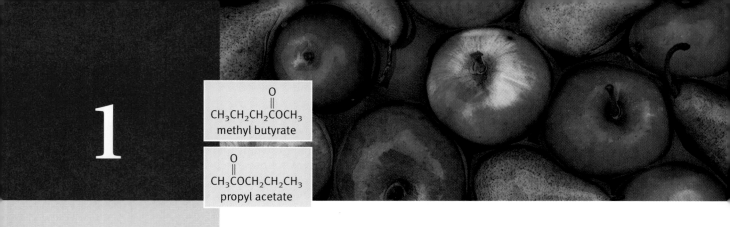

$$\overset{\text{O}}{\underset{\|}{}}$$

CH₃CH₂CH₂COCH₃
methyl butyrate

CH₃COCH₂CH₂CH₃
propyl acetate

1

BONDING AND ISOMERISM

Why does sucrose (table sugar) melt at 185°C, while sodium chloride (table salt) melts at a much higher temperature, 801°C? Why do both of these substances dissolve in water, while olive oil does not? Why does the molecule methyl butyrate smell like apples, while the molecule propyl acetate, which contains the same atoms, smells like pears? To answer questions such as these, you must understand how atoms bond with one another. Bonding is the key to the structure, physical properties, and chemical behavior of different kinds of matter.

Perhaps you have already studied bonding and related concepts in a beginning chemistry course. Browse through each section of this chapter to see whether it is familiar, and try to work the problems. If you can work the problems, you can safely skip that section. But if you have difficulty with any of the problems within or at the end of this chapter, study the entire chapter carefully because we will use the ideas developed here throughout the rest of the book.

1.1 How Electrons Are Arranged in Atoms

Atoms contain a small, dense **nucleus** surrounded by **electrons.** The nucleus is positively charged and contains most of the mass of the atom. The nucleus consists of **protons,** which are positively charged, and **neutrons,** which are neutral. (The only exception is hydrogen, whose nucleus consists of only a single proton.) In a neutral atom the positive charge of the nucleus is exactly balanced by the negative charge of the electrons that surround it. The **atomic number** of an element is equal to the number of protons in the nucleus (and to the number of electrons around the nucleus in a neutral atom). The **atomic weight** is approximately equal to the sum of the number of protons and the number of neutrons in the nucleus; the electrons are not counted because they are very light by comparison. The periodic table on the inside back cover of this book shows all the elements with their atomic numbers and weights.

▲ Methyl butyrate and propyl acetate, organic flavor and fragrance molecules found in apples and pears, respectively, are structural isomers (Sec. 1.8).

Table 1.1	Numbers of orbitals and electrons in the first three shells			
	Number of orbitals of each type			Total number of electrons when shell is filled
Shell number	s	p	d	
1	1	0	0	2
2	1	3	0	8
3	1	3	5	18

An **atom** consists of a small dense **nucleus** containing positively charged **protons** and neutral **neutrons** and surrounded by negatively charged **electrons**. The **atomic number** of an element equals the number of protons in its nucleus; its **atomic weight** is the sum of the number of protons and neutrons in its nucleus.

We are concerned here mainly with the atom's electrons because their number and arrangement provide the key to how a particular atom reacts with other atoms to form molecules. Also, we will deal only with electron arrangements in the lighter elements because these elements are the most important in organic molecules.

Electrons are concentrated in certain regions of space around the nucleus called **orbitals.** Each orbital can contain a maximum of two electrons. The orbitals, which differ in shape, are designated by the letters *s, p,* and *d.* In addition, orbitals are grouped in **shells** designated by the numbers 1, 2, 3, and so on. Each shell contains different types and numbers of orbitals, corresponding to the shell number. For example, shell 1 contains only one type of orbital, designated the 1*s* orbital. Shell 2 contains two types of orbitals, 2*s* and 2*p*, and shell 3 contains three types, 3*s*, 3*p*, and 3*d*. Within a particular shell, the number of *s, p,* and *d* orbitals is 1, 3, and 5, respectively (Table 1.1). These rules permit us to count how many electrons each shell will contain when it is filled (last column in Table 1.1). Table 1.2 shows how the electrons of the first 18 elements are arranged.

Electrons are located in **orbitals.** Orbitals are grouped in **shells.** An orbital can hold a maximum of two electrons.

Table 1.2	Electron arrangements of the first 18 elements					
		Number of electrons in each orbital				
Atomic number	Element	1s	2s	2p	3s	3p
1	H	1				
2	He	2				
3	Li	2	1			
4	Be	2	2			
5	B	2	2	1		
6	C	2	2	2		
7	N	2	2	3		
8	O	2	2	4		
9	F	2	2	5		
10	Ne	2	2	6		
11	Na	2	2	6	1	
12	Mg	2	2	6	2	
13	Al	2	2	6	2	1
14	Si	2	2	6	2	2
15	P	2	2	6	2	3
16	S	2	2	6	2	4
17	Cl	2	2	6	2	5
18	Ar	2	2	6	2	6

Table 1.3	Valence electrons of the first 18 elements						
Group	**I**	**II**	**III**	**IV**	**V**	**VI**	**VII**	**VIII**
	H·							He :
	Li·	Be·	·B·	·Ċ·	·N̈:	·Ö:	:F̈:	:N̈e:
	Na·	Mg·	·Al·	·S̈i·	·P̈:	·S̈:	:C̈l:	:Är:

The first shell is filled for helium (He) and all elements beyond, and the second shell is filled for neon (Ne) and all elements beyond. Filled shells play almost no role in chemical bonding. Rather, the outer electrons, or **valence electrons,** are mainly involved in chemical bonding, and we will focus our attention on them.

Table 1.3 shows the valence electrons, the electrons in the outermost shell, for the first 18 elements. The element's symbol stands for the **kernel** of the element (the nucleus plus the filled electron shells), and the dots represent the valence electrons. The elements are arranged in groups according to the periodic table, and (except for helium) these group numbers correspond to the number of valence electrons.

Armed with this information about atomic structure, we are now ready to tackle the problem of how elements combine to form chemical bonds.

Valence electrons are located in the outermost shell. The **kernel** of the atom contains the nucleus and the inner electrons.

1.2   Ionic and Covalent Bonding

An early, but still useful, theory of chemical bonding was proposed in 1916 by Gilbert Newton Lewis, then a professor at the University of California in Berkeley. Lewis noticed that the **inert gas** helium had only two electrons surrounding its nucleus and that the next inert gas neon had ten such electrons (2 + 8; see Table 1.2). He concluded that atoms of these gases must have very stable electron arrangements *because these elements do not combine with other atoms.* He further suggested that other atoms might react in such a way as to achieve these stable arrangements. This stability could be achieved in one of two ways: by complete transfer of electrons from one atom to another or by sharing of electrons between atoms.

An **inert gas** has a stable electron configuration.

1.2.a Ionic Compounds

Ionic bonds are formed by the transfer of one or more valence electrons from one atom to another. Because electrons are negatively charged, the atom that gives up the electron(s) becomes positively charged, a **cation.** The atom that receives the electron(s) becomes negatively charged, an **anion.** The reaction between sodium and chlorine atoms to form sodium chloride (ordinary table salt) is a typical electron-transfer reaction.*

Ionic compounds are composed of positively charged **cations** and negatively charged **anions.**

---

*The curved arrow in eq. 1.1 shows the movement of one electron from the valence shell of the sodium atom to the valence shell of the chlorine atom. The use of curved arrows to show the movement of electrons is explained in greater detail in Section 1.13.

$$\text{Na} \cdot + \cdot \overset{\cdot\cdot}{\underset{\cdot\cdot}{\text{Cl}}} : \longrightarrow \text{Na}^+ + : \overset{\cdot\cdot}{\underset{\cdot\cdot}{\text{Cl}}} :^- \tag{1.1}$$

sodium   chlorine   sodium   chloride

atom   atom   cation   anion

The sodium atom has only one valence electron (it is in the third shell; see Table 1.2). By giving up that electron it achieves the electron arrangement of neon. At the same time, it becomes positively charged, a sodium cation. The chlorine atom has seven valence electrons. By accepting an additional electron, it achieves the electron arrangement of argon and becomes negatively charged, a chloride anion. Atoms, such as sodium, that tend to give up electrons are said to be **electropositive.** Often such atoms are metals. Atoms, such as chlorine, that tend to accept electrons are said to be **electronegative.** Often such atoms are nonmetals.

Electropositive atoms give up electrons and form cations.
Electronegative atoms accept electrons and form anions.

EXAMPLE 1.1

Write an equation for the reaction of magnesium (Mg) with fluorine atoms (F).

Solution
$$\overset{\cdot\cdot}{\text{Mg}} \cdot + \cdot \overset{\cdot\cdot}{\underset{\cdot\cdot}{\text{F}}} : + \cdot \overset{\cdot\cdot}{\underset{\cdot\cdot}{\text{F}}} : \longrightarrow \text{Mg}^{2+} + 2 : \overset{\cdot\cdot}{\underset{\cdot\cdot}{\text{F}}} :^-$$

Magnesium has two valence electrons. Since each fluorine atom can accept only one electron (from the magnesium) to complete its valence shell, two fluorine atoms are needed to react with one magnesium atom.

PROBLEM 1.1 Write an equation for the reaction of lithium atoms (Li) with bromine atoms (Br).

The product of eq. 1.1 is sodium chloride, an ionic compound made up of equal numbers of sodium and chloride ions. In general, ionic compounds form when strongly electropositive atoms and strongly electronegative atoms interact. The ions in a crystal of an ionic substance are held together by the attractive force between their opposite charges, as shown in Figure 1.1 for a sodium chloride crystal.

In a sense, the ionic bond is not really a bond at all. Being oppositely charged, the ions attract one another like the opposite poles of a magnet. In the crystal, the ions are packed in a definite arrangement, but we cannot say that any particular ion is bonded or connected to any other particular ion. And, of course, when the substance is dissolved, the ions separate and are able to move about in solution relatively freely.

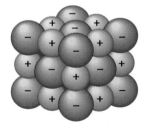

Figure 1.1
Sodium chloride, Na^+Cl^-, is an ionic crystal. The purple spheres represent sodium ions, Na^+, and the green spheres are chloride ions, Cl^-. Each ion is surrounded by six oppositely charged ions, except for those ions that are at the surface of the crystal.

EXAMPLE 1.2

What charge will a beryllium ion carry?

Solution As seen in Table 1.3, beryllium (Be) has two valence electrons. To achieve the filled-shell electron arrangement of helium, it must lose both its valence electrons. Thus, the beryllium cation will carry two positive charges and is represented by Be^{2+}.

PROBLEM 1.2 Using Table 1.3, determine what charge the ion will carry when each of the following elements reacts to form an ionic compound: Al, Li, S, H.

In general, within a given horizontal row in the periodic table, the more electropositive elements are those farthest to the left, and the more electronegative elements are those farthest to the right. Within a given vertical column, the more electropositive elements are those toward the bottom, and the more electronegative elements are those toward the top.

EXAMPLE 1.3

Which atom is more electropositive:

a. lithium or beryllium?   b. lithium or sodium?

Solution
a. The lithium nucleus has less positive charge (+3) to attract electrons than the beryllium nucleus (+4). It takes less energy, therefore, to remove an electron from lithium than it does to remove one from beryllium. Since lithium loses an electron more easily than beryllium, lithium is the more electropositive atom.
b. The valence electron in the sodium atom is shielded from the positive charge of the nucleus by two inner shells of electrons, whereas the valence electron of lithium is shielded by only one inner shell. It takes less energy, therefore, to remove an electron from sodium, so sodium is the more electropositive element.

PROBLEM 1.3   Using Table 1.3, determine which is the more electropositive element: sodium or aluminum, boron or carbon, boron or aluminum.

PROBLEM 1.4   Using Table 1.3, determine which is the more electronegative element: oxygen or fluorine, oxygen or nitrogen, fluorine or chlorine.

PROBLEM 1.5   Judging from its position in Table 1.3, do you expect carbon to be electropositive or electronegative?

1.2.b The Covalent Bond

A **covalent bond** is formed when two atoms share one or more electron pairs. A **molecule** consists of two or more atoms joined by covalent bonds.

Elements that are neither strongly electronegative nor strongly electropositive, or that have similar electronegativities, tend to form bonds by sharing electron pairs instead of completely transferring electrons. A **covalent bond** involves the mutual sharing of one or more electron pairs between atoms. Two (or more) atoms joined by covalent bonds constitute a **molecule.** When the two atoms are identical or have equal electronegativities, the electron pairs are shared equally. The hydrogen molecule is an example.

$$\text{H}\cdot + \text{H}\cdot \longrightarrow \text{H}:\text{H} + \text{heat} \qquad (1.2)$$

$$\underset{\text{hydrogen atoms}}{} \qquad \underset{\text{hydrogen molecule}}{}$$

Each hydrogen atom can be considered to have filled its first electron shell by the sharing process. That is, each atom is considered to "own" all the electrons it shares with the other atom, as shown by the loops in these structures.

$$\widehat{\text{H}:}\text{H} \qquad \text{H}\widehat{:\text{H}}$$

EXAMPLE 1.4

Write an equation similar to eq. 1.2 for the formation of a chlorine molecule from two chlorine atoms.

Solution
$$: \overset{..}{\underset{..}{Cl}} \cdot + \cdot \overset{..}{\underset{..}{Cl}} : \longrightarrow : \overset{..}{\underset{..}{Cl}} : \overset{..}{\underset{..}{Cl}} : + \text{ heat}$$

One electron pair is shared by the two chlorine atoms. In that way, each chlorine completes its valence shell with eight electrons (three unshared pairs and one shared pair).

PROBLEM 1.6   Write an equation similar to eq. 1.2 for the formation of a fluorine molecule from two fluorine atoms.

When two hydrogen atoms combine to form a molecule, heat is liberated. Conversely, this same amount of heat (energy) has to be supplied to a hydrogen molecule to break it apart into atoms. To break apart 1 mole of hydrogen molecules (2 g) into atoms requires 104 kcal (or 435 kJ*) of heat, quite a lot of energy. This energy is called the **bond energy,** or **BE,** and is different for bonds between different atoms. (See Table A in the Appendix.)

> Bond energy (BE) is the energy necessary to break a mole of covalent bonds. The amount of energy depends on the type of bond broken.

The H—H bond is a very strong bond. The main reason for this is that the shared electron pair is attracted to *both* hydrogen nuclei, whereas in a hydrogen atom, the valence electron is associated with only one nucleus. But other forces in the hydrogen molecule tend to counterbalance the attraction between the electron pair and the nuclei. These forces are the repulsion between the two like-charged nuclei and the repulsion between the two like-charged electrons. A balance is struck between the attractive and the repulsive forces. The hydrogen atoms neither fly apart nor fuse together. Instead, they remain connected, or bonded, and vibrate about some equilibrium distance, which we call the **bond length.** For a hydrogen molecule, the bond length (that is, the average distance between the two hydrogen nuclei) is 0.74 Å.** The length of a covalent bond depends on the atoms that are bonded and the number of electron pairs shared between the atoms. Bond lengths for some typical covalent bonds are given in Table B in the Appendix.

> The **bond length** is the average distance between two covalently bonded atoms.

1.3   Carbon and the Covalent Bond

Now let us look at carbon and its bonding. We represent atomic carbon by the symbol $\cdot \overset{\cdot}{C} \cdot$ where the letter C stands for the kernel (the nucleus plus the two 1s electrons) and the dots represent the valence electrons.

---

*Although most organic chemists use the kilocalorie as the unit of heat energy, the currently used international unit is the kilojoule; 1 kcal = 4.184 kJ. In this text, the kilocalorie will be used. If your instructor prefers to use kJ, multiply kcal $\times$ 4.184 (or $\times$ 4 for a rough estimate) to convert to kJ.

**1 Å, or angstrom unit, is 10^{-8} cm, so the H—H bond length is 0.74×10^{-8} cm. Although the angstrom is commonly used by organic chemists, another unit often used for bond lengths is the picometer (pm; 1 Å = 100 pm). To convert the H—H bond length from Å to pm, multiply 0.74×100. The H—H bond length is 74 pm. In this text, the angstrom will be used as the unit for bond lengths.

With four valence electrons, the valence shell of carbon is half filled (or half empty). Carbon atoms have neither a strong tendency to lose all their electrons (and become C^{4+}) nor a strong tendency to gain four electrons (and become C^{4-}). Being in the middle of the periodic table, *carbon is neither strongly electropositive nor strongly electronegative.* Instead, it usually forms covalent bonds with other atoms by sharing electrons. For example, carbon combines with four hydrogen atoms (each of which supplies one valence electron) by sharing four electron pairs.* The substance formed is known as methane. Carbon can also share electron pairs with four chlorine atoms, forming tetrachloromethane.**

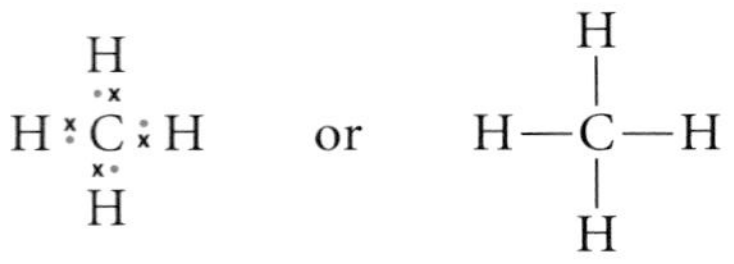

methane

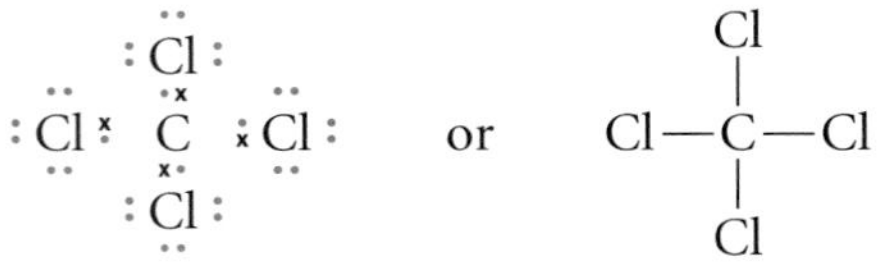

tetrachloromethane
(carbon tetrachloride)

By sharing electron pairs, the atoms complete their valence shells. In both examples, carbon has eight valence electrons around it. In methane, each hydrogen atom completes its valence shell with two electrons, and in tetrachloromethane each chlorine atom fills its valence shell with eight electrons. In this way, all valence shells are filled and the compounds are quite stable.

The shared electron pair is called a covalent bond because it bonds or links the atoms (by its attraction to both nuclei). The single bond is usually represented by a dash, or single line, as shown in the structures above for methane and tetrachloromethane.

EXAMPLE 1.5

Draw the structure for chloromethane (also called methyl chloride), CH_3Cl.

Solution

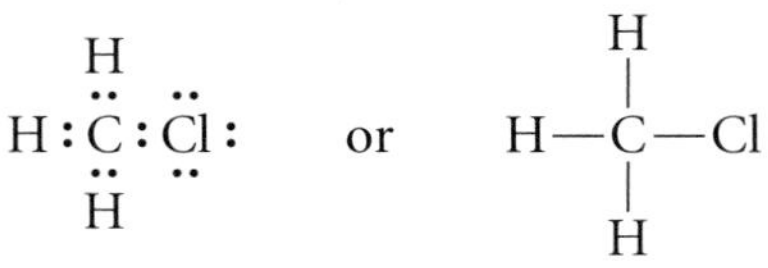

PROBLEM 1.7   Draw the structures for dichloromethane (also called methylene chloride), CH_2Cl_2, and trichloromethane (chloroform), $CHCl_3$.

---

*To designate electrons from different atoms, the symbols · and x are often used. But the electrons are, of course, identical.

**Tetrachloromethane is the systematic name, but carbon tetrachloride is the common name. We discuss how to name organic compounds later.

1.4 Carbon–Carbon Single Bonds

The unique property of carbon atoms—that is, the property that makes it possible for millions of organic compounds to exist—is their ability to share electrons not only with different elements but also with other carbon atoms. For example, two carbon atoms may be bonded to one another, and each of these carbon atoms may be linked to other atoms. In ethane and hexachloroethane, each carbon is connected to the other carbon *and* to three hydrogen atoms or three chlorine atoms. Although they have two carbon atoms instead of one, these compounds have chemical properties similar to those of methane and tetrachloromethane, respectively.

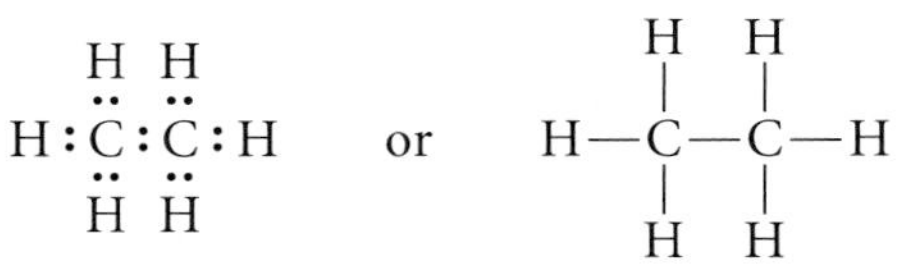

ethane 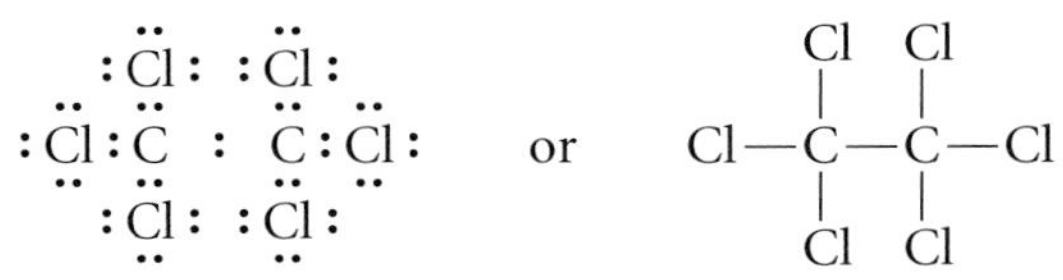   hexachloroethane

The carbon–carbon bond in ethane, like the hydrogen–hydrogen bond in a hydrogen molecule, is a purely covalent bond, with the electrons shared *equally* between the two identical carbon atoms. As with the hydrogen molecule, heat is required to break the carbon–carbon bond of ethane to give two CH_3 fragments (called methyl radicals). A **radical** is a molecular fragment with an odd number of unshared electrons.

A **radical** is a molecular fragment with an odd number of unshared electrons.

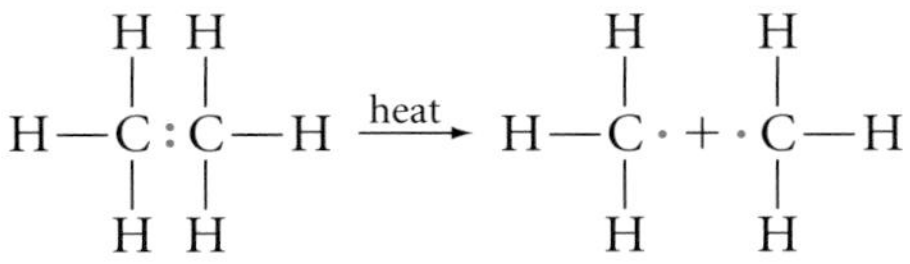

ethane   two methyl radicals   **(1.3)**

However, less heat is required to break the carbon–carbon bond in ethane than is required to break the hydrogen–hydrogen bond in a hydrogen molecule. The actual amount is 88 kcal (or 368 kJ) per mole of ethane. The carbon–carbon bond in ethane is longer (1.54 Å) than the hydrogen–hydrogen bond (0.74 Å) and also somewhat weaker. Breaking carbon–carbon bonds by heat, as represented in eq. 1.3, is the first step in the *cracking* of petroleum, an important process in the manufacture of gasoline (see "A Word About Petroleum, Gasoline, and Octane Number" on pages 106–107).

EXAMPLE 1.6

What do you expect the length of a C—H bond (as in methane or ethane) to be?

Solution It should measure somewhere between the H—H bond length in a hydrogen molecule (0.74 Å) and the C—C bond length in ethane (1.54 Å). The actual value is about 1.09 Å, close to the average of the H—H and C—C bond lengths.

PROBLEM 1.8 The Cl—Cl bond length is 1.98 Å. Which bond will be longer, the C—C bond in ethane or the C—Cl bond in chloromethane?

Catenation is the ability of an element to form chains of its own atoms through covalent bonding.

There is almost no limit to the number of carbon atoms that can be linked, and some molecules contain as many as 100 or more carbon–carbon bonds. This ability of an element to form chains as a result of bonding between the same atoms is called **catenation.**

PROBLEM 1.9   Using the structure of ethane as a guide, draw the structure for propane, C_3H_8.

1.5   Polar Covalent Bonds

A **polar covalent bond** is a covalent bond in which the electron pair is not shared equally between two atoms.

As we have seen, covalent bonds can be formed not only between identical atoms (H—H, C—C) but also between different atoms (C—H, C—Cl), provided that the atoms do not differ too greatly in electronegativity. However, if the atoms are different from one another, the electron pair may not be shared equally between them. Such a bond is sometimes called a **polar covalent bond** because the atoms that are linked carry a partial negative and a partial positive charge.

The hydrogen chloride molecule provides an example of a polar covalent bond. Chlorine atoms are more electronegative than hydrogen atoms, but even so, the bond that they form is covalent rather than ionic. However, the shared electron pair is attracted more toward the chlorine, which therefore is slightly negative with respect to the hydrogen. This bond polarization is indicated by an arrow whose head is negative and whose tail is marked with a plus sign. Alternatively, a partial charge, written as $\delta+$ or $\delta-$ (read as "delta plus" or "delta minus"), may be shown:

$$\text{H} :\!\overset{..}{\underset{..}{\text{Cl}}}: \quad \text{or} \quad \overset{\delta+}{\text{H}} \overset{\delta-}{:\!\overset{..}{\underset{..}{\text{Cl}}}:} \quad \text{or} \quad \overset{\delta+}{\text{H}}\!-\!\overset{\delta-}{\overset{..}{\underset{..}{\text{Cl}}}:}$$

The bonding electron pair, which is shared *unequally,* is displaced toward the chlorine.

You can usually rely on the periodic table to determine which end of a polar covalent bond is more negative and which end is more positive. As we proceed from left to right across the table within a given period, the elements become *more* electronegative, owing to increasing atomic number or charge on the nucleus. The increasing nuclear charge attracts valence electrons more strongly. As we proceed from the top to the bottom of the table within a given group (down a column), the elements become *less* electronegative because the valence electrons are shielded from the nucleus by an increasing number of inner-shell electrons. From these generalizations, we can safely predict that the atom on the right in each of the following bonds will be negative with respect to the atom on the left:

$$\begin{array}{cccc} \text{C}\!-\!\text{N} & \text{C}\!-\!\text{Cl} & \text{H}\!-\!\text{O} & \text{Br}\!-\!\text{Cl} \\ \text{C}\!-\!\text{O} & \text{C}\!-\!\text{Br} & \text{H}\!-\!\text{S} & \text{Si}\!-\!\text{C} \end{array}$$

The carbon–hydrogen bond, which is so common in organic compounds, requires special mention. Carbon and hydrogen have nearly identical electronegativities, so the C—H bond is almost purely covalent. The electronegativities of some common elements are listed in Table 1.4.

Table 1.4	Electronegativities of some common elements					

Group

I	II	III	IV	V	VI	VII
H 2.2						
Li 1.0	Be 1.6	B 2.0	C 2.5	N 3.0	O 3.4	F 4.0
Na 0.9	Mg 1.3	Al 1.6	Si 1.9	P 2.2	S 2.6	Cl 3.2
K 0.8	Ca 1.0					Br 3.0
						I 2.7

	< 1.0		1.5–1.9		2.5–2.9
	1.0–1.4		2.0–2.4		3.0–4.0

EXAMPLE 1.7

Indicate any bond polarization in the structure of tetrachloromethane.

Solution

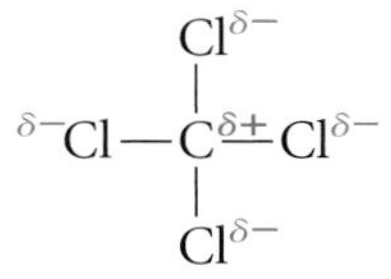

$$Cl^{\delta-}$$
$$|$$
$$^{\delta-}Cl-\overset{}{\underset{|}{C}}^{\delta\pm}Cl^{\delta-}$$
$$Cl^{\delta-}$$

Chlorine is more electronegative than carbon. The electrons in each C—Cl bond are therefore displaced toward the chlorine.

PROBLEM 1.10 Predict the polarity of the P—Cl bond and of the S—O bond.

PROBLEM 1.11 Draw the structure of the refrigerant dichlorodifluoromethane, CCl_2F_2 (CFC-12), and indicate the polarity of the bonds.

PROBLEM 1.12 Draw the formula for methanol, CH_3OH, and (where appropriate) indicate bond polarity with an arrow, $\longmapsto$.

1.6 Multiple Covalent Bonds

To complete their valence shells, atoms may sometimes share more than one electron pair. Carbon dioxide, CO_2, is an example. The carbon atom has four valence electrons, and each oxygen has six valence electrons. A structure that allows each atom to complete its valence shell with eight electrons is

$$\overset{+}{\underset{+}{O}} \overset{x}{\underset{x}{\vdots}} C \overset{x}{\underset{x}{\vdots}} \overset{+}{\underset{+}{O}} \quad \text{or} \quad \overset{xx}{\underset{xx}{O}} = C = \overset{xx}{\underset{xx}{O}} \quad \text{or} \quad O = C = O$$

A   B   C

In structure A, the dots represent the electrons from carbon, and the x's are the electrons from the oxygens. Structure B shows the bonds and oxygen's unshared electrons, and structure C shows only the covalent bonds. Two electron pairs are shared between carbon and oxygen. Consequently, the bond is called a **double bond.** Each oxygen atom also has two pairs of **nonbonding electrons,** or **unshared electron pairs.** The loops in the following structures show that each atom in carbon dioxide has a complete valence shell of eight electrons:

In a **double bond,** two electron pairs are shared between two atoms.

Nonbonding electrons, or **unshared electron pairs,** reside on one atom.

In a **triple bond,** three electron pairs are shared between two atoms.

Hydrogen cyanide, HCN, is an example of a simple compound with a **triple bond,** a bond in which three electron pairs are shared.

$$H \overset{x}{{}} C \overset{x}{\underset{x}{\vdots}} \overset{x}{N} \overset{x}{{}} \quad \text{or} \quad H - C \equiv N \overset{x}{{}} \quad \text{or} \quad H - C \equiv N$$

hydrogen cyanide

PROBLEM 1.13 Show with loops how each atom in hydrogen cyanide completes its valence shell.

EXAMPLE 1.8

Determine what, if anything, is wrong with the following electron arrangement for carbon dioxide:

$$: O ::: C :: \overset{..}{O} :$$

Solution The formula contains the correct total number of valence electrons (16), and each oxygen is surrounded by 8 valence electrons, which is correct. What is wrong is that the carbon atom has 10 valence electrons, 2 more than is allowable.

PROBLEM 1.14 Show what is wrong with each of the following electron arrangements for carbon dioxide:

a. $: O ::: C ::: O :$ b. $: \overset{..}{\underset{..}{O}} : \overset{..}{C} : \overset{..}{\underset{..}{O}} :$ c. $: \overset{..}{O} : C ::: O :$

PROBLEM 1.15 Methanal (formaldehyde) has the formula H_2CO. Draw a structure that shows how the valence electrons are arranged.

PROBLEM 1.16 Draw an electron-dot structure for carbon monoxide, CO.

Hydrocarbons are compounds composed of just hydrogen and carbon atoms.

Carbon atoms can be connected to one another by double bonds or triple bonds, as well as by single bonds. Thus there are three **hydrocarbons** (compounds with just carbon and hydrogen atoms) that have two carbon atoms per molecule: ethane, ethene, and ethyne.

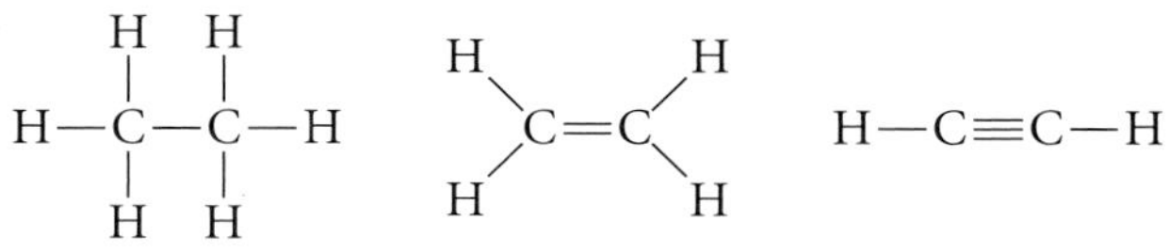

$$
\begin{array}{ccc}
\text{H} \quad \text{H} & \text{H} \qquad\qquad \text{H} & \\
\text{H}-\overset{\displaystyle|}{\underset{\displaystyle|}{\text{C}}}-\overset{\displaystyle|}{\underset{\displaystyle|}{\text{C}}}-\text{H} & \text{C}=\text{C} & \text{H}-\text{C}\equiv\text{C}-\text{H} \\
\text{H} \quad \text{H} & \text{H} \qquad\qquad \text{H} & \\
\text{ethane} & \text{ethene} & \text{ethyne} \\
 & \text{(ethylene)} & \text{(acetylene)}
\end{array}
$$

They differ in that the carbon–carbon bond is single, double, or triple, respectively. They also differ in number of hydrogens. As we will see later, these compounds have different chemical reactivities because of the different types of bonds between the carbon atoms.

EXAMPLE 1.9

Draw the structure for C_3H_6 having one carbon–carbon double bond.

Solution   First, draw the three carbons with one double bond.

$$C=C-C$$

Then add the hydrogens in such a way that each carbon has eight electrons around it (or in such a way that each carbon has four bonds).

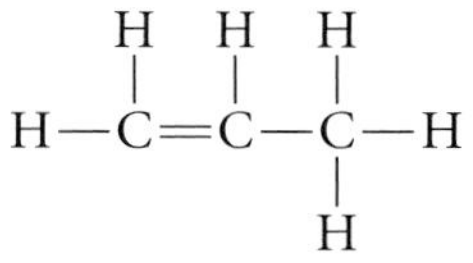

$$
\begin{array}{ccc}
\text{H} & \text{H} & \text{H} \\
| & | & | \\
\text{H}-\text{C}=\text{C}-\text{C}-\text{H} \\
 & & | \\
 & & \text{H}
\end{array}
$$

PROBLEM 1.17   Draw three different structures that have the formula C_4H_8 and have one carbon–carbon double bond.

1.7 Valence

The **valence** of an element is simply the number of bonds that an atom of the element can form. The number is usually equal to the *number of electrons needed to fill the valence shell*. Table 1.5 gives the common valences of several elements. Notice the difference between the number of valence electrons and the valence. Oxygen, for example, has six valence electrons but a valence of only 2. The *sum* of the two numbers is equal to the number of electrons in the filled shell.

The valences in Table 1.5 apply whether the bonds are single, double, or triple. For example, carbon has four bonds in each of the structures we have written so far:

The **valence** of an element is the number of bonds that an atom of the element can form.

Table 1.5	Valences of common elements					
Element	H·	·Ç·	·N̈:	·Ö:	:F̈:	:C̈l:
Valence	1	4	3	2	1	1

methane, tetrachloromethane, ethane, ethene, ethyne, carbon dioxide, and so on. These common valences are worth remembering, because they will help you to write correct structures.

EXAMPLE 1.10

Using dashes for bonds, draw a structure for C_3H_4 that has the proper valence of 1 for each hydrogen and 4 for each carbon.

Solution   There are three possibilities:

$$
\begin{array}{ccc}
\mathrm{H} & & \\
| & & \\
\mathrm{H-C-C}\equiv\mathrm{C-H} & & \\
| & & \\
\mathrm{H} & &
\end{array}
$$

A compound that corresponds to each of these three different arrangements of the atoms is known.

PROBLEM 1.18   Use dashes for bonds and use the valences given in Table 1.5 to write a structure for each of the following:

a. CH_5N b. CH_4O

PROBLEM 1.19   Does C_2H_5 represent a stable molecule?

In Example 1.10, we saw that three carbon atoms and four hydrogen atoms can be connected to one another in three different ways, each of which satisfies the valences of both kinds of atoms. Let us take a closer look at this phenomenon.

1.8 Isomerism

The **molecular formula** of a substance gives the number of different atoms present; the **structural formula** indicates how those atoms are arranged.

The **molecular formula** of a substance tells us the numbers of different atoms present, but a **structural formula** tells us how those atoms are arranged. For example, H_2O is the molecular formula for water. It tells us that each water molecule contains two hydrogen atoms and one oxygen atom. But the structural formula H—O—H tells us more than that. It tells us that the hydrogens are connected to the oxygen (and not to each other).

It is sometimes possible to arrange the same atoms in more than one way and still satisfy their valences. Molecules that have the same kinds and numbers of atoms but different arrangements are called **isomers,** a term that comes from the Greek (*isos,* equal, and *meros,* part). **Structural (or constitutional) isomers** are compounds that have the same molecular formula but different structural formulas. Let us look at a particular pair of isomers.

Isomers are molecules with the same number and kinds of atoms but different arrangements of the atoms. **Structural (or constitutional) isomers** have the same molecular formula but different structural formulas.

Two very different chemical substances are known, each with the molecular formula C_2H_6O. One of these substances is a colorless liquid that boils at 78.5°C, whereas the other is a colorless gas at ordinary temperatures (bp −23.6°C). The only possible explanation is that the atoms must be arranged differently in the molecules of each substance and that these arrangements are somehow responsible for the fact that one substance is a liquid and the other, a gas.

For the molecular formula C_2H_6O, two (and only two) structural formulas are possible that satisfy the valence requirement of 4 for carbon, 2 for oxygen, and 1 for hydrogen. They are

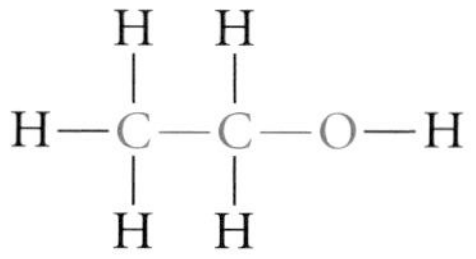

ethanol
(ethyl alcohol)
bp 78.5°C

and

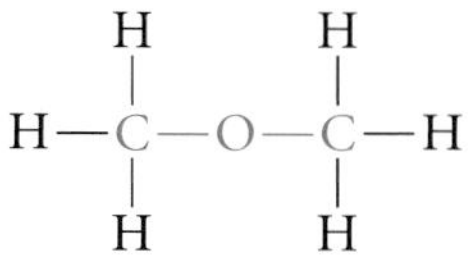

methoxymethane
(dimethyl ether)
bp −23.6°C

In one formula, the two carbons are connected to one another by a single covalent bond; in the other formula, each carbon is connected to the oxygen. When we complete the valences by adding hydrogens, each arrangement requires six hydrogens. Many kinds of experimental evidence verify these structural assignments. We leave for later (Chapters 7 and 8) an explanation of why these arrangements of atoms produce substances that are so different from one another.

Ethanol and methoxymethane are structural isomers. They have the same molecular formula but different structural formulas. Ethanol and methoxymethane differ in physical and chemical properties as a consequence of their different molecular structures. In general, structural isomers are different compounds. They differ in physical and chemical properties as a consequence of their different molecular structures.

PROBLEM 1.20 Draw structural formulas for the three possible isomers of C_3H_8O.

1.9 Writing Structural Formulas

You will be writing structural formulas throughout this course. Perhaps a few hints about how to do so will be helpful. Let's look at another case of isomerism. Suppose we want to write out all possible structural formulas that correspond to the molecular formula C_5H_{12}. We begin by writing all five carbons in a **continuous chain.**

C—C—C—C—C
a continuous chain

In a **continuous chain,** atoms are bonded one after another; in a **branched chain** some atoms form branches from the longest continuous chain.

This chain uses up one valence for each of the end carbons and two valences for the carbons in the middle of the chain. Each end carbon therefore has three valences left for bonds to hydrogens. Each middle carbon has only two valences for bonds to hydrogens. As a consequence, the structural formula in this case is written

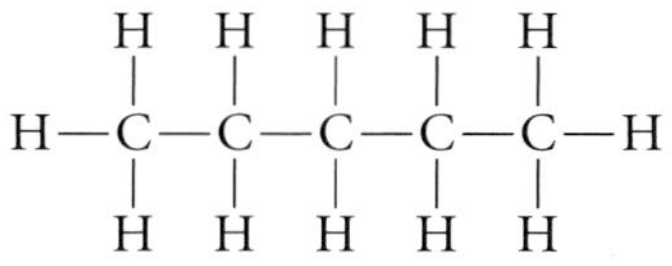

pentane, bp 36°C

To find structural formulas for the other isomers, we must consider **branched chains.** For example, we can reduce the longest chain to only four carbons and connect the fifth carbon to one of the middle carbons, as in the following structural formula:

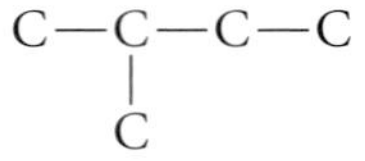

$$C—C—C—C$$
$$|$$
$$C$$

a branched chain

If we add the remaining bonds so that each carbon has a valence of 4, we see that three of the carbons have three hydrogens attached, but the other carbons have only one or two hydrogens. The molecular formula, however, is still C_5H_{12}.

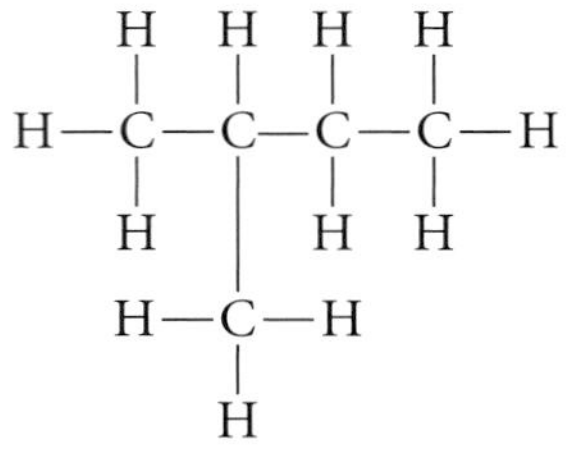

$$
\begin{array}{cccc}
H & H & H & H \\
| & | & | & | \\
H—C—C—C—C—H \\
| & | & | & | \\
H & & H & H \\
\end{array}
$$
$$H—C—H$$
$$|$$
$$H$$

2-methylbutane, bp 28°C
(isopentane)

Suppose we keep the chain of four carbons and try to connect the fifth carbon somewhere else. Consider the following chains:

Ch. 1, Refs. 1–3. *View and manipulate 3D images of pentane, isopentane, and neopentane.*

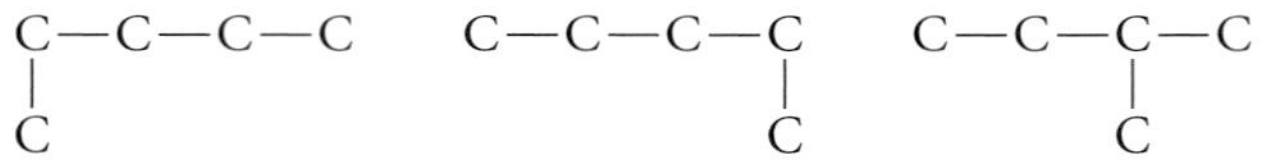

$$
\begin{array}{ccc}
C—C—C—C & C—C—C—C & C—C—C—C \\
| & | & | \\
C & C & C \\
\end{array}
$$

Do we have anything new here? *No!* The first two structures have five-carbon chains, exactly as in the formula for pentane, and the third structure is identical to the branched chain we have already drawn for 2-methylbutane—a four-carbon chain with a one-carbon branch attached to the second carbon in the chain (counting now from the right instead of from the left). Notice that for every drawing of pentane you can draw a line through all five carbon atoms without lifting your pencil from the paper. For every drawing of 2-methylbutane, a continuous line can be drawn through exactly four carbon atoms.*

But there is a third isomer of C_5H_{12}. We can find it by reducing the longest chain to only three carbons and connecting two one-carbon branches to the middle carbon.

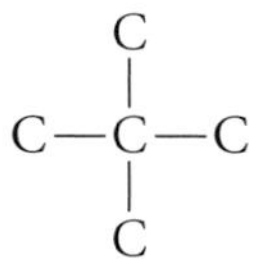

$$C$$
$$|$$
$$C—C—C$$
$$|$$
$$C$$

---

*Using a molecular model kit (see note on p. 41) to construct the carbon chains as drawn will help you to see which representations are identical and which are different.

If we fill in the hydrogens, we see that the middle carbon has no hydrogens attached to it.

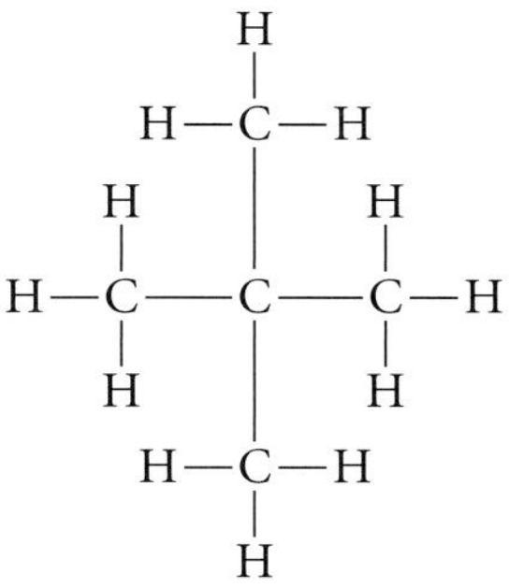

$$
\begin{array}{c}
\text{H} \\
| \\
\text{H}-\text{C}-\text{H} \\
\text{H} \quad | \quad \text{H} \\
| \quad\quad | \quad\quad | \\
\text{H}-\text{C}-\!\!-\!\!-\text{C}-\!\!-\!\!-\text{C}-\text{H} \\
| \quad\quad | \quad\quad | \\
\text{H} \quad | \quad \text{H} \\
\text{H}-\text{C}-\text{H} \\
| \\
\text{H}
\end{array}
$$

2,2-dimethylpropane, bp 10°C
(neopentane)

So we can draw three (and only three) different structural formulas that correspond to the molecular formula C_5H_{12}, and in fact we find that only three different chemical substances with this formula exist. They are commonly called *n*-pentane (*n* for normal, with an unbranched carbon chain), isopentane, and neopentane.

PROBLEM 1.21   To which isomer of C_5H_{12} does each of the following structural formulas correspond?

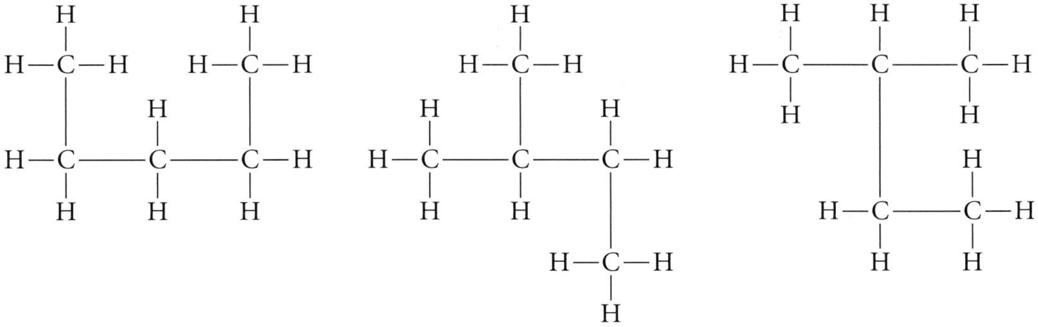

1.10   Abbreviated Structural Formulas

Structural formulas like the ones we have written so far are useful, but they are also somewhat cumbersome. They take up a lot of space and are tiresome to write out. Consequently, we often take some shortcuts that still convey the meaning of structural formulas. For example, we may abbreviate the structural formula of ethanol (ethyl alcohol) from

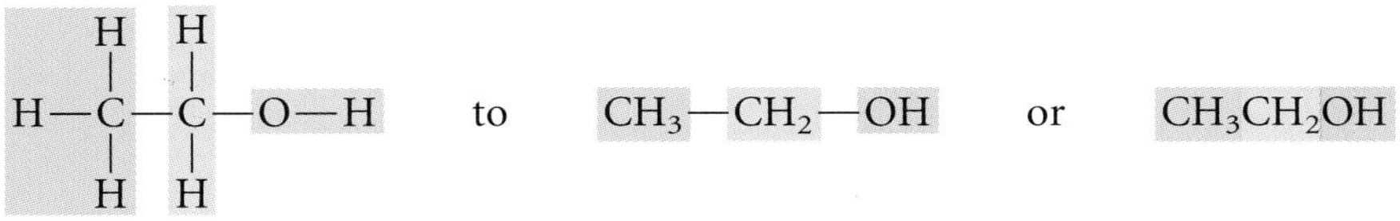

$$
\begin{array}{c}
\text{H} \quad \text{H} \\
| \quad\quad | \\
\text{H}-\text{C}-\text{C}-\text{O}-\text{H} \\
| \quad\quad | \\
\text{H} \quad \text{H}
\end{array}
\qquad \text{to} \qquad
\text{CH}_3-\text{CH}_2-\text{OH}
\qquad \text{or} \qquad
\text{CH}_3\text{CH}_2\text{OH}
$$

Each formula clearly represents ethanol rather than its isomer methoxymethane (dimethyl ether), which can be represented by any of the following structures:

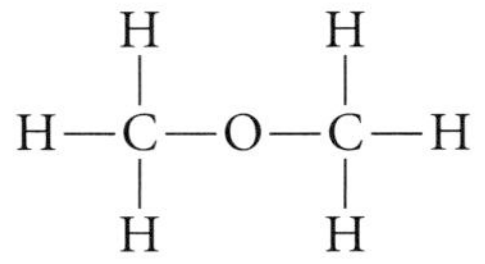

$$
\begin{array}{ccc}
\text{H} & & \text{H} \\
| & & | \\
\text{H—C—O—C—H} & & \\
| & & | \\
\text{H} & & \text{H}
\end{array}
\quad \text{to} \quad \text{CH}_3\text{—O—CH}_3 \quad \text{or} \quad \text{CH}_3\text{OCH}_3
$$

The structural formulas for the three pentanes can be abbreviated in a similar fashion.

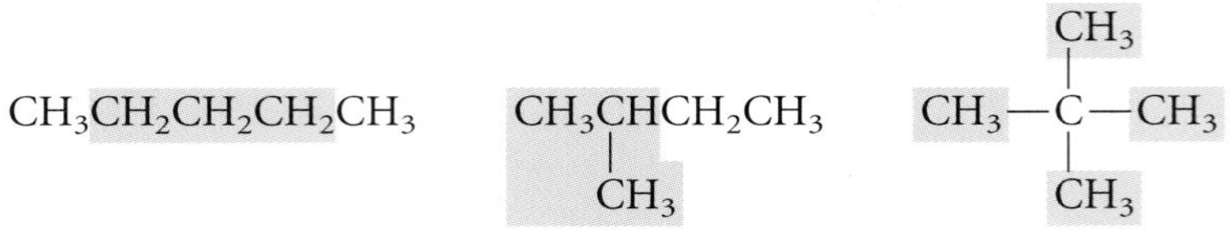

$$
\text{CH}_3\text{CH}_2\text{CH}_2\text{CH}_2\text{CH}_3 \qquad
\begin{array}{c}
\text{CH}_3\text{CHCH}_2\text{CH}_3 \\
| \\
\text{CH}_3
\end{array}
\qquad
\begin{array}{c}
\text{CH}_3 \\
| \\
\text{CH}_3\text{—C—CH}_3 \\
| \\
\text{CH}_3
\end{array}
$$

n-pentane \qquad isopentane \qquad neopentane

Sometimes these formulas are abbreviated even further. For example, they can be printed on a single line in the following ways:

$$
\text{CH}_3(\text{CH}_2)_3\text{CH}_3 \qquad (\text{CH}_3)_2\text{CHCH}_2\text{CH}_3 \qquad (\text{CH}_3)_4\text{C}
$$

n-pentane \qquad isopentane \qquad neopentane

EXAMPLE 1.11

Write a structural formula that shows all bonds for each of the following:

a. $\text{CH}_3\text{CCl}_2\text{CH}_3$ b. $(\text{CH}_3)_2\text{C}(\text{CH}_2\text{CH}_3)_2$

Solution

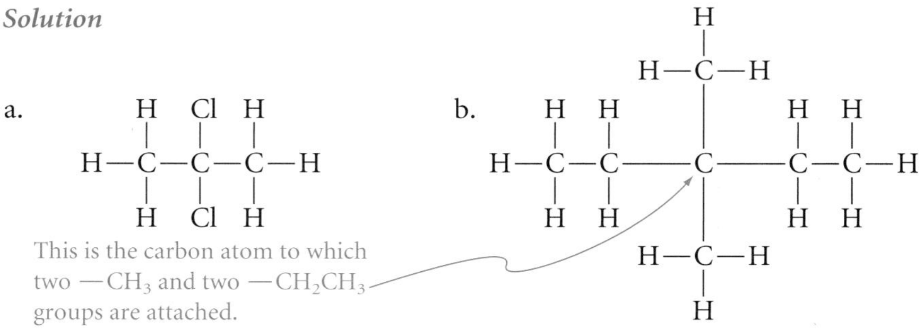

a.
$$
\begin{array}{ccc}
\text{H} & \text{Cl} & \text{H} \\
| & | & | \\
\text{H—C—C—C—H} \\
| & | & | \\
\text{H} & \text{Cl} & \text{H}
\end{array}
$$

This is the carbon atom to which two —CH$_3$ and two —CH$_2$CH$_3$ groups are attached.

b.
$$
\begin{array}{c}
\text{H} \\
| \\
\text{H—C—H} \\
\\
\text{H H} \quad | \quad \text{H H} \\
|~~| \quad\quad |~~| \\
\text{H—C—C——C——C—C—H} \\
|~~| \quad | \quad |~~| \\
\text{H H} \quad\quad \text{H H} \\
\\
\text{H—C—H} \\
| \\
\text{H}
\end{array}
$$

PROBLEM 1.22 Write a structural formula that shows all bonds for each of the following:

a. $(\text{CH}_3)_2\text{CHCH}_2\text{OH}$ b. $\text{Cl}_2\text{C}=\text{CCl}_2$

Perhaps the ultimate abbreviation of structures is the use of lines to represent the carbon framework:

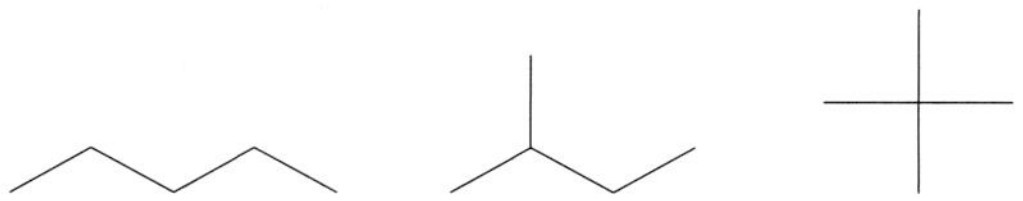

n-pentane \qquad isopentane \qquad neopentane

In these formulas, *each line segment is understood to have a carbon atom at each end.* The hydrogens are omitted, but we can quickly find the number of hydrogens on each carbon by subtracting from four (the valence of carbon) the number of line segments that emanates from any point. Multiple bonds are represented by multiple line segments. For example, the hydrocarbon with a chain of five carbon atoms and a double bond between the second and third carbon atoms (that is, $CH_3CH = CHCH_2CH_3$) is represented as follows:

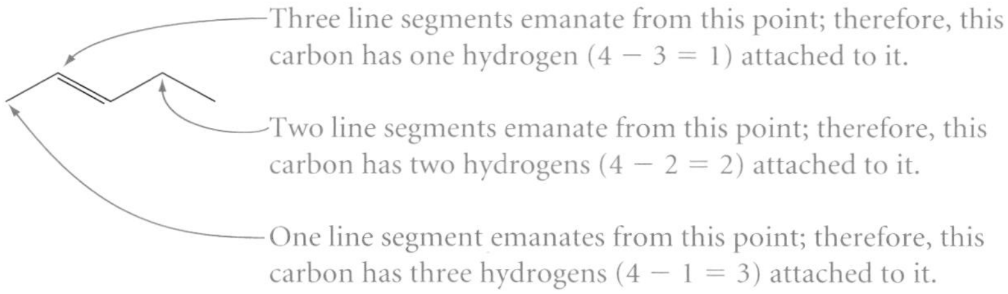

Three line segments emanate from this point; therefore, this carbon has one hydrogen $(4 - 3 = 1)$ attached to it.

Two line segments emanate from this point; therefore, this carbon has two hydrogens $(4 - 2 = 2)$ attached to it.

One line segment emanates from this point; therefore, this carbon has three hydrogens $(4 - 1 = 3)$ attached to it.

EXAMPLE 1.12

Write a more detailed structural formula for

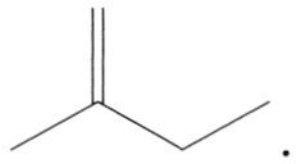

Solution

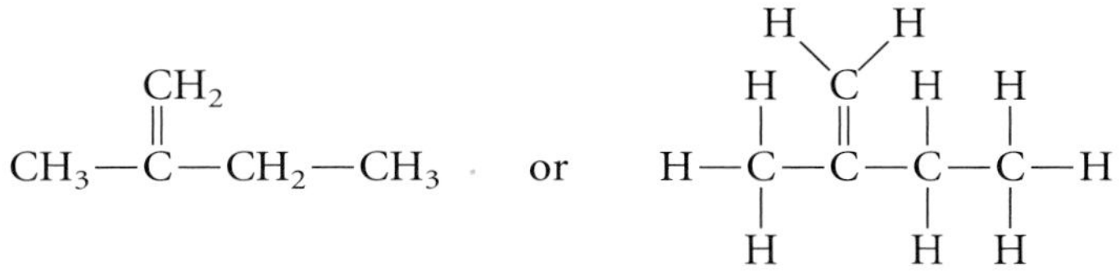

$$CH_3 - \overset{\overset{\textstyle CH_2}{\|}}{C} - CH_2 - CH_3 \quad \text{or} \quad H - \overset{\overset{\textstyle H}{|}}{\underset{\overset{\textstyle |}{H}}{C}} - \overset{\overset{\textstyle H \quad H}{\diagdown \quad /}}{\underset{\overset{\textstyle \|}{C}}{C}} - \overset{\overset{\textstyle H}{|}}{\underset{\overset{\textstyle |}{H}}{C}} - \overset{\overset{\textstyle H}{|}}{\underset{\overset{\textstyle |}{H}}{C}} - H$$

PROBLEM 1.23 Write a more detailed structural formula for

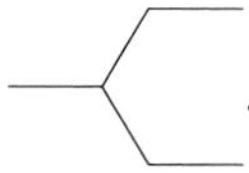

EXAMPLE 1.13

Write a line-segment formula for $CH_3CH_2CH = CHCH_2CH(CH_3)_2$

Solution

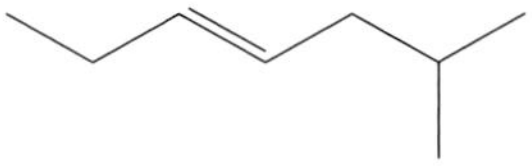

PROBLEM 1.24 Write a line-segment formula for $(CH_3)_2CHCH_2CH(CH_3)_2$.

1.11 Formal Charge

So far we have considered only molecules whose atoms are neutral. But in some compounds one or more atoms may be charged, either positively or negatively. Because such charges usually affect the chemical reactions of such molecules, it is important to know how to tell where the charge is located.

Consider the formula for hydronium ion, H_3O^+, the product of the reaction of a water molecule with a proton.

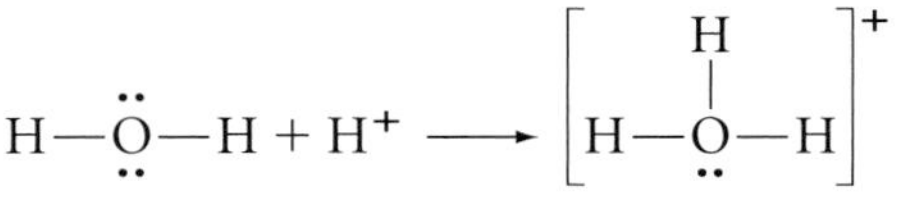

$$H - \overset{..}{\underset{..}{O}} - H + H^+ \longrightarrow \left[H - \overset{\overset{\textstyle H}{|}}{\underset{..}{O}} - H \right]^+ \qquad (1.4)$$

hydronium ion

The structure has eight electrons around the oxygen and two electrons around each hydrogen, so that all valence shells are complete. Note that there are eight valence electrons altogether. Oxygen contributes six, and each hydrogen contributes one, for a total of nine, but the ion has a single positive charge, so one electron must have been given away, leaving eight. Six of these eight electrons are used to form three O—H single bonds, leaving one unshared electron pair on the oxygen.

Although the entire hydronium ion carries a positive charge, we can ask, "Which atom, in a formal sense, bears the charge?" To determine **formal charge,** we consider each atom to "own" *all* of its unshared electrons plus only *half* of its shared electrons (one electron from each covalent bond). We then subtract this total from the number of valence electrons in the neutral atom to get the formal charge. This definition can be expressed in equation form as follows:

The **formal charge** on an atom in a covalently bonded molecule or ion is the number of valence electrons in the neutral atom minus the number of covalent bonds to the atom and the number of unshared electrons on the atom.

$$\begin{array}{ll}\text{Formal} \\ \text{charge}\end{array} = \begin{array}{l}\text{number of valence electrons} \\ \text{in the neutral atom}\end{array} - \left(\begin{array}{l}\text{unshared} \\ \text{electrons}\end{array} + \begin{array}{l}\text{half the shared} \\ \text{electrons}\end{array}\right) \qquad (1.5)$$

or, in a simplified form,

$$\begin{array}{ll}\text{Formal} \\ \text{charge}\end{array} = \begin{array}{l}\text{number of valence electrons} \\ \text{in the neutral atom}\end{array} - (\text{dots} + \text{bonds})$$

Let us apply this definition to the hydronium ion.

For each hydrogen atom:
Number of valence electrons in the neutral atom $= 1$
Number of unshared electrons $\qquad\qquad = 0$
Half the number of the shared electrons $\qquad = 1$
Therefore, the formal charge $\qquad\qquad = 1 - (0 + 1) = 0$

For the oxygen atom:
Number of valence electrons in the neutral atom $= 6$
Number of unshared electrons $\qquad\qquad = 2$
Half the number of the shared electrons $\qquad = 3$
Therefore, the formal charge $\qquad\qquad = 6 - (2 + 3) = +1$

It is the oxygen atom that formally carries the $+1$ charge in the hydronium ion.

EXAMPLE 1.14

On which atom is the formal charge in the hydroxide ion, OH^-?

Solution The electron-dot formula is

$$\left[:\ddot{O}:H\right]^-$$

Oxygen contributes six electrons, hydrogen contributes one, and there is one more for the negative charge, for a total of eight electrons. The formal charge on oxygen is $6 - (6 + 1) = -1$, so the oxygen carries the negative charge. The hydrogen is neutral.

PROBLEM 1.25 Calculate the formal charge on the nitrogen atom in ammonia, NH_3; in the ammonium ion, NH_4^+; and in the amide ion, NH_2^-.

Now let us look at a slightly more complex situation involving electron-dot structures and formal charge.

1.12 Resonance

Sometimes an electron pair is involved with more than two atoms in the process of forming bonds. As an example, consider the structure of the carbonate ion, CO_3^{2-}.

The total number of valence electrons in the carbonate ion is 24 (4 from the carbon, $3 \times 6 = 18$ from the three oxygens, *plus* 2 more electrons that give the ion its negative charge; these 2 electrons presumably have been donated by some metal, perhaps one each from two sodium atoms). An electron-dot structure that completes the valence shell of eight electrons around the carbon and each oxygen is

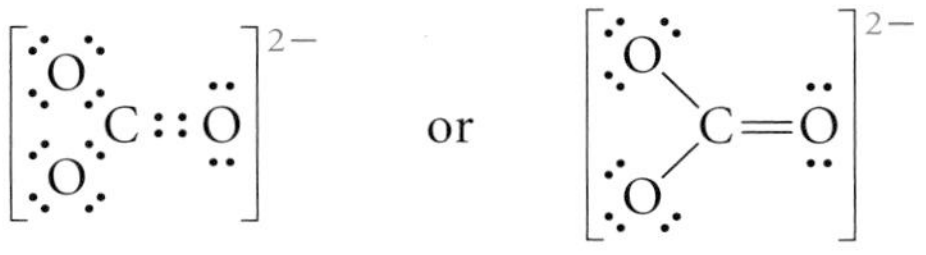

carbonate ion, CO_3^{2-}

The structure contains two carbon–oxygen *single* bonds and one carbon–oxygen *double* bond. Application of the definition for formal charge shows that the carbon is formally neutral, each singly bonded oxygen has a formal charge of -1, and the doubly bonded oxygen is formally neutral.

PROBLEM 1.26   Show that the last sentence of the preceding paragraph is correct.

When we wrote the electron-dot structure for the carbonate ion, our choice of which oxygen atom would be doubly bonded to the carbon atom was purely arbitrary. There are in fact *three exactly equivalent* structures that we might write.

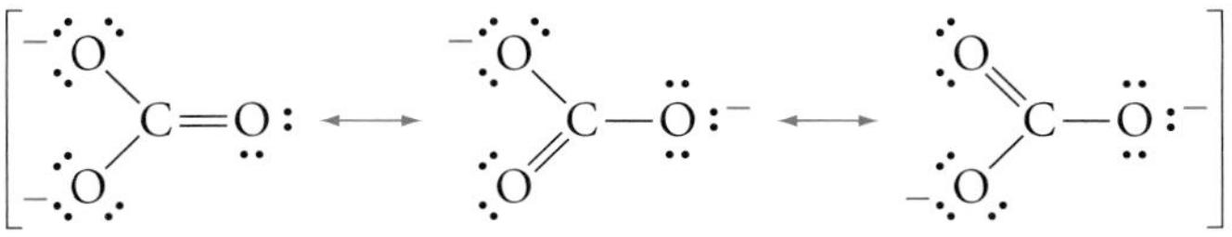

three equivalent structures for the carbonate ion

In each structure there is one C=O bond and there are two C—O bonds. These structures have the same arrangement of the atoms. They differ from one another *only* in the arrangement of the electrons.

The three structures for the carbonate ion are redrawn below, with curved arrows to show how electron pairs can be moved to interconnect the structures:

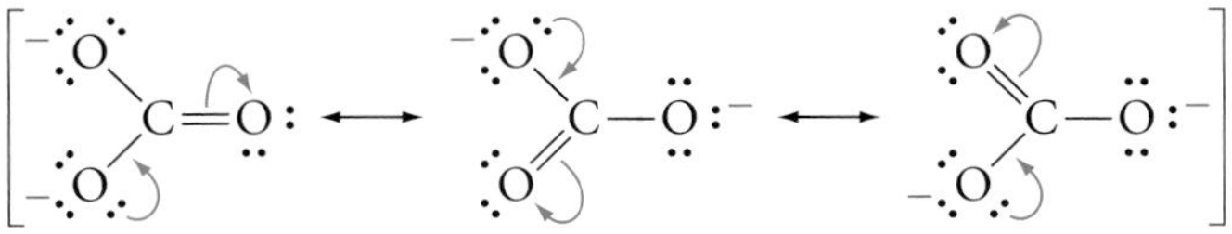

Chemists use curved arrows to keep track of a change in the location of electrons. A detailed explanation of the use of curved arrows is given in Section 1.13.

Physical measurements tell us that *none of the foregoing structures accurately describes the real carbonate ion.* For example, although each structure shows two different types of bonds between carbon and oxygen, we find experimentally that *all three carbon–oxygen bond lengths are identical: 1.31 Å.* This distance is intermediate between the normal C=O (1.20 Å) and C—O (1.41 Å) bond lengths. To explain this fact, we usually say that the real carbonate ion has a structure that is a **resonance hybrid** of the three contributing **resonance structures.** It is as if we could take an average of the

Resonance structures of a molecule or ion are two or more structures with identical arrangements of the atoms but different arrangements of the electrons. If resonance structures can be written, the structure of the molecule or ion is a **resonance hybrid** of the contributing resonance structures.

three structures. In the real carbonate ion, the two formal negative charges are spread *equally* over the three oxygen atoms, so that each oxygen atom carries two-thirds of a negative charge. It is important to note that the carbonate ion does not physically alternate among three resonance structures but has in fact one structure—a *hybrid* of the three resonance structures.

Resonance arises whenever we can write two or more structures for a molecule with different arrangements of the electrons but identical arrangements of the atoms. Resonance is very different from isomerism, for which the atoms themselves are arranged differently. When resonance is possible, the substance is said to have a structure that is a resonance hybrid of the various contributing structures. We use a double-headed arrow ($\longleftrightarrow$) between contributing structures to distinguish resonance from an equilibrium between different compounds, for which we use $\rightleftharpoons$.

Each carbon–oxygen bond in the carbonate ion is neither single nor double, but something in between—perhaps a one-and-one-third bond (any particular carbon–oxygen bond is single in two contributing structures and double in one). Sometimes we represent a resonance hybrid with one formula by writing a solid line for each full bond and a dotted line for each partial bond (in the carbonate ion, the dots represent one-third of a bond).

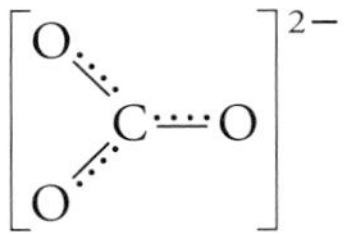

carbonate ion
resonance hybrid

PROBLEM 1.27   Draw the three equivalent contributing resonance structures for the nitrate ion, $NO_3{}^-$. What is the formal charge on the nitrogen atom and on each oxygen atom in the individual structures? What is the charge on the oxygens and on the nitrogen in the resonance hybrid structure? Show with curved arrows how the structures can be interconverted.

1.13   Arrow Formalism

Arrows in chemical drawings have specific meanings. For example, in Section 1.12 we used curved arrows to move electrons to show the relatedness of the three resonance structures of the carbonate ion. Just as it is important to learn the structural representations and names of molecules, it is important to learn the language of arrow formalism in organic chemistry.

Curved arrows show the movement of electrons in resonance structures and in reactions.

1. **Curved arrows** are used to show the movement of electrons in resonance structures and in reactions. Therefore, curved arrows always start at the initial position of electrons and end at their final position. In the example given below, the arrow that points from the C=O bond to the oxygen atom in the structure on the left indicates that the two electrons in one of the covalent bonds between carbon and oxygen move onto the oxygen atom:

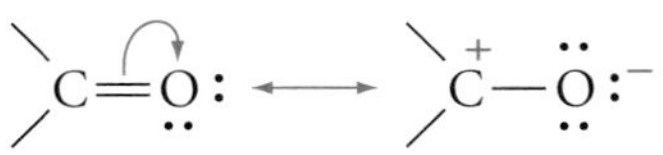

Note that the carbon atom in the structure on the right now has a formal positive charge, and the oxygen has a formal negative charge. Notice also that when a pair of electrons in a polar covalent bond is moved to one of the bonded atoms, *it is moved to the more electronegative atom,* in this case oxygen. In the following example, the arrow that points from the unshared pair of electrons on the oxygen atom to a point between the carbon and oxygen atoms in the structure on the left indicates that the unshared pair of electrons on the oxygen atom moves between the oxygen and carbon atoms to form a covalent bond:

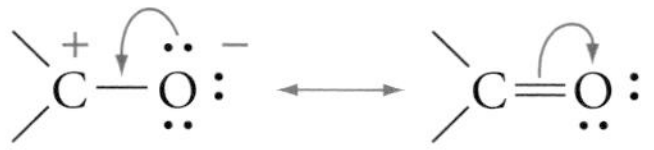

Note that both carbon and oxygen have formal charges of 0 in the structure on the right.

A curved arrow with half a head is called a **fishhook.** This kind of arrow is used to indicate the movement of a single electron. In eq. 1.6, two fishhooks are used to show the movement of each of the two electrons in the C—C bond of ethane to a carbon atom, forming two methyl radicals (see eq. 1.3):

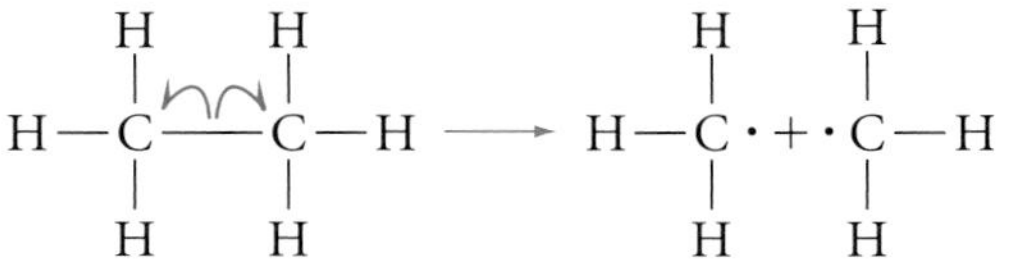

(1.6)

Fishhook arrows indicate the movement of single electrons. **Straight arrows** point from reactants to products in chemical reaction equations. A **double-headed straight arrow** between two structures indicates resonance structures.

2. **Straight arrows** point from reactants to products in chemical reaction equations. An example is the straight arrow pointing from ethane to the two methyl radicals in eq. 1.6. Straight arrows with half-heads are commonly used in pairs to indicate that a reaction is *reversible.*

$$A + B \rightleftharpoons C + D$$

A **double-headed straight arrow** between two structures indicates that they are resonance structures. Such an arrow does not indicate the occurrence of a chemical reaction. The double-headed arrows between resonance structures (Sec. 1.12) for the C=O bond are shown above.

EXAMPLE 1.15

Using correct arrow formalism, write the contributors to the resonance hybrid structure of the acetate ion, $CH_3CO_2^-$. Indicate any formal charges.

Solution   There are two equivalent resonance structures for the acetate ion. Each one has a formal negative charge on one of the oxygen atoms.

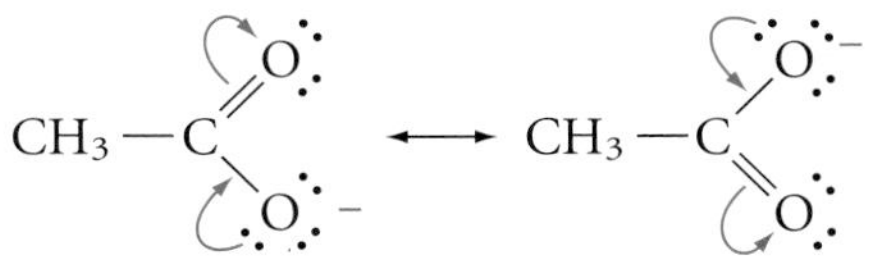

Notice that when one pair of electrons from oxygen is moved to form a covalent bond with carbon, a pair of electrons in a covalent bond between carbon and the other oxygen atom is moved to oxygen. This is necessary to ensure that the carbon atom does not exceed its valence of 4.

PROBLEM 1.28   Using correct arrow formalism, write the contributors to the resonance hybrid of azide ion, a linear ion with three connected nitrogens, N_3^-. Indicate the formal charge on each nitrogen atom.

We will use curved arrows throughout this text as a way of keeping track of electron movement. Several curved-arrow problems are included at the end of this chapter to help you get used to drawing them.

1.14 The Orbital View of Bonding; the Sigma Bond

Although electron-dot structures are often useful, they have some limitations. The Lewis theory of bonding itself has some limitations, especially in explaining the three-dimensional geometries of molecules. For this purpose in particular, we will discuss how another theory of bonding, involving orbitals, is more useful.

The atomic orbitals named in Section 1.1 have definite shapes. The s orbitals are spherical. The electrons that fill an *s* orbital confine their movement to a spherical region of space around the nucleus. The three *p* orbitals are dumbbell shaped and mutually perpendicular, oriented along the three coordinate axes, *x, y,* and *z.* Figure 1.2 shows the shapes of these orbitals.

In the orbital view of bonding, atoms approach each other in such a way that their atomic orbitals can *overlap* to form a bond. For example, if two hydrogen atoms form a hydrogen molecule, their two spherical 1*s* orbitals combine to form a new orbital that encompasses both of the atoms (see Figure 1.3). This orbital contains both valence electrons (one from each hydrogen). Like atomic orbitals, each **molecular orbital** can contain no more than two electrons. In the hydrogen molecule these electrons mainly occupy the space between the two nuclei.

The orbital in the hydrogen molecule is cylindrically symmetric along the H—H internuclear axis. Such orbitals are called **sigma (σ) orbitals,** and the bond is referred

⊚ **Ch. 1, Ref. 4.** *See animation of formation of s-s molecular orbital.*

A **molecular orbital** is the space occupied by electrons in a molecule.

A **sigma (σ) orbital** lies along the axis between two bonded atoms; a pair of electrons in a sigma orbital is called a **sigma bond.**

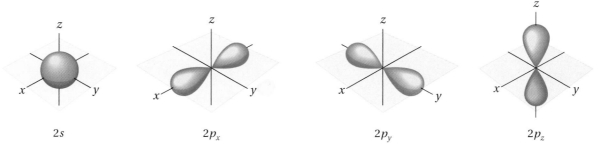

$2s$ $\qquad$ $2p_x$ $\qquad$ $2p_y$ $\qquad$ $2p_z$

Figure 1.2
The shapes of the *s* and *p* orbitals used by the valence electrons of carbon. The nucleus is at the origin of the three coordinate axes.

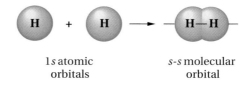

1*s* atomic orbitals $\qquad$ *s-s* molecular orbital

Figure 1.3
The molecular orbital representation of covalent bond formation between two hydrogen atoms.

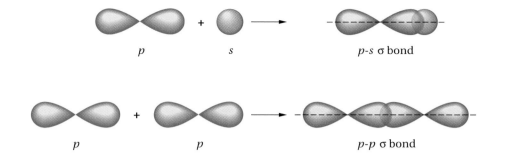

Figure 1.4
Orbital overlap to form σ bonds.

to as a **sigma bond.** Sigma bonds may also be formed by the overlap of an *s* and a *p* orbital or of two *p* orbitals, as shown in Figure 1.4.*

Let us see how these ideas apply to bonding in carbon compounds.

1.15 Carbon *sp*³ Hybrid Orbitals

In a carbon atom, the six electrons are arranged as shown in Figure 1.5 (compare with carbon in Table 1.2). The 1*s* shell is filled, and the four valence electrons are in the 2*s* orbital and two different 2*p* orbitals. There are a few things to notice about Figure 1.5. The energy scale at the left represents the energy of electrons in the various orbitals. The farther the electron is from the nucleus, the greater its potential energy, because it takes energy to keep the electron (negatively charged) and the nucleus (positively charged) apart. The 2*s* orbital has a slightly lower energy than the three 2*p* orbitals, which have equal energies (they differ from one another only in orientation around the nucleus, as shown in Figure 1.2). The two highest energy electrons are placed in different 2*p* orbitals rather than in the same orbital, because this keeps them farther apart and thus reduces the repulsion between these like-charged particles. One *p* orbital is vacant.

We might get a misleading idea about the bonding of carbon from Figure 1.5. For example, we might think that carbon should form only two bonds (to complete the partially filled 2*p* orbitals) or perhaps three bonds (if some atom donated two electrons to the empty 2*p* orbital). But we know from experience that this picture is wrong. Carbon usually forms *four* single bonds, and often these bonds are all equivalent, as in CH_4 or CCl_4. How can this discrepancy between theory and fact be resolved?

Figure 1.5
Distribution of the six electrons in a carbon atom. Each dot stands for an electron.

---

*Two properly aligned *p* orbitals can also overlap to form another type of bond, called a π (pi) bond. We discuss this type of bond in Chapter 3.

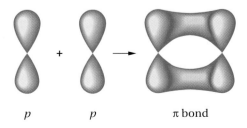

Figure 1.6
Unhybridized vs sp^3 hybridized orbitals on carbon. The dots stand for electrons. (Only the electrons in the valence shell are shown; the electrons in the 1s orbital are omitted because they are not involved in bonding.)

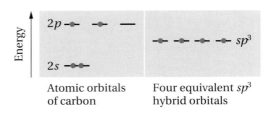

An sp^3 hybrid orbital is a p-shaped orbital that is one part s and three parts p in character.

One solution, illustrated in Figure 1.6, is to mix or combine the four atomic orbitals of the valence shell to form four identical hybrid orbitals, each containing one valence electron. In this model, the hybrid orbitals are called **sp^3 hybrid orbitals** because each one has one part s character and three parts p character. As shown in Figure 1.6, each sp^3 orbital has the same energy: less than that of the 2p orbitals but greater than that of the 2s orbital. The shape of sp^3 orbitals resembles the shape of p orbitals, except that the dumbbell is lopsided, and the electrons are more likely to be found in the lobe that extends out the greater distance from the nucleus, as shown in Figure

Figure 1.7
An sp^3 orbital extends mainly in one direction from the nucleus and forms bonds with other atoms in that direction. The four sp^3 orbitals of any particular carbon atom are directed toward the corners of a regular tetrahedron, as shown in the right-hand part of the figure (in this part of the drawing, the small "back" lobes of the orbitals have been omitted for simplification, although they can be important in chemical reactions).

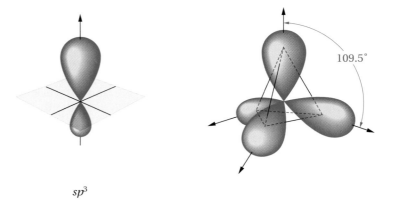

1.7. The four sp^3 hybrid orbitals of a single carbon atom are directed toward the corners of a regular tetrahedron, also shown in Figure 1.7. This particular geometry puts each orbital as far from the other three orbitals as it can be and thus minimizes repulsion when the orbitals are filled with electron pairs. The angle between any two of the four bonds formed from sp^3 orbitals is approximately 109.5°, the angle made by lines drawn from the center to the corners of a regular tetrahedron.

Hybrid orbitals can form sigma bonds by overlap with other hybrid orbitals or with nonhybridized atomic orbitals. Figure 1.8 shows some examples.

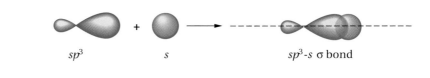

Figure 1.8
Examples of sigma (σ) bonds formed from sp^3 hybrid orbitals.

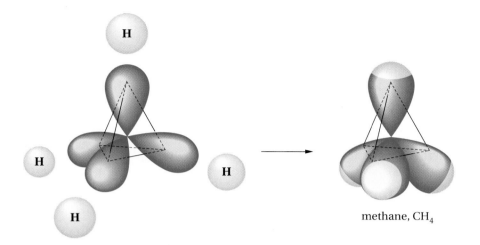

Figure 1.9
A molecule of methane, CH_4, is formed by the overlap of the four sp^3 carbon orbitals with the $1s$ orbitals of four hydrogen atoms. The resulting molecule has the geometry of a regular tetrahedron and contains four sigma bonds of the sp^3-s type.

methane, CH_4

1.16 Tetrahedral Carbon; the Bonding in Methane

We can now describe how a carbon atom combines with four hydrogen atoms to form methane. This process is pictured in Figure 1.9. The carbon atom is joined to each hydrogen atom by a sigma bond, which is formed by the overlap of a carbon sp^3 orbital with a hydrogen $1s$ orbital. The four sigma bonds are directed from the carbon nucleus to the corners of a regular tetrahedron. In this way, the electron pair in any one bond experiences minimum repulsion from the electrons in the other bonds. Each H—C—H **bond angle** is the same, 109.5°. To summarize, in methane there are four sp^3-s C—H sigma bonds, each directed from the carbon atom to one of the four corners of a regular tetrahedron.

A **bond angle** is the angle made by two covalent bonds to the same atom.

PROBLEM 1.29 Considering the repulsion that exists between electrons in different bonds, give a reason why a planar geometry for methane would be less stable than the tetrahedral geometry.

Because the tetrahedral geometry of carbon plays such an important role in organic chemistry, it is a good idea to become familiar with the features of a regular tetrahedron. One feature is that *the center and any two corners of a tetrahedron form a plane that is the perpendicular bisector of a similar plane formed by the center and the other two corners.* In methane, for example, any two hydrogens and the carbon form a plane that perpendicularly bisects the plane formed by the carbon and the other two hydrogens. These planes are illustrated in Figure 1.10.

In a 3D structure, *solid lines* lie in the plane of the page (C and H in C—H lie in the plane). *Dashed wedges* extend behind the plane (H in C⚊H lies behind the plane). *Solid wedges* project out toward you (H in C◀H is in front of the plane).

Figure 1.10
The carbon and two of the hydrogens in methane form a plane that perpendicularly bisects the plane formed by the carbon and the other two hydrogens.

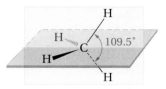

The geometry of carbon with four single bonds, as in methane, can be represented as

Ch. 1, Ref. 5. *3D model of methane. Measure the C—H bond length and the HCH bond angle.*

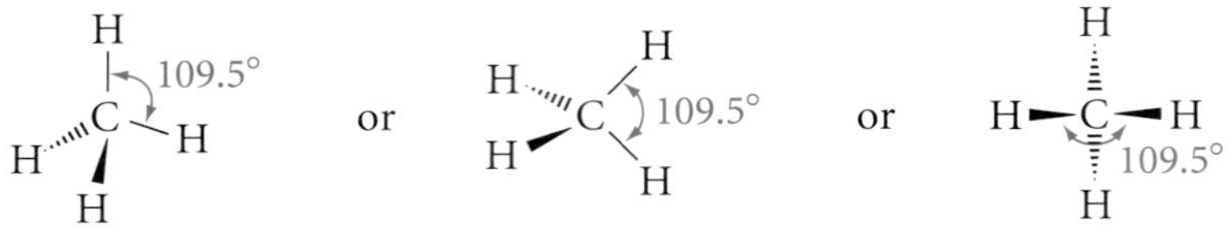

where the **solid lines** lie in the plane of the page, the **dashed wedge** goes behind the plane of the paper, and the **solid wedge** extends out of the plane of the paper toward you. Structures drawn in this way are sometimes called **3D** (that is, three-dimensional) **structures.**

Now that we have described single covalent bonds and their geometry, we are ready to tackle, in the next chapter, the structure and chemistry of saturated hydrocarbons. But before we do that, we present a brief overview of organic chemistry, so that you can see how the subject will be organized for study.

Because carbon atoms can be linked to one another or to other atoms in so many different ways, the number of possible organic compounds is almost limitless. Literally millions of organic compounds have been characterized, and the number grows daily. How can we hope to study this vast subject systematically? Fortunately, organic compounds can be classified according to their structures into a relatively small number of groups. Structures can be classified both according to the carbon framework (sometimes called the carbon *skeleton*) and according to the groups that are attached to that framework.

1.17 Classification According to Molecular Framework

The three main classes of molecular frameworks for organic structures are acyclic, carbocyclic, and heterocyclic compounds.

1.17.a Acyclic Compounds

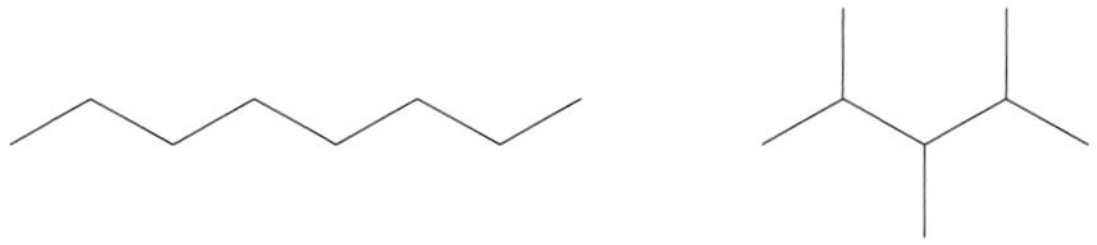

Acyclic compounds contain no rings. **Carbocyclic compounds** contain rings of carbon atoms. **Heterocyclic compounds** have rings containing at least one atom that is *not* carbon.

By **acyclic** (pronounced a′-cyclic) we mean *not cyclic.* Acyclic organic molecules have chains of carbon atoms but no rings. As we have seen, the chains may be unbranched or branched.

unbranched chain of eight carbon atoms   branched chain of eight carbon atoms

Pentane is an example of an acyclic compound with an unbranched carbon chain, whereas isopentane and neopentane are also acyclic but have branched carbon frameworks (Sec. 1.9). Figure 1.11 shows the structures of a few acyclic compounds that occur in nature.

geraniol
(oil of roses)
bp 229–230°C

A branched chain
compound used in
perfumes

$CH_3(CH_2)_5CH_3$

heptane
(petroleum)
bp 98.4°C

A hydrocarbon
present in petroleum,
used as a standard in
testing the octane
rating of gasoline.

$CH_3\overset{O}{\overset{\|}{C}}(CH_2)_4CH_3$

2-heptanone
(oil of cloves)
bp 151.5°C

A colorless liquid
with a fruity odor,
in part responsible
for the "peppery"
odor of blue cheese

Figure 1.11
Examples of natural acyclic
compounds, their sources (in
parentheses), and selected
characteristics.

1.17.b Carbocyclic Compounds

Carbocyclic compounds contain rings of carbon atoms. The smallest possible carbocyclic ring has three carbon atoms, but carbon rings come in many sizes and shapes. The rings may have chains of carbon atoms attached to them and may contain multiple bonds. Many compounds with more than one carbocyclic ring are known. Figure 1.12 shows the structures of a few carbocyclic compounds that occur in nature. Five- and six-membered rings are most common, but smaller and larger rings are also found.

1.17.c Heterocyclic Compounds

Heterocyclic compounds make up the third and largest class of molecular frameworks for organic compounds. In heterocyclic compounds, at least one atom in the ring must be a heteroatom, an atom that is *not* carbon. The most common heteroatoms are oxygen, nitrogen, and sulfur, but heterocyclics with other elements are also known. More than one heteroatom may be present and, if so, the heteroatoms may be alike or different. Heterocyclic rings come in many sizes, may contain multiple bonds, may have carbon chains or rings attached to them, and in short may exhibit a great variety of structures. Figure 1.13 shows the structures of a few natural products that contain heterocyclic rings. In these abbreviated structural formulas, the symbols for the heteroatoms are shown, but the carbons are indicated using lines only.

The structures in Figures 1.11 through 1.13 show not only the molecular frameworks, but also various groups of atoms that may be part of or attached to the frameworks. Fortunately, these groups can also be classified in a way that helps simplify the study of organic chemistry.

Lemons, source of limonene.

1.18   Classification According to Functional Group

Certain groups of atoms have chemical properties that depend only moderately on the molecular framework to which they are attached. These groups of atoms are called **functional groups.** The hydroxyl group, —OH, is an example of a functional group, and compounds with this group attached to a carbon framework are called alcohols. In most organic reactions, some chemical change occurs at the functional group, but the rest of the molecule keeps its original structure. This maintenance of most of the structural formula throughout a chemical reaction greatly simplifies our study of

Functional groups are groups of atoms that have characteristic chemical properties regardless of the molecular framework to which they are attached.

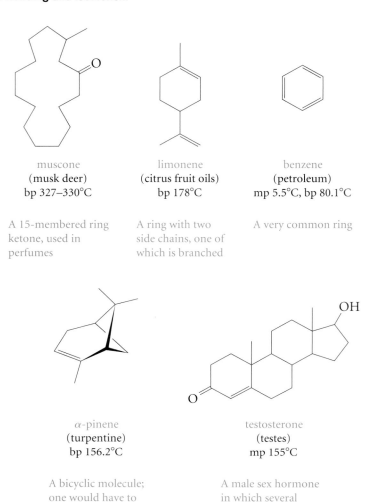

muscone
(musk deer)
bp 327–330°C

A 15-membered ring
ketone, used in
perfumes

limonene
(citrus fruit oils)
bp 178°C

A ring with two
side chains, one of
which is branched

benzene
(petroleum)
mp 5.5°C, bp 80.1°C

A very common ring

α-pinene
(turpentine)
bp 156.2°C

A bicyclic molecule;
one would have to
break *two* bonds to
make it acyclic

testosterone
(testes)
mp 155°C

A male sex hormone
in which several
rings of common sizes
are *fused* together;
that is, they share
two adjacent carbon
atoms

Figure 1.12
Examples of natural carbocyclic compounds with rings of various sizes and shapes. The source and special features of each structure are indicated below it.

organic chemistry. It allows us to focus attention on the chemistry of the various functional groups. We can study classes of compounds instead of having to learn the chemistry of each individual compound.

Some of the main functional groups that we will study are listed in Table 1.6, together with a typical compound of each type. Although we will describe these classes of compounds in greater detail in later chapters, it would be a good idea for you to become familiar with their names and structures now. If a particular functional group is mentioned before its chemistry is discussed in detail, and you forget what it is, you can refer to Table 1.6 or to the inside front cover of this book.

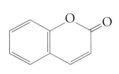

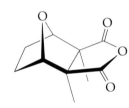

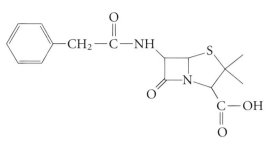

nicotine
bp 246°C

adenine
mp 360–365°C
(decomposes)

penicillin-G
(amorphous solid)

Present in tobacco,
nicotine has two
heterocyclic rings
of different sizes,
each containing
one nitrogen.

One of the four hetero-
cyclic bases of DNA,
adenine contains two
fused heterocyclic
rings, each of
which contains two
heteroatoms (nitrogen).

One of the most widely
used antibiotics,
penicillin has two
heterocyclic rings, the
smaller of which is
crucial to biological
activity.

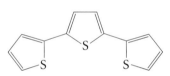

coumarin
mp 71°C

α-terthienyl
mp 92–93°C

cantharidin
mp 218°C

Found in clover and
grasses, coumarin
produces the
pleasant odor of
new-mown hay.

This compound, with
three linked sulfur-
containing rings, is
present in certain
marigold species.

This compound, an
oxygen heterocycle, is
the active principle in
cantharis (also known as
Spanish fly), a material
isolated from certain
dried beetles of the
species *Cantharis vesi-
catoria* and incorrectly
thought by some to
increase sexual desire.

Figure 1.13
Examples of natural heterocyclic compounds having a variety of heteroatoms and ring sizes.

PROBLEM 1.30 What functional groups can you find in the following natural products? (Their formulas are given in Figures 1.11 and 1.12.)

a. geraniol   b. 2-heptanone   c. limonene   d. testosterone

Table 1.6 The main functional groups

	Structure	Class of compound	Specific example	Common name of the specific example
A. Functional groups that are a part of the molecular framework	$-\overset{\mid}{\underset{\mid}{C}}-\overset{\mid}{\underset{\mid}{C}}-$	alkane	CH_3-CH_3	ethane, a component of natural gas
	$\overset{/}{\underset{\backslash}{C}}=\overset{\backslash}{\underset{/}{C}}$	alkene	$CH_2=CH_2$	ethylene, used to make polyethylene
	$-C\equiv C-$	alkyne	$HC\equiv CH$	acetylene, used in welding
	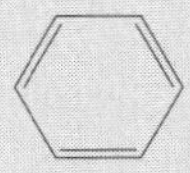 ⬡	arene	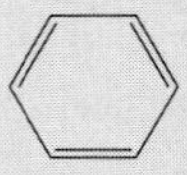 ⬡	benzene, raw material for polystyrene and phenol
B. Functional groups containing oxygen				
1. With carbon–oxygen single bonds	$-\overset{\mid}{\underset{\mid}{C}}-OH$	alcohol	CH_3CH_2OH	ethyl alcohol, found in beer, wines, and liquors
	$-\overset{\mid}{\underset{\mid}{C}}-O-\overset{\mid}{\underset{\mid}{C}}-$	ether	$CH_3CH_2OCH_2CH_3$	diethyl ether, once a common anesthetic
2. With carbon–oxygen double bonds*	$-\overset{O}{\overset{\|}{C}}-H$	aldehyde	$CH_2=O$	formaldehyde, used to preserve biological specimens
	$-\overset{\mid}{\underset{\mid}{C}}-\overset{O}{\overset{\|}{C}}-\overset{\mid}{\underset{\mid}{C}}-$	ketone	$CH_3\overset{O}{\overset{\|}{C}}CH_3$	acetone, a solvent for varnish and rubber cement
3. With single and double carbon–oxygen bonds	$-\overset{O}{\overset{\|}{C}}-OH$	carboxylic acid	$CH_3\overset{O}{\overset{\|}{C}}-OH$	acetic acid, a component of vinegar
	$-\overset{O}{\overset{\|}{C}}-O-\overset{\mid}{\underset{\mid}{C}}-$	ester	$CH_3\overset{O}{\overset{\|}{C}}-OCH_2CH_3$	ethyl acetate, a solvent for nail polish and model airplane glue
C. Functional groups containing nitrogen**	$-\overset{\mid}{\underset{\mid}{C}}-NH_2$	primary amine	$CH_3CH_2NH_2$	ethylamine, smells like ammonia
	$-C\equiv N$	nitrile	$CH_2=CH-C\equiv N$	acrylonitrile, raw material for making Orlon
D. Functional group with oxygen and nitrogen	$-\overset{O}{\overset{\|}{C}}-NH_2$	primary amide	$H-\overset{O}{\overset{\|}{C}}-NH_2$	formamide, a softener for paper
E. Functional group with halogen	$-X$	alkyl or aryl halide	CH_3Cl	methyl chloride, refrigerant and local anesthetic

Table 1.6	Continued				
		Structure	**Class of compound**	**Specific example**	**Common name of the specific example**
F. Functional groups containing sulfur†		—C—SH	thiol (also called mercaptan)	CH_3SH	methanethiol, has the odor of rotten cabbage
		—C—S—C—	thioether (also called sulfide)	$(CH_2{=}CHCH_2)_2S$	diallyl sulfide, has the odor of garlic

*The $\diagdown C{=}O$ group, present in several functional groups, is called a **carbonyl group**. The $-\overset{\overset{\text{O}}{\|}}{C}-OH$ group of acids is called a **carboxyl group** (a contraction of *carb*onyl and hyd*roxyl*).

The $-NH_2$ group is called an **amino group.

†Thiols and thioethers are the sulfur analogs of alcohols and ethers.

ADDITIONAL PROBLEMS

Valence, Bonding, and Lewis Structures

1.31 Show the number of valence electrons in each of the following atoms. Let the element's symbol represent its kernel, and use dots for the valence electrons.

    a. magnesium   **b.** carbon   **c.** oxygen
    d. aluminum   **e.** phosphorus   **f.** chlorine

1.32 When a solution of salt (sodium chloride) in water is treated with a silver nitrate solution, a white precipitate forms immediately. When tetrachloromethane is shaken with aqueous silver nitrate, no such precipitate is produced. Explain these facts in terms of the types of bonds present in the two chlorides.

1.33 Use the relative positions of the elements in the periodic table (Table 1.3 or inside back cover) to classify the following substances as ionic or covalent:

    a. Br_2 **b.** KCl   **c.** P_2O_5
    d. $MgBr_2$ **e.** $SiCl_4$ **f.** BrCl
    g. LiBr   **h.** PCl_3

1.34 For each of the following elements, determine (1) how many valence electrons it has and (2) what its common valence is:

    a. oxygen   **b.** hydrogen   **c.** fluorine
    d. carbon   **e.** nitrogen   **f.** sulfur

= Concept connections. Problems marked by this symbol require you to use two or more concepts. These concepts may be present in the chapter you are studying, or some may come from earlier chapters that you have read. You may find these problems more challenging than other problems because they require you figure out how two or more concepts are connected to each other. The payoff for working these problems is that they will help you to see the big picture. Seeing the big picture helps you to remember concepts without doing so much memorizing. You may find it helpful to team up with one or more of your classmates to work on these problems.

1.35 Write a structural formula for each of the following compounds, using a line to represent each single bond and dots for any unshared electron pairs:

    a. CH_3OH **b.** CH_3Cl **c.** C_3H_8
    d. $CH_3CH_2NH_2$ **e.** C_2H_5Cl **f.** CH_2O

1.36 Draw a structural formula for each of the following covalent molecules. Which bonds are polar? Indicate the polarity by proper placement of the symbols $\delta+$ and $\delta-$.

    a. Br_2 **b.** CH_3Cl **c.** CO_2
    d. HCl **e.** SF_6 **f.** CH_4
    g. SO_2 **h.** CH_3OH

1.37 Consider the X—H bond, in which X is an atom other than H. The H in a polar bond is more acidic (more easily re-

moved) than the H in a nonpolar bond. Considering bond polarity, which hydrogen in acetic acid, $CH_3\overset{\displaystyle O}{\overset{\|}{C}}-OH$, do you expect to be most acidic? Write an equation for the reaction between acetic acid and sodium hydroxide.

Structural Isomers

1.38 Draw structural formulas for all possible isomers having the following molecular formulas:

    a. C_3H_8 **b.** C_3H_7Br **c.** $C_2H_4Cl_2$
    d. $C_3H_6F_2$ **e.** C_4H_9F **f.** $C_2H_2I_2$
    g. C_3H_6 **h.** $C_4H_{10}O$

1.39 Draw structural formulas for the five isomers of C_6H_{14}. As you write them out, try to be systematic, starting with a consecutive chain of six carbon atoms.

Structural Formulas

1.40 For each of the following abbreviated structural formulas, write a structural formula that shows all the bonds:

    a. $CH_3(CH_2)_3CH_3$ **b.** $(CH_3)_3CCH_2CH_3$ **c.** $(CH_3)_2CHOH$
    d. $CH_3CH_2SCH_3$ **e.** CH_2ClCH_2OH **f.** $(CH_3CH_2)_2NH$

1.41 Write structural formulas that correspond to the following abbreviated structures, and show the correct number of hydrogens on each carbon:

    a. 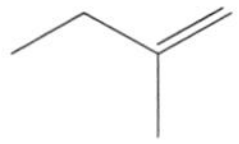   **b.**    **c.**

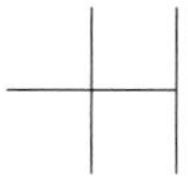

    d. 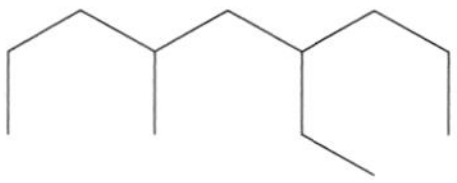   **e.** 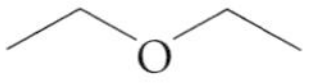   **f.**

    g. 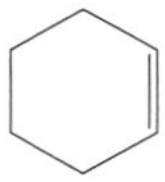   **h.**

1.42 For each of the following abbreviated structural formulas, write a line-segment formula (like those in Problem 1.41).

    a. $CH_3(CH_2)_4CH_3$ **b.** $(CH_3)_2CHCH_2CH_2\overset{\displaystyle O}{\overset{\|}{C}}CH_3$

    c. $CH_3CHCH_2C(CH_3)_3$ **d.** 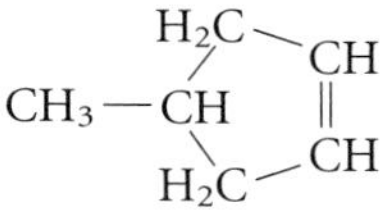 $CH_3-\underset{}{CH}\overset{\displaystyle H_2C\diagdown_{CH}}{\underset{H_2C\diagup^{CH}}{\Big|\Big|}}$
                $|$
              OH

1.43 An abbreviated formula of geraniol is shown in Figure 1.11.

    a. How many carbons does geraniol have?
    b. What is its molecular formula?
    c. Write a more detailed structural formula for it.

1.44 What is the *molecular formula* for each of the following compounds? Consult Figures 1.12 and 1.13 for the abbreviated structural formulas.

    a. benzene   **b.** muscone   **c.** testosterone
    d. adenine   **e.** coumarin   **f.** nicotine

Formal Charge, Resonance, and Curved-Arrow Formalism

1.45 Write electron-dot formulas for the following species. Show where the formal charges, if any, are located.

    a. nitrous acid, HONO   **b.** nitric acid, $HONO_2$
    c. formaldehyde, H_2CO **d.** ammonium ion, NH_4^+
    e. cyanide ion, CN^- **f.** carbon monoxide, CO
    g. sulfate ion, SO_4^{2-} **h.** boron trichloride, BCl_3
    i. hydrogen peroxide, H_2O_2 **j.** bicarbonate ion, HCO_3^-

1.46 Consider each of the following highly reactive carbon species. What is the formal charge on carbon in each species?

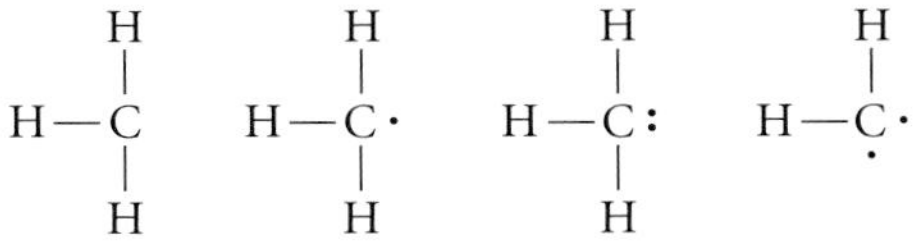

$$
\begin{array}{cccc}
\text{H} & \text{H} & \text{H} & \text{H} \\
| & | & | & | \\
\text{H—C} & \text{H—C·} & \text{H—C:} & \text{H—C·} \\
| & | & | & \cdot \\
\text{H} & \text{H} & \text{H} &
\end{array}
$$

1.47 Draw electron-dot formulas for the two contributors to the resonance hybrid structure of the nitrite ion, NO_2^-. (Each oxygen is connected to the nitrogen.) What is the charge on each oxygen in each contributor and in the hybrid structure? Show by curved arrows how the electron pairs can relocate to interconvert the two structures.

1.48 Write the structure obtained when electrons move as indicated by the curved arrows in the following structure:

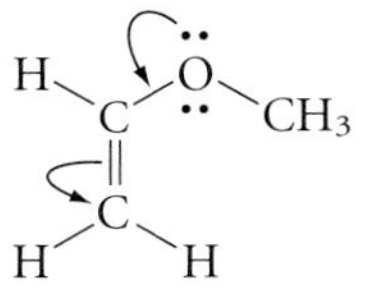

Does each atom in the resulting structure have a complete valence shell of electrons? Locate any formal charges in each structure.

1.49 Add curved arrows to the following structures to show how electron pairs must be moved to interconvert the structures, and locate any formal charges.

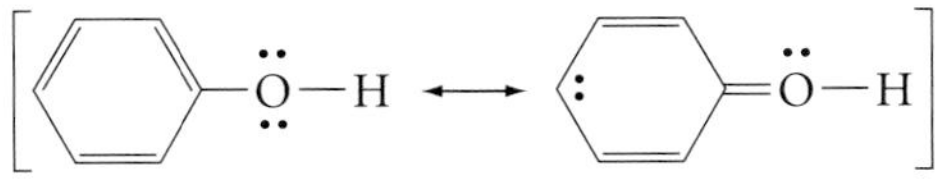

1.50 Add curved arrows to show how electrons must move to form the product from the reactants in the following equation, and locate any formal charges.

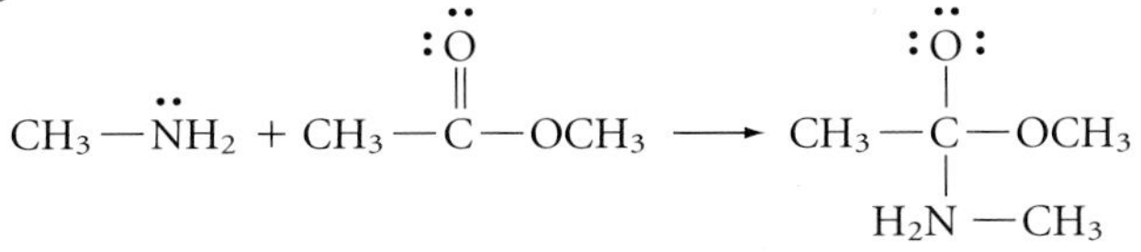

Electronic Structure and Molecular Geometry

1.51 Each of the following substances contains ionic and covalent bonds. Draw their electron-dot formulas.

    a. CH_3ONa **b.** NH_4Cl

1.52 Fill in any unshared electron pairs that are missing from the following formulas:

$$O$$
$$\|$$

    a. $(CH_3)_2NH$ **b.** CH_3C-OH

    c. CH_3CH_2SH **d.** $CH_3OCH_2CH_2OH$

1.53 Make a drawing (similar to the right-hand part of Figure 1.6) of the electron distribution that will be expected in nitrogen atoms if the s and p orbitals are hybridized to sp^3. Based on this model, predict the geometry of the ammonia molecule, NH_3.

1.54 The ammonium ion, NH_4^+, has a tetrahedral geometry analogous to that of methane. Explain this structure in terms of atomic and molecular orbitals.

1.55 Use lines, dashed wedges, and solid wedges to show the geometry of CCl_4 and CH_3OH.

1.56 Silicon is just below carbon in the periodic table. Predict the geometry of silicon tetrachloride, $SiCl_4$.

Classification of Organic Compounds

1.57 Write a structural formula that corresponds to the molecular formula C_4H_8O and is

    a. acyclic   **b.** carbocyclic   **c.** heterocyclic

1.58 Divide the following compounds into groups that might be expected to exhibit similar chemical behavior:

    a. C_5H_{12} **b.** C_4H_9OH **c.** $HOCH_2CH_2OH$

    d. CH_3OCH_3 **e.** C_8H_{18} **f.** CH_3OH

    g. C_3H_7OH **h.** $CH_3OCH_2CH_3$ **i.** C_6H_{14}

1.59 Using Table 1.6, write a structural formula for each of the following:

    a. an alcohol, $C_4H_{10}O$ **b.** an ether, C_3H_8O

    c. an aldehyde, C_3H_6O **d.** a ketone, C_4H_8O

    e. a carboxylic acid, $C_4H_8O_2$ **f.** an ester, $C_5H_{10}O_2$

    g. an alcohol that is an isomer of the one in part a   **h.** an amine, C_3H_9N

1.60 Many organic compounds contain more than one functional group. An example is glycine (shown below), one of the simple building blocks of proteins (Chapter 17).

$$O$$
$$\|$$
$$HO-C-CH_2NH_2$$

    glycine

    a. What functional groups are present in glycine?

    b. Redraw the structure, adding all unshared electron pairs.

    c. What is the molecular formula of glycine?

    d. Draw another structural isomer that has this formula. What functional groups does this isomer have?

ALKANES AND CYCLOALKANES; CONFORMATIONAL AND GEOMETRIC ISOMERISM

T he main components of petroleum and natural gas, resources that now supply most of our fuel for energy, are **hydrocarbons,** compounds that contain only carbon and hydrogen. There are three main classes of hydrocarbons, based on the types of carbon–carbon bonds present. **Saturated hydrocarbons** contain only carbon–carbon *single* bonds. **Unsaturated hydrocarbons** contain carbon–carbon *multiple* bonds—double bonds, triple bonds, or both. **Aromatic hydrocarbons** are a special class of cyclic compounds related in structure to benzene.*

Saturated hydrocarbons are known as **alkanes** if they are acyclic, or as **cycloalkanes** if they are cyclic. Let us look at their structures and properties.

2.1 The Structures of Alkanes

The simplest alkane is methane. Its tetrahedral three-dimensional structure was described in the previous chapter (see Figure 1.9). Additional alkanes are constructed by lengthening the carbon chain and adding an appropriate number of hydrogens to complete the carbon valences (for examples, see Figure 2.1** and Table 2.1).

---

*Unsaturated and aromatic hydrocarbons are discussed in Chapters 3 and 4, respectively.

**Molecular models can help you visualize organic structures in three dimensions. They will be extremely useful to you throughout this course, especially when we consider various types of isomerism. Relatively inexpensive sets are usually at stores that sell textbooks, and your instructor can suggest which kind to buy. If you cannot locate or afford a set, you can create models that are adequate for most purposes from toothpicks (for bonds) and marshmallows, gum drops, or jelly beans (for atoms).

▲ Refining of petroleum, a major natural source of alkanes (see "A Word About Petroleum, Gasoline, and Octane Number," pp. 106–107.)

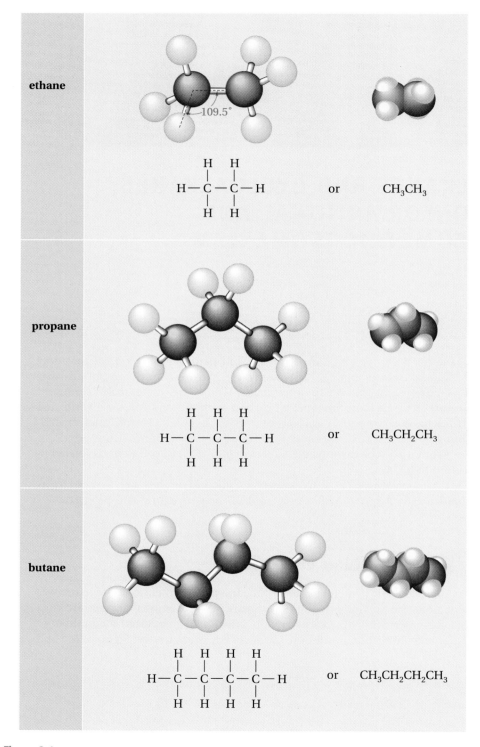

◎ **Ch. 2, Ref. 1.** *Examine and manipulate 3D models of ethane, propane, and butane.*

Figure 2.1

Three-dimensional models of ethane, propane, and butane. The stick-and-ball models at the left show the way in which the atoms are connected and depict the correct bond angles. The space-filling models at the right are constructed to scale and give a better idea of the molecular shape, though some of the hydrogens may appear hidden.

Table 2.1	Names and formulas of the first ten unbranched alkanes			
Name	Number of carbons	Molecular formula	Structural formula	Number of structural isomers
methane	1	CH_4	CH_4	1
ethane	2	C_2H_6	CH_3CH_3	1
propane	3	C_3H_8	$CH_3CH_2CH_3$	1
butane	4	C_4H_{10}	$CH_3CH_2CH_2CH_3$	2
pentane	5	C_5H_{12}	$CH_3(CH_2)_3CH_3$	3
hexane	6	C_6H_{14}	$CH_3(CH_2)_4CH_3$	5
heptane	7	C_7H_{16}	$CH_3(CH_2)_5CH_3$	9
octane	8	C_8H_{18}	$CH_3(CH_2)_6CH_3$	18
nonane	9	C_9H_{20}	$CH_3(CH_2)_7CH_3$	35
decane	10	$C_{10}H_{22}$	$CH_3(CH_2)_8CH_3$	75

Alkanes are **saturated hydrocarbons,** containing only carbon−carbon single bonds. **Cycloalkanes** contain rings. **Unsaturated hydrocarbons** contain carbon−carbon double or triple bonds. **Aromatic hydrocarbons** are cyclic compounds structurally related to benzene.

All alkanes fit the general molecular formula C_nH_{2n+2}, where n is the number of carbon atoms. Alkanes with carbon chains that are unbranched (Table 2.1) are called **normal alkanes.** Each member of this series differs from the next higher and the next lower member by a —CH_2— group (called a **methylene group**). A series of compounds in which the members are built up in a regular, repetitive way like this is called a **homologous series.** Members of such a series have similar chemical and physical properties, which change gradually as carbon atoms are added to the chain.

Unbranched alkanes are called **normal alkanes,** or *n*-alkanes.

Compounds of a **homologous series** differ by a regular unit of structure and share similar properties.

EXAMPLE 2.1

What is the molecular formula of an alkane with six carbon atoms?

Solution   If $n = 6$, then $2n + 2 = 14$. The formula is C_6H_{14}.

PROBLEM 2.1   What is the molecular formula of an alkane with 14 carbon atoms?

PROBLEM 2.2   Which of the following are alkanes?

   a. C_7H_{18} b. C_7H_{16} c. C_8H_{16} d. $C_{27}H_{56}$

2.2   Nomenclature of Organic Compounds

In the early days of organic chemistry, each new compound was given a name that was usually based on its source or use. Examples (Figures 1.12 and 1.13) include limonene (from lemons), α-pinene (from pine trees), coumarin (from the tonka bean, known to South American natives as *cumaru*), and penicillin (from the mold that produces it, *Penicillium notatum*). Even today, this method of naming can be used to give a short, simple name to a molecule with a complex structure. For example, cubane (p. 3) was named after its shape.

It became clear many years ago, however, that one could not rely only on common or trivial names and that a systematic method for naming compounds was needed. Ideally, the rules of the system should result in a unique name for each compound. Knowing the rules and seeing a structure, one should be able to write the systematic name. Seeing the systematic name, one should be able to write the correct structure.

Eventually, internationally recognized systems of nomenclature were devised by a commission of the International Union of Pure and Applied Chemistry; they are known as the IUPAC (pronounced "eye-you-pack") systems. In this book, we will use mainly IUPAC names. However, in some cases, the common name is so widely used that we will ask you to learn it (for example, formaldehyde [common] is used in preference to methanal [systematic], and cubane is much easier to remember than its systematic name, pentacyclo[4.2.0.0^{2,5}.0^{3,8}.0^{4,7}]octane).

2.3 IUPAC Rules for Naming Alkanes

1. The general name for acyclic saturated hydrocarbons is *alkanes*. The *-ane* ending is used for all saturated hydrocarbons. This is important to remember because later other endings will be used for other functional groups.

2. Alkanes without branches are named according to the *number of carbon atoms*. These names, up to ten carbons, are given in the first column of Table 2.1.

3. For alkanes with branches, the **root name** is that of the longest continuous chain of carbon atoms. For example, in the structure

$$CH_3-\overset{\overset{\displaystyle CH_3}{|}}{C}H-\overset{\overset{\displaystyle CH_3}{|}}{C}H-CH_3-CH_3 \qquad or \qquad CH_3-\overset{\overset{\displaystyle CH_3}{|}}{C}H-\overset{\overset{\displaystyle CH_3}{|}}{C}H-CH_2-CH_3$$

the longest continuous chain (in color) has five carbon atoms. The compound is therefore named as a substituted *pent*ane, even though there are seven carbon atoms altogether.

> The **root name** of an alkane is that of the longest continuous chain of carbon atoms.

4. Groups attached to the main chain are called **substituents.** Saturated substituents that contain only carbon and hydrogen are called **alkyl groups.** An alkyl group is named by taking the name of the alkane with the same number of carbon atoms and changing the *-ane* ending to *-yl.*

   In the example above, each substituent has only one carbon. Derived from methane by removing one of the hydrogens, a one-carbon substituent is called a **methyl group.**

> **Substituents** are groups attached to the main chain of a molecule. Saturated substituents containing only C and H are called **alkyl groups.**
>
> The one-carbon alkyl group derived from methane is called a **methyl group.**

$$H-\overset{\overset{\displaystyle H}{|}}{\underset{\underset{\displaystyle H}{|}}{C}}-H \qquad\qquad H-\overset{\overset{\displaystyle H}{|}}{\underset{\underset{\displaystyle H}{|}}{C}}- \quad or \quad CH_3- \quad or \quad Me-$$

methane   methyl group

The names of substituents with more than one carbon atom will be described in Section 2.4.

5. The main chain is numbered in such a way that the first substituent encountered along the chain receives the lowest possible number. Each substituent is then located by its name and by the number of the carbon atom to which it is attached.

When two or more identical groups are attached to the main chain, prefixes such as *di-*, *tri-*, and *tetra-* are used. *Every substituent must be named and numbered, even if two identical substituents are attached to the same carbon of the main chain.* The compound

$$\begin{array}{ccccc} & \overset{CH_3}{\underset{2}{|}} & \overset{CH_3}{\underset{3}{|}} & & \\ \overset{1}{CH_3}-CH-CH-\overset{4}{CH_2}-\overset{5}{CH_3} \end{array}$$

is correctly named 2,3-dimethylpentane. The name tells us that there are two methyl substituents, one attached to carbon-2 and one attached to carbon-3 of a five-carbon saturated chain.

6. If two or more different types of substituents are present, they are listed alphabetically, except that prefixes such as *di-* and *tri-* are not considered when alphabetizing.

7. Punctuation is important when writing IUPAC names. IUPAC names for hydrocarbons are written as one word. Numbers are separated from each other by commas and are separated from letters by hyphens. There is no space between the last named substituent and the name of the parent alkane that follows it.

To summarize and amplify these rules, we take the following steps to find an acceptable IUPAC name for an alkane:

1. Locate the longest continuous carbon chain. This gives the name of the parent hydrocarbon. For example,

$$\begin{array}{ccc} \overset{C-C}{\underset{|}{}} & & \overset{C-C}{\underset{|}{}} \\ C-C-C-C & not & C-C-C-C \end{array}$$

2. Number the longest chain beginning at the end nearest the first branch point. For example,

$$\begin{array}{cc} \overset{C}{\underset{|}{}}\quad\overset{C}{\underset{|}{}} & \overset{C}{\underset{|}{}}\quad\overset{C}{\underset{|}{}} \\ \underset{6\;5\;4\;3\;2\;1}{C-C-C-C-C-C} \quad not \quad \underset{1\;2\;3\;4\;5\;6}{C-C-C-C-C-C} \end{array}$$

If there are two equally long continuous chains, select the one with the most branches. For example,

$$\begin{array}{cc} \overset{C}{\underset{|}{}} & \overset{C}{\underset{|}{}} \\ \underset{1\;2\;3\;4\;5\;6}{C-C-C-C-C-C} \quad not \quad \underset{3\;4\;5\;6}{C-C-C-C-C-C} \\ \overset{|}{\underset{C-C}{}} & \overset{|}{\underset{\underset{2\;1}{C-C}}{}} \end{array}$$

two branches   one branch

If there is a branch equidistant from each end of the longest chain, begin numbering nearest to a third branch:

$$\begin{array}{cc} \overset{C}{\underset{|}{}}\;\overset{C}{\underset{|}{}}\qquad\overset{C}{\underset{|}{}} & \overset{C}{\underset{|}{}}\;\overset{C}{\underset{|}{}}\qquad\overset{C}{\underset{|}{}} \\ \underset{1\;2\;3\;4\;5\;6\;7}{C-C-C-C-C-C-C} \quad not \quad \underset{7\;6\;5\;4\;3\;2\;1}{C-C-C-C-C-C-C} \end{array}$$

2,3,6-trimethylheptane   2,5,6-trimethylheptane

If there is no third branch, begin numbering nearest the substituent whose name has alphabetic priority:

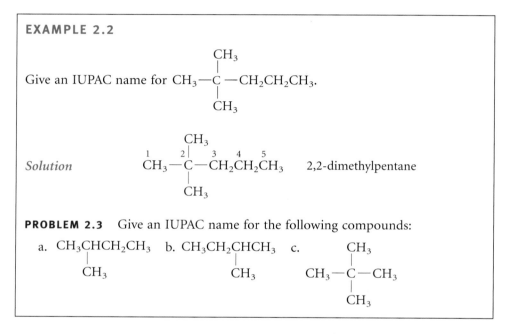

3. Write the name as one word, placing substituents in alphabetic order and using proper punctuation.

EXAMPLE 2.2

Give an IUPAC name for
$$CH_3-\overset{\overset{\displaystyle CH_3}{|}}{\underset{\underset{\displaystyle CH_3}{|}}{C}}-CH_2CH_2CH_3.$$

Solution
$$\overset{1}{CH_3}-\overset{\overset{\displaystyle CH_3}{|}}{\underset{\underset{\displaystyle CH_3}{|}}{\overset{2}{C}}}-\overset{3}{CH_2}\overset{4}{CH_2}\overset{5}{CH_3}$$ 2,2-dimethylpentane

PROBLEM 2.3   Give an IUPAC name for the following compounds:

a. $CH_3\overset{|}{\underset{\underset{\displaystyle CH_3}{|}}{CH}}CH_2CH_3$ b. $CH_3CH_2\overset{|}{\underset{\underset{\displaystyle CH_3}{|}}{CH}}CH_3$ c. $CH_3-\overset{\overset{\displaystyle CH_3}{|}}{\underset{\underset{\displaystyle CH_3}{|}}{C}}-CH_3$

2.4 Alkyl and Halogen Substituents

As illustrated for the methyl group, alkyl substituents are named by changing the -*ane* ending of alkanes to -*yl*. Thus the two-carbon alkyl group is called the **ethyl group,** from ethane.

> The two-carbon alkyl group is the **ethyl group**. The **propyl group** and the **isopropyl group** are three-carbon groups attached to the main chain by the first and second carbons, respectively.

$$CH_3CH_3 \qquad CH_3CH_2- \quad \text{or} \quad C_2H_5- \quad \text{or} \quad Et-$$

ethane   ethyl group

When we come to propane, there are two possible alkyl groups, depending on which type of hydrogen is removed. If a *terminal* hydrogen is removed, the group is called a **propyl group.**

$$H-\overset{\overset{\displaystyle H}{|}}{\underset{\underset{\displaystyle H}{|}}{C}}-\overset{\overset{\displaystyle H}{|}}{\underset{\underset{\displaystyle H}{|}}{C}}-\overset{\overset{\displaystyle H}{|}}{\underset{\underset{\displaystyle H}{|}}{C}}-H \qquad H-\overset{\overset{\displaystyle H}{|}}{\underset{\underset{\displaystyle H}{|}}{C}}-\overset{\overset{\displaystyle H}{|}}{\underset{\underset{\displaystyle H}{|}}{C}}-\overset{\overset{\displaystyle H}{|}}{\underset{\underset{\displaystyle H}{|}}{C}}- \quad \text{or} \quad CH_3CH_2CH_2- \quad \text{or} \quad Pr-$$

propyl group

But if a hydrogen is removed from the *central* carbon atom, we get a different isomeric propyl group, called the **isopropyl** (or 1-methylethyl) **group.**

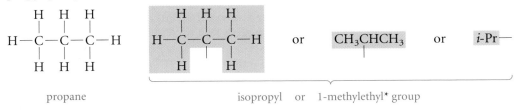

propane   isopropyl   or   1-methylethyl* group

There are four different butyl groups. The butyl and *sec*-butyl groups are based on *n*-butane, while the isobutyl and *tert*-butyl groups come from isobutane.

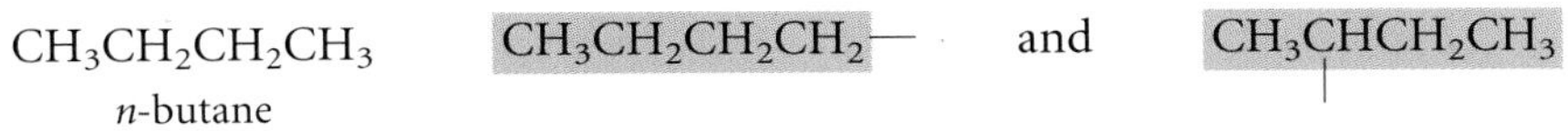

$CH_3CH_2CH_2CH_3$ $CH_3CH_2CH_2CH_2$—   and $CH_3CHCH_2CH_3$

n-butane

butyl

sec-butyl
(or 1-methylpropyl)

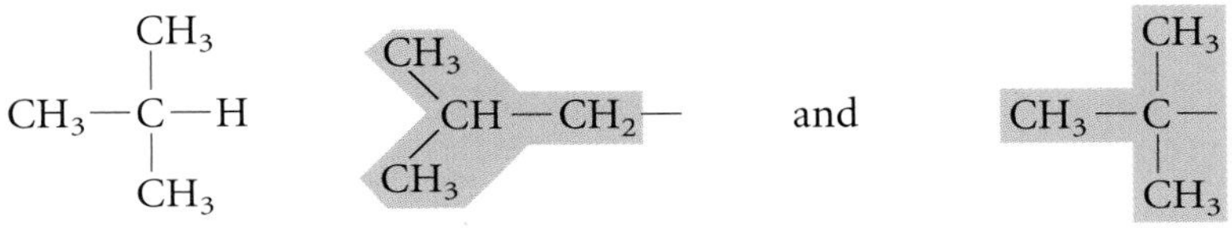

isobutane   isobutyl
(or 2-methylpropyl)   *tert*-butyl
(or 1,1-dimethylethyl)

These names for the alkyl groups with up to four carbon atoms are very commonly used, so you should memorize them.

The letter **R** is used as a general symbol for an alkyl group. The formula R—H therefore represents any alkane, and the formula R—Cl stands for any alkyl chloride (methyl chloride, ethyl chloride, and so on).

R is the general symbol for an alkyl group.

Halogen substituents are named by changing the -*ine* ending of the element to -*o*.

F—   Cl—   Br—   I—
fluoro-   chloro-   bromo-   iodo-

EXAMPLE 2.3

Give the common and IUPAC names for $CH_3CH_2CH_2Br$.

Solution   The common name is propyl bromide (the common name of the alkyl group is followed by the name of the halide). The IUPAC name is 1-bromopropane, the halogen being named as a substituent on the three-carbon chain.

PROBLEM 2.4   Give an IUPAC name for CH_2ClF.

PROBLEM 2.5   Write the formula for each of the following compounds:

a. propyl chloride   b. isopropyl iodide
c. 2-chloropropane   d. *tert*-butyl iodide
e. isobutyl bromide   f. general formula for an alkyl fluoride

*The name 1-methylethyl for this group comes about by regarding it as a substituted ethyl group.

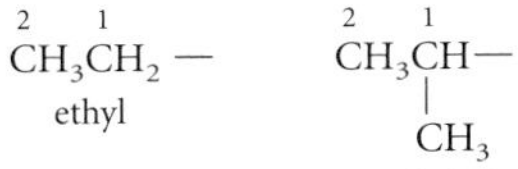

CH_3CH_2—
ethyl

CH_3CH—
CH_3
1-methylethyl

2.5 Use of the IUPAC Rules

The examples given in Table 2.2 illustrate how the IUPAC rules are applied for particular structures. Study each example to see how a correct name is obtained and how to avoid certain pitfalls.

It is important not only to be able to write a correct IUPAC name for a given structure, but also to do the converse: Write the structure given the IUPAC name. In this case, first write the longest carbon chain and number it, then add the substituents to the correct carbon atoms, and finally fill in the formula with the correct number of hydrogens at each carbon. For example, to write the formula for 2,2,4-trimethylpentane, we go through the following steps:

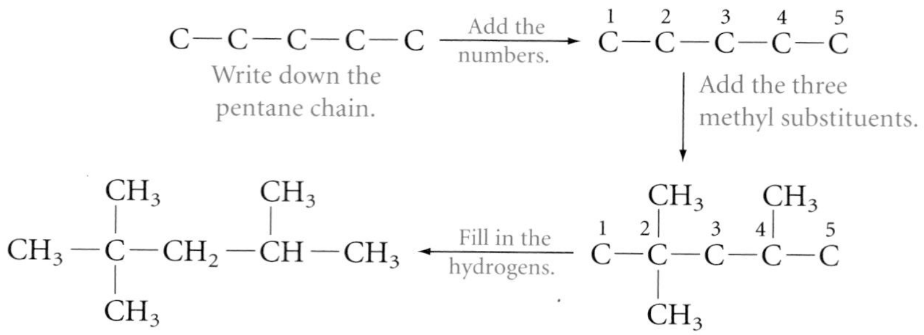

C—C—C—C—C $\xrightarrow[\text{numbers.}]{\text{Add the}}$ $\overset{1}{C}-\overset{2}{C}-\overset{3}{C}-\overset{4}{C}-\overset{5}{C}$

Write down the
pentane chain.

Add the three
methyl substituents.

$$CH_3-\underset{\underset{CH_3}{|}}{\overset{\overset{CH_3}{|}}{C}}-CH_2-\overset{\overset{CH_3}{|}}{CH}-CH_3 \xleftarrow[\text{hydrogens.}]{\text{Fill in the}} \overset{1}{C}-\overset{2}{\underset{\underset{CH_3}{|}}{\overset{\overset{CH_3}{|}}{C}}}-\overset{3}{C}-\overset{4}{\overset{\overset{CH_3}{|}}{C}}-\overset{5}{C}$$

2,2,4-trimethylpentane

Table 2.2 **Examples of use of the IUPAC rules**	
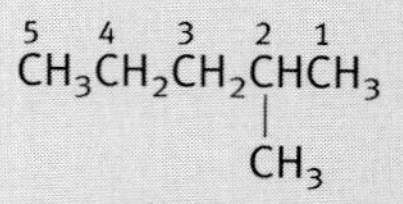 $\overset{5}{C}H_3\overset{4}{C}H_2\overset{3}{C}H_2\overset{2}{C}H\overset{1}{C}H_3$ $\underset{CH_3}{\vert}$ 2-methylpentane (*not* 4-methylpentane)	The ending -*ane* tells us that all the carbon–carbon bonds are single; *pent-* indicates five carbons in the longest chain. We number them from right to left, starting closest to the branch point.

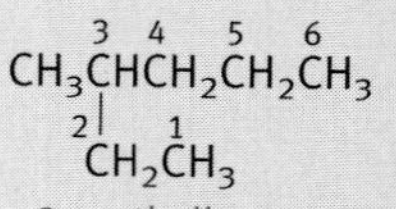 $\overset{3}{C}H_3\overset{4}{C}H\overset{5}{C}H_2\overset{6}{C}H_2CH_3$
  $\overset{2}{\underset{1}{\vert}}$
  CH_2CH_3
 	
3-methylhexane
(*not* 2-ethylpentane
or 4-methylhexane)
	This is a six-carbon saturated chain with a methyl group on the third carbon. We would usually write the structure as $CH_3CH_2CHCH_2CH_2CH_3$.
	    $\underset{CH_3}{\vert}$

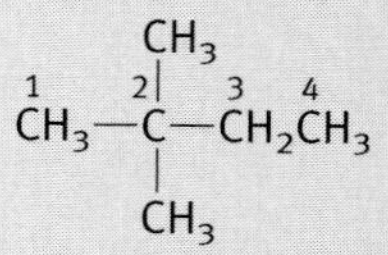 CH_3 $\overset{1}{C}H_3-\overset{2}{\underset{\underset{CH_3}{\vert}}{\overset{\overset{\vert}{}}{C}}}-\overset{3}{C}H_2\overset{4}{C}H_3$ 2,2-dimethylbutane (*not* 2,2-methylbutane or 2-dimethylbutane)	There must be a number for each substituent, and the prefix *di-* says that there are two methyl substituents.
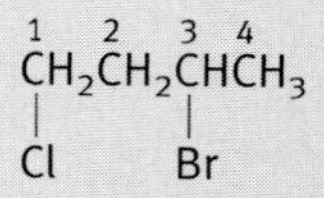 $\overset{1}{C}H_2\overset{2}{C}H_2\overset{3}{C}H\overset{4}{C}H_3$ $\underset{Cl}{\vert}\quad\underset{Br}{\vert}$ 3-bromo-1-chlorobutane (*not* 1-chloro-3-bromobutane or 2-bromo-4-chlorobutane)	First, we number the butane chain from the end closest to the first substituent. Then we name the substituents in alphabetical order, regardless of position number.

PROBLEM 2.6   Name the following compounds by the IUPAC system:

a. CH_3CHFCH_3 b. $(CH_3)_3CCH_2CHClCH_3$

PROBLEM 2.7   Write the structure for 3,3-dimethylpentane.

PROBLEM 2.8   Explain why 1,3-dichlorobutane is a correct IUPAC name, but 1,3-dimethylbutane is *not* a correct IUPAC name.

2.6   Sources of Alkanes

The two most important natural sources of alkanes are **petroleum** and **natural gas.** Petroleum is a complex liquid mixture of organic compounds, many of which are alkanes or cycloalkanes. For more details about how petroleum is refined to obtain gasoline, fuel oil, and other useful substances, read "A Word About Petroleum, Gasoline, and Octane Number" on pages 106–107.

Natural gas, often found associated with petroleum deposits, consists mainly of methane (about 80%) and ethane (5 to 10%), with lesser amounts of some higher alkanes. Propane is the major constituent of liquefied petroleum gas (LPG), a domestic fuel used mainly in rural areas and mobile homes. Butane is the gas of choice in some areas. Natural gas is becoming an energy source that can compete with and possibly surpass oil. In the United States, there are about a million miles of natural gas pipelines distributing this energy source to all parts of the country. Natural gas is also distributed worldwide via huge tankers. To conserve space, the gas is liquefied ($-160°C$), because 1 cubic meter (m^3) of liquefied gas is equivalent to about 600 m^3 of gas at atmospheric pressure. Large tankers can carry more than 100,000 m^3 of liquefied gas.

> **Petroleum** and **natural gas** are the two most important natural sources of alkanes.

A CLOSER LOOK AT ...

Natural Gas

Log on to http://www.college.hmco.com and follow the links to the student text web site. Use the Web Link section for Chapter 2 to answer the following questions and find more information on the topics below.

U.S. Department of Energy—Fossil Fuels

1. What are the physical properties of natural gas? Why does natural gas used for home heating (or from the Bunsen burner in your chemistry lab) have an odor?
2. How is natural gas produced in nature? What keeps it from escaping to the surface of the earth? How is it obtained from its natural source?
3. How many miles (or kilometers) of pipeline for transporting natural gas currently exist in the United States?

What percent of U.S. energy needs are supplied by natural gas? What are the major geographic locations of natural gas in the United States and Canada?

World Map of Petroleum Basins

1. Where are the greatest potential sources of natural gas located?
2. Using the map, go to your region of the world to explore where potential natural gas resources are located. Report on your findings.
3. In the region you have selected, view the natural gas resources data in tabular form. How many trillion cubic feet of natural gas can potentially be found in your country according to the table you are viewing?

(a)

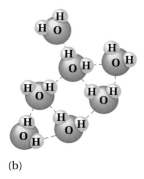

(b)

Figure 2.2
Hydrogen bonding: (a) the polar water molecule and (b) hydrogen bonding between water molecules.

2.7 Physical Properties of Alkanes and Nonbonding Intermolecular Interactions

Alkanes are insoluble in water. This is because water molecules are *polar*, whereas alkanes are *nonpolar* (all the C—C and C—H bonds are nearly purely covalent). The O—H bond in a water molecule is strongly polarized by the high electronegativity of oxygen (Sec. 1.5). This polarization places a partial positive charge on the hydrogen atom and a partial negative charge on the oxygen atom. As a result, the hydrogen atoms in one water molecule are strongly attracted to the oxygen atoms in other water molecules, and the small size of the H atoms allows the molecules to approach each other very closely. This special attraction is called **hydrogen bonding** (Figure 2.2).* To intersperse alkane and water molecules, we would have to break up the hydrogen bonding interactions between water molecules, which would require considerable energy. Alkanes, with their nonpolar C—H bonds, cannot replace hydrogen bonding among water molecules with attractive alkane–water interactions that are comparable in strength, so mixing alkane molecules and water molecules is *not* an energetically favored process.

The mutual insolubility of alkanes and water is used to advantage by many plants. Alkanes often constitute part of the protective coating on leaves and fruits. If you have ever polished an apple, you know that the skin, or cuticle, contains waxes. Among them are the normal alkanes $C_{27}H_{56}$ and $C_{29}H_{60}$. The leaf wax of cabbage and broccoli is mainly *n*-$C_{29}H_{60}$, and the main alkane of tobacco leaves is *n*-$C_{31}H_{64}$. Similar hydrocarbons are found in beeswax. The major function of plant waxes is to prevent water loss from the leaves or fruit.

Alkanes have lower boiling points for a given molecular weight than most other organic compounds. This is because they are nonpolar molecules. Because they are constantly moving, the electrons in a nonpolar molecule can become unevenly distributed within the molecule, causing the molecule to have partially positive and partially negative ends. The *temporarily* polarized molecule causes its neighbor to become temporarily polarized as well, and these molecules are weakly attracted to each other. Such interactions between molecules are called **van der Waals attractions.**

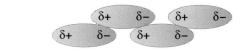

Figure 2.3
As shown by the curve, the boiling points of the normal alkanes rise smoothly as the length of the carbon chain increases. Note from the table, however, that chain branching causes a decrease in boiling point (each compound in the table has the same number of carbons and hydrogens, C_5H_{12}).

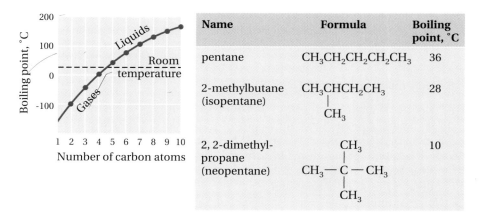

Name	Formula	Boiling point, °C
pentane	$CH_3CH_2CH_2CH_2CH_3$	36
2-methylbutane (isopentane)	$CH_3CHCH_2CH_3$ $\quad\quad\;\mid$ $\quad\quad CH_3$	28
2,2-dimethyl-propane (neopentane)	$\quad\quad\quad CH_3$ $\quad\quad\quad\mid$ CH_3-C-CH_3 $\quad\quad\quad\mid$ $\quad\quad\quad CH_3$	10

*Molecules that contain N—H and F—H covalent bonds also have hydrogen bonding interactions with molecules containing N, O, or F atoms.

A CLOSER LOOK AT ...

Hydrogen Bonding

Log on to http://www.college.hmco.com and follow the links to the student text web site. Use the Web Link section for Chapter 2 to answer the following questions and find more information on the topics below.

Water and Ice

1. Use the section on the Water and Ice link called Water Is a Polar Molecule to answer the following questions.
   a. Identify the oxygen atom and the hydrogen atoms in the model of a water molecule. What is the partial charge on the O atom? On each H atom?
   b. If the water molecule were linear (H—O—H) instead of (H—O—H with O apex) would it still be polar?

2. Now, read the second section, Water Is Highly Cohesive and answer the questions below.
   a. Describe what you see in the movie. What is the average length (in angstroms) of a hydrogen bond between two water molecules?
   b. How is a hydrogen bond different from a covalent bond?

3. Read the third section found on the Water and Ice link, and examine the pictures to answer the questions below.
   a. How many hydrogen bonds does a water molecule make in liquid water? In ice?
   b. Why is ice less dense than liquid water? How is this related to hydrogen bonding?

4. One property of alkanes is that they are not soluble in water. This is related to interactions between molecules. Let us explore the connection to hydrogen bonding.
   a. What is required for hydrogen bonding between molecules?
   b. Can alkane molecules form hydrogen bonds to each other or to water molecules?
   c. When the oil tanker Exxon Valdez broke apart near the Alaskan coast, the oil (composed mostly of alkanes) that spilled did not dissolve in the water. It floated on top. Explain these facts in terms of intermolecular interactions and the properties of alkanes and water molecules.

Because they are weak attractions, the process of separating molecules from one another (which is what we do when we convert a liquid to a gas) requires relatively little energy, and the boiling points of these compounds are relatively low. Figure 2.3 shows the boiling points of some alkanes. Since these attractive forces can only operate over short distances between the surfaces of molecules, *the boiling points of alkanes rise as the chain length increases and fall as the chains become branched and more nearly spherical in shape.* Figure 2.4 illustrates the effect of molecular shape on van der Waals attractions.

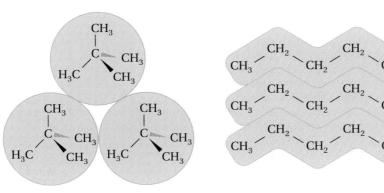

2, 2-dimethylpropane
bp 10°C

pentane
bp 36°C

Figure 2.4
2,2-Dimethylpropane and pentane have the same molecular weight, but the rod-shaped pentane molecules have more surface area available for contact between them than the spherical 2,2-dimethylpropane molecules. Pentane molecules, therefore, experience more van der Waals attractions than 2,2-dimethylpropane molecules.

Hydrogen bonding and **van der Waals attractions** are nonbonding intermolecular interactions.

Hydrogen bonding and van der Waals attractions are examples of **nonbonding intermolecular interactions.** These kinds of interactions have important consequences for the properties and behavior of molecules, and we will encounter more examples as we continue to explore the chemistry of different classes of organic compounds.

2.8   Conformations of Alkanes

The shapes of molecules often affect their properties. A simple molecule like ethane, for example, can have an infinite number of shapes as a consequence of rotating one carbon atom (and its attached hydrogens) with respect to the other carbon atom. These arrangements are called **conformations** or **conformers.** Conformers are **stereoisomers,** isomers in which the atoms are connected in the same order but are arranged differently in space. Two possible conformers for ethane are shown in Figure 2.5.*

Different **conformations** (shapes) of the same molecule that are interconvertible by rotation around a single bond are called **conformers** or **rotamers.** Conformers are **stereoisomers,** isomers with the same atom connectivity but different spatial arrangements of atoms.

In the **staggered conformation** of ethane, each C—H bond on one carbon bisects an H—C—H angle on the other carbon. In the **eclipsed conformation,** C—H bonds on the front and back carbons are aligned. By rotating one carbon 60° with respect to the other, we can interconvert staggered and eclipsed conformations. Between these two extremes are an infinite number of intermediate conformations of ethane.

⊚ **Ch. 2, Ref. 2.** *View animation of C—C bond rotation in ethane.*

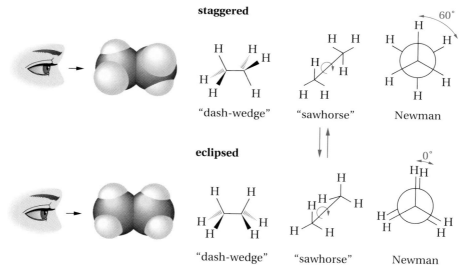

Figure 2.5
Two of the possible conformations of ethane: staggered and eclipsed. Interconversion is easy via a 60° rotation about the C—C bond, as shown by the curved arrows. The structures at the left are space-filling models. In each case, the next structure is a "dash-wedge" structure, which, if viewed as shown by the eyes, converts to the "sawhorse" drawing, or the Newman projection at the right, an end-on view down the C—C axis. In the Newman projection, the circle represents two connected carbon atoms. Bonds on the "front" carbon go to the center of the circle, and bonds on the "rear" carbon go only to the edge of the circle.

*Build a 3D model of ethane (using a molecular model kit) and use it as you read this section to model the staggered and eclipsed conformations shown in Figure 2.5.

The staggered and eclipsed conformations of ethane can be regarded as **rotamers** because each is convertible to the other by rotation about the carbon–carbon bond. Such rotation about a single bond occurs easily because the amount of overlap of the sp^3 orbitals on the two carbon atoms is unaffected by rotation about the sigma bond (see Figure 1.8). Indeed, there is enough energy available at room temperature for the staggered and eclipsed conformers of ethane to interconvert rapidly. Consequently, the conformers cannot be separated from one another. We know from various types of physical evidence, however, that both forms are not equally stable. The staggered conformation is the most stable (has the lowest potential energy) of all ethane conformations, while the eclipsed conformation is the least stable (has the highest potential energy). At room temperature, the staggered conformation is practically the only conformation present.

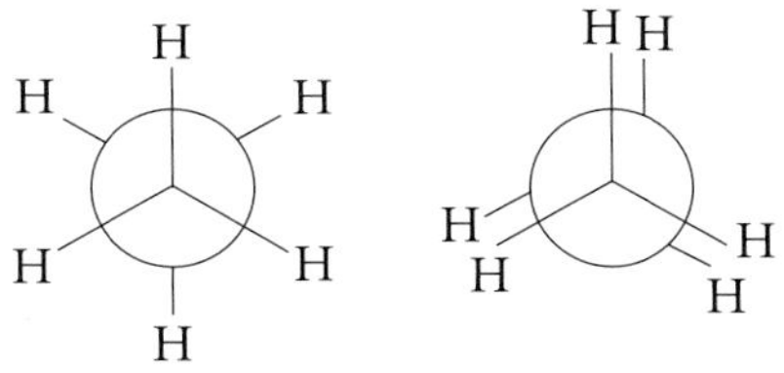

staggered   eclipsed

(2.1)

EXAMPLE 2.4

Draw the Newman projections for the staggered and eclipsed conformations of propane.

Solution

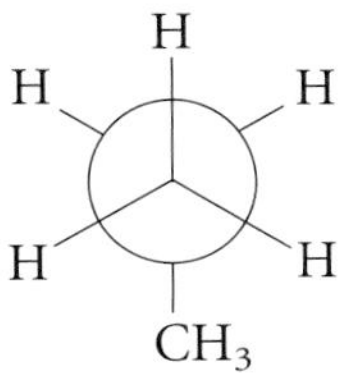

staggered

The projection formula is similar to that of ethane, except for the replacement of one hydrogen with methyl.

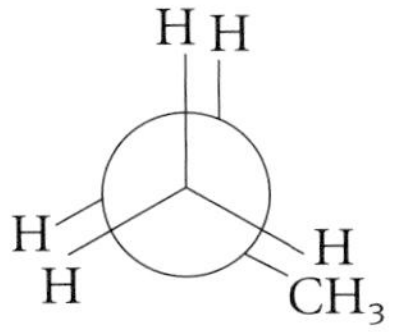

eclipsed

Rotation of the "rear" carbon of the staggered conformation by 60° gives the eclipsed conformation shown.

We are looking down the C_1—C_2 bond.

PROBLEM 2.9 Draw Newman projections for two different *staggered* conformations of butane (looking end-on at the bond between carbon-2 and carbon-3), and predict which of the two conformations is more stable.

The most important thing to remember about conformers is that they are just different forms of a single molecule that can be interconverted by rotational motions about single (sigma) bonds. More often than not, there is sufficient thermal energy for this rotation at room temperature. Consequently, at room temperature it is usually not possible to separate conformers from one another.

Now let us look at the structures of cycloalkanes and their conformations.

2.9 Cycloalkane Nomenclature and Conformation

Cycloalkanes are saturated hydrocarbons that have at least one ring of carbon atoms. A common example is cyclohexane.

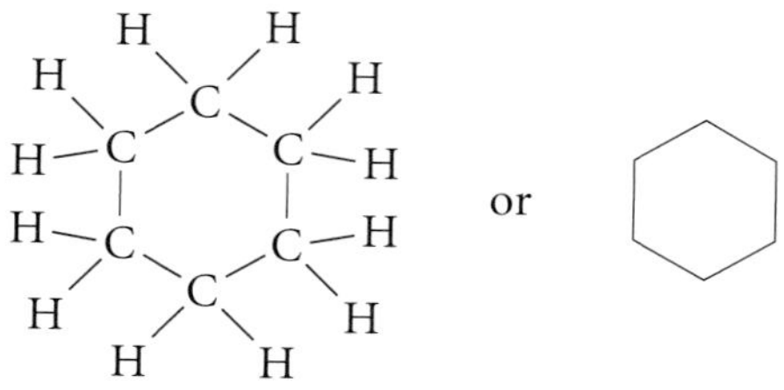

Structural and abbreviated structural
formulas for cyclohexane

Cycloalkanes are named by placing the prefix *cyclo-* before the alkane name that corresponds to the number of carbon atoms in the ring. The structures and names of the first six unsubstituted cycloalkanes are as follows:

| cyclopropane | cyclobutane | cyclopentane | cyclohexane | cycloheptane | cyclooctane |
| bp −32.7°C | bp 12°C | bp 49.3°C | bp 80.7°C | bp 118.5°C | bp 149°C |

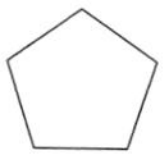

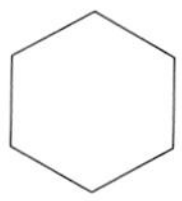

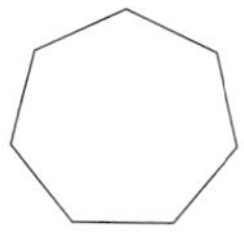

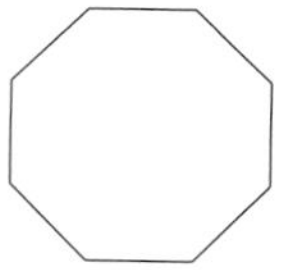

Alkyl or halogen substituents attached to the rings are named in the usual way. If only one substituent is present, no number is needed to locate it. If there are several substituents, numbers are required. One substituent is always located at ring carbon number 1, and the remaining ring carbons are then numbered consecutively in a way that gives the other substituents the lowest possible numbers. With different substituents, the one with highest alphabetic priority is located at carbon 1. The following examples illustrate the system:

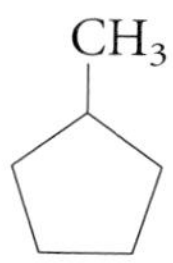

methylcyclopentane
(*not* 1-methylcyclopentane)

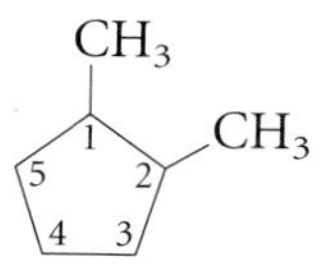

1,2-dimethylcyclopentane
(*not* 1,5-dimethylcyclopentane)

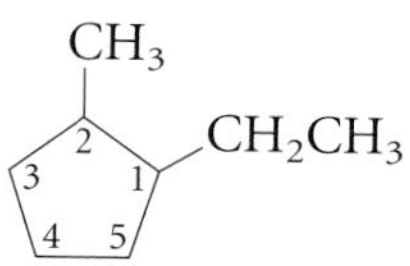

1-ethyl-2-methylcyclopentane
(*not* 2-ethyl-1-methylcyclopentane)

PROBLEM 2.10   The general formula for an alkane is C_nH_{2n+2}. What is the corresponding formula for a cycloalkane with one ring?

PROBLEM 2.11   Draw the structural formulas for

a. 1,3-dimethylcyclohexane.
b. 1,2,3-trichlorocyclopropane.

PROBLEM 2.12   Give IUPAC names for

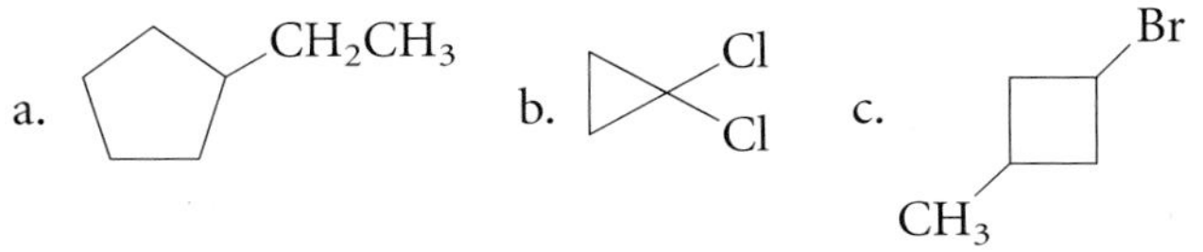

a. (cyclopentane with CH_2CH_3 substituent)   b. (cyclopropane with two Cl substituents)   c. (cyclobutane with Br and CH_3 substituents)

What are the conformations of cycloalkanes? Cyclopropane, with only three carbon atoms, is necessarily planar (because three points determine a plane). The C—C—C angle is only 60° (the carbons form an equilateral triangle), much less than the usual tetrahedral angle of 109.5°. The hydrogens lie above and below the carbon plane, and hydrogens on adjacent carbons are eclipsed.

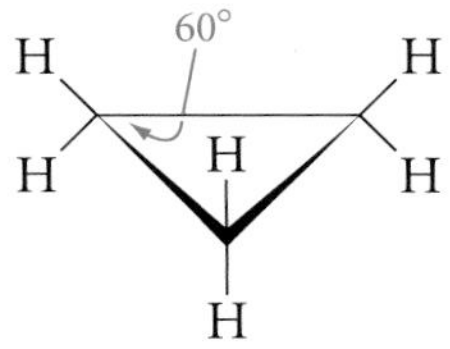

cyclopropane

EXAMPLE 2.5

Explain why the hydrogens in cyclopropane lie above and below the carbon plane.

Solution   Refer to Figure 1.10. The carbons in cyclopropane have a geometry similar to that shown there, except that the C—C—C angle is "squeezed" and is smaller than tetrahedral. In compensation, the H—C—H angle is expanded and is larger than tetrahedral, approximately 120°.

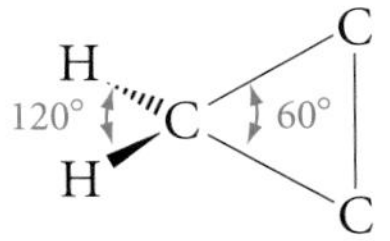

The H—C—H plane perpendicularly bisects the C—C—C plane, which, as drawn here, lies in the plane of the paper.

Cycloalkanes with more than three carbon atoms are nonplanar and have "puckered" conformations. In cyclobutane and cyclopentane, puckering allows the molecule to adopt the most stable conformation (with the least strain energy). Puckering introduces strain by making the C—C—C angles a little smaller than they would be if the molecules were planar; however, less eclipsing of the adjacent hydrogens compensates for this.

C—C—C angle for planar molecule	cyclobutane	cyclopentane
	90°	108°
observed experimentally	88°	105°

Six-membered rings are rather special and have been studied in great detail because they are so common in nature. If cyclohexane were planar, the internal C—C—C angles would be those of a regular hexagon, 120°—quite a bit larger than the normal tetrahedral angle (109.5°). The resulting strain prevents cyclohexane from being planar (flat). Its most favored conformation is the **chair conformation,** an arrangement in which all the C—C—C angles are 109.5° and all the hydrogens on adjacent carbon atoms are perfectly staggered. Figure 2.6 shows models of the cyclohexane chair conformation.* (If a set of molecular models is available, it would be a good idea for you to construct a cyclohexane model to better visualize the concepts discussed in this and the next two sections.)

> In the **chair conformation** of cyclohexane, the six **axial** hydrogen atoms lie above and below the mean plane of the ring, while the six **equatorial** hydrogens lie in the plane.

PROBLEM 2.13   How are the H—C—H and C—C—C planes at any one carbon atom in cyclohexane related? (Refer, if necessary, to Example 2.5.)

In the chair conformation, the hydrogens in cyclohexane fall into two sets, called **axial** and **equatorial.** Three axial hydrogens lie above and three lie below the average plane of the carbon atoms; the six equatorial hydrogens lie approximately in that plane. By a motion in which alternate ring carbons (say, 1, 3, and 5) move in one direction (down) and the other three ring carbons move in the opposite direction (up), one chair conformation can be converted into another chair conformation in which

---

*****Diamond** is one naturally occurring form of carbon. In the diamond crystal, the carbon atoms are connected to one another in a structure similar to the chair form of cyclohexane, except that all of the hydrogens are replaced by carbon atoms, resulting in a continuous network of carbon atoms. The hydrocarbons **adamantane** and **diamantane** show the beginnings of the diamond structure in their fusing of chair cyclohexanes. For a fascinating article on diamond structure, see "Diamond Cleavage" by M. F. Ansell in *Chemistry in Britain*, **1984,** 1017–1021.

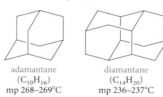

adamantane
($C_{10}H_{16}$)
mp 268–269°C

diamantane
($C_{14}H_{20}$)
mp 236–237°C

all axial hydrogens have become equatorial, and all equatorial hydrogens have become axial.

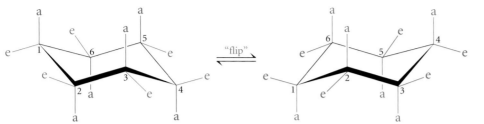

(2.2)

Axial bonds (red) in the left structure become equatorial bonds
(red) in the right structure when the ring "flips."

◎ **Ch. 2, Ref. 3.** *View animation of cyclohexane ring flip.*

At room temperature this flipping process is rapid, but at low temperatures (say, $-90°C$), it slows down enough that the two different types of hydrogens can actually be detected by proton nuclear magnetic resonance (NMR) spectroscopy (see Chapter 12).

Cyclohexane conformations have another important feature. If you look carefully at the space-filling model of cyclohexane (Figure 2.6), you will notice that *the three axial hydrogens on the same face of the ring nearly touch each other.* If an axial hydrogen is replaced by a larger substituent (such as a methyl group), the axial crowding is even worse. Therefore, the preferred conformation is the one in which the larger substituent, in this case the methyl group, is equatorial.

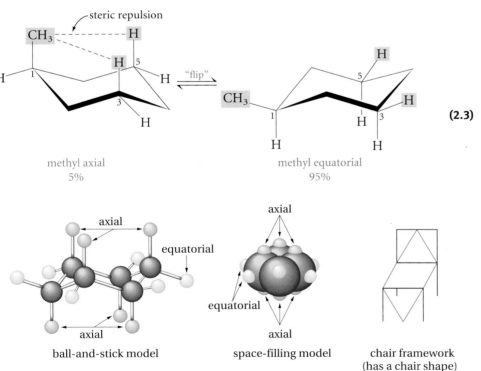

(2.3)

Figure 2.6
The chair conformation of cyclohexane, shown in ball-and-stick and space-filling models. The axial hydrogens lie above or below the mean plane of the carbons, and the six equatorial hydrogens lie approximately in that mean plane. The origin of the chair terminology is illustrated at the right.

PROBLEM 2.14   Another puckered conformation for cyclohexane, one in which all C—C—C angles are the normal 109.5°, is the boat conformation.

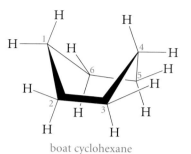

boat cyclohexane

Explain why this conformation is very much less stable than the chair conformation. (*Hint:* Note the arrangement of hydrogens as you sight along the bond between carbon-2 and carbon-3; a molecular model will help you answer this problem.)

PROBLEM 2.15   For *tert*-butylcyclohexane only one conformation, with the *tert*-butyl group equatorial, is detected. Explain why this conformational preference is greater than that for methylcyclohexane (see eq. 2.3).

The six-membered ring in the chair conformation is a common structural feature of many organic molecules, including sugar molecules (Sec. 16.7) like glucose, where one ring carbon is replaced by an oxygen atom.

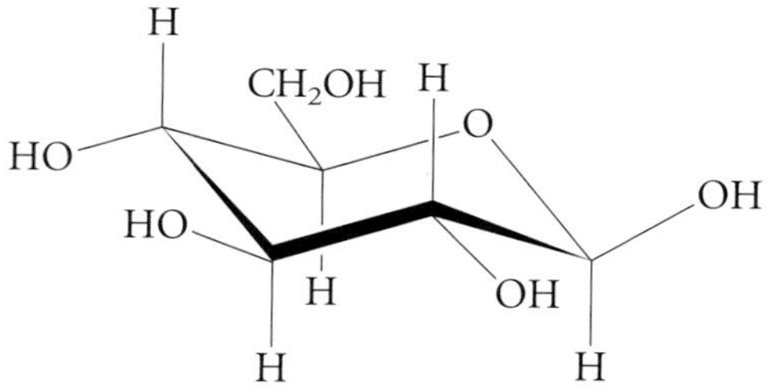

glucose
(β-D-glucopyranose)

Notice that the bulkier group on each carbon is in the equatorial position. The conformations of sugars will be studied in greater detail in Chapter 16.

Before we proceed to reactions of alkanes and cycloalkanes, we need to consider a type of isomerism that may arise when two or more carbon atoms in a cycloalkane have substituents.

2.10   *Cis–trans* Isomerism in Cycloalkanes

Cis–trans isomers of cycloalkanes are a type of stereoisomer, also called **geometric** stereoisomers, in which substituents are on the same side (*cis*) or on opposite sides (*trans*) of the ring.

Stereoisomerism deals with molecules that have the same order of attachment of the atoms, but different arrangements of the atoms in space. *Cis–trans* isomerism (sometimes called **geometric isomerism**) is one kind of stereoisomerism, and it is most easily understood with a specific case. Consider, for example, the possible structures of 1,2-dimethylcyclopentane. For simplicity, let us neglect the slight puckering of the ring and draw it as if it were planar. The two methyl groups may be on the same side of the ring plane or they may be on opposite sides.

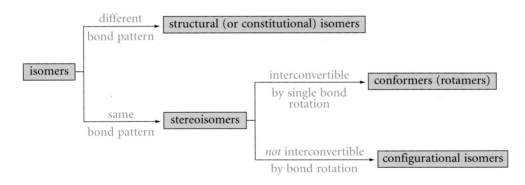

cis-1,2-dimethylcyclopentane
bp 99°C

trans-1,2-dimethylcyclopentane
bp 92°C

The methyl groups are said to be *cis* (Latin, on the same side) or *trans* (Latin, across) to each other.

Cis–trans isomers differ from one another only in the way the atoms or groups are positioned in space. Yet this difference is sufficient to give them different physical and chemical properties. (Note, for example, the boiling points under the 1,2-dimethylcyclopentane structures.) *Cis–trans* isomers are separate and unique compounds. Unlike conformers, they are not readily interconverted by rotation around carbon–carbon bonds. In this example, the cyclic structure limits rotation about the ring bonds. To interconvert these dimethylcyclopentanes, one would have to break open, rotate, and re-form the ring, or carry out some other bond-breaking process.

Cis–trans isomers can be separated from each other and kept separate, usually without interconversion at room temperature. *Cis–trans* isomerism can be important in determining the biological properties of molecules. For example, a molecule in which two reactive groups are *cis* will interact differently with an enzyme or biological receptor site than will its isomer with the same two groups *trans*.

PROBLEM 2.16 Draw the structure for the *cis* and *trans* isomers of

a. 1-bromo-2-chlorocyclopropane
b. 1,3-dichlorocyclobutane

2.11 Summary of Isomerism

At this point, it may be useful to summarize the relationships of the several types of isomers we have discussed so far. These relationships are outlined in Figure 2.7.

Figure 2.7
The relationships of the various types of isomers.

A WORD ABOUT ...

Isomers, Possible and Impossible

Table 2.1 shows that there are 75 structural isomers of the alkane $C_{10}H_{22}$. How many such isomers do you think there might be if we double the number of carbons ($C_{20}H_{42}$)? The answer is 366,319! And if we double the number of carbons again ($C_{40}H_{82}$)? Exactly 62,481,801,147,341. Of course, no one sits down with pencil and paper or molecular models and determines these numbers by constructing all the possibilities; it could take a lifetime. Complex mathematical formulas have been developed to compute these numbers.

Although we can write some isomers' formulas on paper, they are structurally impossible and cannot be synthesized. Consider, for example, the series of alkanes obtained by replacing the hydrogens of methane with methyl groups and then repeating that process on the product indefinitely. You can see from the drawings below, even though they are only two dimensional, that in this way we build up molecules with a central core of carbon atoms and a surface of hydrogen atoms.

$$CH_4 \longrightarrow C(CH_3)_4 \longrightarrow$$
$$C[C(CH_3)_3]_4 \longrightarrow C\{C[C(CH_3)_3]_3\}_4$$
$$CH_4 \longrightarrow C_5H_{12} \longrightarrow C_{17}H_{36} \longrightarrow C_{53}H_{108}$$

In three dimensions, the molecules are nearly spherical. Of these compounds, only the first two are known (methane and 2,2-dimethylpropane). The $C_{17}H_{36}$ hydrocarbon (tetra-*t*-butylmethane or, more accurately, 3,3-di-*t*-butyl-2,2,4,4-tetramethylpentane) has not yet been synthesized, and if it ever is, it will be an exceptionally strained molecule. The reason is simply that there is not enough room for all the methyl groups on the surface of the molecule.

The first thing to look at in a pair of isomers is their bonding patterns (or atom connectivities). If the bonding patterns are *different*, the compounds are **structural** (or **constitutional**) **isomers.** But if the bonding patterns are the *same*, the compounds are stereoisomers. Examples of structural isomers are ethanol and methoxymethane (pages 21–22) or the three isomeric pentanes (page 22). Examples of stereoisomers are the staggered and eclipsed forms of ethane (page 55) or the *cis* and *trans* isomers of 1,2-dimethylcyclopentane (page 59).

If compounds are stereoisomers, we can make a further distinction as to isomer type. If *single-bond rotation* easily interconverts the two stereoisomers (as with staggered and eclipsed ethane), we call them conformers. If the two stereoisomers can be interconverted only by breaking and remaking bonds (as with *cis-* and *trans*-1,2-dimethylcyclopentane), we call them **configurational isomers.***

Configurational isomers (such as *cis–trans* isomers) are stereoisomers that can only be interconverted by breaking and remaking bonds.

PROBLEM 2.17 Classify each of the following isomer pairs according to the scheme in Figure 2.7.

a. 1-iodopropane and 2-iodopropane
b. *cis-* and *trans*-1,2-dimethylcyclohexane
c. chair and boat forms of cyclohexane

---

*Remember that conformers are different conformations of the same molecule, whereas configurational isomers are different molecules. Geometric isomers (*cis–trans* isomers) are one type of configurational isomer. As we will see in Chapter 3, geometric isomerism also occurs in alkenes. Also, we will see other types of configurational isomers in Chapter 5.

Therefore, if this $C_{17}H_{36}$ isomer is ever prepared, its bond angles and bond lengths are likely to be severely distorted from the normal. There is almost no possibility of synthesizing the $C_{53}H_{108}$ isomer in this series; its structure is too strained. It is interesting to note that, like these hydrocarbons, the growth of trees, sponges, and other biological structures is similarly limited by the ratio of surface area to volume. (For more on this subject, see the article by R. E. Davies and P. J. Freyd, *J. Chem. Educ.* **1989,** *66,* 278–281.)

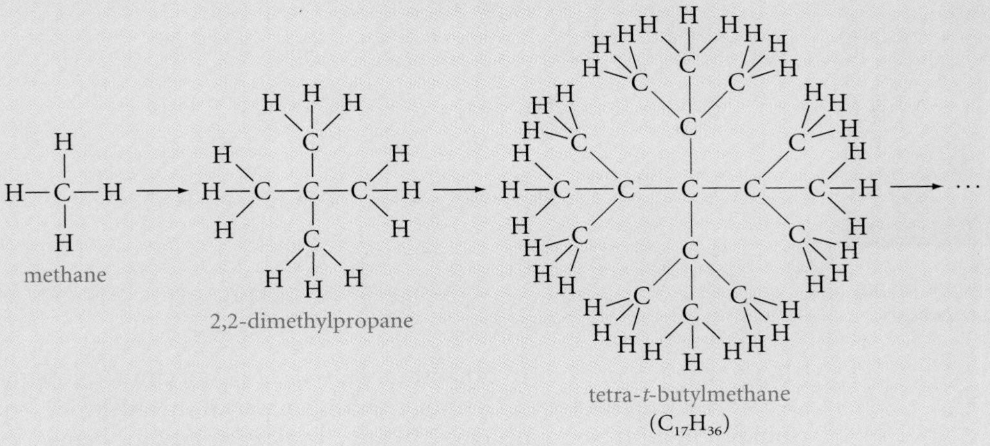

methane → 2,2-dimethylpropane → tetra-*t*-butylmethane ($C_{17}H_{36}$) → ···

2.12 Reactions of Alkanes

All the bonds in alkanes are single, covalent, and nonpolar. Hence alkanes are relatively inert. Alkanes ordinarily do not react with most common acids, bases, or oxidizing and reducing agents. Because of this inertness, alkanes can be used as solvents for extraction or crystallization or for carrying out chemical reactions of other substances. However, alkanes do react with some reagents, such as oxygen and the halogens. We will discuss those reactions here.

2.12.a Oxidation and Combustion; Alkanes as Fuels

The most important use of alkanes is as fuel. With excess oxygen, alkanes burn to form carbon dioxide and water. Most important, the reactions evolve considerable heat (that is, the reactions are **exothermic**).

Exothermic reactions evolve heat.

$$CH_4 \; + \; 2\,O_2 \longrightarrow CO_2 + 2\,H_2O + \text{heat (212.8 kcal/mol)} \qquad \textbf{(2.4)}$$

methane

$$C_4H_{10} + \tfrac{13}{2}\,O_2 \longrightarrow 4\,CO_2 + 5\,H_2O + \text{heat (688.0 kcal/mol)} \qquad \textbf{(2.5)}$$

butane

These combustion reactions are the basis for the use of hydrocarbons for heat (natural gas and heating oil) and for power (gasoline). An initiation step is required—usually ignition by a spark or flame. Once initiated, the reaction proceeds spontaneously and exothermically.

Combustion of hydrocarbons is an **oxidation reaction** in which C—H bonds are replaced with C—O bonds.

Auto tailpipe exhaust fumes contain condensed water.

In methane, all four bonds to the carbon atom are C—H bonds. In carbon dioxide, its combustion product, all four bonds to the carbon are C—O bonds. **Combustion** is an **oxidation reaction,** the replacement of C—H bonds by C—O bonds. In methane, carbon is in its most reduced form. In carbon dioxide, carbon is in its most oxidized form. Intermediate oxidation states of carbon are also known, in which only one, two, or three of the C—H bonds are converted to C—O bonds. It is not surprising, then, that if insufficient oxygen is available for complete combustion of a hydrocarbon, *partial* oxidation may occur, as illustrated in eqs. 2.6 through 2.9.

$$2 \text{ CH}_4 + 3 \text{ O}_2 \longrightarrow \underset{\text{carbon monoxide}}{2 \text{ CO}} + 4 \text{ H}_2\text{O} \qquad \textbf{(2.6)}$$

$$\text{CH}_4 + \text{O}_2 \longrightarrow \underset{\text{carbon}}{\text{C}} + 2 \text{ H}_2\text{O} \qquad \textbf{(2.7)}$$

$$\text{CH}_4 + \text{O}_2 \longrightarrow \underset{\text{formaldehyde}}{\text{CH}_2\text{O}} + \text{H}_2\text{O} \qquad \textbf{(2.8)}$$

$$2 \text{ C}_2\text{H}_6 + 3 \text{ O}_2 \longrightarrow \underset{\text{acetic acid}}{2 \text{ CH}_3\text{CO}_2\text{H}} + 2 \text{ H}_2\text{O} \qquad \textbf{(2.9)}$$

Toxic carbon monoxide in exhaust fumes (eq. 2.6), soot emitted copiously from trucks with diesel engines (eq. 2.7), smog resulting in part from aldehydes (eq. 2.8), and acid buildup in lubricating oils (eq. 2.9) are all prices we pay for being a motorized society.* However, incomplete hydrocarbon combustion is occasionally useful, as in the manufacture of carbon blacks (eq. 2.7) used for automobile tires, and lampblack, a pigment used in ink.

EXAMPLE 2.6

In which compound is carbon more oxidized, formaldehyde (CH_2O) or formic acid (HCO_2H)?

Solution   Draw the structures:

$$\underset{\text{formaldehyde}}{\overset{\displaystyle H}{\underset{\displaystyle H}{>}}C=O} \qquad \underset{\text{formic acid}}{H-C\overset{\displaystyle O}{\underset{\displaystyle OH}{<}}}$$

Formic acid is the more oxidized form (three C—O and one C—H bond, compared to two C—O and two C—H bonds in formaldehyde).

PROBLEM 2.18   Which of the following represents the more oxidized form of carbon?

a. methanol (CH_3OH) or formaldehyde
b. methanol or dimethyl ether (CH_3OCH_3)

---

*You may have noticed white exhaust fumes coming from car tailpipes in cold weather. Combustion of hydrocarbons produces water (eqs. 2.4–2.9), so what you see is condensed water from the combustion of gasoline.

2.12.b Halogenation of Alkanes

When a mixture of an alkane and chlorine gas is stored at low temperatures in the dark, no reaction occurs. In sunlight or at high temperatures, however, an exothermic reaction occurs. One or more hydrogen atoms of the alkane are replaced by chlorine atoms. This reaction can be represented by the general equation

$$R\text{—}H + Cl\text{—}Cl \xrightarrow[\text{heat}]{\text{light or}} R\text{—}Cl + H\text{—}Cl \qquad (2.10)$$

or, specifically for methane:

$$CH_4 + Cl\text{—}Cl \xrightarrow[\text{or heat}]{\text{sunlight}} CH_3Cl + HCl \qquad (2.11)$$

methane

chloromethane
(methyl chloride)
bp −24.2°C

The reaction is called **chlorination.** It is a **substitution reaction;** a chlorine is substituted for a hydrogen.

An analogous reaction, called **bromination,** occurs when the halogen is bromine.

$$R\text{—}H + Br\text{—}Br \xrightarrow[\text{heat}]{\text{light or}} R\text{—}Br + HBr \qquad (2.12)$$

Chlorination of hydrocarbons is a **substitution reaction** in which a chlorine atom is substituted for a hydrogen atom. Likewise in **bromination** reactions, a bromine atom is substituted.

If excess halogen is present, the reaction can continue further to give polyhalogenated products. Thus, methane and excess chlorine can give products with two, three, or four chlorines.*

$$CH_3Cl \xrightarrow{Cl_2} CH_2Cl_2 \xrightarrow{Cl_2} CHCl_3 \xrightarrow{Cl_2} CCl_4 \qquad (2.13)$$

dichloromethane
(methylene chloride)
bp 40°C

trichloromethane
(chloroform)
bp 61.7°C

tetrachloromethane
(carbon tetrachloride)
bp 76.5°C

By controlling the reaction conditions and the ratio of chlorine to methane, we can favor formation of one or another of the possible products.

PROBLEM 2.19   Write the names and structures of all possible products of bromination of methane.

With longer chain alkanes, mixtures of products may be obtained even at the first step.** For example, with propane,

$$CH_3CH_2CH_3 + Cl_2 \xrightarrow[\text{or heat}]{\text{light}} CH_3CH_2CH_2Cl + CH_3CHCH_3 + HCl \qquad (2.14)$$
$$\underset{\displaystyle Cl}{|}$$

propane

1-chloropropane
(*n*-propyl chloride)

2-chloropropane
(isopropyl chloride)

---

*Note that we sometimes write the formula of one of the reactants (in this case Cl_2) over the arrow for convenience, as in eq. 2.13. We also sometimes omit obvious inorganic products (in this case HCl).

**Note that we often do not write a balanced equation, especially when more than one product is formed from a single organic reactant. Instead, we show on the right side of the equation the structures of *all* the important organic products, as in eq. 2.14.

A WORD ABOUT ...

Methane, Marsh Gas, and Miller's Experiment

Methane is commonly found in nature wherever bacteria decompose organic matter in the absence of oxygen, as in marshes, swamps, or the muddy sediment of lakes—hence, its common name, *marsh gas.* In China, methane has been collected from the mud at the bottom of swamps for use in domestic cooking and lighting. Methane is similarly formed from bacteria in the digestive tracts of certain ruminant animals, such as cows.

Coal mining in Appalachia.

The scale of methane production by bacteria is considerable. The earth's atmosphere contains an average of 1 part per million of methane. Because our planet is small and because methane is light compared to most other air constituents (O_2, N_2), one would expect most of the methane to have escaped from our atmosphere, and it has been calculated that the equilibrium concentration should be very much less than is observed. The reason, then, for the relatively high observed concentration is that at the same time that methane escapes from the atmosphere, it is constantly replenished by bacterial decay of plant matter.

In cities, the amount of methane in the atmosphere reaches much higher levels, up to several parts per million. The peak concentrations come in the early morning and late afternoon, directly correlated with the peaks of automobile traffic. Fortunately, methane, which constitutes about 50% of urban atmospheric hydrocarbon pollutants, seems to have no direct harmful effect on human health.

Methane can accumulate in coal mines, where it is a hazard because, when mixed with 5 to 14% of air, it is explosive. Also, miners can be asphyxiated by it (due to lack of sufficient oxygen). Dangerous concentrations of methane can be detected readily by a variety of safety devices.

Hydrogen is the most common element in the solar system (it constitutes about 87% of the sun's mass). It therefore seemed reasonable to think that, when the planets were formed, other elements should have been present in reduced (not oxidized) forms: carbon as methane, nitrogen as ammonia, and oxygen as water. Indeed, some of the outer planets (Saturn and Jupiter) still have atmospheres that are rich in methane and ammonia.

A now-famous experiment by Stanley L. Miller (working in the laboratory of H. C. Urey at Columbia University) supports the idea that life could have arisen in a reducing atmosphere. Miller found that when mixtures of methane, ammonia, water, and hydrogen were subjected to electric discharges (to simulate lightning), some organic compounds were formed (amino acids, for example) that are important to biology and necessary for life. Similar results have since been obtained using heat or ultraviolet light in place of electric discharges (it seems likely that the earth's early atmosphere was subjected to much more ultraviolet radiation than it is now). When oxygen was added to these simulated primeval atmospheres, no amino acids were produced—strong evidence that the earth's original atmosphere did *not* contain free oxygen.

In the years since Miller's experiment, ideas about the chemistry of life's origin have become more precise as a consequence of much experimentation and of exploration in outer space. We now know that the earth's primary atmosphere was formed mainly by degassing the molten interior rather than by accretion from the solar nebula. It seems likely that the main carbon sources in the earth's early atmosphere were CO_2 and CO, *not* methane as assumed by Miller, and that nitrogen was present mainly as N_2 rather than ammonia. Repetition of Miller-type experiments with these assumed primordial atmospheres again gave biomolecules.

Miller's experiment provided a model for much work in the branch of a science now called **chemical evolution** or **prebiotic chemistry,** the study of chemical events that may have taken place on earth or elsewhere in the universe leading to the appearance of the first living cell. For additional reading you can consult *Chemical Evolution* by Stephen F. Mason, Clarendon Press, Oxford, 1991.

When larger alkanes are halogenated, the mixture of products becomes even more complex; individual isomers become difficult to separate and obtain pure, so halogenation tends not to be a useful way to synthesize specific alkyl halides. With unsubstituted *cycloalkanes*, however, where all the hydrogens are equivalent, a single pure organic product can be obtained:

$$\text{cyclopentane} + Br_2 \xrightarrow{\text{light}} \text{bromocyclopentane (cyclopentyl bromide)} + HBr \qquad (2.15)$$

PROBLEM 2.20   Write the structures of all possible products of *mono*bromination of pentane. Note the complexity of the product mixture, compared to that from the corresponding reaction with *cyclo*pentane (eq. 2.15).

PROBLEM 2.21   How many organic products can be obtained from the monochlorination of octane? of cyclooctane?

PROBLEM 2.22   Do you think that the chlorination of 2,2-dimethylpropane might be synthetically useful?

2.13   The Free-Radical Chain Mechanism of Halogenation

One may well ask how halogenation occurs. Why is light or heat necessary? Equations 2.10 and 2.11 express the *overall* reaction for halogenation. They describe the structures of the reactants and the products, and they show necessary reaction conditions or catalysts over the arrow. But they do *not* tell us exactly how the products are formed from the reactants.

A **reaction mechanism** is a step-by-step description of the bond-breaking and bond-making processes that occur when reagents react to form products. In the case of halogenation, various experiments show that this reaction occurs in several steps, not in one. Halogenation occurs via a **free-radical chain** of reactions.

The **chain-initiating step** is the breaking of the halogen molecule into two halogen atoms.

$$\textit{initiation} \qquad :\!\overset{..}{\underset{..}{Cl}}\!:\!\overset{..}{\underset{..}{Cl}}\!: \xrightarrow[\text{or heat}]{\text{light}} :\!\overset{..}{\underset{..}{Cl}}\!\cdot + :\!\overset{..}{\underset{..}{Cl}}\!\cdot \qquad (2.16)^*$$

chlorine molecule   chlorine atoms

The Cl—Cl bond is weaker than either the C—H bond or the C—C bond (compare the bond energies, Table A in the Appendix), and is therefore the easiest bond to break by supplying heat energy. When light is the energy source, chlorine absorbs visible light but alkanes do not, so again it is the Cl—Cl bond that breaks.

A **reaction mechanism** is a step-by-step description of the bond-breaking and bond-making processes that occur when reagents react to form products.

A **free-radical chain reaction** includes a **chain-initiating step, chain-propagating steps,** and **chain-terminating steps.**

---

*Recall from Section 1.13 that we use a "fishhook," or half-headed arrow, ⌒, to show the movement of only *one* electron, whereas we use a complete arrow, ⌒⟩, to describe the movement of an electron *pair*.

The **chain-propagating steps** are

$$\text{propagation} \begin{cases} R{-}H + \cdot \ddot{C}l : \longrightarrow \underset{\substack{\text{alkyl} \\ \text{radical}}}{R \cdot} + H{-}Cl & \textbf{(2.17)} \\[2em] R \cdot + Cl{-}Cl \longrightarrow \underset{\substack{\text{alkyl} \\ \text{chloride}}}{R{-}Cl} + \cdot \ddot{C}l : & \textbf{(2.18)} \end{cases}$$

Chlorine atoms are very reactive, because they have an incomplete valence shell (seven electrons instead of the required eight). They may either recombine to form chlorine molecules (the reverse of eq. 2.16) or, if they collide with an alkane molecule, abstract a hydrogen atom to form hydrogen chloride and an alkyl radical $R \cdot$. Recall from Section 1.4 that a radical is a fragment with an odd number of unshared electrons. The space-filling models in Figure 2.1 show that alkanes seem to have an exposed surface of hydrogens covering the carbon skeleton. So it is most likely that, if a halogen atom collides with an alkane molecule, it will hit the hydrogen end of a C—H bond.

Like a chlorine atom, the alkyl radical formed in the first step of the chain (eq. 2.17) is very reactive (incomplete octet). If it were to collide with a chlorine molecule, it could form an alkyl chloride molecule and a chlorine atom (eq. 2.18). The chlorine atom formed in this step can then react to repeat the sequence. When you add eq. 2.17 and eq. 2.18, you get the overall equation for chlorination (eq. 2.10). In each chain-propagating step, a radical (or atom) is consumed, but another radical (or atom) is formed and can continue the chain. Almost all of the reactants are consumed, and almost all of the products are formed in these steps.

Were it not for **chain-terminating steps,** all of the reactants could, in principle, be consumed by initiating a single reaction chain. However, because many chlorine molecules react to form chlorine atoms in the chain-initiating step, many chains are started simultaneously. Quite a few radicals are present as the reaction proceeds. If any two radicals combine, the chain will be terminated. Three possible chain-terminating steps are

$$\text{termination} \begin{cases} 2 : \ddot{C}l \cdot \longrightarrow Cl{-}Cl & \textbf{(2.19)} \\[1em] 2 R \cdot \longrightarrow R{-}R & \textbf{(2.20)} \\[1em] R \cdot + : \ddot{C}l \cdot \longrightarrow R{-}Cl & \textbf{(2.21)} \end{cases}$$

No new radicals are formed in these reactions, so the chain is broken or, as we say, terminated.

PROBLEM 2.23   Show that when eq. 2.17 and eq. 2.18 are added, the overall equation for chlorination (eq. 2.10) results.

PROBLEM 2.24   Write equations for all the steps (initiation, propagation, termination) in the free-radical chlorination of methane to form methyl chloride.

PROBLEM 2.25   Account for the experimental observation that small amounts of ethane and chloroethane are produced during the monochlorination of methane. (*Hint:* Consider the possible chain-terminating steps.)

REACTION SUMMARY

1. Reactions of Alkanes and Cycloalkanes

a. Combustion (Sec. 2.12a)

$$C_nH_{2n+2} + \left(\frac{3n+1}{2}\right)O_2 \longrightarrow nCO_2 + (n+1)H_2O$$

b. Halogenation (Sec. 2.12b)

$$R\!-\!H + X_2 \xrightarrow[\text{or light}]{\text{heat}} R\!-\!X + H\!-\!X \quad (X = Cl, Br)$$

MECHANISM SUMMARY

1. Halogenation (Sec. 2.13)

initiation $:\!\ddot{X}\!-\!\ddot{X}\!: \longrightarrow 2 :\!\ddot{X}\!\cdot$

termination $2 :\!\ddot{X}\!\cdot \longrightarrow :\!\ddot{X}\!-\!\ddot{X}\!:$

$2\ R\!\cdot \longrightarrow R\!-\!R$

$R\!\cdot + :\!\ddot{X}\!\cdot \longrightarrow R\!-\!\ddot{X}\!:$

propagation $R\!-\!H + :\!\ddot{X}\!\cdot \longrightarrow R\!\cdot + H\!-\!\ddot{X}\!:$

$R\!\cdot + :\!\ddot{X}\!-\!\ddot{X}\!: \longrightarrow R\!-\!\ddot{X}\!: + :\!\ddot{X}\!\cdot$

ADDITIONAL PROBLEMS

Alkane Nomenclature and Structural Formulas

2.26 Write structural formulas for the following compounds:

a. 3-methylpentane
b. 2,2-dimethylbutane
c. 4-ethyl-2,2-dimethylhexane
d. 2-bromo-3-methylpentane
e. 1,1-dichlorocyclopropane
f. 2-iodopropane
g. 1,1,3-trimethylcyclohexane
h. 1,1,3,3-tetrachloropropane

2.27 Write expanded formulas for the following compounds, and name them using the IUPAC system:

a. $CH_3(CH_2)_2CH_3$
b. $(CH_3)_2CHCH_2CH_2CH_3$
c. $(CH_3)_3CCH_2CH_2CH_3$
d. $(CH_2)_4$
e. $CH_3CH_2CHFCH_3$
f. $CH_3CCl_2CBr_3$
g. $i\text{-PrCl}$
h. MeBr
i. CH_2ClCH_2Cl
j. $(CH_3CH_2)_2CHCH(CH_3)CH_2CH_3$

2.28 Give both common and IUPAC names for the following compounds:

a. CH_3I
b. CH_3CH_2Br
c. CH_2Cl_2
d. CHI_3
e. $(CH_3)_2CHBr$
f. $CH_3CH_2CH_2F$
g. $(CH_3)_3CBr$

2.29 Write a structure for each of the compounds listed. Explain why the name given here is incorrect, and give a correct name in each case.

a. 1-methylbutane
b. 2,3-dibromopropane
c. 2-ethylbutane
d. 4-chloro-3-methylbutane
e. 1,3-dimethylcyclopropane
f. 1,1,3-trimethylpentane

2.30 Chemical substances used for communication in nature are called *pheromones*. The pheromone used by the female tiger moth to attract the male is the 18-carbon-atom alkane 2-methylheptadecane. Write its structural formula.

2.31 Write the structural formula for all isomers of each of the following compounds, and name each isomer by the IUPAC system (The number of isomers is indicated in parentheses.):

    a. C_4H_{10} (2)   **b.** $C_3H_6F_2$ (4)   **c.** $C_2H_2ClBr_3$ (3)
    d. C_5H_{12} (3)   **e.** C_4H_9Cl (4)   **f.** C_3H_6BrCl (5)

2.32 Write structural formulas and names for all possible cycloalkanes having each of the following molecular formulas. Be sure to include *cis–trans* isomers when appropriate. Name each compound by the IUPAC system.

    a. C_5H_{10} (there are 6)   **b.** C_6H_{12} (there are 16)

Alkane Properties and Intermolecular Interactions

2.33 Without referring to tables, arrange the following five hydrocarbons in order of increasing boiling point. (*Hint:* Draw structures or make models of the five hydrocarbons to see their shapes and sizes.)

    a. 2-methylhexane   **b.** heptane   **c.** 3,3-dimethylpentane
    d. hexane   **e.** 2-methylpentane

Explain your answer in terms of intermolecular interactions.

2.34 Arrange the following liquids in order from least soluble in hexane to most soluble in hexane:

    a. $CH_3(CH_2)_6CH_3$ **b.** H_2O **c.** CH_3OH

Explain your answer in terms of intermolecular interactions.

Conformations of Alkanes

2.35 In Problem 2.9 you drew two staggered conformations of butane (looking end-on down the bond between carbon-2 and carbon-3). There are also two eclipsed conformations around this bond. Draw Newman projections for them. Arrange all four conformations in order of decreasing stability.

2.36 Draw all possible staggered and eclipsed conformations of 1-bromo-2-chloroethane, using Newman projections. Underneath each, draw the corresponding "dash-wedge" and "sawhorse" structures. Rank the conformations in order of decreasing stability.

Conformations of Cycloalkanes; *Cis–Trans* Isomerism

2.37 Draw the formula for the preferred conformation of

    a. bromocyclohexane.   **b.** *trans*-1,4-dimethylcyclohexane.
    c. *cis*-1-ethyl-3-methylcyclohexane.   **d.** 1,1-dichlorocyclohexane.

2.38 Name the following *cis–trans* pairs:

 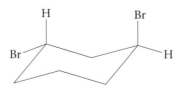

    a.

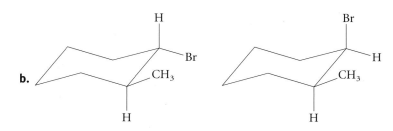

2.39 Explain with the aid of conformational structures why *cis*-1,3-dimethylcyclohexane is more stable than *trans*-1,3-dimethyl-cyclohexane, whereas the reverse order of stability is observed for the 1,2 and 1,4 isomers.

2.40 Which will be more stable, *cis*- or *trans*-1,4-di-*tert*-butylcyclohexane? Explain your answer by drawing conformational structures for each compound.

2.41 Examine the relationships of isomers as described in Figure 2.7 (p. 59). Then classify the following pairs of structures as structural isomers, conformers, configurational (*cis–trans*) isomers, or identical.

a. the pairs of compounds in Problem 2.38.

b.

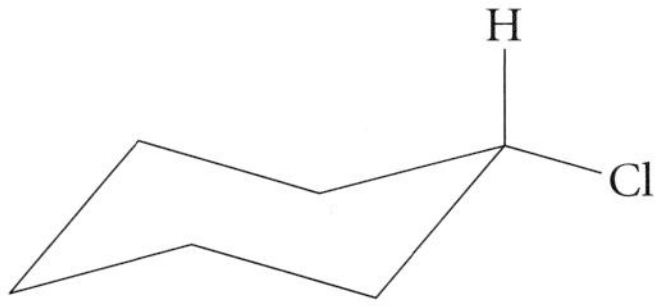

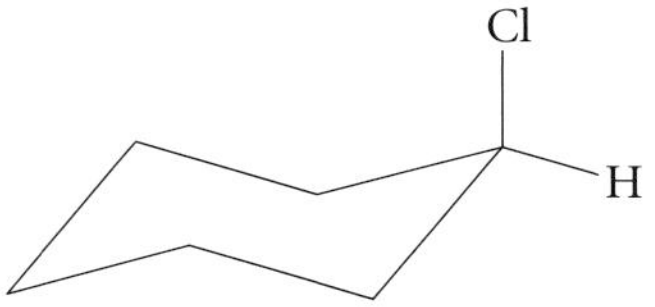

c.

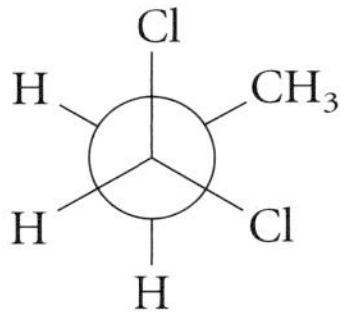

and

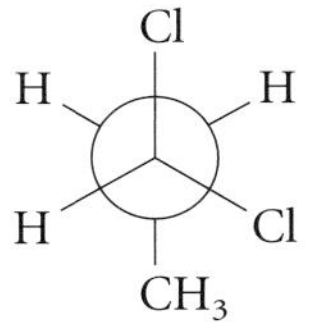

d.

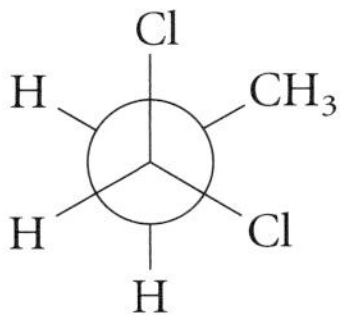

and

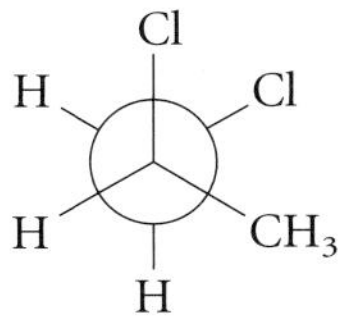

e. CH₃CHCH₂CH₂CH₃   and   CH₃CH₂CH₂CHCH₃   (careful!)
         | |
         CH₃   CH₃

2.42 Draw structural formulas for all possible dichlorocyclohexanes. Include *cis–trans* isomers.

Reactions of Alkanes: Combustion and Halogenation

2.43 How many monochlorination products can be obtained from each of the following polycyclic alkanes?

a. 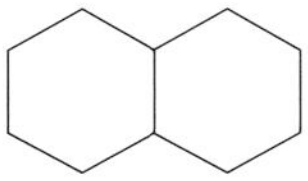   **b.** 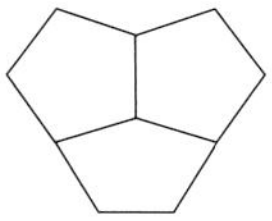   **c.**

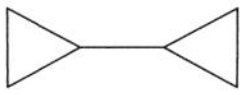

2.44 Using structural formulas, write equations for each of the following combustion reactions (see Reaction Summary 1.a, p. 67):

   a. the complete combustion of pentane
   b. the complete combustion of hexane
   c. the complete combustion of cyclohexane

2.45 Using structural formulas, write equations for the following halogenation reactions (see Reaction Summary 1.b, p. 67), and name each organic product:

   a. the monobromination of butane
   b. the monochlorination of cyclopentane
   c. the complete chlorination of propane

2.46 From the dichlorination of propane, four isomeric products with the formula $C_3H_6Cl_2$ were isolated and designated A, B, C, and D. Each was separated and further chlorinated to give one or more trichloropropanes, $C_3H_5Cl_3$. A and B gave three trichloro compounds, C gave one, and D gave two. Deduce the structures of C and D. One of the products from A was identical to the product from C. Deduce structures for A and B. (*Hint:* Start by drawing the structures of all four dichlorinated propane isomers.)

2.47 Write all the steps in the free-radical chain mechanism for the monochlorination of ethane (see Mechanism Summary, p. 67).

$$CH_3CH_3 + Cl_2 \longrightarrow CH_3CH_2Cl + HCl$$

What trace by-products would you expect to be formed as a consequence of the chain-terminating steps?

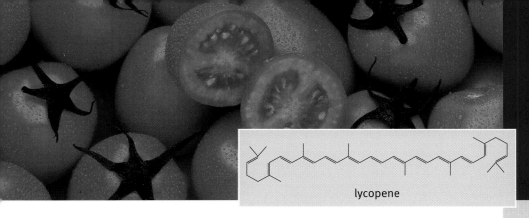

lycopene

3

ALKENES AND ALKYNES

Alkenes are compounds containing carbon–carbon double bonds. The simplest alkene, ethene, is a plant hormone (see A Word About Ethylene, p. 102–103) and an important starting material for the manufacture of other organic compounds (see Figure 3.12, p. 103). The alkene functional group is found in sources as varied as citrus fruits (limonene, Fig. 1.12), steroids (cholesterol, Sec. 15.9), and insect pheromones (muscalure; see A Word About the Gypsy Moth's Epoxide, p. 242). Alkenes have physical properties similar to those of alkanes (Sec. 2.7). They are less dense than water and, being nonpolar, are not very soluble in it. As with alkanes, compounds with four or fewer carbons are colorless gases, whereas higher homologs are volatile liquids.

Alkynes, compounds containing carbon–carbon triple bonds, are similar to alkenes in their physical properties and chemical behavior. In this chapter, we will examine the structure and chemical reactions of these two classes of compounds. We will also examine briefly the relationship between chemical reactions and energy.

3.1 Definition and Classification

Hydrocarbons that contain a carbon–carbon double bond are called **alkenes;** those with a carbon–carbon triple bond are **alkynes.*** Their general formulas are

$$C_nH_{2n} \qquad C_nH_{2n-2}$$
alkenes   alkynes

---

*An old but still used synonym for alkenes is *olefins.* Alkynes are also called *acetylenes,* after the first member of the series.

▲ Ripe, red tomatoes contain lycopene, a conjugated alkene (Sec. 3.15).

Alkenes and alkynes are **unsaturated** hydrocarbons containing carbon–carbon double bonds and carbon–carbon triple bonds respectively.

Both of these classes of hydrocarbons are **unsaturated,** because they contain fewer hydrogens per carbon than alkanes (C_nH_{2n+2}). Alkanes can be obtained from alkenes or alkynes by adding 1 or 2 moles of hydrogen.

$$RCH{=}CHR \xrightarrow[\text{catalyst}]{H_2}$$
alkene

$$RC{\equiv}CR \xrightarrow[\text{catalyst}]{2H_2}$$
alkyne

$$\longrightarrow RCH_2CH_2R$$
alkane

(3.1)

Compounds with more than one double or triple bond exist. If two double bonds are present, the compounds are called **alkadienes** or, more commonly, **dienes.** There are also trienes, tetraenes, and even polyenes (compounds with *many* double bonds, from the Greek *poly,* many). Polyenes are responsible for the color of carrots (β-carotene, p. 81) and tomatoes (lycopene, p. 71). Compounds with more than one triple bond, or with double and triple bonds, are also known.

Alkadienes, or **dienes,** contain two C—C double bonds which can be **cumulated** (next to each other), **conjugated** (separated by one C—C single bond), or **nonconjugated** (separated by more than one C—C single bond).

EXAMPLE 3.1

What are all the structural possibilities for the compound C_3H_4?

Solution   The formula C_3H_4 corresponds to the general formula C_nH_{2n-2}. The compound could have one triple bond, two double bonds, or one ring and one double bond. For their structures, see the solution to Example 1.10 on page 18.

PROBLEM 3.1   What are all the structural possibilities for C_4H_6? (Nine compounds, four acyclic and five cyclic, are known.)

When two or more multiple bonds are present in a molecule, it is useful to classify the structure further, depending on the relative positions of the multiple bonds. Double bonds are said to be **cumulated** when they are right next to one another. When multiple bonds *alternate* with single bonds, they are called **conjugated.** When more than one single bond comes between multiple bonds, the latter are **isolated** or **nonconjugated.**

C=C=C	C=C—C=C	C=C—C—C=C
C=C=C=C	C=C—C≡C	C≡C—C—C—C≡C
cumulated	conjugated	nonconjugated (isolated)

PROBLEM 3.2   Which of the following compounds have conjugated multiple bonds?

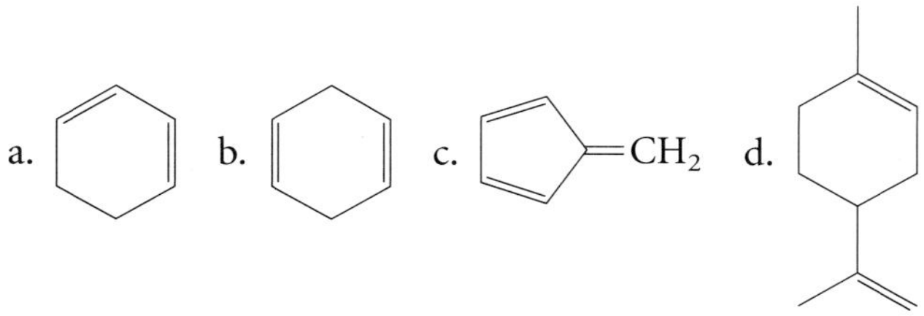

3.2 Nomenclature

The IUPAC rules for naming alkenes and alkynes are similar to those for alkanes (Sec. 2.3), but a few rules must be added for naming and locating the multiple bonds.

1. The ending *-ene* is used to designate a carbon–carbon double bond. When more than one double bond is present, the ending is *-diene, -triene,* and so on. The ending *-yne* (rhymes with wine) is used for a triple bond (*-diyne* for two triple bonds, and so on). Compounds with a double *and* a triple bond are *-enynes.*

2. Select the longest chain that includes *both* carbons of the double or triple bond. For example,

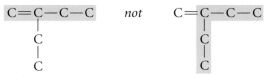

named as a butene, not as a pentene

3. Number the chain from the end nearest the multiple bond, so that the carbon atoms in that bond have the lowest possible numbers.

$$\overset{1}{C}-\overset{2}{C}=\overset{3}{C}-\overset{4}{C}-\overset{5}{C} \quad not \quad \overset{5}{C}-\overset{4}{C}=\overset{3}{C}-\overset{2}{C}-\overset{1}{C}$$

If the multiple bond is equidistant from both ends of the chain, number from the end nearest the first branch point.

$$\overset{1}{C}-\overset{2}{C}=\overset{3}{C}-\overset{4}{C} \quad not \quad \overset{4}{C}-\overset{3}{C}=\overset{2}{C}-\overset{1}{C}$$
$$\underset{C}{\mid} \qquad\qquad\qquad \underset{C}{\mid}$$

4. Indicate the position of the multiple bond using the *lower numbered carbon atom* of that bond. For example,

$$\overset{1}{CH_2}=\overset{2}{CH}\overset{3}{CH_2}\overset{4}{CH_3} \qquad \text{1-butene, } not \text{ 2-butene}$$

5. If more than one multiple bond is present, number from the end nearest the first multiple bond.

$$\overset{1}{C}=\overset{2}{C}-\overset{3}{C}=\overset{4}{C}-\overset{5}{C} \quad not \quad \overset{5}{C}=\overset{4}{C}-\overset{3}{C}=\overset{2}{C}-\overset{1}{C}$$

If a double and a triple bond are equidistant from the end of the chain, the *double* bond receives the lowest numbers. For example,

$$\overset{1}{C}=\overset{2}{C}-\overset{3}{C}\equiv\overset{4}{C} \quad not \quad \overset{4}{C}=\overset{3}{C}-\overset{2}{C}\equiv\overset{1}{C}$$

Let us see how these rules are applied. The first two members of each series are

$$CH_3CH_3 \qquad\qquad CH_2=CH_2 \qquad\qquad HC\equiv CH$$
ethane ethene ethyne

$$CH_3CH_2CH_3 \qquad\qquad CH_2=CHCH_3 \qquad\qquad HC\equiv CCH_3$$
propane propene propyne

The root of the name (*eth-* or *prop-*) tells us the number of carbons, and the ending (*-ane, -ene,* or *-yne*) tells us whether the bonds are single, double, or triple. No number is necessary in these cases, because in each instance, only one structure is possible. With four carbons, a number is necessary to locate the double or triple bond.

$$\overset{1}{C}H_2=\overset{2}{C}H\overset{3}{C}H_2\overset{4}{C}H_3 \qquad \overset{1}{C}H_3\overset{2}{C}H=\overset{3}{C}H\overset{4}{C}H_3 \qquad H\overset{1}{C}\equiv\overset{2}{C}\overset{3}{C}H_2\overset{4}{C}H_3 \qquad \overset{1}{C}H_3\overset{2}{C}\equiv\overset{3}{C}\overset{4}{C}H_3$$

1-butene   2-butene   1-butyne   2-butyne

Branches are named in the usual way.

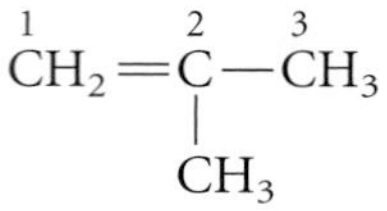

$$\overset{1}{C}H_2=\overset{2}{\underset{\underset{CH_3}{|}}{C}}-\overset{3}{C}H_3 \qquad \overset{1}{C}H_2=\overset{2}{\underset{\underset{CH_3}{|}}{C}}-\overset{3}{C}H_2\overset{4}{C}H_3 \qquad \overset{1}{C}H_3-\overset{2}{\underset{\underset{CH_3}{|}}{C}}=\overset{3}{C}H\overset{4}{C}H_3 \qquad \overset{1}{C}H_2=\overset{2}{\underset{\underset{CH_3}{|}}{C}}-\overset{3}{C}H=\overset{4}{C}H_2$$

methylpropene   2-methyl-1-butene 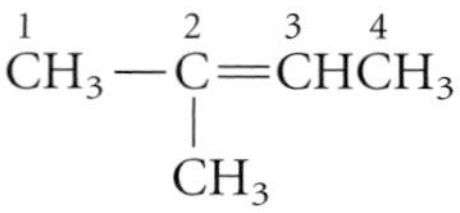   2-methyl-2-butene 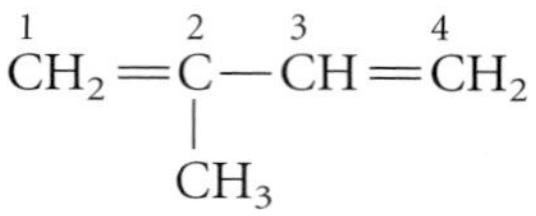   2-methyl-1,3-butadiene
(isobutylene)   (isoprene)

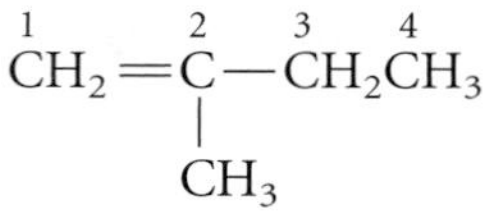

Note how the rules are applied in the following examples:

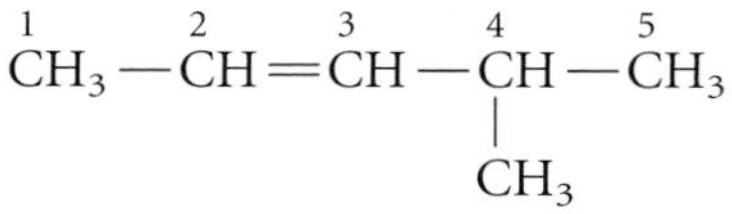

$$\overset{1}{C}H_3-\overset{2}{C}H=\overset{3}{C}H-\overset{4}{\underset{\underset{CH_3}{|}}{C}H}-\overset{5}{C}H_3 \qquad \overset{1}{C}H_2=\overset{2}{\underset{\underset{CH_2CH_3}{|}}{C}}-\overset{3}{C}H_2\overset{4}{C}H_3 \qquad \overset{1}{C}H_2=\overset{2}{C}H-\overset{3}{C}H=\overset{4}{C}H_2$$

4-methyl-2-pentene
(*Not* 2-methyl-3-pentene; the chain is numbered so that the double bond gets the lower number.)

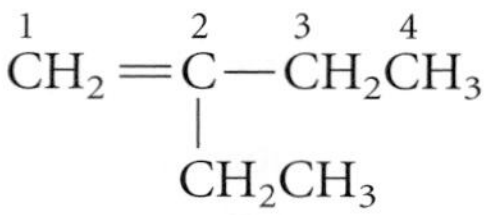

2-ethyl-1-butene
(Named this way, even though there is a five-carbon chain present, because that chain does not include both carbons of the double bond.)

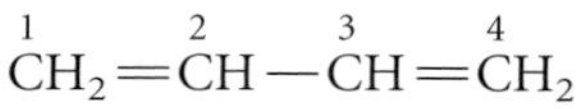

1,3-butadiene
(Note the *a* inserted in the name, to help in pronunciation.)

With cyclic hydrocarbons, we start numbering the ring with the carbons of the multiple bond.

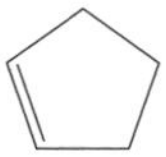

cyclopentene
(No number is necessary, because there is only one possible structure.)

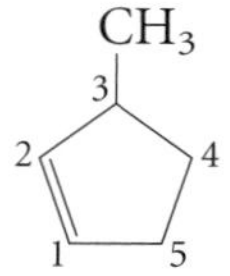

3-methylcyclopentene
(Start numbering at, and number through the double bond; 5-methylcyclopentene and 1-methyl-2-cyclopentene are incorrect names.)

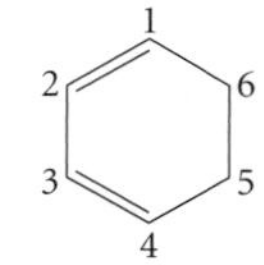

1,3-cyclohexadiene

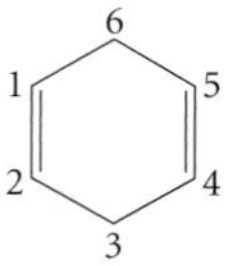

1,4-cyclohexadiene

PROBLEM 3.3 Name each of the following structures by the IUPAC system:

a. $CH_2=C(Br)CH_3$ b. $(CH_3)_2C=C(CH_3)_2$ c. $ClCH=CHCH_3$

d. 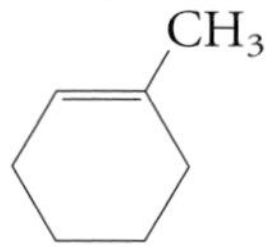 CH_3 e. $CH_2=C(CH_3)CH=CH_2$ f. $CH_3(CH_2)_3C\equiv CH$

EXAMPLE 3.2

Write the structural formula for 3-methyl-2-pentene.

Solution To get the structural formula from the IUPAC name, first write the longest chain or ring, number it, and locate the multiple bond. In this case, note that the chain has five carbons and that the double bond is located between carbon-2 and carbon-3:

$$\overset{1}{C}-\overset{2}{C}=\overset{3}{C}-\overset{4}{C}-\overset{5}{C}$$

Next, add the substituent:

$$\overset{1}{C}-\overset{2}{C}=\overset{3}{\underset{\underset{CH_3}{|}}{C}}-\overset{4}{C}-\overset{5}{C}$$

Finally, fill in the hydrogens:

$$CH_3-CH=\underset{\underset{CH_3}{|}}{C}-CH_2-CH_3$$

PROBLEM 3.4 Write structural formulas for the following:

a. 2,4-dimethyl-2-pentene
b. 3-hexyne
c. 1,2-dichlorocyclobutene
d. 2-chloro-1,3-butadiene

In addition to the IUPAC rules, it is important to learn a few common names. For example, the simplest members of the alkene and alkyne series are frequently referred to by their older common names, **ethylene, acetylene,** and **propylene.**

$$CH_2=CH_2 \quad HC\equiv CH \quad CH_3CH=CH_2$$

ethylene   acetylene   propylene
(ethene)   (ethyne)   (propene)

Two important groups also have common names. They are the **vinyl** and **allyl** groups (their IUPAC names are in parentheses below), shown on the left. These groups are used in common names, illustrated in the examples on the right.

$$CH_2=CH- \qquad\qquad CH_2=CHCl$$

vinyl   vinyl chloride
(ethenyl)   (chloroethene)

$$CH_2=CH-CH_2- \qquad CH_2=CH-CH_2Cl$$

allyl   allyl chloride
(2-propenyl)   (3-chloropropene)

PROBLEM 3.5 Write the structural formula for

a. vinylcyclopentane.
b. allylcyclopropane.

◎ **Ch. 3, Ref. 1.** *Examine and manipulate 3D model of ethylene.*

A **trigonal** carbon atom is bonded to only three other atoms.

3.3 Some Facts About Double Bonds

Carbon–carbon double bonds have some special features that are different from those of single bonds. For example, each carbon atom of a double bond is connected to only *three* other atoms (instead of four atoms, as with tetrahedral carbon). We speak of such a carbon as being **trigonal.** Furthermore, the two carbon atoms of a double bond and the four atoms that are attached to them lie in a single plane. This planarity is shown in Figure 3.1 for ethylene. The H—C—H and H—C=C angles in ethylene are approximately 120°. Although rotation occurs freely around single bonds, *rotation around double bonds is restricted.* Ethylene does not adopt any other conformation except the planar one. The doubly bonded carbons with two attached hydrogens do not rotate with respect to each other. Finally, carbon–carbon double bonds are shorter than carbon–carbon single bonds.

Figure 3.1
Three models of ethylene, each showing that the four atoms attached to a carbon–carbon double bond lie in a single plane.

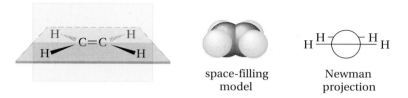

space-filling model   Newman projection

These differences between single and double bonds are summarized in Table 3.1. Let us see how the orbital model for bonding can explain the structure and properties of double bonds.

Table 3.1 Comparison of C—C and C=C bonds		
Property	**C—C**	**C=C**
1. Number of atoms attached to a carbon	4 (tetrahedral)	3 (trigonal)
2. Rotation	relatively free	restricted
3. Geometry	many conformations are possible; staggered is preferred	planar
4. Bond angle	109.5°	120°
5. Bond length	1.54 Å	1.34 Å

sp^2-**Hybridized orbitals** are one part *s* and two parts *p* in character and are directed toward the three vertices of an equilateral triangle. The angle between two *sp^2* orbitals is 120°.

3.4 The Orbital Model of a Double Bond; the Pi Bond

Figure 3.2 shows what must happen with the atomic orbitals of carbon to accommodate trigonal bonding, bonding to only three other atoms. The first part of this figure is exactly the same as Figure 1.6 (p. 30). But now we combine only *three* of the orbitals, to make *three equivalent sp^2*-**hybridized orbitals** (called *sp^2* because they are

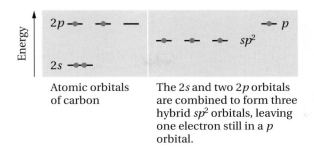

Figure 3.2
Unhybridized vs. sp^2-hybridized orbitals on carbon.

formed by combining one s and two p orbitals). These orbitals lie in a plane and are directed to the corners of an equilateral triangle. The angle between them is 120°. This angle is preferred because repulsion between electrons in each orbital is minimized. Three valence electrons are placed in the three sp^2 orbitals. The fourth valence electron is placed in the remaining $2p$ orbital, whose axis is perpendicular to the plane formed by the three sp^2 hybrid orbitals (see Figure 3.3).

Now let us see what happens when two sp^2-hybridized carbons are brought together to form a double bond. The process can be imagined as occurring stepwise (Figure 3.4). One of the two bonds, formed by *end-on* overlap of two sp^2 orbitals, is a **sigma (σ) bond.** The second bond of the double bond is formed differently. If the two carbons are aligned with the p orbitals on each carbon parallel, lateral overlap can occur, as shown at the bottom of Figure 3.4. The bond formed by lateral p-orbital overlap is called a **pi (π) bond.** The bonding in ethylene is summarized in Figure 3.5.

The orbital model explains the facts about double bonds listed in Table 3.1. Rotation about a double bond is restricted because, for rotation to occur, we would have

A **pi (π) bond** is formed by lateral overlap of p orbitals on adjacent atoms.

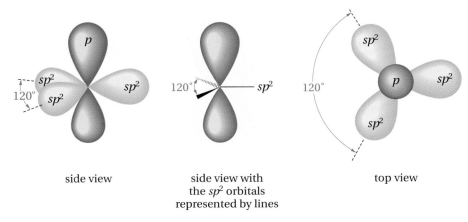

side view

side view with
the sp^2 orbitals
represented by lines

top view

Figure 3.3
A trigonal carbon showing three sp^2 hybrid orbitals in a plane with a 120° angle between them. The remaining p orbital is perpendicular to the sp^2 orbitals. There is a small back lobe to each sp^2 orbital, which has been omitted for ease of representation.

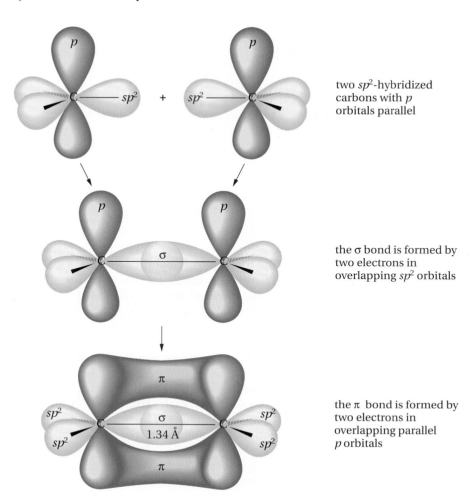

two sp^2-hybridized carbons with p orbitals parallel

the σ bond is formed by two electrons in overlapping sp^2 orbitals

the π bond is formed by two electrons in overlapping parallel p orbitals

Figure 3.4
Schematic formation of a carbon–carbon double bond. Two sp^2 carbons form a sigma (σ) bond (end-on overlap of two sp^2 orbitals) and a pi (π) bond (lateral overlap of two properly aligned p orbitals).

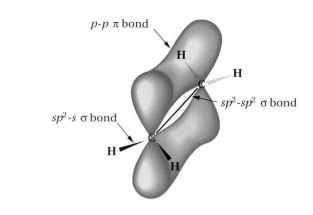

Figure 3.5
The bonding in ethylene consists of one sp^2–sp^2 carbon–carbon σ bond, four sp^2–s carbon–hydrogen σ bonds, and one p–p π bond.

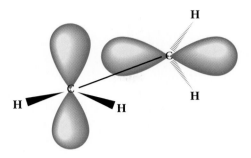

Figure 3.6
Rotation of one *sp²* carbon 90° with respect to another orients the *p* orbitals perpendicular to one another so that no overlap (and therefore no π bond) is possible.

to "break" the pi bond, as seen in Figure 3.6. For ethylene, it takes about 62 kcal/mol (259 kJ/mol) to break the pi bond, much more thermal energy than is available at room temperature. With the pi bond intact, the *sp²* orbitals on each carbon lie in a single plane. The 120° angle between those orbitals minimizes repulsion between the electrons in them. Finally, the carbon–carbon double bond is shorter than the carbon–carbon single bond because the two shared electron pairs draw the nuclei closer together than a single pair does.

To recap, according to the orbital model, the carbon–carbon double bond consists of one sigma bond and one pi bond. The two electrons in the sigma bond lie along the internuclear axis; the two electrons in the pi bond lie in a region of space above and below the plane formed by the two carbons and the four atoms attached to them. The π electrons are more exposed than the σ electrons and, as we will see, can be attacked by various electron-seeking reagents.

But before we consider reactions at the double bond, let us examine an important result of the restricted rotation around double bonds.

3.5 *Cis–trans* Isomerism in Alkenes

Because rotation at carbon–carbon double bonds is restricted, *cis–trans* isomerism (geometric isomerism) is possible in appropriately substituted alkenes. For example, 1,2-dichloroethene exists in two different forms:

cis-1,2-dichloroethene
bp 60°C, mp −80°C

trans-1,2-dichloroethene
bp 47°C, mp −50°C

These stereoisomers are *not* readily interconverted by rotation around the double bond at room temperature. Like *cis–trans* isomers of cycloalkanes, they are configurational stereoisomers and can be separated from one another by distillation, taking advantage of the difference in their boiling points.

A W O R D A B O U T ...

The Chemistry of Vision

Color in organic molecules is usually associated with extended conjugated systems of double bonds. A good example is **β-carotene,** a yellow-orange pigment found in carrots and many other plants. This $C_{40}H_{56}$ hydrocarbon has 11 carbon–carbon double bonds in conjugation. It is the biological precursor of the C_{20} unsaturated alcohol **vitamin A** (also called retinol), which in turn leads to the key substance involved in vision, **11-cis-retinal.** Notice in Figure 3.7 that the conversion of vitamin A to 11-cis-retinal involves not only oxidation of the alcohol group ($-CH_2OH$) to an aldehyde ($-CH=O$), but also trans → cis isomerism at the $C_{11}-C_{12}$ double bond.

Carrots contain β-carotene.

Cis–trans isomerism plays a key role in the process of vision. The rod cells in the retina of the eye contain a red, light-sensitive pigment called **rhodopsin.** This pigment consists of the protein **opsin** combined at its active site with 11-cis-retinal. When visible light with the appropriate energy is absorbed by rhodopsin, the complexed cis-retinal is isomerized to the trans isomer. This process is fantastically fast, occurring in only pico-seconds (10^{-12} seconds). As you can see from their structures, the shapes of the cis and trans isomers are very different.

trans-retinal

The trans-retinal complex with opsin (called metarhodopsin-II) is less stable than the cis-retinal complex, and it dissociates into opsin and trans-retinal. This change in geometry triggers a response in the rod nerve cells that is transmitted to the brain and perceived as vision.

If this were all that happened, we would be able to see for only a few moments, because all of the 11-cis-retinal present in the rod cells would be quickly consumed. Fortunately, the enzyme retinal isomerase, in the presence of light, converts the trans-retinal back to the 11-cis isomer, so that the cycle can be repeated. Calcium ions in the cell and its membrane control how fast the visual system recovers after exposure to light. They also mediate the way in which cells adapt to various light

---

EXAMPLE 3.3

Are cis–trans isomers possible for 1-butene and 2-butene?

Solution 2-Butene has cis–trans isomers, but 1-butene does not.

cis-2-butene
bp 3.7°C, mp −139°C

trans-2-butene
bp 0.3°C, mp −106°C

levels. The following sequence summarizes the visual cycle:

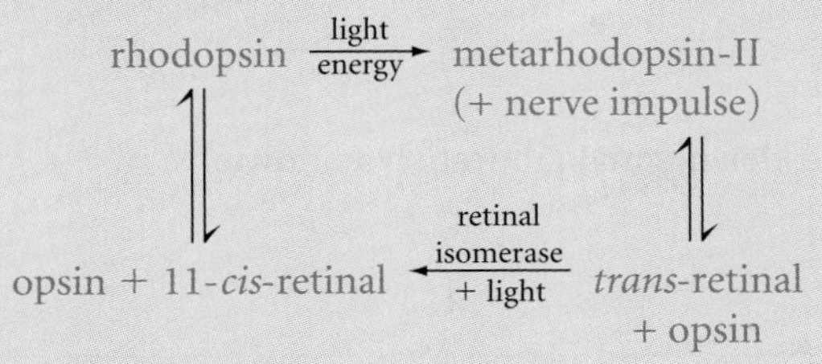

This representation is simplified because there are actually several additional intermediates between rhodopsin and the fully dissociated *trans*-retinal and opsin.

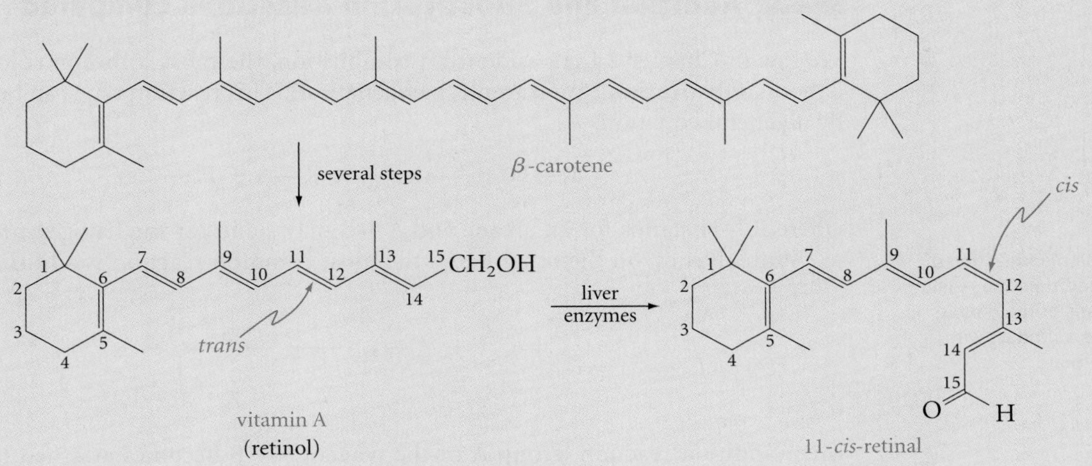

Figure 3.7
In the liver, *β*-carotene is converted into vitamin A first and then into 11-*cis*-retinal.

For 1-butene, carbon-1 has two identical hydrogen atoms attached to it; therefore, only one structure is possible.

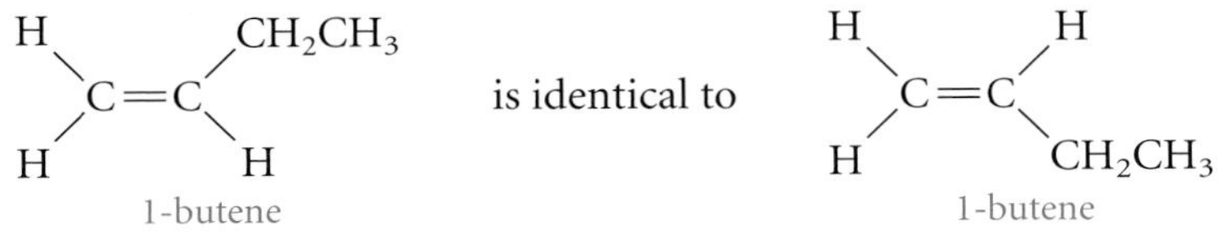

For *cis–trans* isomerism to occur in alkenes, *each* carbon of the double bond must have two different atoms or groups attached to it.

PROBLEM 3.6   Which of the following compounds can exist as *cis–trans* isomers? Draw their structures.

a. propene   b. 3-hexene   c. 2-methyl-2-butene   d. 2-hexene

Geometric isomers of alkenes can be interconverted if sufficient energy is supplied to break the pi bond and allow rotation about the remaining, somewhat stronger, sigma bond (eq. 3.2). The required energy may take the form of light or heat.

$$\underset{cis}{\overset{A}{\underset{B}{\diagdown}}C=C\overset{A}{\underset{B}{\diagup}}} \;\underset{\text{light}}{\overset{\text{heat or}}{\rightleftharpoons}}\; \overset{A}{\underset{B}{\diagup}}\dot{C}\!\ominus\!\dot{C}\overset{A}{\underset{B}{\diagdown}} \;\rightleftharpoons\; \overset{A}{\underset{B}{\diagup}}\dot{C}\!\ominus\!\dot{C}\overset{B}{\underset{A}{\diagdown}} \;\rightleftharpoons\; \overset{A}{\underset{B}{\diagup}}C\!-\!C\overset{B}{\underset{A}{\diagdown}} \;\rightleftharpoons\; \underset{trans}{\overset{A}{\underset{B}{\diagup}}C=C\overset{B}{\underset{A}{\diagdown}}} \qquad \textbf{(3.2)}$$

This conversion does not occur under normal laboratory conditions.

3.6 Addition and Substitution Reactions Compared

We saw in Chapter 2 that, aside from combustion, the most common reaction of alkanes is **substitution** (for example, halogenation). This reaction type can be expressed by a general equation.

$$R\!-\!H + A\!-\!B \longrightarrow R\!-\!A + H\!-\!B \qquad \textbf{(3.3)}$$

where R—H stands for an alkane and A—B may stand for the halogen molecule.

The most common reaction of alkenes is **addition** of a reagent to the carbons of the double bond to give a product with a C—C single bond.

With alkenes, on the other hand, the most common reaction is **addition:**

$$\overset{}{\underset{}{\diagdown}}C=C\overset{}{\underset{}{\diagup}} + A\!-\!B \longrightarrow \underset{A\quad B}{-\overset{|}{C}\!-\!\overset{|}{C}-} \qquad \textbf{(3.4)}$$

In an addition reaction, group A of the reagent A—B becomes attached to one carbon atom of the double bond, group B becomes attached to the other carbon atom, and the product has only a single bond between the two carbon atoms.

What bond changes take place in an addition reaction? The pi bond of the alkene is broken and the sigma bond of the reagent is also broken. Two new sigma bonds are formed. In other words, we break a pi and a sigma bond, and we make two sigma bonds. Because sigma bonds are usually stronger than pi bonds, the net reaction is favorable.

> **PROBLEM 3.7** Why, in general, is a sigma bond between two atoms stronger than a pi bond between the same two atoms?

3.7 Polar Addition Reactions

Several reagents add to double bonds by a two-step polar process. In this section we will describe examples of this reaction type, after which we will consider details of the reaction mechanism.

3.7.a Addition of Halogens

Alkenes readily add chlorine or bromine.

$$CH_3CH{=}CHCH_3 + Cl_2 \longrightarrow CH_3CH{-}CHCH_3 \qquad\qquad \textbf{(3.5)}$$
$$\underset{Cl}{|} \quad \underset{Cl}{|}$$

2-butene
bp 1–4°C

2,3-dichlorobutane
bp 117–119°C

$$CH_2{=}CH{-}CH_2{-}CH{=}CH_2 + 2\,Br_2 \longrightarrow CH_2{-}CH{-}CH_2{-}CH{-}CH_2 \qquad \textbf{(3.6)}$$
$$\underset{Br}{|}\ \ \underset{Br}{|} \qquad\quad \underset{Br}{|}\ \ \underset{Br}{|}$$

1,4-pentadiene
bp 26.0°C

1,2,4,5-tetrabromopentane
mp 85–86°C

Usually the halogen is dissolved in some inert solvent such as tri- or tetrachloromethane, and then this solution is added dropwise to the alkene. Reaction is nearly instantaneous, even at room temperature or below. No light or heat is required, as in the case of substitution reactions.

PROBLEM 3.8 Write an equation for the reaction of bromine at room temperature with

  a. 1-butene.
  b. cyclohexene.

   The addition of bromine can be used as a chemical test for the presence of unsaturation in an organic compound. Bromine solutions in tetrachloromethane are dark reddish-brown, and the unsaturated compound and its bromine adduct are usually both colorless. As the bromine solution is added to the unsaturated compound, the bromine color disappears. If the compound being tested is saturated, it will not react with bromine under these conditions, and the color will persist.

3.7.b Addition of Water (Hydration)

If an acid catalyst is present, water adds to alkenes. It adds as H—OH, and the products are alcohols.

$$CH_2{=}CH_2 + H{-}OH \xrightarrow{\text{H}^+} CH_2{-}CH_2 \quad (\text{or } CH_3CH_2OH) \qquad \textbf{(3.7)}$$
$$\underset{H}{|}\quad \underset{OH}{|}$$

ethanol

Bromine solution (red-brown) is added to a saturated hydrocarbon (left) and an unsaturated hydrocarbon (right).

$$\text{cyclohexene} + H{-}OH \xrightarrow{\text{H}^+} \text{cyclohexanol} \qquad\qquad \textbf{(3.8)}$$

cyclohexene
bp 83.0°C

cyclohexanol
bp 161.1°C

An acid catalyst is required in this case, because the neutral water molecule is not acidic enough to provide protons to start the reaction. The stepwise mechanism for this reaction is given later in eq. 3.20. Hydration is used industrially and occasionally in the laboratory to synthesize alcohols from alkenes.

PROBLEM 3.9 Write an equation for the acid-catalyzed addition of water to

a. cyclopentene. b. 2-butene.

3.7.c Addition of Acids

A variety of acids add to the double bond of alkenes. The hydrogen ion (or proton) adds to one carbon of the double bond, and the remainder of the acid becomes connected to the other carbon.

$$\begin{array}{c} \diagup \\ C=C \\ \diagdown \end{array} + \overset{\delta+}{H}-\overset{\delta-}{A} \longrightarrow \begin{array}{c} | \quad | \\ -C-C- \\ | \quad | \\ H \quad A \end{array} \tag{3.9}$$

Acids that add in this way are the hydrogen halides (HF, HCl, HBr, HI) and sulfuric acid ($H-OSO_3H$). Here are two typical examples:

$$CH_2=CH_2 + H-Cl \longrightarrow \underset{\substack{| \quad | \\ H \quad Cl}}{CH_2-CH_2} \quad \text{(or } CH_3CH_2Cl\text{)} \tag{3.10}$$

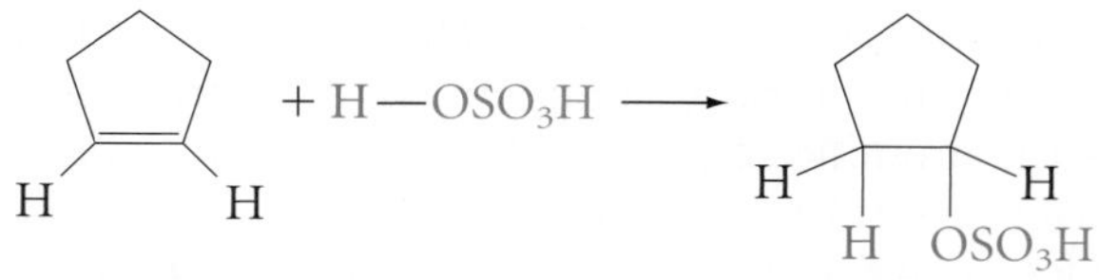

$$\underset{\text{ethene}}{} \quad \underset{\substack{\text{hydrogen} \\ \text{chloride}}}{} \quad \underset{\substack{\text{chloroethane} \\ \text{(ethyl chloride)}}}{}$$

$$\tag{3.11}$$

cyclopentene   sulfuric acid   cyclopentyl hydrogen sulfate

PROBLEM 3.10 Write an equation for each of the following reactions:

a. 2-butene + HI
b. cyclohexene + HBr

Before we discuss the mechanism of addition reactions, we must introduce a complication that we have carefully avoided in all the examples given so far.

The products of addition of **unsymmetric reagents** to **unsymmetric alkenes** are called **regioisomers. Regiospecific** additions produce only one regioisomer. **Regioselective** additions produce mainly one regioisomer.

3.8 Addition of Unsymmetric Reagents to Unsymmetric Alkenes; Markovnikov's Rule

Reagents and alkenes can be classified as either **symmetric** or **unsymmetric** with respect to addition reactions. Table 3.2 illustrates what this means. If a reagent and/or an alkene is symmetric, only one addition product is possible. If you check back

Table 3.2	Classification of reagents and alkenes by symmetry with regard to addition reactions	
	Symmetric	**Unsymmetric**
Reagents	Br$-$Br	H$-$Br
	Cl$-$Cl	H$-$OH
	H$-$H	H$-$OSO$_3$H
Alkenes	CH$_2$=CH$_2$	CH$_3$CH=CH$_2$
	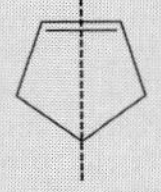	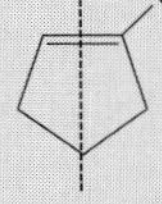 CH$_3$
	mirror plane	not a mirror plane

through all the equations and problems for addition reactions up to now, you will see that either the alkene or the reagent (or both) was symmetric. But if *both* the reagent *and* the alkene are *unsymmetric,* two products are, in principle, possible.

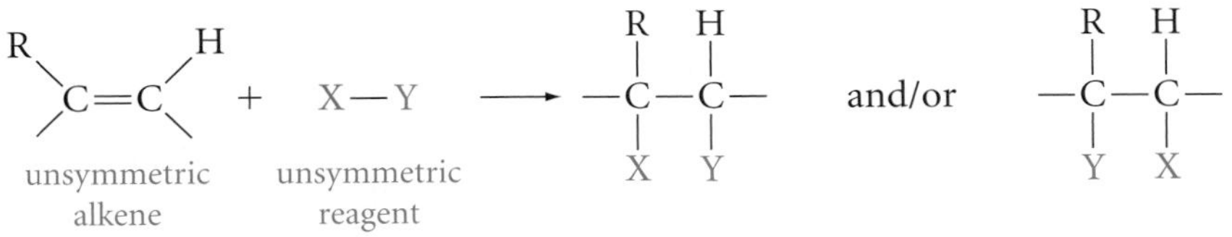

$$\underset{\substack{\text{unsymmetric}\\\text{alkene}}}{\overset{R}{\underset{}{}}\!\!\!\overset{}{\underset{}{}}C=C\overset{H}{\underset{}{}}} \ + \ \underset{\substack{\text{unsymmetric}\\\text{reagent}}}{X-Y} \ \longrightarrow \ \overset{R\quad H}{\underset{X\quad Y}{-C-C-}} \quad \text{and/or} \quad \overset{R\quad H}{\underset{Y\quad X}{-C-C-}} \qquad \textbf{(3.12)}$$

The products of eq. 3.12 are sometimes called **regioisomers.** If a reaction of this type gives *only one* of the two possible regioisomers, it is said to be **regiospecific.** If it gives *mainly one* product, it is said to be **regioselective.**

Let us consider as a specific example the acid-catalyzed addition of water to propene. In principle, two products could be formed: 1-propanol or 2-propanol.

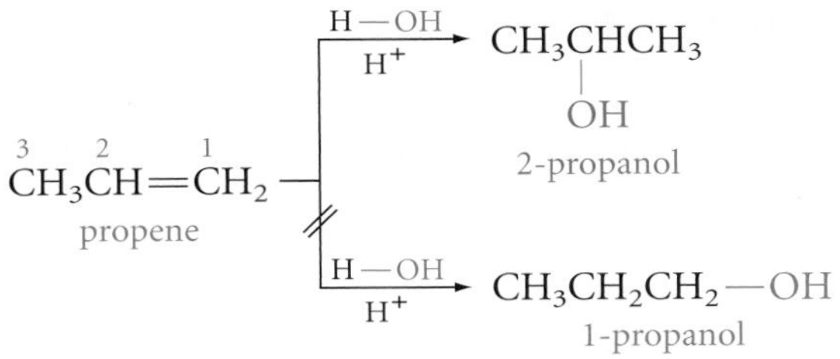

$$\underset{\substack{\text{propene}}}{\overset{3\quad2\quad1}{CH_3CH=CH_2}}\ \begin{cases} \xrightarrow[H^+]{H-OH} & \underset{\substack{|\\OH\\ \text{2-propanol}}}{CH_3CHCH_3} \\[3em] \xrightarrow[H^+]{H-OH} & \underset{\text{1-propanol}}{CH_3CH_2CH_2-OH} \end{cases} \qquad \textbf{(3.13)}$$

That is, the hydrogen of the water could add to C-1 and the hydroxyl group to C-2 of propene, or vice versa. When the experiment is carried out, *only one product is observed. The addition is regiospecific, and the only product is 2-propanol.*

Most addition reactions of alkenes show a similar preference for the formation of only (or mainly) one of the two possible addition products. Here are some examples.

$$CH_3CH{=}CH_2 + \overset{\delta+}{H}{-}\overset{\delta-}{Cl} \longrightarrow CH_3\underset{\underset{Cl}{|}}{C}HCH_3 \qquad (CH_3CH_2CH_2Cl) \qquad \textbf{(3.14)}$$

not observed

$$CH_3\underset{\underset{CH_3}{|}}{C}{=}CH_2 + \overset{\delta+}{H}{-}\overset{\delta-}{OH} \xrightarrow{H^+} CH_3\underset{\underset{CH_3}{|}}{\overset{\overset{OH}{|}}{C}}CH_3 \qquad (CH_3\underset{\underset{CH_3}{|}}{C}HCH_2OH) \qquad \textbf{(3.15)}$$

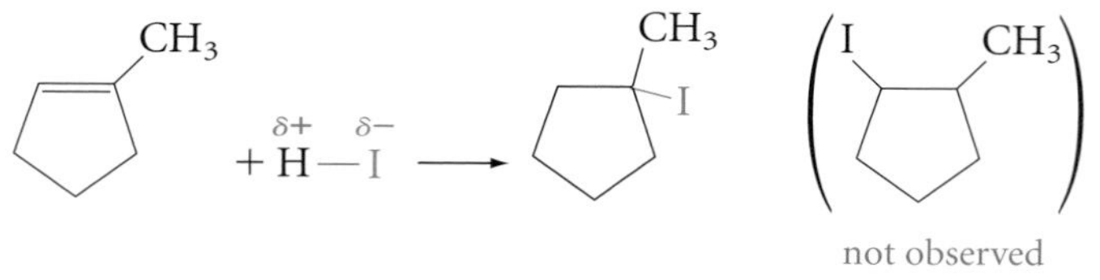

not observed

$$\textbf{(3.16)}$$

not observed

Notice that the reagents are all polar, with a positive and a negative end. After studying a number of such addition reactions, the Russian chemist Vladimir Markovnikov formulated the following rule more than 100 years ago: *When an unsymmetric reagent adds to an unsymmetric alkene, the electropositive part of the reagent bonds to the carbon of the double bond that has the greater number of hydrogen atoms attached to it.*[*]

PROBLEM 3.11 Use Markovnikov's rule to predict which regioisomer predominates in each of the following reactions:

a. 1-butene + HCl
b. 2-methyl-2-butene + H_2O (H^+ catalyst)

PROBLEM 3.12 What two products are *possible* from the addition of HCl to 2-pentene? Would you expect the reaction to be regiospecific?

Let us now develop a rational explanation for Markovnikov's rule in terms of modern chemical theory.

3.9 Mechanism of Electrophilic Addition to Alkenes

The pi electrons of a double bond are more exposed to an attacking reagent than are the sigma electrons. The π bond is also weaker than the σ bond. It is the pi electrons, then, that are involved in additions to alkenes. The double bond can act as a supplier of pi electrons to an electron-seeking reagent.

Polar reactants can be classified as either **electrophiles** or **nucleophiles.** Electrophiles (literally, electron lovers) are electron-poor reagents; in reactions with some other molecule, they seek electrons. They are often positive ions (cations) or

Electrophiles are electron-poor reactants; they seek electrons. **Nucleophiles** are electron-rich reactants; they form bonds by donating electrons to electrophiles.

---

[*]Actually, Markovnikov stated the rule a little differently. The form given here is easier to remember and apply. For an interesting historical article on what he actually said, when he said it, and how his name is spelled, see J. Tierney, *J. Chem. Educ.* **1988,** *65,* 1053–54.

otherwise electron-deficient species. Nucleophiles (literally, nucleus lovers), on the other hand, are electron rich; they form bonds by donating electrons to an electrophile.

$$E^+ \quad + \quad :Nu^- \quad \longrightarrow \quad E:Nu \qquad \qquad \textbf{(3.17)}$$
$$\text{electrophile} \quad \text{nucleophile}$$

Let us now consider the mechanism of polar addition to a carbon–carbon double bond, specifically the addition of acids to alkenes. The carbon–carbon double bond, because of its pi electrons, is a nucleophile. The proton (H^+) is the attacking electrophile. As the proton approaches the π bond, the two pi electrons are used to form a σ bond between the proton and one of the two carbon atoms. Because this bond uses *both* pi electrons, the other carbon acquires a positive charge, producing a **carbocation**.

A **carbocation** is a positively charged carbon atom bonded to three other atoms.

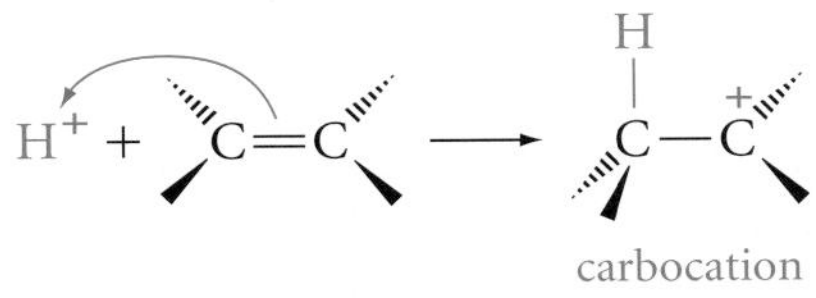

$$\textbf{(3.18)}$$

carbocation

The resulting carbocations are, however, extremely reactive because there are only six electrons (instead of the usual eight) around the positive carbon. The carbocation rapidly combines with some species that can supply it with two electrons, a nucleophile.

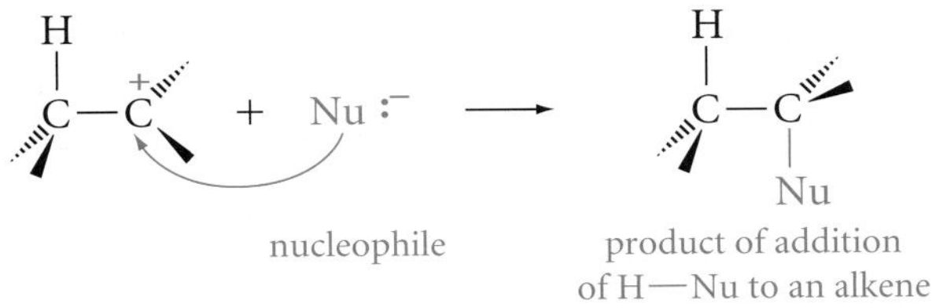

$$\textbf{(3.19)}$$

nucleophile   product of addition
of H—Nu to an alkene

Examples include the addition of HCl, HOSO$_3$H, and HOH to alkenes:

Ch. 3, Ref. 2. View animation of addition of H-Cl to ethylene.

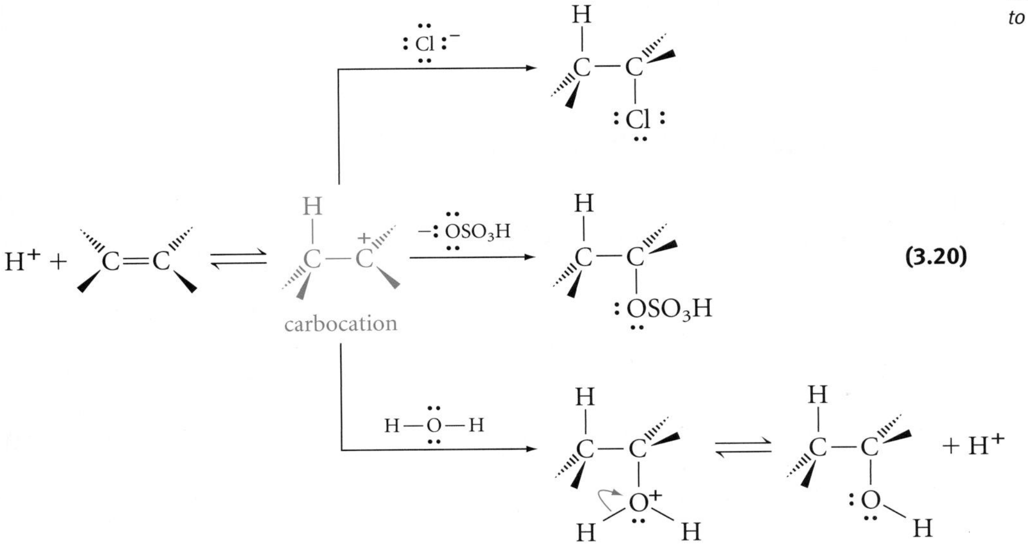

$$\textbf{(3.20)}$$

carbocation

In these reactions, the electrophile H^+ first adds to the alkene to give a carbocation. Then the carbocation combines with a nucleophile, in these examples, a chloride ion, a bisulfate ion, or a water molecule.

With most alkenes, the first step in this process—the formation of the carbocation—is the slower of the two steps. The resulting carbocation is usually so reactive that combination with the nucleophile is extremely rapid. *Since the first step in these additions is attack by the electrophile, the whole process is called an* **electrophilic addition reaction.**

Electrophilic addition of the halogens Cl_2 and Br_2 to alkenes occurs in a similar manner. Although the mechanism is not identical to that for acids, the end results are the same. For example, when a molecule of Br_2 approaches the pi bond of an alkene, the Br—Br bond becomes polarized: $\overset{\delta+}{Br}$—$\overset{\delta-}{Br}$. The Br atom closest to the pi bond develops a partial positive charge and thus becomes an electrophile, while the other Br atom develops a partial negative charge and becomes the nucleophile. Although it is impossible to tell from the products, the addition occurs in Markovnikov fashion.*

A reaction in which an electrophile is added to an alkene is called an **electrophilic addition reaction.**

EXAMPLE 3.4

Since carbocations are involved in the electrophilic addition reactions of alkenes, it is important to understand the bonding in these chemical intermediates. Describe the bonding in carbocations in orbital terms.

Solution   The carbon atom is positively charged and therefore has only three valence electrons to use in bonding. Each of these electrons is in an sp^2 orbital. The three sp^2 orbitals lie in one plane with 120° angles between them, an arrangement that minimizes repulsion between the electrons in the three bonds. The remaining p orbital is perpendicular to that plane and vacant.

carbocation

3.10   **Markovnikov's Rule Explained**

To explain Markovnikov's rule, let us consider a specific example, the addition of H—Cl to propene. The first step is addition of a proton to the double bond. This can occur in two ways, to give either an isopropyl cation or a propyl cation.

$$\overset{3}{CH_3}-\overset{2}{CH}=\overset{1}{CH_2} \xrightarrow{\ H^+\ }$$
propene

adds to C-1 → $CH_3\overset{+}{C}HCH_3$
isopropyl cation

adds to C-2 → $CH_3CH_2\overset{+}{C}H_2$
propyl cation

(3.21)

At this stage of the reaction, the structure of the product is already determined; when combining with chloride ion, the isopropyl cation can give only 2-chloropropane, and the propyl

---

*Consult your instructor if you are curious about the detailed mechanism.

cation can give only 1-chloropropane. The only observed product is 2-chloropropane, so we must conclude that the *proton adds to C-1 to form only the isopropyl cation.* Why?

Carbocations can be classified as **tertiary, secondary,** or **primary,** depending on whether the positive carbon atom has attached to it three organic groups, two groups, or only one group. From many studies, it has been established that the stability of carbocations decreases in the following order:

$$
\underset{\substack{\text{tertiary }(3°)\\ \text{most stable}}}{R-\overset{\displaystyle R}{\underset{\displaystyle R}{C^+}}} \;>\; \underset{\substack{\text{secondary }(2°)}}{R-\overset{+}{\underset{\displaystyle R}{CH}}} \;>>\; \underset{\substack{\text{primary }(1°)}}{R-\overset{+}{CH_2}} \;>\; \underset{\substack{\text{methyl (unique)}\\ \text{least stable}}}{\overset{+}{CH_3}}
$$

most stable ⟶ least stable

One reason for this order is the following: A carbocation will be more stable when the positive charge can be spread out, or delocalized, over several atoms in the ion, instead of being concentrated on a single carbon atom. In alkyl cations, this delocalization occurs by drift of electron density to the positive carbon from C—H and C—C sigma bonds that can align themselves with the empty *p* orbital on the positively charged carbon atom (Figure 3.8). If the positive carbon is surrounded by other carbon atoms (alkyl groups), instead of by hydrogen atoms, more C—H and C—C bonds will be available to provide electrons to help delocalize the charge. This is the main reason for the observed stability order of carbocations.

Markovnikov's rule can now be restated in modern and more generally useful terms: *The electrophilic addition of an unsymmetric reagent to an unsymmetric double bond proceeds in such a way as to involve the most stable carbocation.*

> **PROBLEM 3.13** Classify each of the following carbocations as primary, secondary, or tertiary:
>
> a. $CH_3CH_2\overset{+}{C}HCH_3$ b. $(CH_3)_2\overset{+}{C}HCH_2$ c. CH_3
>
> **PROBLEM 3.14** Which carbocation in Problem 3.13 is most stable? least stable?
>
> **PROBLEM 3.15** Write the steps in the electrophilic additions in eqs. 3.15 and 3.16, and in each case, show that reaction occurs via the more stable carbocation.

Carbocations are classified as **primary, secondary,** and **tertiary** when one, two, and three R groups, respectively, are attached to the positively charged carbon atom.

C-H σ-*p* overlap

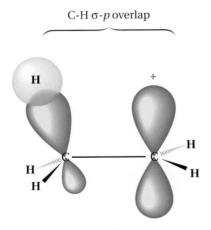

Figure 3.8
Alkyl groups stabilize carbocations by donating electron density from C—H and C—C sigma bonds that can line up with the empty *p* orbital on the positively charged carbon atom.

Discussion of Markovnikov's rule raises two important general questions about chemical reactions: (1) Under what conditions is a reaction likely to proceed? (2) How rapidly will a reaction occur? We will consider these questions briefly in the next two sections before continuing our survey of the reactions of alkenes.

3.11 Reaction Equilibrium: What Makes a Reaction Go?

A chemical reaction can proceed in two directions. Reactant molecules can form product molecules, and product molecules can react to re-form the reactant molecules. For the reaction*

$$aA + bB \rightleftharpoons cC + dD \qquad (3.22)$$

we describe the chemical equilibrium for the forward and backward reactions by the following equation:

$$K_{eq} = \frac{[C]^c[D]^d}{[A]^a[B]^b} \qquad (3.23)$$

The **equilibrium constant, K_{eq},** indicates the direction that is favored for a reaction.

In this equation, K_{eq}, the **equilibrium constant,** is equal to the product of the concentrations of the products divided by the product of the concentrations of the reactants. (The small letters a, b, c, and d are the numbers of molecules of reactants and products in the balanced reaction equation.)

The equilibrium constant tells us the direction that is favored for the reaction. If K_{eq} is greater than 1, the formation of products C and D will be favored over the formation of reactants A and B. The preferred direction for the reaction is from left to right. Conversely, if K_{eq} is less than 1, the preferred direction for the reaction is from right to left.

What determines whether a reaction will proceed to the right, toward products? A reaction will occur when the products are lower in energy (more stable) than the reactants. A reaction in which products are higher in energy than reactants will proceed to the left, toward reactants. When products are lower in energy than reactants, heat is given off in the course of the reaction. For example, heat is given off when an acid such as HBr is added to ethene (eq. 3.24). Such a reaction is **exothermic.**

$$\begin{array}{c} H \\ \diagdown \\ C=C \\ \diagup \\ H \end{array} \begin{array}{c} H \\ \diagup \\ \diagdown \\ H \end{array} + HBr \rightleftharpoons CH_3CH_2Br \qquad (3.24)$$

An **exothermic** reaction evolves heat energy; an **endothermic** reaction takes in heat energy. The chemists' term for heat energy is **enthalpy, H.**

On the other hand, heat must be added to ethane to produce two methyl radicals (eq. 1.3). This reaction is **endothermic** (takes in heat). The term used by chemists for heat energy is **enthalpy** and is designated by the symbol **H.** The difference in enthalpy between products and reactants is designated by the symbol ΔH (pronounced "delta H").

For the addition of HBr to ethene, the product (bromoethane) is more stable than the reactants (ethene and HBr), and the reaction proceeds to the right. For this reaction ΔH is negative (heat is given off), and K_{eq} is much greater than 1 (Figure 3.9a). For the formation of two methyl radicals from ethane, ΔH is positive (heat is absorbed), and K_{eq} is much less than 1 (Figure 3.9b).**

---

*The double arrow indicates that this reaction goes both ways and reaches chemical equilibrium.

Actually, enthalpy is not the only factor that contributes to the energy difference between products and reactants. A factor called **entropy, S, also contributes to the total energy difference, which is known as the **Gibbs free-energy difference, ΔG,** in the equation $\Delta G = \Delta H - T\Delta S$. For most organic reactions, however, the entropy contribution is very small compared to the enthalpy contribution.

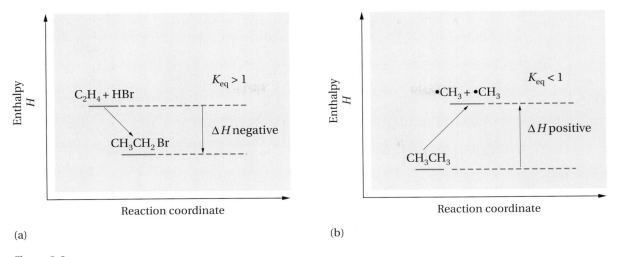

Figure 3.9
(a) The addition of HBr to ethene; the reaction equilibrium lies to the right.
(b) The formation of methyl radicals from ethane; the reaction equilibrium lies to the left.

3.12   Reaction Rates: How Fast Does a Reaction Go?

The equilibrium constant for a reaction tells us whether or not products are more stable than reactants. However, *the equilibrium constant does not tell us anything about the rate of a reaction.* For example, the equilibrium constant for the reaction of gasoline with oxygen is very large, but gasoline can be safely handled in air because the reaction is very slow unless a spark is used to initiate it. The rate of addition of HBr to ethene is also very slow, although the reaction is exothermic.

In order to react, molecules must collide with each other with enough energy and with the right orientation so that the breaking and making of bonds can occur. The energy required for this process is a barrier to reaction. The higher the barrier, the slower the reaction.

Chemists use **reaction energy diagrams** to show the changes in energy that occur in the course of a reaction. Figure 3.10 shows the reaction energy diagram for the polar addition of the acid HBr to ethene (eq. 3.24). This reaction occurs in two steps. In the first step, as a proton adds to the double bond, the π bond of the alkene is broken and a C—H σ bond is formed, giving a carbocation intermediate product. The reactants start with the energy shown at the left of the diagram. As the π bond begins to break and the new σ bond begins to form, the structure formed by the reactants reaches a maximum energy. This structure with maximum energy is called the **transition state** for the first step. This structure cannot be isolated and continues to change until the carbocation product of the first step is fully formed.

The difference in energy between the transition state and the reactants is called the **activation energy, E_a.** It is this energy that determines the rate of the reaction. If E_a is great, the reaction will be slow. A small E_a means that the reaction will proceed rapidly.

In the second step of the reaction, a new carbon–bromine σ bond is formed. Again, the approach of the bromide ion to the positively charged carbon of the carbocation intermediate causes a rise in energy to a maximum. The structure at this energy

A **reaction energy diagram** shows the changes in energy that occur in the course of a reaction. A **transition state** is a structure with maximum energy for a particular reaction step. **Activation energy, E_a,** is the difference in energy between reactants and the transition state and it determines the **reaction rate.**

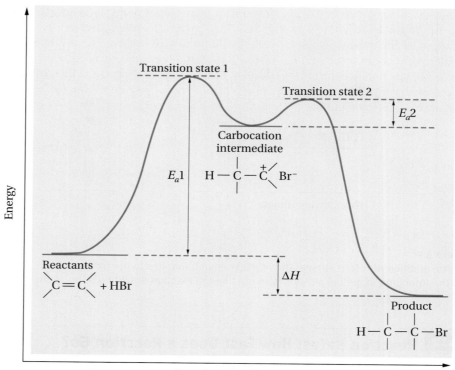

Figure 3.10
Reaction energy diagram for the addition of HBr to ethene.

maximum is the transition state for the second step. The difference in energy between the carbocation and this transition state is the activation energy E_a for this step. This structure cannot be isolated and continues to change until the σ bond is fully formed, completing the formation of the product.

Notice in Figure 3.10 that although the final product of the reaction is lower in energy (ΔH) than the reactants, the reactants must surmount two energy barriers (E_a1 and E_a2), one for each step of the reaction. Between the two transition states, the carbocation intermediate is at an energy minimum that is higher than reactants or

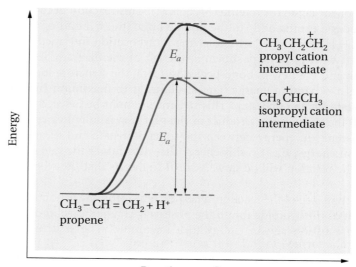

Figure 3.11
Reaction energy diagram for formation of the isopropyl and propyl cations from propene (Eq. 3.21).

products. The first step of the reaction is *endothermic,* because the carbocation intermediate product is higher in energy than the reactants. The second step is exothermic, because the product is lower in energy than the carbocation. The overall reaction is *exothermic,* because the product is lower in energy than the reactants. However, the rate of the reaction is determined by the highest energy barrier, E_a1. The second activation energy, E_a2, is very low compared to the activation energy for the first step. Therefore, as described in Section 3.9, the first step is the slower of the two steps, and the rate of the reaction is determined by the rate of this first step.

EXAMPLE 3.5

Sketch a reaction energy diagram for a one-step reaction that is very slow and slightly exothermic.

Solution   A very slow reaction has a large E_a, and a slightly exothermic reaction has a small negative ΔH. Therefore, the diagram will look like this:

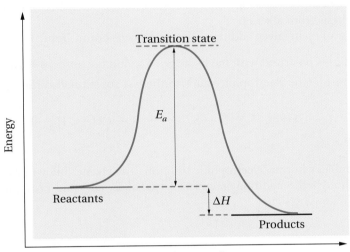

PROBLEM 3.16   Draw a reaction energy diagram for a one-step reaction that is very fast and very exothermic.

PROBLEM 3.17   Draw a reaction energy diagram for a one-step reaction that is very slow and slightly endothermic.

PROBLEM 3.18   Draw a reaction energy diagram for a two-step reaction that has an endothermic first step and an exothermic second step. Label the reactants, transition states, reaction intermediate, activation energies, and enthalpy differences.

Let us see how reaction rates are related to Markovnikov's rule. In electrophilic addition reactions, more stable carbocations are formed more rapidly than less stable carbocations. This is because more stable carbocations are lower in energy than less stable carbocations and it follows that the activation energy E_a for the formation of more stable carbocations is also lower. For example, both isopropyl and propyl cations could be formed from propene and H^+ (eq. 3.21), but the isopropyl cation is much lower in energy than the propyl cation (Figure 3.11). Formation of the isopropyl cation therefore has a lower activation energy E_a and this carbocation is formed more rapidly than the propyl cation. The regioselectivity of electrophilic additions is thus the result of competing first steps, in which the more stable carbocation is formed at a faster rate.

Increasing **temperature** or using a **catalyst** increases reaction rates.

Other factors that affect reaction rates are **temperature** and **catalysts.** Heating a reaction generally increases the rate at which the reaction occurs by providing the reactant molecules with more energy to surmount activation energy barriers. Catalysts speed up a reaction by providing an alternative pathway or mechanism for the reaction, one in which the activation energy is lower. Enzymes play this role in biochemical reactions.

In the next five sections, we will continue our survey of the reactions of alkenes.

3.13 Hydroboration of Alkenes

Hydroboration is the addition of H—B⟨ to an alkene.

Hydroboration was discovered by Professor Herbert C. Brown (Purdue University). This reaction is so useful in synthesis that Brown's work earned him a Nobel Prize in 1979. We will describe here only one practical example of hydroboration, a two-step alcohol synthesis from alkenes.

Hydroboration involves addition of a hydrogen–boron bond to an alkene. The H—B⟨ bond is polarized with the hydrogen $\delta-$ and the boron $\delta+$. Addition occurs so that the boron (the electrophile) adds to the less substituted carbon.

$$R-CH=CH_2 + \overset{\delta-}{H}-\overset{\delta+}{B}\diagdown \longrightarrow R-\underset{H}{CH}-CH_2-B\diagdown \tag{3.25}$$

Thus, it resembles a normal electrophilic addition to an alkene, following Markovnikov's rule, even though the addition is concerted (that is, all bond-breaking and bond-making occurs in one step).

transition state for hydroboration

Because it has three B—H bonds, one molecule of borane, BH_3, can react with three molecules of an alkene. For example, propene gives tri-*n*-propylborane.

$$3\ CH_3CH=CH_2 + BH_3 \longrightarrow CH_3CH_2CH_2-B\diagup^{CH_2CH_2CH_3}_{\diagdown CH_2CH_2CH_3} \tag{3.26}$$

propene   borane   tri-*n*-propylborane

The trialkylboranes made in this way are usually not isolated but are treated with some other reagent to obtain the desired final product. For example, trialkylboranes can be oxidized by hydrogen peroxide and base to give alcohols.

$$(CH_3CH_2CH_2)_3B + 3\ H_2O_2 + 3\ NaOH \longrightarrow$$
tri-*n*-propylborane

$$3\ CH_3CH_2CH_2OH + Na_3BO_3 + 3\ H_2O \tag{3.27}$$
n-propyl alcohol   sodium borate

One great advantage of this hydroboration–oxidation sequence is that it provides a route to alcohols that *cannot* be obtained by the acid-catalyzed hydration of alkenes (review eq. 3.13).

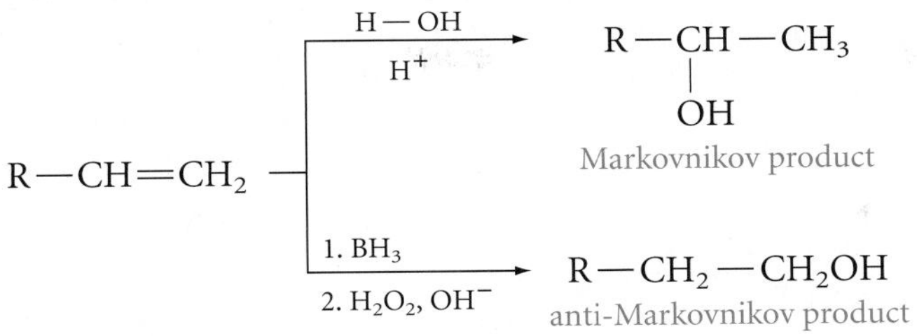

$$R-CH=CH_2 \begin{cases} \xrightarrow[H^+]{H-OH} & R-\underset{\underset{OH}{|}}{CH}-CH_3 \quad \text{Markovnikov product} \\[2em] \xrightarrow[2.\ H_2O_2,\ OH^-]{1.\ BH_3} & R-CH_2-CH_2OH \quad \text{anti-Markovnikov product} \end{cases}$$

(3.28)

The overall result of the two-step hydroboration sequence *appears* to be the addition of water to the carbon–carbon double bond in the reverse of the usual Markovnikov sense.

EXAMPLE 3.6

What alcohol is obtained from this sequence?

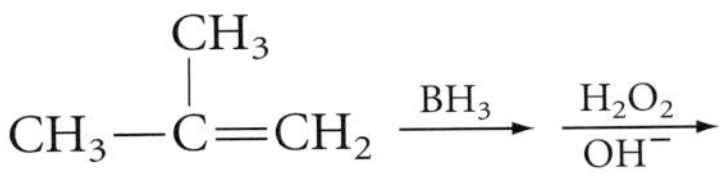

$$CH_3-\underset{\underset{CH_3}{|}}{C}=CH_2 \xrightarrow{BH_3} \xrightarrow[OH^-]{H_2O_2}$$

Solution   The boron adds to the less substituted carbon; oxidation gives the corresponding alcohol. Compare this result with that of eq. 3.15.

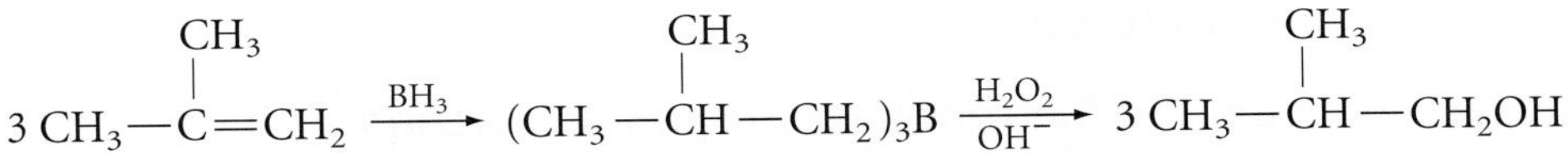

$$3\ CH_3-\underset{\underset{CH_3}{|}}{C}=CH_2 \xrightarrow{BH_3} (CH_3-\underset{\underset{CH_3}{|}}{CH}-CH_2)_3B \xrightarrow[OH^-]{H_2O_2} 3\ CH_3-\underset{\underset{CH_3}{|}}{CH}-CH_2OH$$

PROBLEM 3.19 What alcohol is obtained by applying the hydroboration–oxidation sequence to 2-methyl-2-butene?

PROBLEM 3.20 What alkene is needed to obtain ⬡—CH₂CH₂OH via the hydroboration–oxidation sequence? What product would this alkene give with acid-catalyzed hydration?

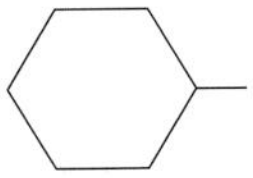

3.14 Addition of Hydrogen

Hydrogen adds to alkenes in the presence of an appropriate catalyst. The process is called **hydrogenation.**

Hydrogenation is the addition of hydrogen to alkenes in the presence of a catalyst.

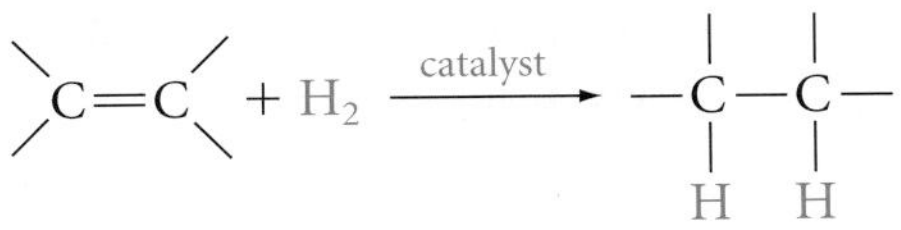

$$\underset{/}{\overset{\backslash}{}}C=C\underset{\backslash}{\overset{/}{}} + H_2 \xrightarrow{\text{catalyst}} -\underset{\underset{H}{|}}{C}-\underset{\underset{H}{|}}{C}-$$

(3.29)

The catalyst is usually a finely divided metal, such as nickel, platinum, or palladium. These metals adsorb hydrogen gas on their surfaces and activate the hydrogen–hydrogen bond. Both hydrogen atoms usually add from the catalyst surface to the same face of the double bond. For example, 1,2-dimethylcyclopentene gives mainly *cis*-1,2-dimethylcyclopentane.

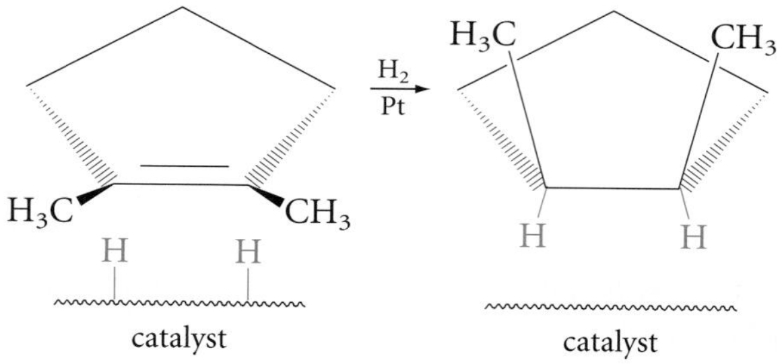

(3.30)

Catalytic hydrogenation of double bonds is used commercially to convert vegetable oils to margarine and other cooking fats (Sec. 15.2).

PROBLEM 3.21 Write an equation for the catalytic hydrogenation of

a. methylpropene.
b. 1,2-dimethylcyclohexene.

3.15 Additions to Conjugated Systems

3.15.a Electrophilic Additions to Conjugated Dienes

Alternate double and single bonds of conjugated systems have special consequences for their addition reactions. When 1 mole of hydrogen bromide adds to 1 mole of 1,3-butadiene, a rather surprising result is obtained. Two products are isolated.

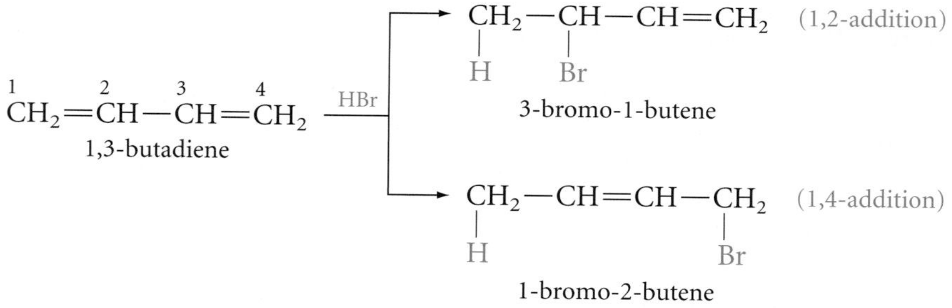

(3.31)

In one of these products, HBr has added to one of the two double bonds, and the other double bond is still present in its original position. We call this the product of **1,2-addition.** The other product may at first seem unexpected. The hydrogen and bromine have added to carbon-1 and carbon-4 of the original diene, and a new double bond has appeared between carbon-2 and carbon-3. This process, known as **1,4-addition,** is quite general for electrophilic additions to conjugated systems. How can we explain it?

In **1,2-addition** a reagent is added to the first and second carbons of a conjugated diene, whereas **1,4-addition** is addition to the first and fourth carbons.

In the first step, the proton adds to the terminal carbon atom, according to Markovnikov's rule.

$$H^+ + CH_2{=}CH{-}CH{=}CH_2 \longrightarrow CH_3{-}\overset{+}{C}H{-}CH{=}CH_2 \qquad \textbf{(3.32)}$$

The resulting carbocation can be stabilized by resonance; in fact, it is a hybrid of two contributing resonance structures (see Sec. 1.12).

$$[CH_3{-}\overset{+}{C}H{-}CH{=}CH_2 \longleftrightarrow CH_3{-}CH{=}CH{-}\overset{+}{C}H_2]$$

The positive charge is delocalized over carbon-2 and carbon-4. When, in the next step, the carbocation reacts with bromide ion (the nucleophile), it can react either at carbon-2 to give the product of 1,2-addition, or at carbon-4 to give the product of 1,4-addition.

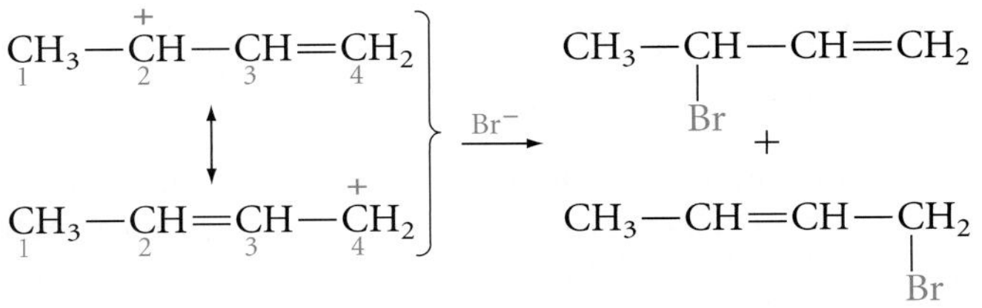

$$\left.\begin{array}{c} CH_3{-}\underset{1}{C}\overset{+}{\underset{2}{H}}{-}\underset{3}{CH}{=}\underset{4}{CH_2} \\ \updownarrow \\ CH_3{-}\underset{1}{CH}{=}\underset{2}{CH}{-}\overset{+}{\underset{3}{C}}\underset{4}{H_2} \end{array}\right\} \xrightarrow{Br^-} \begin{array}{c} CH_3{-}CH{-}CH{=}CH_2 \\ | \\ Br \\ + \\ CH_3{-}CH{=}CH{-}CH_2 \\ | \\ Br \end{array} \qquad \textbf{(3.33)}$$

PROBLEM 3.22 Explain why, in the first step of this reaction, the proton adds to C-1 (eq. 3.32) and not to C-2.

The carbocation intermediate in these reactions is a single species, a resonance hybrid. This type of carbocation, with a carbon–carbon double bond adjacent to the positive carbon, is called an **allylic cation.** The parent allyl cation, shown below as a resonance hybrid, is a primary carbocation, but it is more stable than simple primary ions (such as propyl), because its positive charge is delocalized over the two end carbon atoms.

In an **allylic cation,** a carbon–carbon double bond is adjacent to the positively charged carbon atom.

$$CH_2{=}CH{-}\overset{+}{C}H_2 \longleftrightarrow \overset{+}{C}H_2{-}CH{=}CH_2$$

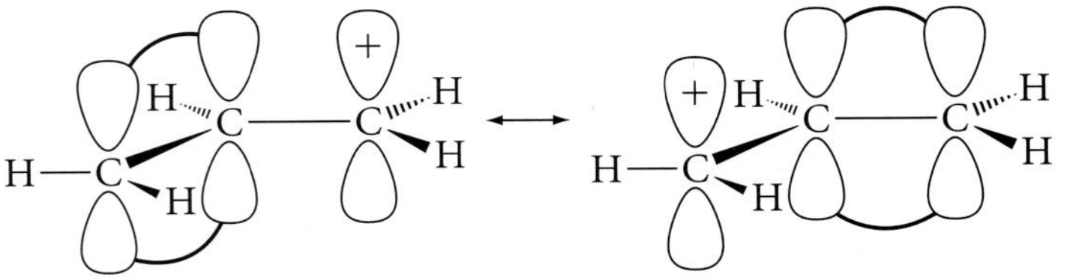

$$\qquad \textbf{(3.34)}$$

the allyl carbocation

PROBLEM 3.23 Draw the contributors to the resonance hybrid structure of the

3-cyclopentenyl cation 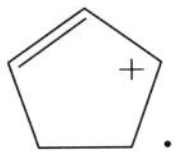 .

PROBLEM 3.24 Write an equation for the expected products of 1,2-addition and 1,4-addition of bromine to 1,3-butadiene.

3.15.b Cycloaddition to Conjugated Dienes: The Diels–Alder Reaction

Ch. 3, Ref. 3. *View animation of Diels–Alder reaction.*

Conjugated dienes undergo another type of 1,4-addition when they react with alkenes (or alkynes). The simplest example is the addition of ethylene to 1,3-butadiene to give cyclohexene.

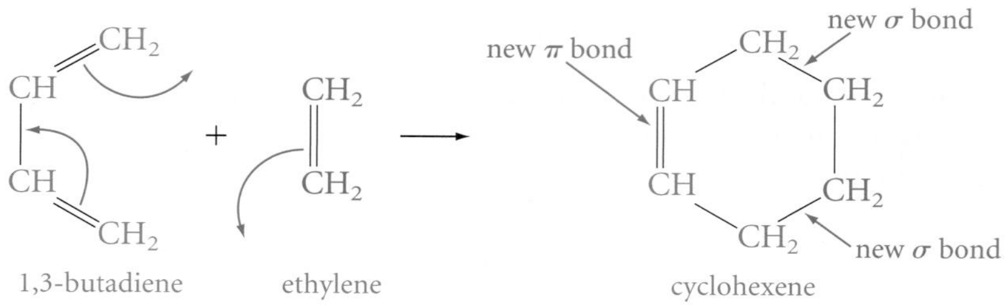

1,3-butadiene   ethylene   cyclohexene   (3.35)

The **Diels–Alder reaction** is the **cycloaddition reaction** of a conjugated **diene** and a **dienophile** to give a cyclic product in which three pi bonds are converted to two sigma bonds and a new pi bond.

This reaction is an example of a **cycloaddition reaction,** an addition that results in a cyclic product. This cycloaddition, which converts three π bonds to two σ bonds and one new π bond, is called the **Diels–Alder reaction,** after its discoverers, Otto Diels and Kurt Alder. It is so useful for making cyclic compounds that it earned the 1950 Nobel Prize in chemistry for its discoverers. As with hydroboration (Sec. 3.13), this reaction is *concerted.* All bond-breaking and bond-making occurs at the same time.

The two reactants are a **diene** and a **dienophile** (diene lover). The simple example in eq. 3.35 is not typical of most Diels–Alder reactions because it proceeds only under pressure and not in good yield. However, this type of reaction gives excellent yields at moderate temperatures if the dienophile has *electron-withdrawing groups** attached, as in the following examples:

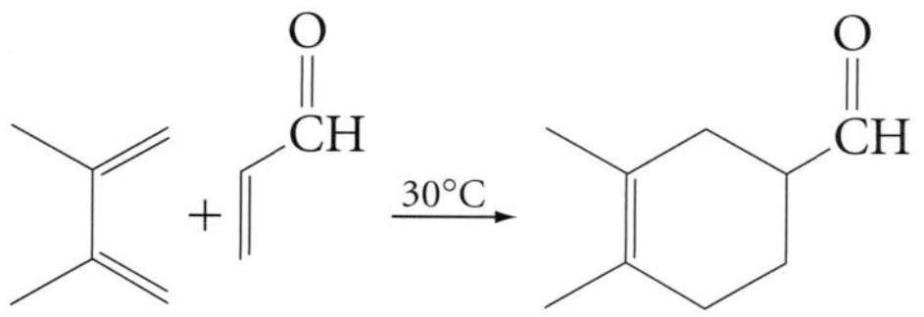

(3.36)

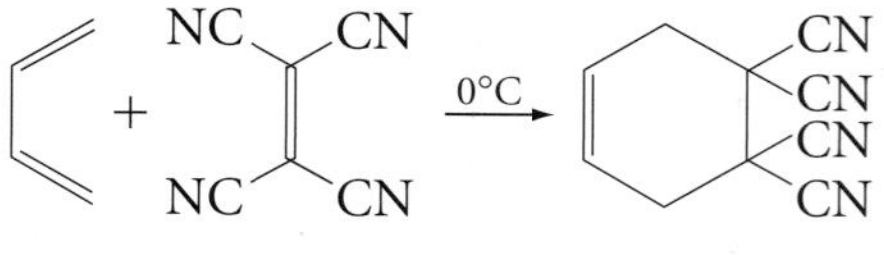

(3.37)

---

*Electron-withdrawing groups are groups of atoms that attract the electrons of the π bond, making the alkene electron poor and therefore more electrophilic toward the diene.

EXAMPLE 3.7

How could a Diels–Alder reaction be used to synthesize the following compound?

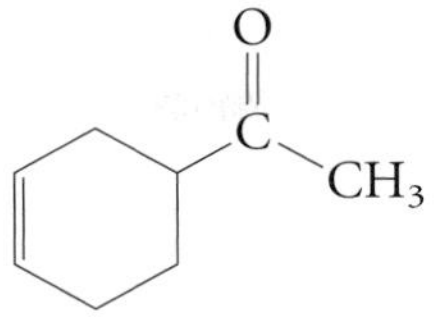

Solution Work backwards. The double bond in the product was a single bond in the starting diene. Therefore,

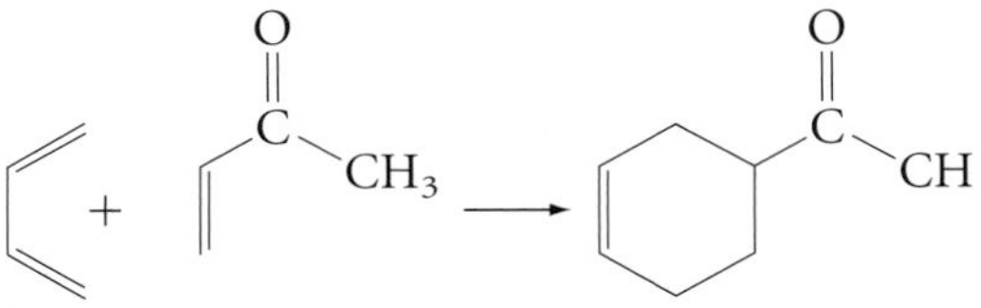

PROBLEM 3.25 Show how limonene (Figure 1.12) could be formed by a Diels–Alder reaction of isoprene (2-methyl-1,3-butadiene) with itself.

PROBLEM 3.26 Draw the structure of the product of each of the following cycloaddition reactions.

a. 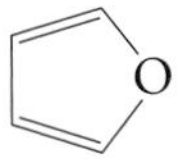 $O + CH_2\text{=}CH\text{—}CN$

b. $CH_2\text{=}CH\text{—}CH\text{=}CH_2 + NC\text{—}C\text{≡}C\text{—}CN$

3.16 Free-Radical Additions; Polyethylene

Some reagents add to alkenes by a free-radical mechanism instead of by an ionic mechanism. From a commercial standpoint, the most important of these free-radical additions are those that lead to polymers.

A **polymer** is a large molecule, usually with a high molecular weight, built up from small repeating units. The simple molecule from which these repeating units are derived is called a **monomer,** and the process of converting a monomer to a polymer is called **polymerization.**

The free-radical polymerization of ethylene gives **polyethylene,** a material that is produced on a very large scale (more than 10 billion pounds annually in the United States alone). The reaction is carried out by heating ethylene under pressure with a catalyst (eq. 3.38). How does this reaction occur?

$$CH_2\text{=}CH_2 \xrightarrow[\text{1000 atm, >100°C}]{\text{ROOR}} \text{(}CH_2\text{—}CH_2\text{)}_{\overline{n}} \qquad \textbf{(3.38)}$$

ethylene   polyethylene
                        (*n* = several thousand)

A **polymer** is a large molecule containing a repeating unit derived from small molecules called **monomers.** The process of polymer formation is called **polymerization.**

One common type of catalyst for polymerization is an organic peroxide. The O—O single bond is weak, and on heating this bond breaks, with one electron going to each of the oxygens.

$$R\!-\!O\!-\!O\!-\!R \xrightarrow{\text{heat}} 2\,R\!-\!O\cdot \qquad\qquad (3.39)$$

organic peroxide   two radicals

A catalyst radical then adds to the carbon–carbon double bond:

$$RO\cdot \quad CH_2\!=\!CH_2 \longrightarrow RO\!-\!CH_2\!-\!\overset{\cdot}{C}H_2 \qquad (3.40)$$

catalyst   a carbon
radical   free radical

The result of this addition is a carbon free radical, which may add to another ethylene molecule, and another, and another, and so on.

$$RO\overset{\cdot}{C}H_2\overset{\cdot}{C}H_2 \xrightarrow{CH_2=CH_2} ROCH_2CH_2CH_2\overset{\cdot}{C}H_2 \xrightarrow{CH_2=CH_2}$$

$$ROCH_2CH_2CH_2CH_2CH_2\overset{\cdot}{C}H_2 \text{ and so on} \qquad (3.41)$$

The carbon chain continues to grow in length until some chain-termination reaction occurs (perhaps a combination of two radicals).

We might think that only a single long chain of carbons will be formed in this way, but this is not always the case. A "growing" polymer chain may abstract a hydrogen atom from its back, so to speak, to cause **chain branching.**

$$\qquad\qquad (3.42)$$

A giant molecule with long and short branches is thus formed:

branched polyethylene

Fish in polyethylene bag.

The degree of chain branching and other features of the polymer structure can often be controlled by the choice of catalyst and reaction conditions.

A polyethylene molecule is mainly saturated despite its name (polyethyl*ene*) and consists mostly of linked CH_2 groups, but with CH groups at the branch points and CH_3 groups at the ends of the branches. It also contains an OR group from the catalyst at one end, but since the molecular weight is very large, this OR group constitutes a minor and, as far as properties go, relatively insignificant fraction of the molecule.

Polyethylene made in this way is transparent and used in packaging and film (for example, for freezer and sandwich bags).

In Chapter 14, we will describe many other polymers, some made by the process just described for polyethylene and some made by other methods.

3.17 Oxidation of Alkenes

In general, alkenes are more easily oxidized than alkanes by chemical oxidizing agents. These reagents attack the pi electrons of the double bond. The reactions may be useful as chemical tests for the presence of a double bond or for synthesis.

3.17.a Oxidation with Permanganate; a Chemical Test

Alkenes react with alkaline potassium permanganate to form **glycols** (compounds with two adjacent hydroxyl groups).

Glycols are compounds with two hydroxyl groups on adjacent carbons.

$$3 \ \underset{\diagup}{\overset{\diagdown}{C}} {=} \underset{\diagdown}{\overset{\diagup}{C}} \ + \ 2 \, K^+MnO_4^- \ + \ 4 \, H_2O \ \longrightarrow \ 3 -\overset{|}{\underset{|}{C}}-\overset{|}{\underset{|}{C}}- \ + \ 2 \, MnO_2 \ + \ 2 \, K^+OH^-$$
$$ OH \ OH$$

(3.43)

alkene   potassium permanganate (purple)   a glycol   manganese dioxide (brown-black)

As the reaction occurs, the purple color of the permanganate ion is replaced by the brown precipitate of manganese dioxide. Because of this color change, the reaction can be used as a chemical test to distinguish alkenes from alkanes, which normally do not react.

> **PROBLEM 3.27** Write an equation for the reaction of 2-butene with potassium permanganate.

Hexane does not react with purple KMnO₄ (left); cyclohexene (right) reacts, producing a brown-black precipitate of MnO₂.

3.17.b Ozonolysis of Alkenes

Alkenes react rapidly and quantitatively with ozone, O_3. Ozone is generated by passing oxygen over a high-voltage electric discharge. The resulting gas stream is then bubbled at low temperature into a solution of the alkene in an inert solvent, such as dichloromethane. The first product, a **molozonide,** is formed by cycloaddition of the oxygen at each end of the ozone molecule to the carbon–carbon double bond. This product then rearranges rapidly to an **ozonide.** Since these products may be explosive if isolated, they are usually treated directly with a reducing agent, commonly zinc and aqueous acid, to give carbonyl compounds as the isolated products.

$$\underset{\diagup}{\overset{\diagdown}{C}}{=}\underset{\diagdown}{\overset{\diagup}{C}} \ \xrightarrow{O_3} \ \left[\begin{array}{c} -C - C - \\ O \qquad O \\ O \end{array} \right] \ \longrightarrow \ \begin{array}{c} C \qquad C \\ O \qquad O \\ O - O \end{array} \ \xrightarrow[H_3O^+]{Zn} \ \underset{\diagup}{\overset{\diagdown}{C}}{=}O + O{=}\underset{\diagdown}{\overset{\diagup}{C}}$$

(3.44)

alkene   molozonide   ozonide   two carbonyl groups

The net result of this reaction is to break the double bond of the alkene and to form two carbon–oxygen double bonds (carbonyl groups), one at each carbon of the original double bond. The overall process is called **ozonolysis.**

Ozonolysis is the oxidation of alkenes with ozone to give carbonyl compounds.

A W O R D A B O U T ...

Ethylene: Raw Material and Plant Hormone

Ethylene, the simplest alkene, ranks first among organic chemicals in industrial production. Current U.S. annual production of ethylene is well over 58 billion pounds. Propene comes in second with about half that amount.

How is all this ethylene produced, and what is it used for? Most hydrocarbons can be "cracked" to give ethylene. (See "A Word About Petroleum, Gasoline, and Octane Number" on pages 106–107.) In the United States, the major raw material for this purpose is ethane.

Cranberry harvest.

$$CH_3—CH_3 \xrightarrow{700–900°C} CH_2=CH_2 + H_2$$

A substantial fraction of industrial ethylene is, of course, converted to polyethylene, as described in Section 3.16; but ethylene is also a key raw material for the manufacture of other industrial organic chemicals, because of the reactivity of the carbon–carbon double bond. Shown in Figure 3.12 are 9 of the top 50 organic chemicals; each is produced from ethylene.

Ethylene is not only the most important industrial source of organic chemicals, it also has some biochemical properties that are crucial to agriculture. Ethylene is a **plant hormone** that can cause seeds to sprout, flowers to bloom, fruit to ripen and fall, and leaves and petals to shrivel and turn brown. It is produced naturally by plants from the amino acid *methionine* via an unusual cyclic amino acid, *1-aminocyclopropane-1-carboxylic acid (ACC)*, which is then, in several steps, converted to ethylene.

$$CH_3—S—CH_2CH_2—\underset{\underset{+}{\overset{|}{NH_3}}}{\overset{|}{CH}}—CO_2^- \xrightarrow{\text{several steps}}$$
methionine

$$\underset{CH_2}{\overset{CH_2}{|}}—\underset{\underset{+}{NH_3}}{\overset{CO_2^-}{C}} \xrightarrow{\text{several steps}} CH_2=CH_2 + CO_2 + HCN$$
ACC   ethylene

The mode by which ethylene functions biologically is still being studied.

Chemists have prepared synthetic compounds that can release ethylene in plants in a controlled manner. One such example is 2-chloroethylphosphonic acid, $ClCH_2CH_2PO(OH)_2$. Sold by Union Carbide under the trade name *ethrel*, it is water soluble and is taken up by plants, where it breaks down to ethylene, chloride, and phosphate. It has been used commercially to induce fruits, such as pineapples and tomatoes, to ripen uniformly so that an entire field can be harvested efficiently, as shown in the photo above. Ethylene has also been used to regulate the growth of other crops, such as wheat, apples, cherries, and cotton. Only a small amount need be used, since plants are very sensitive to ethylene and respond to concentrations lower than 0.1 part per million of the gas.

Ozonolysis can be used to locate the position of a double bond. For example, ozonolysis of 1-butene gives two different aldehydes, whereas 2-butene gives a single aldehyde.

$$CH_2=CHCH_2CH_3 \xrightarrow[\text{2. Zn, H}^+]{\text{1. O}_3} CH_2=O + O=CHCH_2CH_3 \quad \textbf{(3.45)}$$
1-butene   formaldehyde   propanal

$$CH_3CH=CHCH_3 \xrightarrow[\text{2. Zn, H}^+]{\text{1. O}_3} 2\ CH_3CH=O \quad \textbf{(3.46)}$$
2-butene   ethanal

Using ozonolysis, one can easily tell which butene isomer is which. By working backward from the structures of ozonolysis products, one can deduce the structure of an unknown alkene.

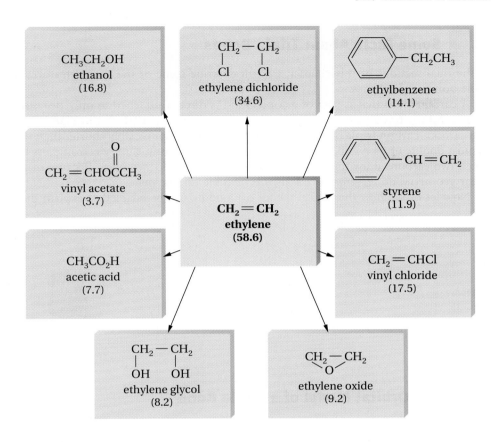

Figure 3.12
Ethylene is central to the manufacture of many industrial organic chemicals. The numbers in parentheses give the recent U.S. production of these chemicals in *billions* of pounds.

EXAMPLE 3.8

Ozonolysis of an alkene produces equal amounts of acetone and formaldehyde, $(CH_3)_2C{=}O$ and $CH_2{=}O$, respectively. Deduce the alkene structure.

Solution   Connect to each other by a double bond the carbons that are bound to oxygen in the ozonolysis products. The alkene is $(CH_3)_2C{=}CH_2$.

PROBLEM 3.28   Which alkene will give only acetone, $(CH_3)_2C{=}O$, as the ozonolysis product?

3.17.c Other Alkene Oxidations

Various reagents can convert alkenes to epoxides (eq. 3.47).

$$\ce{>C=C< -> >C-C<}$$

alkene   epoxide

(3.47)

This reaction and the chemistry of epoxides are detailed in Chapter 8.

Like alkanes (and all other hydrocarbons), alkenes can be used as fuels. Complete combustion gives carbon dioxide and water.

$$C_nH_{2n} + \tfrac{3n}{2}O_2 \longrightarrow nCO_2 + nH_2O$$

(3.48)

◎ **Ch. 3, Ref. 4.** *Examine and manipulate 3D model of acetylene.*

3.18 Some Facts About Triple Bonds

In the final sections of this chapter, we will describe some of the special features of triple bonds and alkynes.

A carbon atom that is part of a triple bond is directly attached to only *two* other atoms, and the bond angle is 180°. Thus, acetylene is linear, as shown in Figure 3.13. The carbon–carbon triple bond distance is about 1.21 Å, appreciably shorter than that of most double (1.34 Å) or single (1.54 Å) bonds. Apparently, three electron pairs between two carbons draw them even closer together than do two pairs. Because of the linear geometry, no *cis–trans* isomerism is possible for alkynes.

Now let us see how the orbital theory of bonding can be adapted to explain these facts.

Figure 3.13
Models of acetylene, showing its linearity.

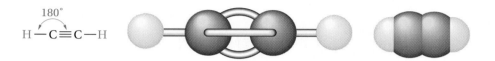

3.19 The Orbital Model of a Triple Bond

sp-**Hybrid orbitals** are half *s* and half *p* in character. The angle between two *sp* orbitals is 180°.

The carbon atom of an acetylene is connected to only *two* other atoms. Therefore, we combine the 2*s* orbital with only one 2*p* orbital to make two ***sp*-hybrid orbitals** (Figure 3.14). These orbitals extend in opposite directions from the carbon atom. The angle between the two hybrid orbitals is 180° so as to minimize repulsion between any electrons placed in them. One valence electron is placed in each *sp*-hybrid orbital. The remaining two valence electrons occupy two different *p* orbitals that are perpendicular to each other and perpendicular to the hybrid *sp* orbitals.

The formulation of a triple bond from two *sp*-hybridized carbons is shown in Figure 3.15. End-on overlap of two *sp* orbitals forms a sigma bond between the two carbons, and lateral overlap of the properly aligned *p* orbitals forms two pi bonds (designated π_1 and π_2 in the figure). This model nicely explains the linearity of acetylenes.

Figure 3.14
Unhybridized versus *sp*-hybridized orbitals on carbon.

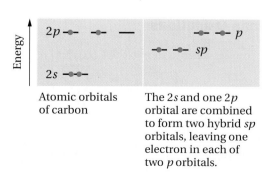

Atomic orbitals of carbon

The 2*s* and one 2*p* orbital are combined to form two hybrid *sp* orbitals, leaving one electron in each of two *p* orbitals.

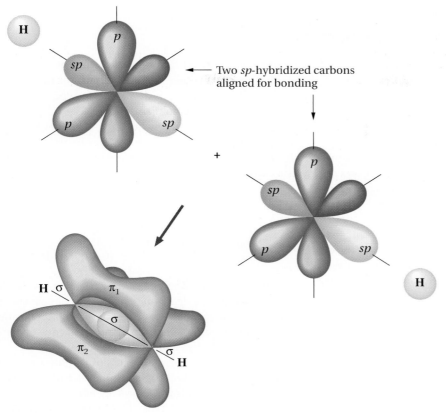

Two *sp*-hybridized carbons aligned for bonding

The resulting carbon–carbon triple bond, with a hydrogen atom attached to each remaining *sp* bond. (The orbitals involved in the C—H bonds are omitted for clarity.)

Figure 3.15
A triple bond consists of the end-on overlap of two *sp*-hybrid orbitals to form a σ bond and the lateral overlap of two sets of parallel-oriented p orbitals to form two mutually perpendicular π bonds.

Addition Reactions of Alkynes

Many addition reactions described for alkenes also occur, though usually more slowly, with alkynes. For example, bromine adds as follows:

$$H-C\equiv C-H \xrightarrow{Br_2} \underset{\text{\textit{trans}-1,2-dibromoethene}}{\overset{H}{\underset{Br}{}}C=C\overset{Br}{\underset{H}{}}} \xrightarrow{Br_2} \underset{\text{1,1,2,2-tetrabromoethane}}{H-\underset{Br}{\overset{Br}{C}}-\underset{Br}{\overset{Br}{C}}-H} \qquad \textbf{(3.49)}$$

In the first step, the addition occurs mainly *trans*.

With an ordinary nickel or platinum catalyst, alkynes are hydrogenated all the way to alkanes (eq. 3.1). However, a special palladium catalyst (called **Lindlar's catalyst**) can control hydrogen addition so that only 1 mole of hydrogen adds. In this case, the

Lindlar's catalyst limits addition of hydrogen to an alkyne to 1 mole and produces a *cis* alkene.

A WORD ABOUT ...

Petroleum, Gasoline, and Octane Number

Petroleum is at present our most important fossil fuel. The need for petroleum to keep our industrial society going follows close behind our need for food, air, water, and shelter. What is black gold, as petroleum has been called, and how do we use it?

Petroleum is a complex mixture of hydrocarbons formed over eons of time through the gradual decay of buried animal and vegetable matter. **Crude oil** is a viscous black liquid that collects in vast underground pockets in sedimentary rock (the word *petroleum* literally means rock oil, from the Latin *petra*, rock, and *oleum*, oil). It must be brought to the surface via drilling and pumping. To be most useful, the crude oil must be refined.

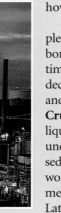

Petroleum refinery, where crude oil is processed.

The first step in petroleum refining is usually **distillation.** The crude oil is heated to about 400°C, and the vapors rise through a tall fractionating column. The lower boiling fractions rise faster and higher in the column before condensing to liquids; higher boiling fractions do not rise as high. By drawing off liquid at various column levels, technicians separate crude oil roughly into the fractions shown in Table 3.3.

The gasoline fraction comprises only about 25% of crude oil. It is the most valuable fraction, however, both as a fuel and as a source material for the petrochemical industry, the industry that furnishes our synthetic fibers, plastics, and many other useful materials. For this reason, many processes have been developed for converting the other fractions into gasoline.

Higher boiling fractions can be "cracked" by heat and catalysts (mainly silica and alumina), to give products with shorter carbon chains and therefore lower boiling points. The carbon chain can break at many points.

$$C_{10}H_{22} \longrightarrow \begin{array}{l} \overset{\text{alkane}}{C_5H_{12}} + \overset{\text{alkene}}{C_5H_{10}} \\ C_8H_{18} + C_2H_4 \\ C_2H_6 + C_8H_{16} \\ C_4H_{10} + (C_4H_8 + C_2H_4) \end{array}$$

To balance the number of hydrogens, any particular alkane must give at least one alkane and one alkene as products. Thus, catalytic cracking converts larger alkanes into a mixture of smaller alkanes and alkenes and increases the yield of gasoline from petroleum.

During cracking, large amounts of the lower gaseous hydrocarbons—ethene, propene, butanes, and butenes—are formed. Some of these, especially ethene, are used as petrochemical raw materials. To obtain more gasoline, scientists sought methods to convert these low-molecular-weight hydrocarbons to somewhat larger hydrocarbons that boil in the gasoline range. One such process is **alkylation,** the combination of an alkane with an alkene to form a higher boiling alkane.

$$C_2H_6 + C_4H_8 \xrightarrow{\text{catalyst}} C_6H_{14}$$
$$\underbrace{C_4H_{10} + C_4H_8}_{\text{gases}} \xrightarrow{\text{catalyst}} \underbrace{C_8H_{18}}_{\text{liquids}}$$

Table 3.3	Common petroleum fractions			
Boiling range, °C	**Name**	**Range of carbon atoms per molecule**	**Use**	
<20	gases	C_1 to C_4	heating, cooking, petrochemical raw material	
20–200	naphtha; straight-run gasoline	C_5 to C_{12}	fuel; lighter fractions (such as petroleum ether, bp 30–60°C) also used as laboratory solvents	
200–300	kerosene	C_{12} to C_{15}	fuel	
300–400	fuel oil	C_{15} to C_{18}	heating homes, diesel fuel	
>400		over C_{18}	lubricating oil, greases, paraffin waxes, asphalt	

These processes, which were developed in the 1930s, were important for producing aviation fuel during World War II and are still used to make high-octane gasoline, now for our much improved automobile engines.

This brings us to **octane number** and why it is important. Some hydrocarbons, especially those with highly branched structures, burn smoothly in an automobile engine and drive the piston forward evenly. Other hydrocarbons, especially those with unbranched carbon chains, tend to explode in the cylinder and drive the piston forward violently. These undesirable explosions produce audible knocks. A scale was set up many years ago to evaluate this important knock property of gasolines. **Isooctane** (2,2,4-trimethylpentane), an excellent fuel with a highly branched structure, was arbitrarily given a rating of 100, and **heptane,** a very poor automotive fuel, was given a rating of 0. A "regular" gasoline with an octane number of 87 has the same "knock" properties as a mixture that is 87% isooctane and 13% heptane.

Small amounts of **tetraethyllead,** $(CH_3CH_2)_4Pb$, in gasoline improve its octane rating but are undesirable for environmental reasons and most gasoline is now unleaded. However, unleaded gasoline must therefore contain hydrocarbons with a high octane rating. It became important to develop methods that convert straight-chain to branched-chain hydrocarbons, since these have higher octane ratings.

Certain catalysts can produce branched-chain alkanes from straight-chain alkanes. This process, called **isomerization,** is carried out on a large scale commercially.

$$CH_3CH_2CH_2CH_3 \xrightarrow[\text{alumina}]{AlCl_3,\ HCl} CH_3\overset{\overset{\displaystyle CH_3}{|}}{C}HCH_3$$

n-butane   isobutane

Aromatic hydrocarbons, such as benzene and toluene, also have a high octane rating. A platinum catalyst used in a process called **platforming** cyclizes and dehydrogenates alkanes to cycloalkanes and to aromatic hydrocarbons. Of course, large amounts of hydrogen gas are also formed during platforming. Millions of gallons of aromatic hydrocarbons are produced daily by such processes, not only to add to unleaded gasoline to improve its octane rating, but also to supply raw materials for many other petrochemically based products, as we will see in the next chapter.

$$CH_3(CH_2)_5CH_3 \xrightarrow[\text{catalyst}]{Pt} \quad \xrightarrow[\text{catalyst}]{Pt}$$

methylcyclohexane

toluene
(an aromatic hydrocarbon)

product is a *cis* alkene, because both hydrogens add to the same face of the triple bond from the catalyst surface.

$$CH_3-C\equiv C-CH_3 \xrightarrow[\text{catalyst)}]{\overset{\displaystyle H-H}{\text{Pd (Lindlar's}}} \quad \text{(3.50)}$$

2-butyne
bp 27°C

cis-2-butene
bp 3.7°C

With unsymmetric triple bonds and unsymmetric reagents, Markovnikov's rule is followed in each step, as shown in the following example:

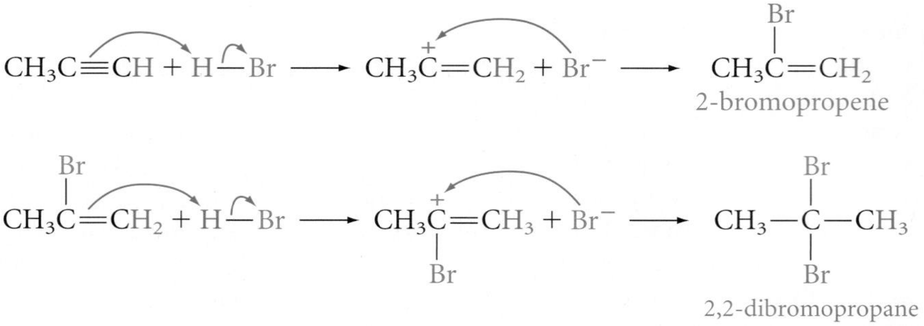

$$CH_3C{\equiv}CH + H{-}Br \longrightarrow CH_3\overset{+}{C}{=}CH_2 + Br^- \longrightarrow CH_3\overset{\overset{\displaystyle Br}{|}}{C}{=}CH_2 \qquad \textbf{(3.51)}$$

2-bromopropene

$$CH_3\overset{\overset{\displaystyle Br}{|}}{C}{=}CH_2 + H{-}Br \longrightarrow CH_3\underset{\underset{\displaystyle Br}{|}}{\overset{+}{C}}{=}CH_3 + Br^- \longrightarrow CH_3{-}\underset{\underset{\displaystyle Br}{|}}{\overset{\overset{\displaystyle Br}{|}}{C}}{-}CH_3$$

2,2-dibromopropane

Addition of water to alkynes requires not only an acid catalyst but mercuric ion as well. The mercuric ion forms a complex with the triple bond and activates it for addition. Although the reaction is similar to that of alkenes, the initial product—a **vinyl alcohol,** or **enol**—rearranges to a carbonyl compound.

A **vinyl alcohol** or **enol** is an alcohol with a carbon–carbon double bond on the carbon that bears the hydroxyl group.

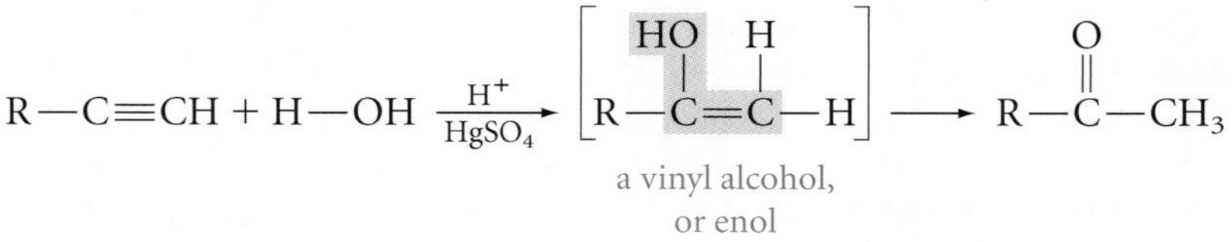

$$R{-}C{\equiv}CH + H{-}OH \xrightarrow[\text{HgSO}_4]{\text{H}^+} \left[R{-}\underset{\underset{\displaystyle H}{|}}{\overset{\overset{\displaystyle HO}{|}}{C}}{=}\overset{\overset{\displaystyle H}{|}}{C}{-}H \right] \longrightarrow R{-}\overset{\overset{\displaystyle O}{\|}}{C}{-}CH_3 \qquad \textbf{(3.52)}$$

a vinyl alcohol, or enol

The product is a methyl ketone or, in the case of acetylene itself (R=H), acetaldehyde. We will have more to say about the chemistry of enols and the mechanism of the second step of eq. 3.52 in Chapter 9.

PROBLEM 3.29 Write equations for the following reactions:

a. $CH_3C{\equiv}CH + Br_2$ (1 mole)
b. $CH_3C{\equiv}CH + Cl_2$ (2 moles)
c. 1-butyne + HBr (1 and 2 moles)
d. 1-pentyne + H_2O (Hg^{2+}, H^+)

3.21 Acidity of Alkynes

A hydrogen atom on a triply bonded carbon is weakly acidic and can be removed by a very strong base. Sodium amide, for example, converts acetylenes to acetylides.

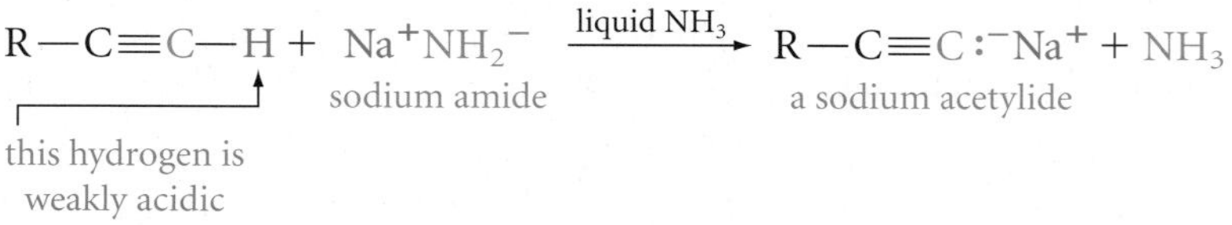

$$R{-}C{\equiv}C{-}H + Na^+NH_2^- \xrightarrow{\text{liquid NH}_3} R{-}C{\equiv}C{:}^-Na^+ + NH_3 \qquad \textbf{(3.53)}$$

sodium amide   a sodium acetylide

this hydrogen is
weakly acidic

A CLOSER LOOK AT ...

Petroleum

Log on to http://www.college.hmco.com and follow the links to the student text web site. Use the Web Link section for Chapter 3 to answer the following questions and find more information on the topics below.

U.S. Department of Energy Educational Site

1. How much oil does the world use in a year? What percent of US energy needs are supplied through oil?
2. What are some of the earliest known examples of oil use by humans? When did the modern oil industry come into being?
3. What are some of the techniques used to extract petroleum (crude oil) from rock? Describe some of the new technologies being developed to extract oil.

Chemical Processing

1. What is cracking? What is the difference between thermal and catalytic cracking?
2. *Unification* and *alteration* are other methods of chemical processing. Define these processes and explain why they are used.

Lots More Information

For more information on a number of related topics, click this link!

This type of reaction occurs easily with a hydrogen attached to a triply bound carbon, but less so when the hydrogen is adjacent to a double or single bond. Why? Consider the hybridization of the carbon atom in each type of C—H bond:

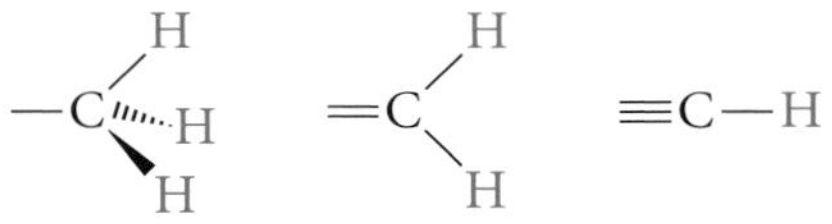

$$sp^3 \qquad sp^2 \qquad sp$$
25% s,   33⅓% s,   50% s,
75% p   66⅔% p   50% p

increasing acidity →

*As the hybridization at carbon becomes more **s**-like and less **p**-like, the acidity of the attached hydrogen increases.* Recall that s orbitals are closer to the nucleus than are p orbitals. Consequently, the bonding electrons are closest to the carbon nucleus in the ≡C—H bond, making it easiest for a base to remove that type of proton. Sodium amide is a sufficiently strong base for this purpose.*

PROBLEM 3.30 Write an equation for the reaction of 1-hexyne with sodium amide in liquid ammonia.

Though acidic, 1-alkynes are much less so than water. Acetylides can therefore be hydrolyzed to alkynes by water. Internal alkynes have no exceptionally acidic hydrogens.

PROBLEM 3.31 Write an equation for the reaction of a sodium acetylide with water.

PROBLEM 3.32 Will 2-butyne react with sodium amide? Explain.

---

*See Table C in the Appendix for the relative acidities of organic functional groups.

REACTION SUMMARY

1. Reactions of Alkenes

a. Addition of Halogens (Sec. 3.7a)

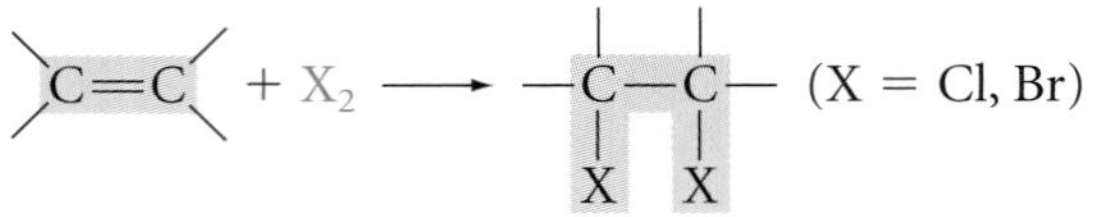

$$\text{C=C} + X_2 \longrightarrow \underset{\underset{X}{|}\ \underset{X}{|}}{-\text{C}-\text{C}-} \quad (X = Cl, Br)$$

b. Addition of Polar Reagents (Sec. 3.7b and Sec. 3.7c)

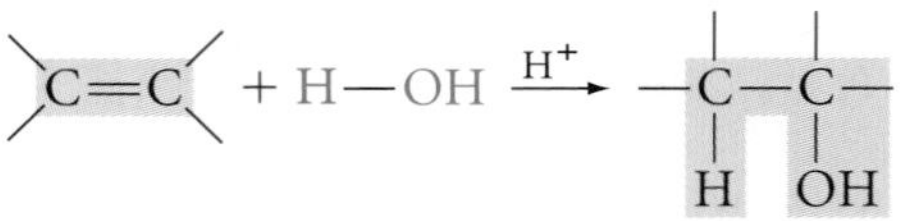

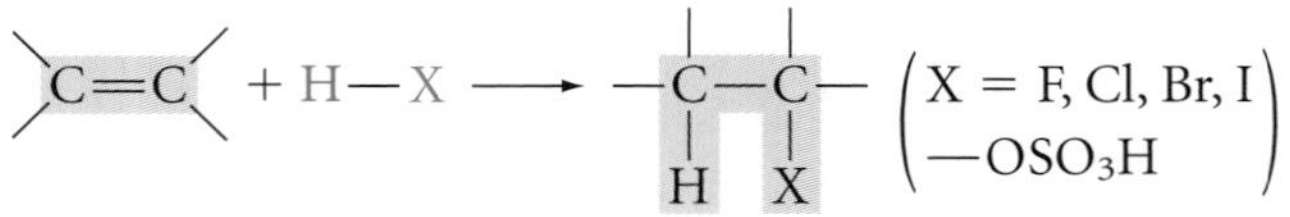

$$\text{C=C} + \text{H—OH} \xrightarrow{\text{H}^+} \underset{\underset{H}{|}\ \underset{OH}{|}}{-\text{C}-\text{C}-}$$

$$\text{C=C} + \text{H—X} \longrightarrow \underset{\underset{H}{|}\ \underset{X}{|}}{-\text{C}-\text{C}-} \quad \left(\begin{array}{l} X = F, Cl, Br, I \\ -OSO_3H \end{array} \right)$$

c. Hydroboration–Oxidation (Sec. 3.13)

$$\text{RCH=CH}_2 \xrightarrow{\text{BH}_3} (\text{RCH}_2\text{CH}_2)_3\text{B} \xrightarrow[\text{HO}^-]{\text{H}_2\text{O}_2} \text{RCH}_2\text{CH}_2\text{OH}$$

d. Addition of Hydrogen (Sec. 3.14)

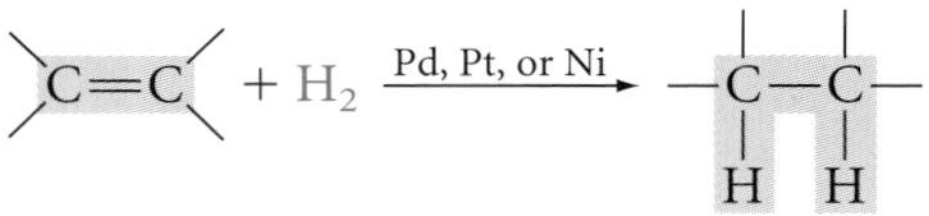

$$\text{C=C} + \text{H}_2 \xrightarrow{\text{Pd, Pt, or Ni}} \underset{\underset{H}{|}\ \underset{H}{|}}{-\text{C}-\text{C}-}$$

e. Addition of X_2 and HX to Conjugated Dienes (Sec. 3.15a)

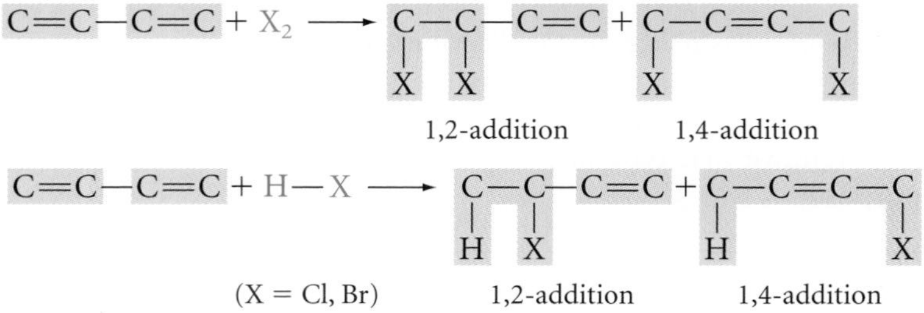

$$\text{C=C—C=C} + X_2 \longrightarrow \underset{\underset{X}{|}\ \underset{X}{|}}{\text{C}-\text{C}}\text{—C=C} + \underset{\underset{X}{|}}{\text{C}}\text{—C=C—}\underset{\underset{X}{|}}{\text{C}}$$

$$ \text{1,2-addition} \qquad \text{1,4-addition}$$

$$\text{C=C—C=C} + \text{H—X} \longrightarrow \underset{\underset{H}{|}\ \underset{X}{|}}{\text{C}-\text{C}}\text{—C=C} + \underset{\underset{H}{|}}{\text{C}}\text{—C=C—}\underset{\underset{X}{|}}{\text{C}}$$

$$(X = Cl, Br) \qquad \text{1,2-addition} \qquad\qquad \text{1,4-addition}$$

f. Cycloaddition to Conjugated Dienes (Sec. 3.15b)

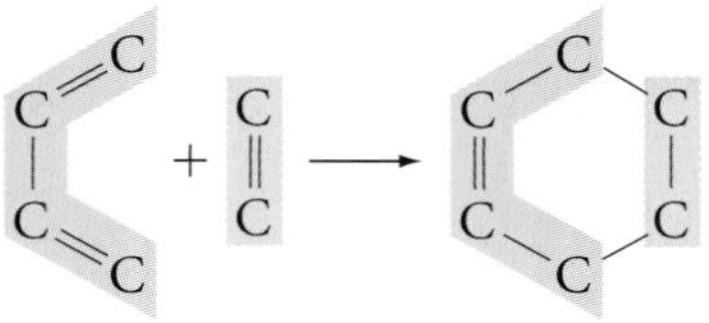

g. **Polymerization of Ethylene (Sec. 3.16)**

$$n\text{H}_2\text{C}=\text{CH}_2 \xrightarrow{\text{catalyst}} -(\text{CH}_2-\text{CH}_2)_n$$

h. **Oxidation to Diols or Carbonyl-Containing Compounds (Sec. 3.17)**

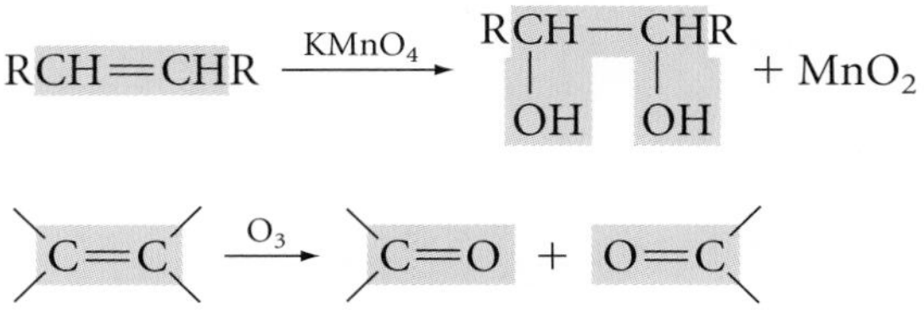

$$\text{RCH}=\text{CHR} \xrightarrow{\text{KMnO}_4} \begin{array}{c}\text{RCH}-\text{CHR}\\ | \quad\quad |\\ \text{OH}\quad\text{OH}\end{array} + \text{MnO}_2$$

$$\text{C}=\text{C} \xrightarrow{\text{O}_3} \text{C}=\text{O} + \text{O}=\text{C}$$

---

2. Reactions of Alkynes

a. **Additions to the Triple Bond (Sec. 3.20)**

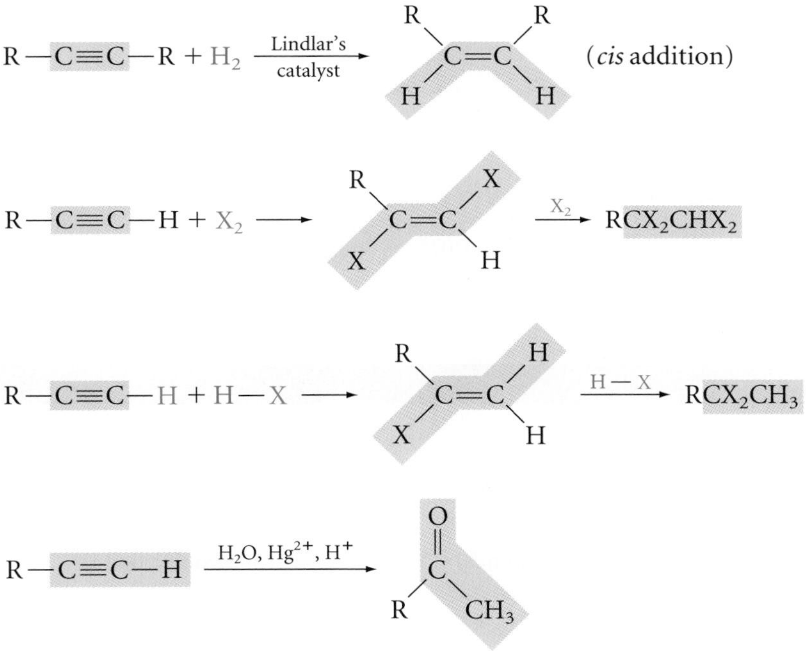

$$\text{R}-\text{C}\equiv\text{C}-\text{R} + \text{H}_2 \xrightarrow[\text{catalyst}]{\text{Lindlar's}} \begin{array}{c}\text{R}\quad\quad\text{R}\\ \text{C}=\text{C}\\ \text{H}\quad\quad\text{H}\end{array} \quad (cis \text{ addition})$$

$$\text{R}-\text{C}\equiv\text{C}-\text{H} + \text{X}_2 \longrightarrow \begin{array}{c}\text{R}\quad\quad\text{X}\\ \text{C}=\text{C}\\ \text{X}\quad\quad\text{H}\end{array} \xrightarrow{\text{X}_2} \text{RCX}_2\text{CHX}_2$$

$$\text{R}-\text{C}\equiv\text{C}-\text{H} + \text{H}-\text{X} \longrightarrow \begin{array}{c}\text{R}\quad\quad\text{H}\\ \text{C}=\text{C}\\ \text{X}\quad\quad\text{H}\end{array} \xrightarrow{\text{H}-\text{X}} \text{RCX}_2\text{CH}_3$$

$$\text{R}-\text{C}\equiv\text{C}-\text{H} \xrightarrow{\text{H}_2\text{O, Hg}^{2+}, \text{H}^+} \begin{array}{c}\quad\text{O}\\ \quad||\\ \quad\text{C}\\ \text{R}\quad\quad\text{CH}_3\end{array}$$

b. **Formation of Acetylide Anions (Sec. 3.21)**

$$\text{R}-\text{C}\equiv\text{C}-\text{H} + \text{Na}^+ \text{NH}_2{}^- \xrightarrow{\text{NH}_3} \text{R}-\text{C}\equiv\text{C}:{}^- \text{Na}^+ + \text{NH}_3$$

MECHANISM SUMMARY

1. Electrophilic Addition (E^+ = electrophile and Nu:$^-$ = nucleophile; Sec. 3.9)

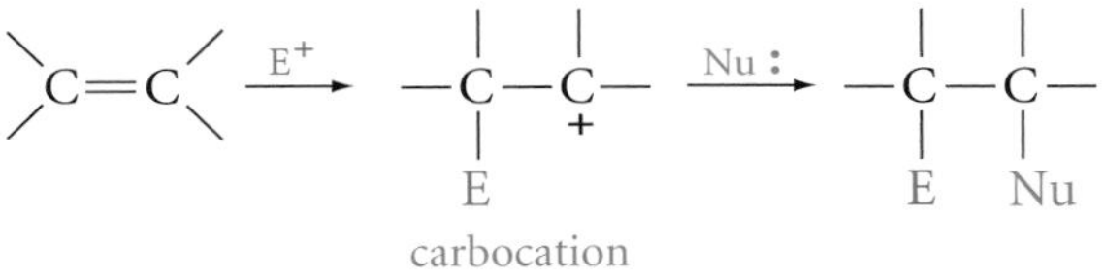

$$\text{C}=\text{C} \xrightarrow{\text{E}^+} \begin{array}{c}| \quad |\\ -\text{C}-\text{C}-\\ | \quad +|\\ \text{E}\end{array} \xrightarrow{\text{Nu}:} \begin{array}{c}| \quad |\\ -\text{C}-\text{C}-\\ | \quad |\\ \text{E}\quad\text{Nu}\end{array}$$

carbocation

2. 1,2-Addition and 1,4-Addition (Sec. 3.15a)

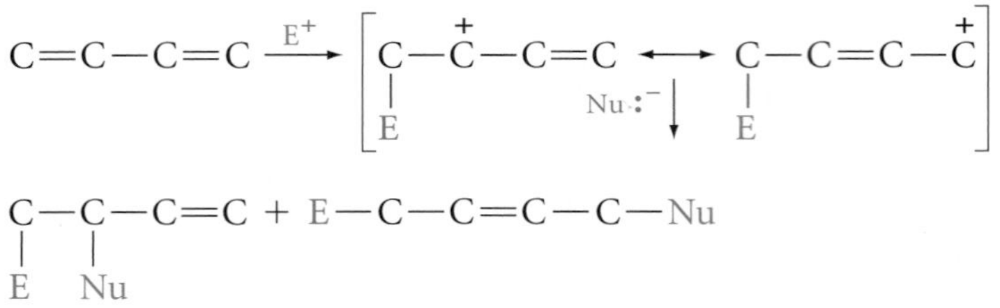

$$C=C-C=C \xrightarrow{E^+} \left[\underset{\overset{|}{E}}{C}-\overset{+}{C}-C=C \longleftrightarrow \underset{\overset{|}{E}}{C}-C=C-\overset{+}{C} \right] \quad \text{allyl carbocation}$$

$$\text{Nu}:^- \downarrow$$

$$\underset{\overset{|}{E} \ \overset{|}{\text{Nu}}}{C}-C-C=C + E-C-C=C-C-\text{Nu}$$

$$\qquad \text{1,2-product} \qquad\qquad \text{1,4-product}$$

3. Cycloaddition (Sec. 3.15b)

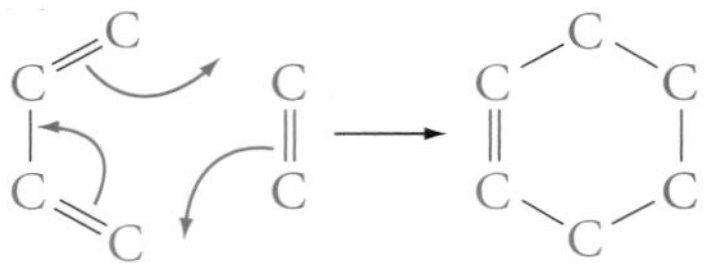

4. Free-Radical Polymerization of Ethylene (Sec. 3.16)

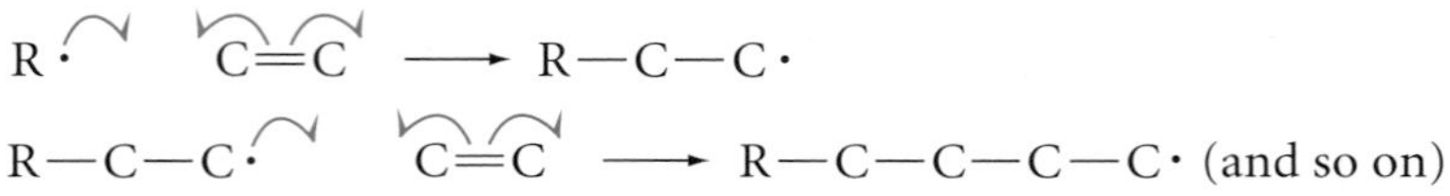

$$R\cdot \quad C=C \longrightarrow R-C-C\cdot$$

$$R-C-C\cdot \quad C=C \longrightarrow R-C-C-C-C\cdot \text{ (and so on)}$$

ADDITIONAL PROBLEMS

Alkenes and Alkynes: Nomenclature and Structure

3.33 For the following compounds, write structural formulas and IUPAC names for all possible isomers having the indicated number of multiple bonds:

    a. C_5H_8 (one triple bond)   **b.** C_5H_8 (two double bonds)   **c.** C_4H_8 (one double bond)

3.34 Name the following compounds by the IUPAC system:

    a. $CH_3CH=C(CH_2CH_3)_2$ **b.** $CH_3CH_2CH=CHCH_3$ **c.**

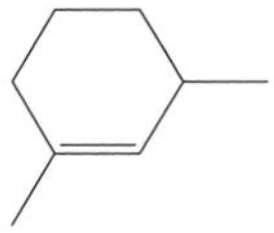

    d. $CH_3CH_2C\equiv CCH_2CH_3$ **e.** $CH_2=CH-CBr=CH_2$ **f.** $H-C\equiv C-CH_2-CH=CH_2$

    g.    **h.**

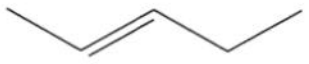

3.35 Write a structural formula for each of the following compounds:

    a. 1-hexene   **b.** cyclobutene   **c.** 1,3-dichloro-2-butene

    d. 4-methyl-2-pentyne   **e.** 1,3-cyclohexadiene   **f.** vinyl chloride

    g. allyl iodide   **h.** vinylcyclopropane   **i.** 3-methylcyclopentene

    j. 2,3-dichloro-1,3-cyclopentadiene

3.36 Explain why the following names are incorrect, and give a correct name in each case:

a. 3-pentene

d. 2-ethyl-1-propene

g. 3-pentyne-1-ene

b. 3-butyne

e. 3-methyl-1,3-butadiene

h. 3-buten-1-yne

c. 2-methylcyclohexene

f. 1-methyl-2-butene

3.37

a. What are the usual lengths for the single (sp^3–sp^3), double (sp^2–sp^2), and triple (sp–sp) carbon–carbon bonds?

 b. The *single* bond in each of the following compounds has the length shown. Suggest a possible explanation for the observed shortening.

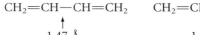

$$CH_2=CH-CH=CH_2 \qquad CH_2=CH-C\equiv CH \qquad HC\equiv C-C\equiv CH$$
$$\uparrow \qquad\qquad\qquad \uparrow \qquad\qquad\qquad \uparrow$$
$$1.47 \text{ Å} \qquad\qquad 1.43 \text{ Å} \qquad\qquad 1.37 \text{ Å}$$

3.38 Which of the following compounds can exist as *cis–trans* isomers? If such isomerism is possible, draw the structures in a way that clearly illustrates the geometry.

a. 2-pentene

d. 3-bromopropene

b. 1-hexene

e. 1,3,5-hexatriene

c. 1-chloropropene

f. 1,2-dichlorocyclodecene

3.39 The mold metabolite and antibiotic *mycomycin* has the formula

$$HC\equiv C-C\equiv C-CH=C=CH-CH=CH-CH=CH-CH_2-\overset{\overset{\displaystyle O}{\|}}{C}-OH$$

Number the carbon chain, starting with the carbonyl carbon.

a. Which multiple bonds are conjugated?

b. Which multiple bonds are cumulated?

c. Which multiple bonds are isolated?

Electrophilic Addition to Alkenes

3.40 Write the structural formula and name of the product when each of the following reacts with 1 mole of bromine:

a. 2-butene

d. 1,3-cyclohexadiene

b. vinyl chloride

e. 2,3-dimethyl-2-butene

c. 1,4-cyclohexadiene

3.41 What reagent will react by addition to what unsaturated hydrocarbon to form each of the following compounds?

 a. $CH_3CHClCHClCH_3$

d.

b. $(CH_3)_2CHOSO_3H$

e. $CH_3CH=CHCH_2Br$

c. $(CH_3)_3COH$

f. $CH_3CCl_2CCl_2CH_3$

g.

3.42 Which of the following reagents are electrophiles? Which are nucleophiles?

a. HCl   **b.** H_3O^+ **c.** Br^- **d.** $AlCl_3$ **e.** HO^-

3.43 Water can act as an electrophile or as a nucleophile. Explain.

 = concept connections

3.44 The acid-catalyzed hydration of 1-methylcyclohexene gives 1-methylcyclohexanol.

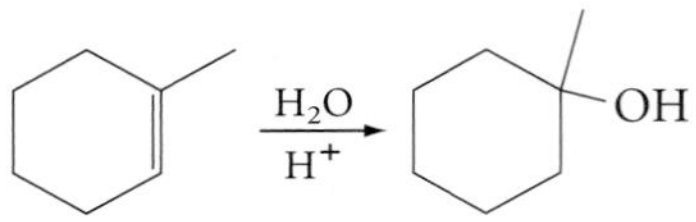

Write every step in the mechanism of this reaction.

3.45 When 2-methylpropene reacts with water and an acid catalyst, only one product alcohol is observed: *tert*-butyl alcohol (2-methyl-2-propanol).

   a. Draw the structures of the two intermediate carbocations that could form from protonation of 2-methylpropene. Which is more stable (has lower energy)? (*Hint:* See eq. 3.21.)

   b. Draw a reaction energy diagram for the formation of the two intermediate carbocations in 3.45.a from protonation of 2-methylpropene. Use your diagram to explain why only one alcohol is formed (see Figure 3.11).

3.46 *Caryophyllene* is an unsaturated hydrocarbon mainly responsible for the odor of oil of cloves. It has the molecular formula $C_{15}H_{24}$. Hydrogenation of caryophyllene gives a saturated hydrocarbon $C_{15}H_{28}$. Does caryophyllene contain any rings? How many? What else can be learned about the structure of caryophyllene from its hydrogenation?

3.47 Predict the structures of the two possible monohydration products of limonene (Figure 1.12). These alcohols are called *ter-pineols*. Predict the structure of the diol (dialcohol) obtained by hydrating both double bonds in limonene. These alcohols are used in the cough medicine "elixir of terpin hydrate."

Reactions of Conjugated Dienes

3.48 Draw the resonance contributors to the carbocation

$$(CH_3)_2CH\overset{+}{C}HCH\!=\!CHCH(CH_3)_2$$

Does the ion have a symmetric structure?

3.49 Adding 1 mole of hydrogen bromide to 1,3-hexadiene gives two products. Give their structures, and write all the steps in a reaction mechanism that explains how each product is formed.

3.50 Predict the product of each of the following Diels–Alder reactions:

   a.

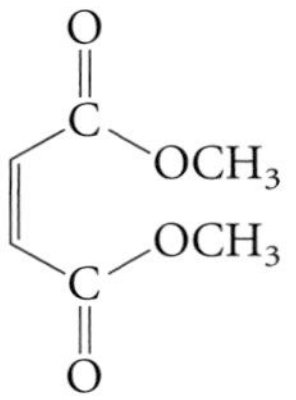

   b. $CH_3\!-\!CH\!=\!CH\!-\!CH\!=\!CH\!-\!CH_3 + NC\!-\!CH\!=\!CH\!-\!CN$

3.51 From what diene and dienophile could each of the following be made?

   a.   **b.**

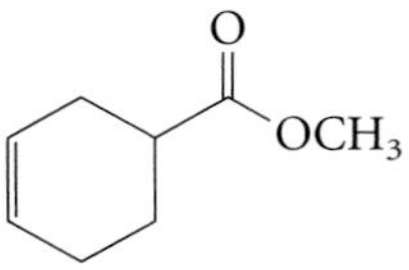

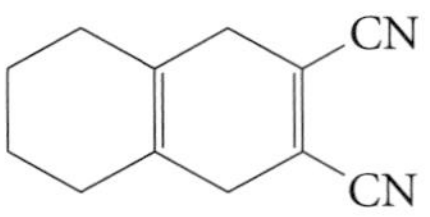

Other Reactions of Alkenes

3.52 Write an equation that clearly shows the structure of the alcohol obtained from the sequential hydroboration and H_2O_2/OH^- oxidation of

a. 3,3-dimethyl-1-butene.   **b.** 1-ethylcyclohexene.

3.53 Write equations to show how ⬠=CH$_2$ could be converted to

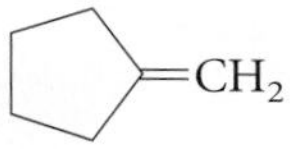

a. 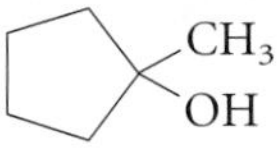 ⬠–CH$_3$/OH   **b.** 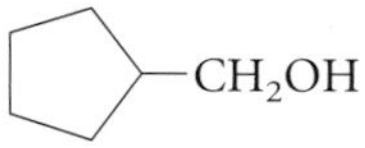 ⬠–CH$_2$OH

3.54 Given the information that free-radical stability follows the same order as carbocation stability ($3° > 2° > 1°$), predict the structure of polypropylene produced by the free-radical polymerization of propene. It should help to write out each step in the mechanism, as in eqs. 3.40 and 3.41.

3.55 Describe two simple chemical tests that could be used to distinguish cyclopentane from cyclopentene. (*Hint:* Both tests produce color changes when alkenes are present.)

3.56 Give the structural formulas of the alkenes that, on ozonolysis, give

a. only $CH_3CH_2CH{=}O$.   **b.** $(CH_3)_2C{=}O$ and $CH_3CH_2CH{=}O$.
c. $CH_2{=}O$ and $(CH_3)_2CHCH{=}O$.   **d.** $O{=}CHCH_2CH_2CH{=}O$.

Reactions of Alkynes

3.57 Write equations for the following reactions:

a. 3-hexyne + H_2 (1 mole, Lindlar's catalyst)
b. 2-pentyne + Cl_2 (2 moles)
c. propyne + sodium amide in liquid ammonia
d. propyne + H_2O (H^+, Hg^{2+} catalyst)

3.58 Determine what alkyne and what reagent will give

a. 2,2-dichlorobutane.   **b.** 2,2,3,3-tetrachlorobutane.

Summary Problems

3.59 Write an equation for the reaction of 1-butene with each of the following reagents:

a. chlorine   **b.** hydrogen chloride
c. hydrogen (Pt catalyst)   **d.** ozone, followed by Zn, H^+
e. BH_3 followed by H_2O_2, OH^- **f.** H_2O, H^+
g. $KMnO_4$, OH^- **h.** oxygen (combustion)

3.60 Write a complete reaction mechanism for the addition of HCl to 1-butene (Problem 3.59b).

3.61 When propyne is treated with $[(CH_3)_2CHCH]_2BH$ followed by H_2O_2 and OH^-, the product isolated is the aldehyde
CH_3

 propanal: $CH_3CH_2{-}\overset{\overset{\displaystyle O}{\|}}{C}{-}H$.

a. What reaction of alkenes is similar to this reaction?
b. Compare the product with the product of Problem 3.57d. Write the structure of an intermediate product (not isolated) that rearranges to form propanal. (*Hint:* See eq. 3.52.)

benzaldehyde

4

AROMATIC COMPOUNDS

Spices and herbs have long played a romantic role in the course of history. They bring to mind frankincense and myrrh and the great explorers of past centuries—Vasco da Gama, Christopher Columbus, Ferdinand Magellan, Sir Francis Drake—whose quest for spices helped open up the Western world. Though not without risk, trade in spices was immensely profitable. It was natural, therefore, that spices and herbs were among the first natural products studied by organic chemists. If one could extract from plants the pure compounds with these desirable fragrances and flavors and determine their structures, perhaps one could synthesize them in large quantity, at low cost, and without danger.

It turned out that many of these aromatic substances have rather simple structures. Many contain a six-carbon unit that passes unscathed through various chemical reactions that alter only the rest of the structure. This group, C_6H_5—, is common to many substances, including **benzaldehyde** (isolated from the oil of bitter almonds), **benzyl alcohol** (isolated from gum benzoin, a balsam resin obtained from certain Southeast Asian trees), and **toluene** (a hydrocarbon isolated from tolu balsam). When any of these three compounds is oxidized, the C_6H_5 group remains intact; the product is **benzoic acid** (another constituent of gum benzoin). The calcium salt of this acid, when heated, yields the parent hydrocarbon C_6H_6 (eq. 4.1).

$$C_6H_5CH{=}O \xrightarrow{\text{oxidize}}$$
benzaldehyde

$$C_6H_5CH_2OH \xrightarrow{\text{oxidize}} C_6H_5CO_2H \xrightarrow[\text{2. heat}]{\text{1. CaO}} C_6H_6 \qquad (4.1)$$
benzyl alcohol   benzoic acid   benzene

$$C_6H_5CH_3 \xrightarrow{\text{oxidize}}$$
toluene

▲ Bitter almonds are the source of the aromatic compound benzaldehyde.

This same hydrocarbon, first isolated from compressed illuminating gas by Michael Faraday in 1825, is now called **benzene.**[*] It is the parent hydrocarbon of a class of substances that we now call **aromatic compounds,** *not because of their aroma,* but because of their special chemical properties, in particular, their stability. Why is benzene unusually stable, and what chemical reactions will benzene and related aromatic compounds undergo? These are the subjects of this chapter.

Benzene, C_6H_6, is the parent hydrocarbon of the especially stable compounds known as **aromatic compounds.**

4.1 Some Facts About Benzene

The carbon-to-hydrogen ratio in benzene, C_6H_6, suggests a highly unsaturated structure. Compare the number of hydrogens, for example, with that in hexane, C_6H_{14}, or in cyclohexane, C_6H_{12}, both of which also have six carbons but are saturated.

PROBLEM 4.1 Draw at least five isomeric structures that have the molecular formula C_6H_6. Note that all are highly unsaturated or contain small, strained rings.

Despite its molecular formula, benzene for the most part does not behave as if it were unsaturated. For instance, it does not decolorize bromine solutions the way alkenes and alkynes do (Sec. 3.7a), nor is it easily oxidized by potassium permanganate (Sec. 3.17a). It does not undergo the typical addition reactions of alkenes or alkynes. Instead, *benzene reacts mainly by substitution.* For example, when treated with bromine in the presence of ferric bromide as a catalyst, benzene gives bromobenzene and hydrogen bromide.

$$C_6H_6 \;+\; Br_2 \;\xrightarrow[\text{catalyst}]{\text{FeBr}_3}\; C_6H_5Br \;+\; HBr \tag{4.2}$$

benzene   bromobenzene

Chlorine, with a ferric chloride catalyst, reacts similarly.

$$C_6H_6 \;+\; Cl_2 \;\xrightarrow[\text{catalyst}]{\text{FeCl}_3}\; C_6H_5Cl \;+\; HCl \tag{4.3}$$

benzene   chlorobenzene

Only *one* monobromobenzene or monochlorobenzene has ever been isolated; that is, no isomers are obtained in either of these reactions. This result requires *all six hydrogens in benzene to be chemically equivalent.* It does not matter which hydrogen is replaced by bromine; we get the same monobromobenzene. This fact has to be accounted for in any structure proposed for benzene.

When bromobenzene is treated with a second equivalent of bromine and the same type of catalyst, *three di*bromobenzenes are obtained.

$$C_6H_5Br \;+\; Br_2 \;\xrightarrow[\text{catalyst}]{\text{FeBr}_3}\; C_6H_4Br_2 \;+\; HBr \tag{4.4}$$

dibromobenzenes
(three isomers)

The isomers are not formed in equal amounts. Two of them predominate, and only a small amount of the third isomer is formed. The important point is that there are

---

[*]Today, benzene is one of the most important commercial organic chemicals. Approximately 16 billion pounds are produced annually in the United States alone. Benzene is obtained mostly from petroleum by catalytic reforming of alkanes and cycloalkanes or by cracking certain gasoline fractions. It is used to make styrene, phenol, acetone, cyclohexane, and other industrial chemicals.

three isomers—no more and no less. Similar results are obtained when chlorobenzene is further chlorinated to give dichlorobenzenes. These facts also have to be explained by any structure proposed for benzene.

The problem of benzene's structure does not sound overwhelming, yet it took decades to solve. Let us examine the main ideas that led to our modern view of its structure.

4.2 The Kekulé Structure of Benzene

In 1865 Kekulé proposed a reasonable structure for benzene.* He suggested that the six carbon atoms are located at the corners of a regular hexagon, with one hydrogen atom attached to each carbon atom. To give each carbon atom a valence of 4, he suggested that single and double bonds alternate around the ring (what we now call a *conjugated* system of double bonds). But this structure is highly unsaturated. To explain benzene's negative tests for unsaturation (that is, its failure to decolorize bromine or to give a permanganate test), Kekulé suggested that the single and double bonds exchange positions around the ring *so rapidly* that the typical reactions of alkenes cannot take place.

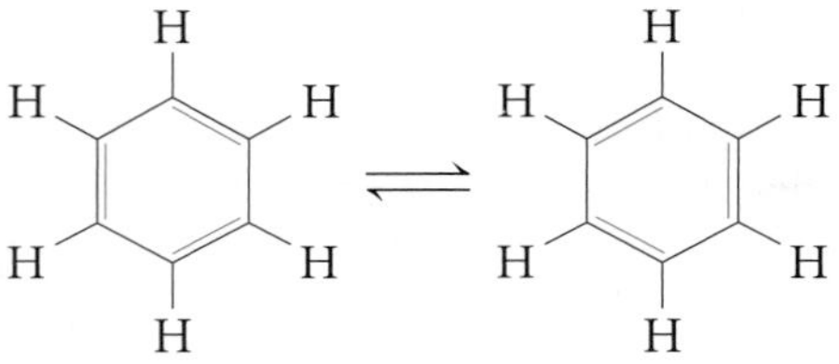

the Kekulé structures for benzene

PROBLEM 4.2 Write out eqs. 4.2 and 4.4 using a Kekulé structure for benzene. Does this model explain the existence of only one monobromobenzene? only three dibromobenzenes?

PROBLEM 4.3 How might Kekulé explain the fact that there is only one dibromobenzene with the bromines on adjacent carbon atoms, even though we can draw two different structures, with either a double or a single bond between the bromine-bearing carbons?

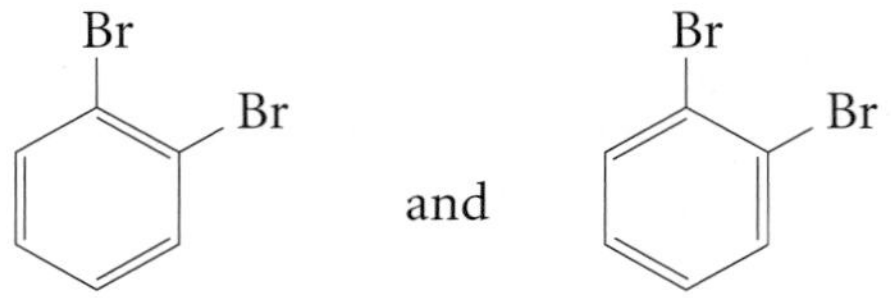

and

---

*Friedrich August Kekulé (1829–1896) was a pioneer in the development of structural formulas in organic chemistry. He was among the first to recognize the tetracovalence of carbon and the importance of carbon chains in organic structures. He is best known for his proposal regarding the structure of benzene and other aromatic compounds. It is interesting that Kekulé first studied architecture, and only later switched to chemistry. Judging from his contributions, he apparently viewed chemistry as molecular architecture.

4.3 Resonance Model for Benzene

Kekulé's model for the structure of benzene is nearly, but not entirely, correct. *Kekulé's two structures for benzene differ only in the arrangement of the electrons;* all the atoms occupy the same positions in both structures. *This is precisely the requirement for resonance* (review Sec. 1.12). Kekulé's formulas represent two identical contributing structures to a *single resonance hybrid* structure of benzene. Instead of writing an equilibrium symbol between them, as Kekulé did, we now write the double-headed arrow used to indicate a resonance hybrid:

Benzene is a resonance hybrid of
these two contributing structures.

To express this model another way, all benzene molecules are identical, and their structure is not adequately represented by either of Kekulé's contributing structures. Being a resonance hybrid, benzene is more stable than its contributing Kekulé structures. There are no single or double bonds in benzene—only one type of carbon–carbon bond, which is of some intermediate type. Consequently, it is not surprising that benzene does not react chemically exactly like alkenes.

Modern physical measurements support this model for the benzene structure. *Benzene is planar, and each carbon atom is at the corner of a regular hexagon. All the carbon–carbon bond lengths are identical;* 1.39 Å, intermediate between typical single (1.54 Å) and double (1.34 Å) carbon–carbon bond lengths. Figure 4.1 shows a space-filling model of the benzene molecule.*

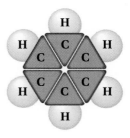

Properties:
colorless liquid
bp 80°C
mp 5.5°C

Figure 4.1
Space-filling model of benzene.

> **PROBLEM 4.4** How does the resonance model for benzene explain the fact that there are only three isomers of dibromobenzene?

4.4 Orbital Model for Benzene

Orbital theory, which is so useful in rationalizing the geometries of alkanes, alkenes, and alkynes, is also useful in explaining the structure of benzene. Each carbon atom in benzene is connected to only *three* other atoms (two carbons and a hydrogen). Each carbon is therefore sp^2-hybridized, as in ethylene. Two sp^2 orbitals of each carbon atom overlap with similar orbitals of adjacent carbon atoms to form the sigma bonds of the hexagonal ring. The third sp^2 orbital of each carbon overlaps with a hydrogen 1s orbital to form the C—H sigma bonds. Perpendicular to the plane of the three sp^2 orbitals at each carbon is a *p* orbital containing one electron, the fourth valence electron. The *p* orbitals on all six carbon atoms can overlap laterally to form pi orbitals that create a ring or cloud of electrons above and below the plane of the ring. The construction of a benzene ring from six sp^2-hybridized carbons is shown schematically in Figure 4.2. This model explains nicely the planarity of benzene. It also explains its hexagonal shape, with H—C—C and C—C—C angles of 120°.

◎ **Ch. 4, Ref. 1.** *View and manipulate 3D model of benzene.*

---

*Notice the difference in the shapes of benzene and cyclohexane (Figure 2.6).

Figure 4.2
An orbital representation of the bonding in benzene. Sigma bonds are formed by the end-on overlap of *sp²* orbitals. In addition, each carbon contributes one electron to the pi system by lateral overlap of its *p* orbital with the *p* orbitals of its two neighbors.

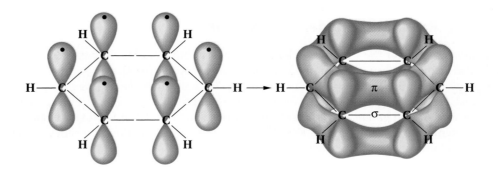

4.5   Symbols for Benzene

Two symbols are used to represent benzene. One is the Kekulé structure, and the other is a hexagon with an inscribed circle, to represent the idea of a delocalized pi electron cloud.

Kekulé   delocalized pi cloud

Regardless of which symbol is used, the hydrogens are usually not written explicitly, but we must remember that one hydrogen atom is attached to the carbon at each corner of the hexagon.

The symbol with the inscribed circle emphasizes the fact that the electrons are distributed evenly around the ring, and in this sense, it is perhaps the more accurate of the two. The Kekulé symbol, however, reminds us very clearly that there are six pi electrons in benzene. For this reason, it is particularly useful in allowing us to keep track of the valence electrons during chemical reactions of benzene. In this book, we will use the Kekulé symbol. However, we must keep in mind that the "double bonds" are not fixed in the positions shown, nor are they really double bonds at all.

4.6   Nomenclature of Aromatic Compounds

Because aromatic chemistry developed in a haphazard fashion many years before systematic methods of nomenclature were developed, common names have acquired historic respectability and are accepted by IUPAC. Examples include

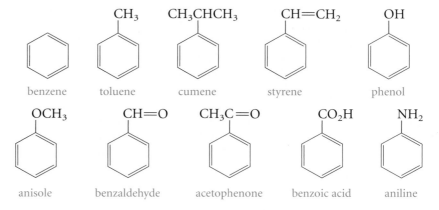

EXAMPLE 4.1

Write the structural formula for benzaldehyde (eq. 4.1).

Solution   One hydrogen in the formula for benzene is replaced by the aldehyde group.

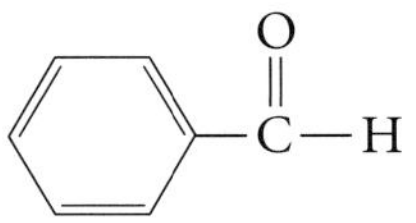

$$\text{\includegraphics} \quad \overset{\displaystyle O}{\overset{\displaystyle \|}{C}} - H$$

PROBLEM 4.5   Write the formulas for benzyl alcohol, toluene, and benzoic acid (eq. 4.1).

Monosubstituted benzenes that do not have common names accepted by IUPAC are named as derivatives of benzene.

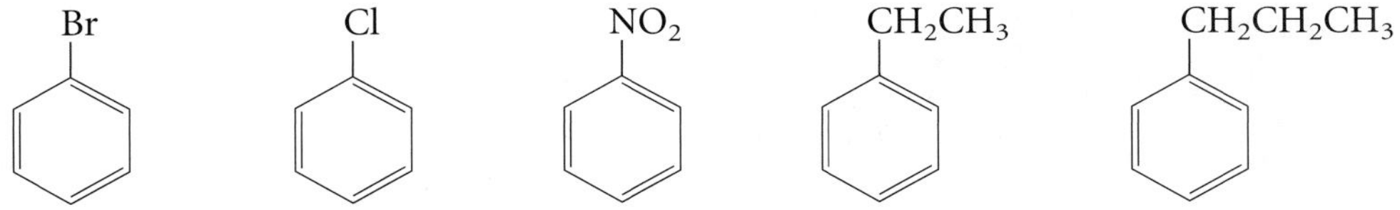

Br	Cl	NO_2	CH_2CH_3	$CH_2CH_2CH_3$
bromobenzene	chlorobenzene	nitrobenzene	ethylbenzene	propylbenzene

When two substituents are present, three isomeric structures are possible. They are designated by the prefixes **ortho-, meta-,** and **para-,** which are usually abbreviated as **o-, m-,** and **p-,** respectively. If substituent X is attached (by convention) to carbon 1, then *o*-groups are on carbons 2 and 6, *m*-groups are on carbons 3 and 5, and *p*-groups are on carbon 4.*

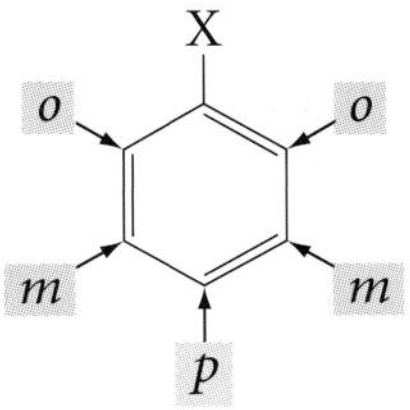

Specific examples are

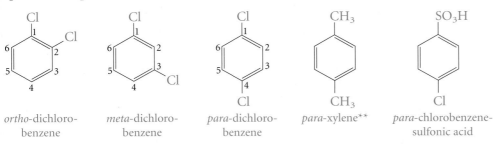

ortho-dichloro-
benzene

meta-dichloro-
benzene

para-dichloro-
benzene

para-xylene**

para-chlorobenzene-
sulfonic acid

---

*Note that X can be on any carbon of the ring. It is the *position of the second substituent relative to that of X* that is important.

**The common and IUPAC name is xylene, *not* p-methyltoluene.

PROBLEM 4.6 Draw the structures for *ortho*-xylene and *meta*-xylene.

The prefixes *ortho-*, *meta-*, and *para-* are used even when the two substituents are not identical.

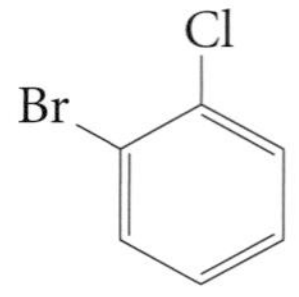

o-bromochlorobenzene
(note alphabetical order) 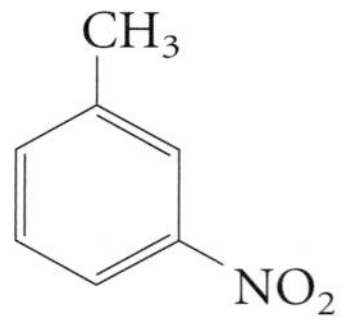   *m*-nitrotoluene 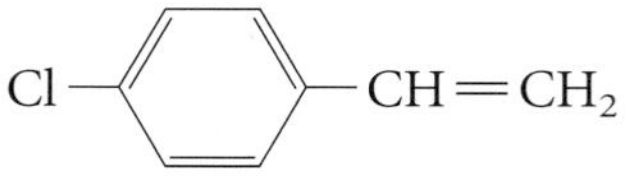   *p*-chlorostyrene 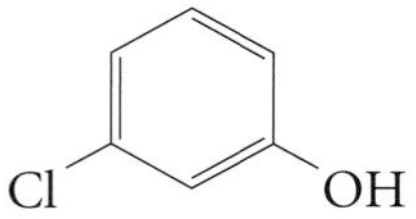   *m*-chlorophenol 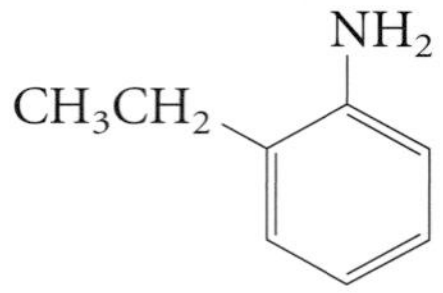   *o*-ethylaniline

Although *o-*, *m-*, and *p-* designations are commonly used in naming disubstituted benzenes, position numbers of substituents can also be used. For example, *o*-dichlorobenzene can also be named 1,2-dichlorobenzene, and *m*-chlorophenol can also be named 2-chlorophenol.

When more than two substituents are present, their positions are designated by numbering the ring.

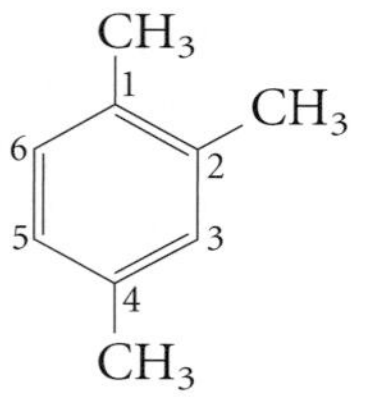

1,2,4-tri-
methylbenzene 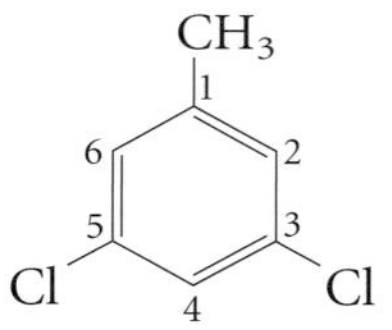   3,5-dichlorotoluene 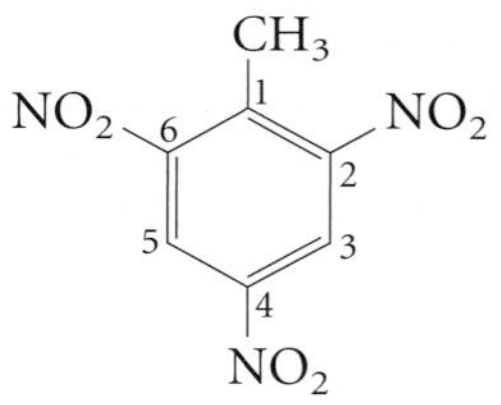   2,4,6-trinitrotoluene
                                                        (TNT)

PROBLEM 4.7 Draw the structure of

a. *o*-nitrophenol.   b. *p*-bromotoluene.
c. 2-chlorophenol.   d. *m*-dinitrobenzene.
e. *p*-divinylbenzene.   f. 1,4-dibromobenzene.

PROBLEM 4.8 Draw the structure of

a. 1,3,5-trimethylbenzene.
b. 4-bromo-2,6-dichlorotoluene.

Aromatic hydrocarbons, are
called **arenes.** An aromatic
substituent is called an **aryl
group, Ar.**

Aromatic hydrocarbons, as a class, are called **arenes.** The symbol **Ar** is used for an **aryl group,** just as the symbol R is used for an alkyl group. The formula Ar-R would therefore represent any arylalkane.

Two groups with special names occur frequently in aromatic compounds. They are the **phenyl group** and the **benzyl group.**

C_6H_5— or

phenyl group

$C_6H_5CH_2$— or
CH₂—

benzyl group

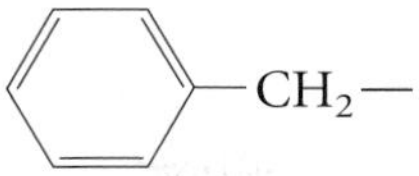

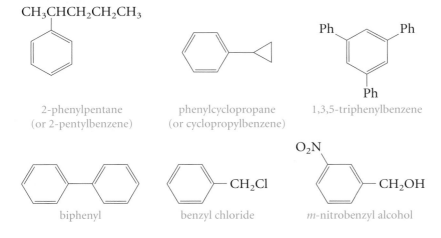

The symbol Ph is sometimes used as an abbreviation for the phenyl group. The use of these group names is illustrated in the following examples:

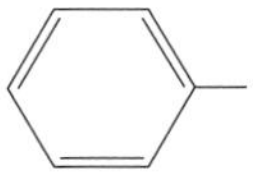

CH3CHCH2CH2CH3

2-phenylpentane
(or 2-pentylbenzene)

phenylcyclopropane
(or cyclopropylbenzene)

Ph
Ph
Ph

1,3,5-triphenylbenzene

biphenyl

CH2Cl

benzyl chloride

O2N
CH2OH

m-nitrobenzyl alcohol

PROBLEM 4.9 Draw the structure of

a. cyclopentylbenzene. b. benzyl bromide.
c. *p*-phenylstyrene. d. dibenzyl.

PROBLEM 4.10 Name the following structures:

a.

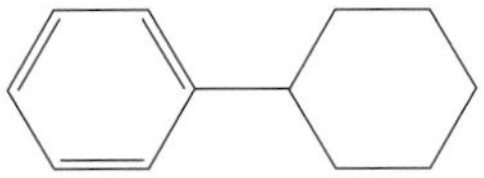

b.
OH
CH2

4.7 The Resonance Energy of Benzene

We have asserted that a resonance hybrid is always more stable than any of its contributing structures. Fortunately, in the case of benzene, this assertion can be proved experimentally, and we can even measure how much more stable benzene is than the hypothetical molecule 1,3,5-cyclohexatriene (the IUPAC name for one Kekulé structure).

Hydrogenation of a carbon–carbon double bond is an exothermic reaction. The amount of energy (heat) released is about 26 to 30 kcal/mol for each double bond (eq. 4.5). (The exact value depends on the substituents attached to the double bond.)

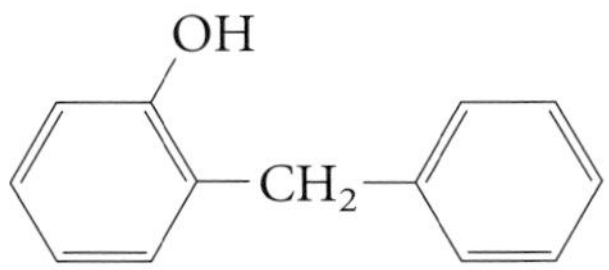

When two double bonds in a molecule are hydrogenated, twice as much heat is evolved, and so on.

$$\diagup C = C \diagup + H-H \longrightarrow \underset{\underset{H\ \ H}{|\ \ \ |}}{-C-C-} + \text{heat (26–30 kcal/mol)} \qquad \textbf{(4.5)}$$

Hydrogenation of cyclohexene releases 28.6 kcal/mol (Figure 4.3). We expect that the complete hydrogenation of 1,3-cyclohexadiene should release twice that amount of heat, or 2 × 28.6 = 57.2 kcal/mol; experimentally the value is close to what we expect (Figure 4.3). It seems reasonable, therefore, to expect that the heat of hydrogenation of a Kekulé structure (the *hypothetical* triene 1,3,5-cyclohexatriene) should correspond to that for *three* double bonds, or about 84 to 86 kcal/mol. However, we find experimentally that benzene is more difficult to hydrogenate than simple alkenes, and the heat evolved when benzene is hydrogenated to cyclohexane is *much lower* than expected: only 49.8 kcal/mol (Figure 4.3).

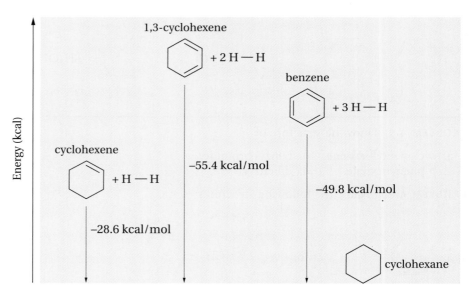

Figure 4.3
Comparison of heat released by the hydrogenation of cyclohexene, 1,3-cyclohexadiene, and benzene to produce cyclohexane.

We conclude that *real benzene molecules are more stable than the contributing resonance structures* (the hypothetical molecule 1,3,5-cyclohexatriene) *by about 36 kcal/mol* (86 − 50 = 36).

We define the **stabilization energy,** or **resonance energy,** of a substance as the difference between the actual energy of the real molecule (the resonance hybrid) and the calculated energy of the most stable contributing structure. For benzene this value is about 36 kcal/mol. This is a substantial amount of energy. Consequently, as we will see, *benzene and other aromatic compounds usually react in such a way as to preserve their aromatic structure and therefore retain their resonance energy.*

The **stabilization energy,** or **resonance energy,** of a substance is the difference between the energy of the real molecule and the calculated energy of the most stable contributing structure.

ADDITIONAL PROBLEMS

ADDITIONAL PROBLEMS

Aromatic Compounds: Nomenclature and Structural Formulas

4.20 Write structural formulas for the following compounds:

a. 1,3,5-trichlorobenzene
b. *p*-chlorotoluene
c. *p*-diethylbenzene
d. isopropylbenzene
e. *o*-chlorophenol
f. benzyl bromide
g. 2,3-diphenylbutane
h. *m*-bromostyrene
i. 2-bromo-4-ethyl-3,5-dinitrotoluene
j. *m*-chlorobenzenesulfonic acid
k. *o*-nitroanisole
l. 2,4,6-trichloroaniline
m. *p*-bromobenzoic acid
n. *o*-fluoroacetophenone

4.21 Name the following compounds:

a. $CH(CH_3)_2$

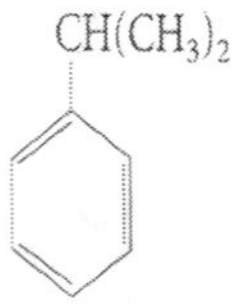

b. $CH{=}O$ ... Br

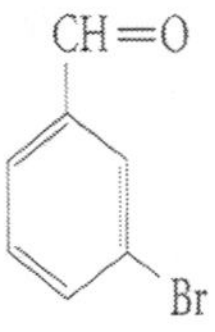

c. Cl   Cl

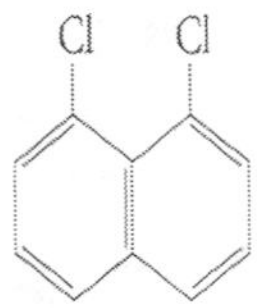

d. Br ... CH_3 ... Br

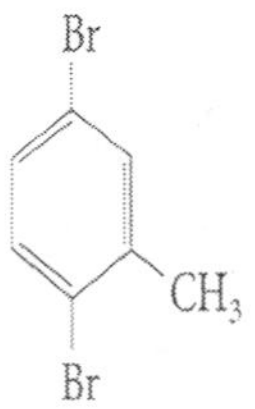

e. $(CH_3)_3C$—⬡—OH

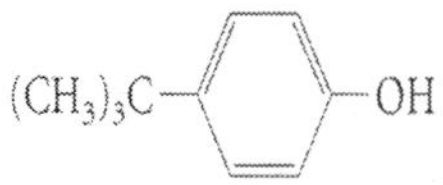

f. CH_3 ... NO_2

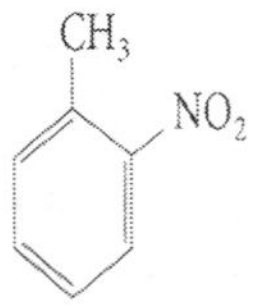

g. F F F F F F

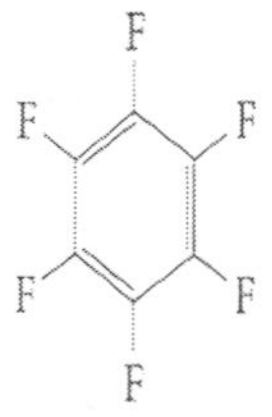

h. $CH{=}CH_2$ ... Cl   Cl

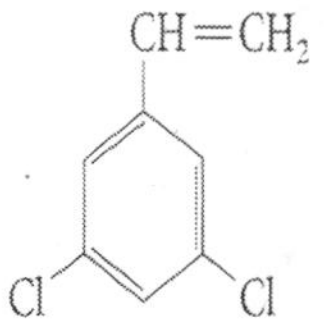

i. CH_3CH_2 ...

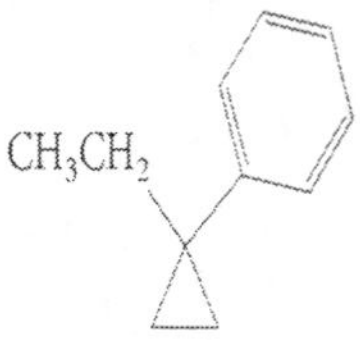

4.22 Give the structures and names for all possible

    a. trimethylbenzenes.

    b. dichloronitrobenzenes.

4.23 Give the structure and name of each of the following aromatic hydrocarbons:

    a. C_8H_{10}; has two possible ring-substituted monobromo derivatives

    b. C_9H_{12}; can give only one mononitro product on nitration

    c. C_9H_{12}; can give four mononitro derivatives on nitration

4.24 There are three dibromobenzenes (*o*-, *m*-, and *p*-). Suppose we have samples of each in separate bottles, but we don't know which is which. Let us call them A, B, and C. On nitration, compound A (mp 87°C) gives only *one* nitrodibromobenzene. What is the structure of A? B and C are both liquids. On nitration, B gives *two* nitrodibromobenzenes, and C gives *three* nitrodibromobenzenes (of course, not in equal amounts). What are the structures of B and C? of their mononitration products? (This method, known as Körner's method, was used years ago to assign structures to isomeric benzene derivatives.) (*Hint:* Start by drawing the structures of the three dibromobenzenes. Then figure out where they can be nitrated.)

Aromaticity and Resonance

4.25 The observed amount of heat evolved when 1,3,5,7-cyclooctatetraene is hydrogenated is 110 kcal/mol. What does this tell you about the possible resonance energy of this compound?

4.26 The structure of the nitro group—NO_2 is usually shown as

yet experiments show that the two nitrogen–oxygen bonds have the same length of 1.21 Å. This length is intermediate between 1.36 Å for the N—O single bond and 1.18 Å for the N=O double bond. Draw structural formulas that explain this observation.

4.27 Draw all reasonable electron-dot formulas for the nitronium ion, $(NO_2)^+$, the electrophile in aromatic nitrations. Show any formal charges. Which structure is favored and why? (See Secs. 1.11 and 1.12.)

Mechanism of Electrophilic Aromatic Substitution

4.28 Write out all steps in the mechanism for the reaction of

    a. *p*-xylene + nitric acid (H_2SO_4 catalyst).

    b. toluene + *t*-butyl chloride + $AlCl_3$.

4.29 Draw all possible contributing structures to the carbocation intermediate in the chlorination of chlorobenzene. Explain why the major products are *o*- and *p*-dichlorobenzene. (*Note: p*-Dichlorobenzene is produced commercially this way, for use against clothes moths.)

4.30 Repeat Problem 4.29 for the chlorination of acetophenone, and explain why the product is *m*-chloroacetophenone.

4.31 Suggest a reason why $FeCl_3$ is used as a catalyst for aromatic chlorinations and $FeBr_3$ for brominations (that is, why the iron halide used has the same halogen as the halogenating agent).

    = concept connections

4.32 When benzene is treated with propene and sulfuric acid, two different monoalkylation products are possible. Draw their structures. Which one do you expect to be the major product? Why? (*Hint:* See eq. 3.21, p. 88.)

4.33 When benzene is treated with excess D_2SO_4 at room temperature, the hydrogens on the benzene ring are gradually replaced by deuterium. Write a mechanism that explains this observation. (*Hint:* D_2SO_4 is a form of the acid H_2SO_4 in which deuterium has been substituted for hydrogen.)

4.34 Draw a molecular orbital picture for the resonance hybrid benzenonium ion shown in eq. 4.16, and describe the hybridization of each ring carbon atom. (*Hint:* Examine the molecular orbital structure of benzene in Figure 4.2.)

Reactions of Substituted Benzenes: Activating and Directing Effects

4.35 Indicate the main *mono*substitution products in each of the following reactions. Keep in mind that certain substituents are *meta* directing and others are *ortho,para* directing.

    a. anisole + bromine (Fe catalyst)
    b. nitrobenzene + chlorine (Fe catalyst)
    c. bromobenzene + concentrated sulfuric acid (heat) + SO_3
    d. toluene + chlorine (Fe catalyst)
    e. benzenesulfonic acid + concentrated nitric acid (heat) (H_2SO_4 catalyst)
    f. iodobenzene + chlorine (Fe catalyst)
    g. toluene + acetyl chloride ($AlCl_3$ catalyst)
    h. ethylbenzene + concentrated nitric acid (H_2SO_4 catalyst)

4.36 Predict whether the following substituents on the benzene ring are likely to be *ortho,para* directing or *meta* directing and whether they are likely to be ring activating or ring deactivating:

    a. $-OCH_2CH_3$
    b. $-\overset{\overset{\displaystyle O}{\|}}{C}-OCH_3$
    c. $-\overset{+}{N}H(CH_3)_2$
    d. $-\overset{\overset{\displaystyle O}{\|}}{N}HCCH_3$

4.37 Which compound is more reactive toward electrophilic substitution (for example, nitration)?

    a. 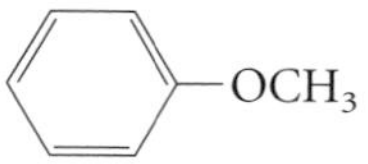 $C_6H_5-OCH_3$ or $C_6H_5-\overset{\overset{\displaystyle O}{\|}}{C}-OH$ **b.** 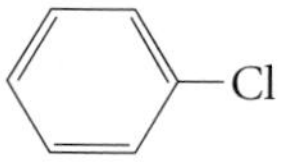 C_6H_5-Cl or $C_6H_5-CH_2CH_3$

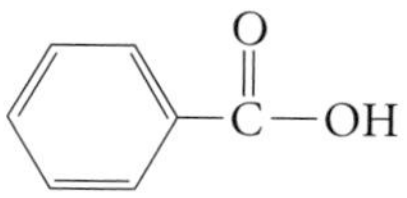

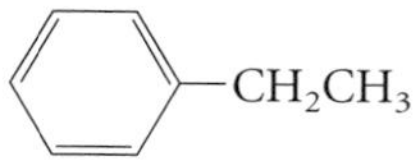

4.38 The explosive TNT (2,4,6-trinitrotoluene) can be made by nitrating toluene with a mixture of nitric and sulfuric acids, but the reaction conditions must gradually be made more severe as the nitration proceeds. Explain why.

Electrophilic Aromatic Substitution Reactions in Synthesis

4.39 Using benzene or toluene as the only aromatic organic starting material, devise a synthesis for each of the following:

    a. *p*-bromonitrobenzene   **b.** *p*-nitroethylbenzene   **c.** *p*-toluenesulfonic acid   **d.** ethylcyclohexane

4.40 Using benzene or toluene as the only aromatic organic starting material, devise a synthesis for each of the following compounds. Name the product.

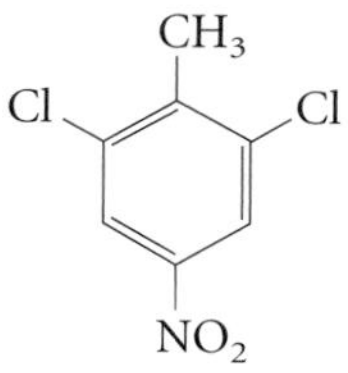

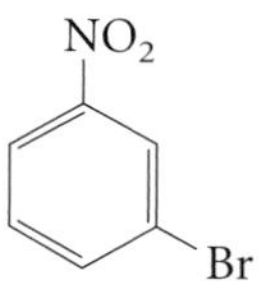

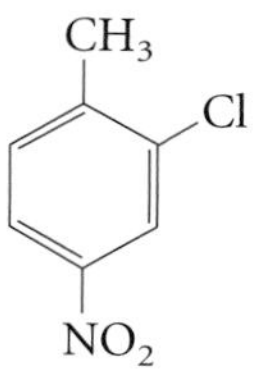

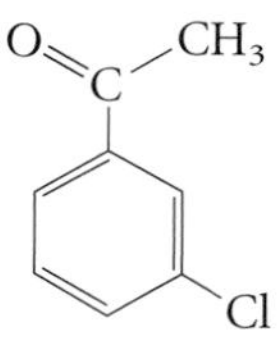

4.41 For a one-step synthesis of 3-bromo-5-nitrobenzoic acid, which is the better starting material, 3-bromobenzoic acid or 3-nitrobenzoic acid? Why?

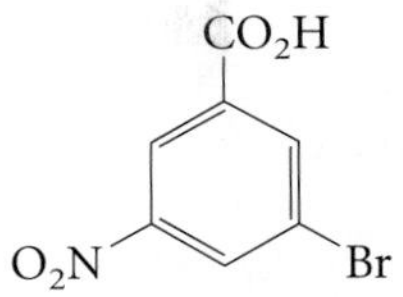

3-bromo-5-nitrobenzoic acid

4.42 Show how pure 3,5-dinitrochlorobenzene can be prepared, starting from a disubstituted benzene.

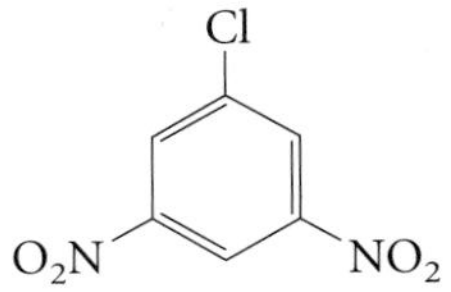

3,5-dinitrochlorobenzene

Polycyclic Aromatic Hydrocarbons

4.43 How many possible monosubstitution products are there for each of the following? (See page 140 for structures.)

    a. anthracene   **b.** phenanthrene

4.44 Bromination of anthracene gives mainly 9-bromoanthracene. Write out the steps in the mechanism of this reaction.

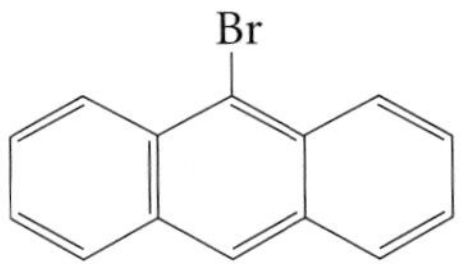

9-bromoanthracene

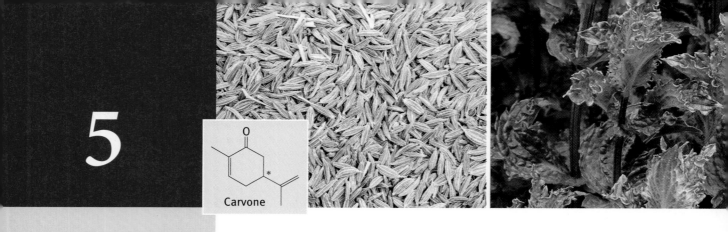

Carvone

5

STEREOISOMERISM

Stereoisomers have the same order of attachment of the atoms, but different arrangements of the atoms in space. The differences between stereoisomers are more subtle than those between structural isomers. Yet stereoisomerism is responsible for significant differences in chemical properties of molecules. The effectiveness of a drug often depends on which stereoisomer is used, as does the presence or absence of side effects (see A Word About Enantiomers and Biological Activity, p. 172). The chemistry of life itself is affected by the natural predominance of particular stereoisomers in biological molecules such as carbohydrates (Chapter 16), amino acids (Chapter 17), and nucleic acids (Chapter 18).

We have already seen that stereoisomers can be characterized according to the ease with which they can be interconverted (see Sec. 2.11 and Figure 2.7). That is, they may be **conformers,** which can be interconverted by rotation about a single bond, or they may be **configurational isomers,** which can be interconverted only by breaking and remaking covalent bonds. Here we will consider other useful ways to categorize stereoisomers, ways that are particularly helpful in describing their properties.

5.1 Chirality and Enantiomers

Consider the difference between a pair of gloves and a pair of socks. A sock, like its partner, can be worn on either the left or the right foot. But a left-hand glove, unlike its partner, cannot be worn on the right hand. Like a pair of gloves, certain molecules possess this property of "handedness," which affects their chemical behavior. Let us examine the idea of molecular handedness.

A molecule (or object) is either **chiral** or **achiral.** The word *chiral*, pronounced "kai-ral" to rhyme with spiral, comes from the Greek χειρ (*cheir*, hand).

▲ The odors of caraway seeds (left) and mint leaves (right) arise from the enantiomers of carvone, which differ in arrangement of the atoms attached to the indicated (*) carbon.

A chiral molecule (or object) is one that exhibits the property of handedness. An achiral molecule does not have this property.

What test can we apply to tell whether a molecule (or object) is chiral or achiral? We examine the molecule (or object) *and its mirror image. The mirror image of a chiral molecule cannot be superimposed on the molecule itself. The mirror image of an achiral molecule, however, is identical with or superimposable on the molecule itself.*

Let us apply this test to some specific examples. Figure 5.1 shows one of the more obvious examples. The mirror image of a left hand is not another left hand, but a right hand. A hand and its mirror image are *not* superimposable. A hand is chiral. But the mirror image of a ball (sphere) is also a ball (sphere), so a ball (sphere) is achiral.

Stereoisomers have the same order of attachment of the atoms but different spatial arrangements of the atoms. **Conformers** and **configurational isomers** are two classes of stereoisomers (Sec. 2.11).

Chiral molecules possess the property of handedness, whereas **achiral** molecules do not.

PROBLEM 5.1 Which of the following objects are chiral and which are achiral?

a. golf club	b. teacup	c. football	d. corkscrew
e. tennis racket	f. shoe	g. portrait	h. pencil

Now let us look at two molecules, 2-chloropropane and 2-chlorobutane, and their mirror images.* Figure 5.2 shows that 2-chloropropane is achiral. Its mirror image is superimposable on the molecule itself. Therefore 2-chloropropane has only one possible structure.

The mirror image of a left hand is not a left hand, but a right hand.

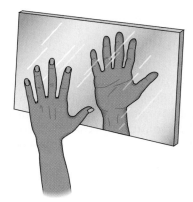

Chiral object

The mirror image of a ball is identical with the object itself.

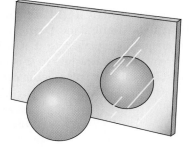

Achiral object

Figure 5.1
The mirror-image relationships of chiral and achiral objects.

---

*Build 3D models of these molecules to visualize them better. In general, using 3D models while reading this chapter will help you understand the concepts.

⊚ **Ch. 5, Ref. 1.** *View anima-tion of superposition of identical molecules.*

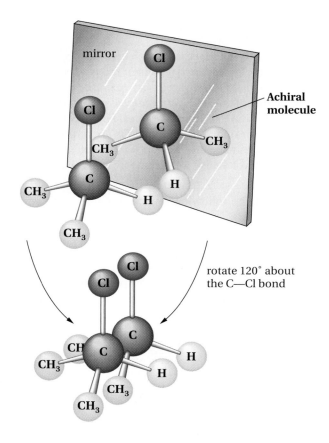

Figure 5.2
Model of 2-chloropropane and its mirror image. The mirror image is superimposable on the original molecule.

Enantiomers are a pair of molecules related as nonsuperimposable mirror images.

A **stereogenic carbon atom** or **stereogenic center** is a carbon atom with four different groups attached to it.

On the other hand, as Figure 5.3 shows, 2-chlorobutane has two possible struc-tures, related to one another as nonsuperimposable mirror images. We call a pair of molecules that are related as nonsuperimposable mirror images **enantiomers.** Every molecule, of course, has a mirror image. Only those that are *nonsuperimposable* are called enantiomers.

5.2   Stereogenic Centers; the Stereogenic Carbon Atom

What is it about their structures that leads to chirality in 2-chlorobutane but not in 2-chloropropane? Notice that, in 2-chlorobutane, carbon atom 2, the one marked with an asterisk, has four different groups attached to it (Cl, H, CH_3, and CH_2CH_3). A carbon atom with four different groups attached to it is called a **stereogenic carbon atom.** This type of carbon is also called a **stereogenic center** because it gives rise to stereoisomers.

$$CH_3 - \overset{Cl}{\underset{H}{\overset{|}{\underset{|}{C}}}} - CH_2CH_3$$

Let us examine the more general case of a carbon atom with any four different groups attached; let us call the groups A, B, D, and E. Figure 5.4 shows such a

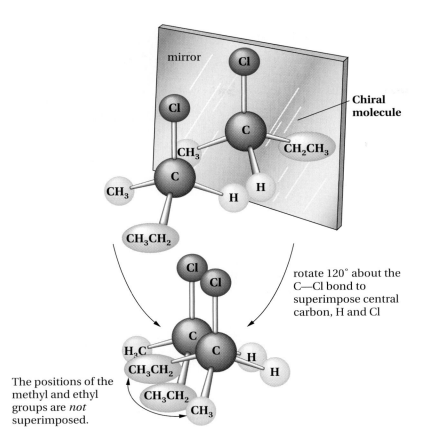

◎ **Ch. 5, Ref. 2.** *View animation of attempted superposition of enantiomers.*

Figure 5.3
Model of 2-chlorobutane and its mirror image. The mirror image is *not* superimposable on the original molecule. The two forms of 2-chlorobutane are enantiomers.

molecule and its mirror image. That the molecules on each side of the mirror in Figure 5.4 are nonsuperimposable mirror images (enantiomers) becomes clear by examining Figure 5.5. (We strongly urge you to use molecular models when studying this chapter. It is sometimes difficult to visualize three-dimensional structures when they are drawn on a two-dimensional surface [this page or a blackboard], though with experience, your ability to do so will improve.) The handedness of these molecules is also illustrated in Figure 5.4, where the clockwise or counterclockwise

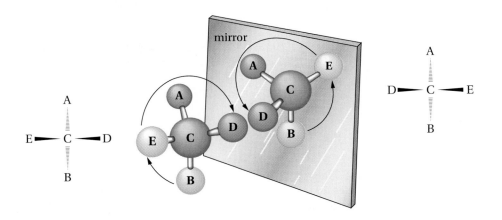

Figure 5.4
The chirality of enantiomers. Looking down the C—A bond, we have to read clockwise to spell BED for the model on the left, but we must read counterclockwise for its mirror image.

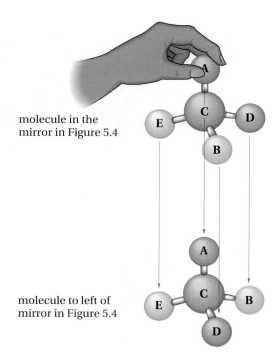

Figure 5.5
When the four different groups attached to a stereogenic carbon atom are arranged to form mirror images, the molecules are not superimposable. The models may be twisted or turned in any direction, but as long as no bonds are broken, only two of the four attached groups can be made to coincide.

molecule in the mirror in Figure 5.4

molecule to left of mirror in Figure 5.4

arrangement of the groups (we might call them right- or left-handed arrangements) is apparent.

What happens when all four of the groups attached to the central carbon atom are *not* different from one another? Suppose two of the groups are identical—say, A, A, B, and D. Figure 5.6 describes this situation. The molecule and its mirror image are now *identical,* and the molecule is achiral. This is exactly the situation with 2-chloropropane, where two of the four groups attached to carbon-2 are identical (CH_3, CH_3, H, and Cl).

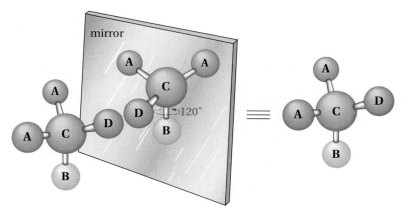

Figure 5.6
The tetrahedral model at the left has two corners occupied by identical groups (A). It has a plane of symmetry that passes through atoms B, C, and D and bisects angle ACA. Its mirror image is identical to itself, seen by a 120° rotation of the mirror image about the C—B bond. Hence the model is achiral.

Notice that the molecule in Figure 5.6 has a plane of symmetry. This plane passes through atoms B, C, and D and bisects the ACA angle. On the other hand, the molecule in Figure 5.4 does *not* have a symmetry plane.

A **plane of symmetry** (sometimes called a mirror plane) is a plane that passes through a molecule (or object) in such a way that what is on one side of the plane is the exact reflection of what is on the other side. *Any molecule with a plane of symmetry is achiral. Chiral molecules do not have a plane of symmetry.* Seeking a plane of symmetry is usually one quick way to tell whether a molecule is chiral or achiral.

What is on one side of a **plane of symmetry,** or mirror plane, is the exact reflection of what is on the other side.

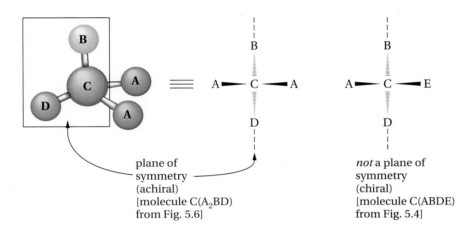

plane of symmetry (achiral) [molecule C(A$_2$BD) from Fig. 5.6]

not a plane of symmetry (chiral) [molecule C(ABDE) from Fig. 5.4]

To summarize, a molecule with a stereogenic center (in our examples, the stereogenic center is a carbon atom with four different groups attached to it) can exist in two stereoisomeric forms, that is, as a pair of enantiomers. Such a molecule does not have a symmetry plane. Compounds with a symmetry plane are achiral.

EXAMPLE 5.1

Locate the stereogenic center in 3-methylhexane.

Solution   Draw the structure, and look for a carbon atom with four different groups attached.

$$\overset{1}{C}H_3\overset{2}{C}H_2\overset{3}{C}H\overset{4}{C}H_2\overset{5}{C}H_2\overset{6}{C}H_3$$
$$|$$
$$CH_3$$

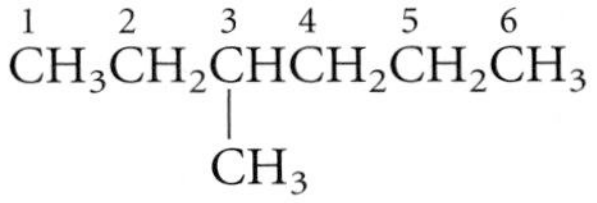

All of the carbons except carbon-3 have at least two hydrogens (two identical groups) and therefore cannot be stereogenic centers. But carbon-3 has four different groups attached (H, CH_3—, CH_3CH_2—, and $CH_3CH_2CH_2$—) and is therefore a stereogenic center. By convention, we sometimes mark such centers with an asterisk.

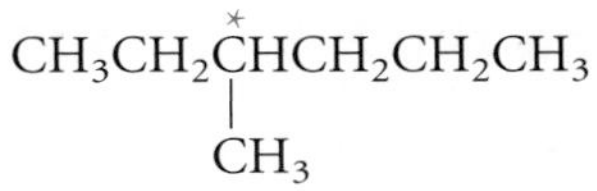

$$CH_3CH_2\overset{*}{C}HCH_2CH_2CH_3$$
$$|$$
$$CH_3$$

EXAMPLE 5.2

Draw the two enantiomers of 3-methylhexane.

Solution   There are many ways to do this. Here are two of them. First draw carbon-3 with four tetrahedral bonds.

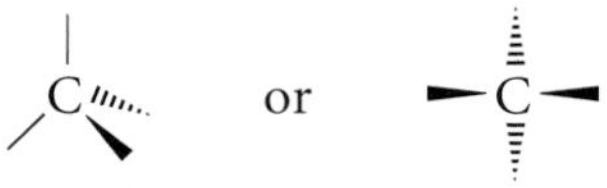

$$\text{or}$$

Then attach the four different groups, in any order.

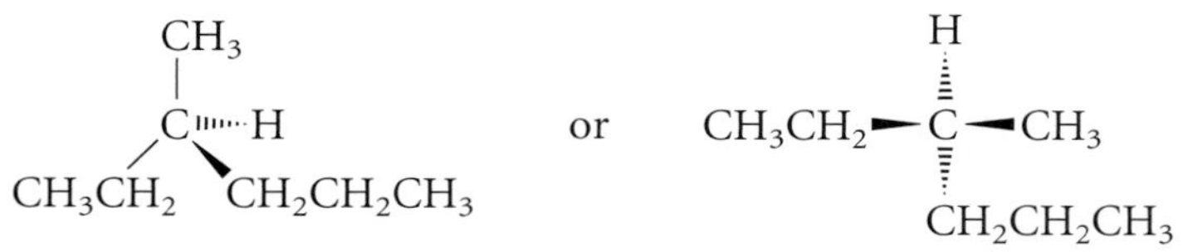

$$\text{CH}_3\text{CH}_2 \quad \text{C----H} \quad \text{CH}_2\text{CH}_2\text{CH}_3 \qquad \text{or} \qquad \text{CH}_3\text{CH}_2\text{---C---CH}_3$$

Now draw the mirror image, or interchange the positions of any two groups.

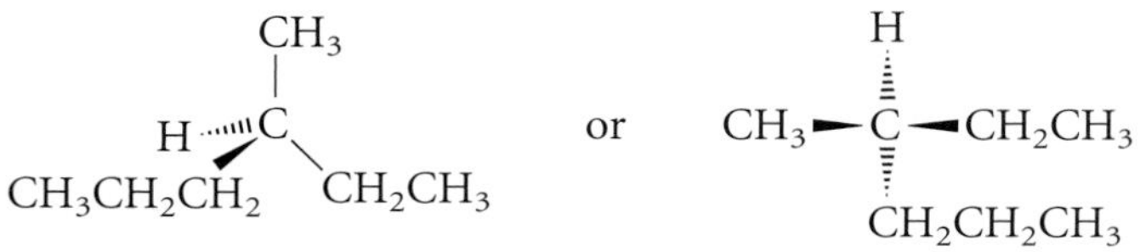

$$\text{H----C} \quad \text{CH}_3\text{CH}_2\text{CH}_2 \quad \text{CH}_2\text{CH}_3 \qquad \text{or} \qquad \text{CH}_3\text{---C---CH}_2\text{CH}_3$$

To convince yourself that the *interchange of any two groups at a stereogenic center produces the enantiomer*, work with molecular models.

PROBLEM 5.2   Find the stereogenic centers in

   a. 3-iodohexane.   b. 2,3-dibromobutane.
   c. 3-methylcyclohexene.   d. 1-bromo-1-fluoroethane.

PROBLEM 5.3   Which of the following compounds is chiral?

   a. 1-bromo-1-phenylethane   b. 1-bromo-2-phenylethane

PROBLEM 5.4   Draw three-dimensional structures for the two enantiomers of the chiral compound in Problem 5.3.

PROBLEM 5.5   Locate the planes of symmetry in the eclipsed conformation of ethane. In this conformation, is ethane chiral or achiral?

PROBLEM 5.6   Does the staggered conformation of ethane have planes of symmetry? In this conformation, is ethane chiral or achiral? *(Careful!)*

PROBLEM 5.7   Locate the planes of symmetry in *cis*- and *trans*-1,2-dichloroethene. Are these molecules chiral or achiral? *(Careful!)*

5.3   **Configuration and the *R-S* Convention**

The arrangement of four groups attached to a stereogenic center is called the **configuration** of that center.

Enantiomers differ in the arrangement of the groups attached to the stereogenic center. This arrangement of groups is called the **configuration** of the stereogenic center. *Enantiomers are another type of configurational isomer; they are said to have opposite configurations.*

When referring to a particular enantiomer, we would like to be able to specify which configuration we mean without having to draw the structure. A convention for doing this is known as the *R-S* or Cahn–Ingold–Prelog (CIP)* system. Here is how it works.

a → b → c clockwise
R

or

a → b → c clockwise
R

a → b → c counterclockwise
S

or

a → b → c counterclockwise
S

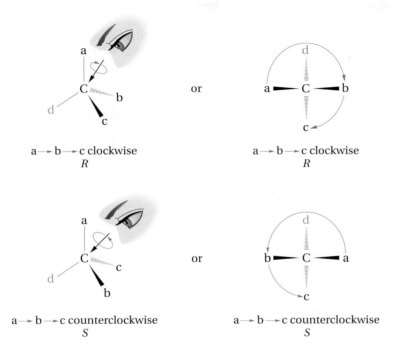

The four groups attached to the stereogenic center are placed in a priority order (by a system we will describe next), $a \rightarrow b \rightarrow c \rightarrow d$. The stereogenic center is then observed *from the side opposite the lowest priority group, d.* If the remaining three groups ($a \rightarrow b \rightarrow c$) form a clockwise array, the configuration is designated *R* (from the Latin *rectus,* right).** If they form a *counterclockwise* array, the configuration is designated as *S* (from the Latin *sinister,* left).

The priority order of the four groups is set in the following way:

Rule 1. The atoms directly attached to the stereogenic center are ranked according to *atomic number:* the higher the atomic number, the higher the priority.

$$Cl > O > C > H$$

high priority ⟶ low priority

*After R. S. Cahn and C. K. Ingold, both British organic chemists, and V. Prelog, a Swiss chemist and Nobel Prize winner.

**More precisely, *rectus* means "right" in the sense of "correct, or proper," and not in the sense of direction (which is *dexter* = right, opposite to left). It may not be entirely coincidental that the initials of one of the inventors of this system are *R. S.*

If one of the four groups is H, it always has the lowest priority, and one views the stereogenic center looking down the C—H bond from C to H.

Rule 2. If a decision cannot be reached with rule 1 (that is, if two or more of the directly attached atoms are the same), work outward from the stereogenic center until a decision is reached. For example, the ethyl group has a higher priority than the methyl group, because at the first point of difference, working outward from the stereogenic center, we come to a *carbon* (higher priority) in the ethyl group and a *hydrogen* (lower priority) in the methyl group.

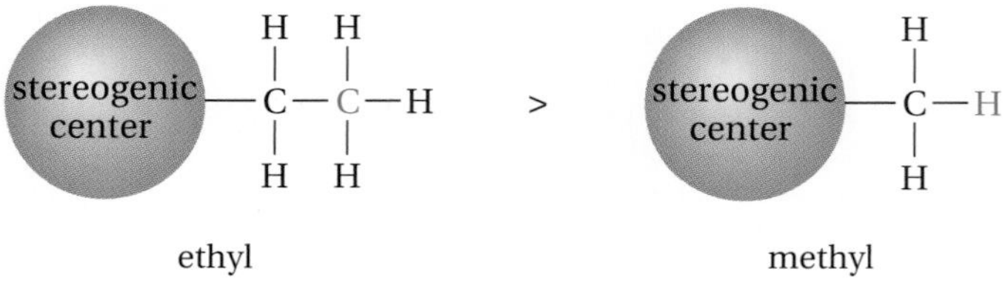

ethyl   methyl

EXAMPLE 5.3

Assign a priority order to the following groups: —H, —Br, —CH$_2$CH$_3$, and —CH$_2$OCH$_3$.

Solution   —Br > —CH$_2$OCH$_3$ > —CH$_2$CH$_3$ > —H

The atomic numbers of the directly attached atoms are ordered Br > C > H. To prioritize the two carbon groups, we must go out until a point of difference is reached.

$$—CH_2OCH_3 > —CH_2CH_3 \quad (O > C)$$

PROBLEM 5.8 Assign a priority order to each of the following sets of groups:

a. —CH$_3$, —CH(CH$_3$)$_2$, —H, —NH$_2$
b. —OH, —F, —CH$_3$, —CH$_2$OH
c. —OCH$_3$, —NHCH$_3$, —CH$_2$NH$_2$, —OH
d. —CH$_2$CH$_3$, —CH$_2$CH$_2$CH$_3$, —C(CH$_3$)$_3$, —CH(CH$_3$)$_2$

For stereogenic centers in cyclic compounds, the same rule for assigning priorities is followed. For example, in 1,1,3-trimethylcyclohexane, the four groups attached to carbon-3 in order of priority are —CH$_2$C(CH$_3$)$_2$CH$_2$ > —CH$_2$CH$_2$ > —CH$_3$ > —H.

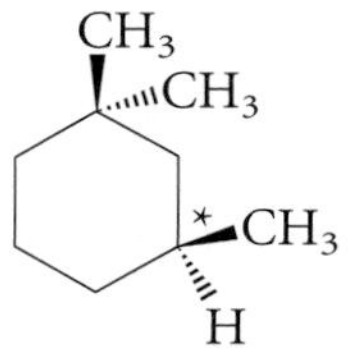

1,1,3-trimethylcyclohexane

A third, somewhat more complicated, rule is required to handle double or triple bonds and aromatic rings (which are written in the Kekulé fashion).

Rule 3.   Multiple bonds are treated as if they were an equal number of single bonds. For example, the vinyl group —CH=CH$_2$ is counted as

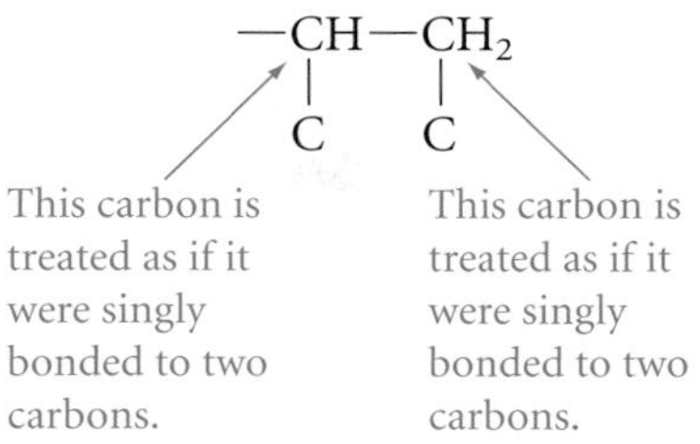

$$-\text{CH}-\text{CH}_2$$

This carbon is treated as if it were singly bonded to two carbons.

This carbon is treated as if it were singly bonded to two carbons.

Similarly,

$$-\text{C}{\equiv}\text{CH} \qquad \text{is treated as} \qquad -\overset{\overset{\text{C}}{|}}{\underset{\underset{\text{C}}{|}}{\text{C}}}-\overset{\overset{\text{C}}{|}}{\underset{\underset{\text{C}}{|}}{\text{C}}}-\text{H}$$

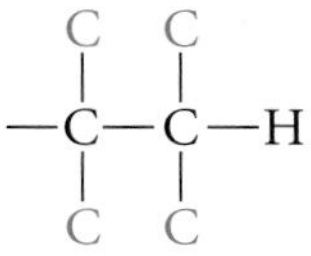

and

$$-\text{CH}{=}\text{O} \qquad \text{is treated as} \qquad -\overset{\overset{\text{H}}{|}}{\underset{\underset{\text{O}}{|}}{\text{C}}}-\overset{}{\underset{\underset{\text{C}}{|}}{\text{O}}}$$

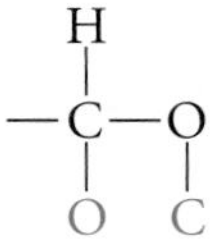

EXAMPLE 5.4

Which group has the higher priority, isopropyl or vinyl?

Solution   The vinyl group has the higher priority. We go out until we reach a difference, shown in color.

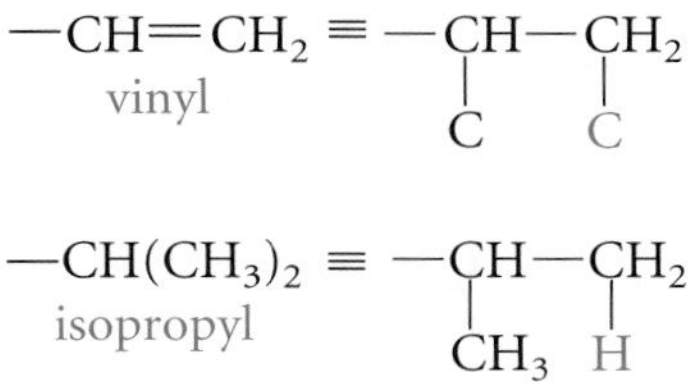

$$-\text{CH}{=}\text{CH}_2 \equiv -\text{CH}-\text{CH}_2$$
$$\text{vinyl} \qquad\qquad \overset{}{\underset{\underset{\text{C}}{|}}{}} \quad \overset{}{\underset{\underset{\text{C}}{|}}{}}$$

$$-\text{CH(CH}_3)_2 \equiv -\text{CH}-\text{CH}_2$$
$$\text{isopropyl} \qquad\qquad \overset{}{\underset{\underset{\text{CH}_3}{|}}{}} \quad \overset{}{\underset{\underset{\text{H}}{|}}{}}$$

PROBLEM 5.9   Assign a priority order to

a. —C≡CH   and —CH=CH$_2$ b. —CH=CH$_2$ and   [benzene ring]

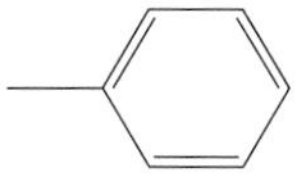

c. —CH=O, —CH=CH$_2$, —CH$_2$CH$_3$,   and   —CH$_2$OH

Now let us see how these rules are applied.

EXAMPLE 5.5

Assign the configuration (R or S) to the following enantiomer of 3-methylhexane (see Example 5.2).

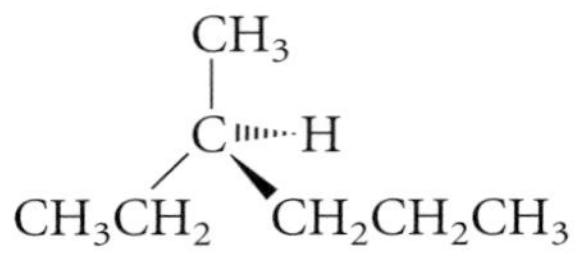

$$\begin{array}{c}
CH_3 \\
| \\
C\text{······}H \\
CH_3CH_2 \quad CH_2CH_2CH_3
\end{array}$$

Solution   First assign the priority order to the four different groups attached to the stereogenic center.

$$-CH_2CH_2CH_3 > -CH_2CH_3 > -CH_3 > -H$$

Now view the molecule *from the side opposite the lowest priority group* (—H) and determine whether the remaining three groups, from high to low priority, form a clockwise (R) or counterclockwise (S) array.

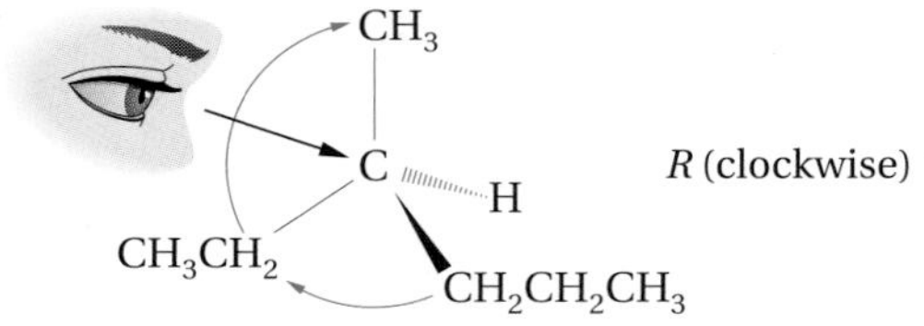

R (clockwise)

We write the name (R)-3-methylhexane.

If we view the other representation of this molecule shown in Example 5.2, we come to the same conclusion.

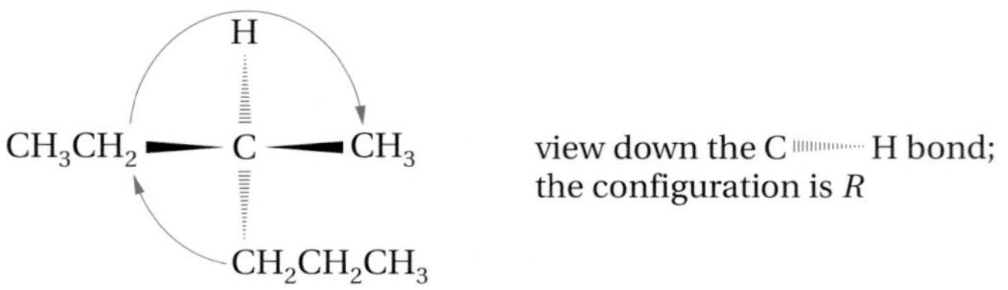

view down the C······H bond;
the configuration is R

PROBLEM 5.10   Determine the configuration (R or S) at the stereogenic center in

a.

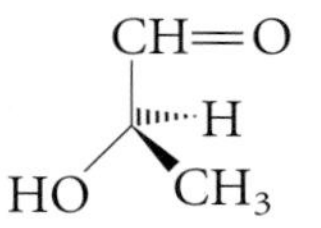

$$\begin{array}{c}
CH{=}O \\
\text{······}H \\
HO \quad CH_3
\end{array}$$

b. 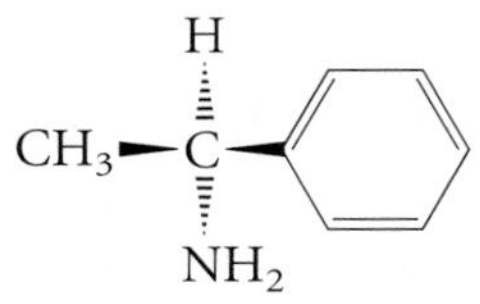
$$CH_3{-}C{-}\phi$$
with H above and NH_2 below

EXAMPLE 5.6

Draw the structure of (R)-2-bromobutane.

Solution   First, write out the structure and prioritize the groups attached to the stereogenic center.

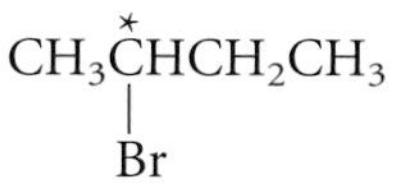

$$CH_3\overset{*}{C}HCH_2CH_3$$
$$|$$
$$Br$$

$$Br— > CH_3CH_2— > CH_3— > H—$$

Now make the drawing with the H (lowest priority group) "away" from you, and place the three remaining groups (Br → CH_3CH_2 → CH_3) in a clockwise (R) array.

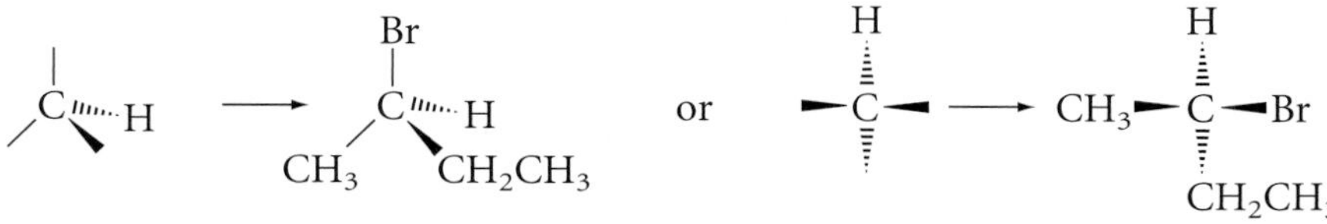

Of course, we could have started with the top-priority group at either of the other two bonds to give the following structures, which are equivalent to those above:

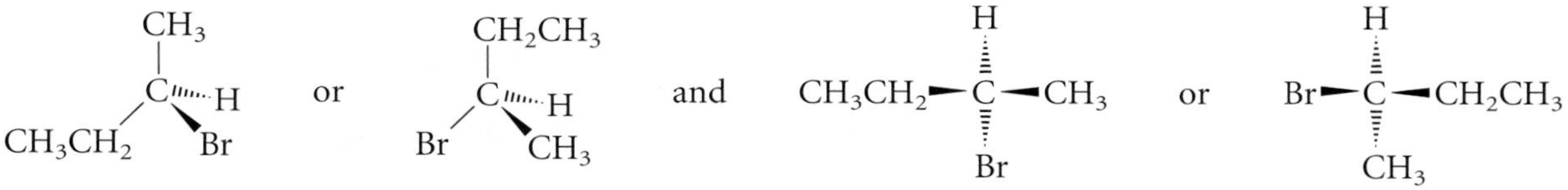

PROBLEM 5.11   Draw the structure of

a. (S)-2-phenylbutane.
b. (R)-3-methyl-1-pentene.
c. (S)-3-methylcyclopentene.

5.4   The *E-Z* Convention for *Cis–Trans* Isomers

Before we continue with other aspects of chirality, let us digress briefly to describe a useful extension of the Cahn–Ingold–Prelog system of nomenclature to *cis–trans* isomers. Although we can easily use *cis–trans* nomenclature for 1,2-dichloroethene or 2-butene (see Sec. 3.5), that system is sometimes ambiguous, as in the following examples:

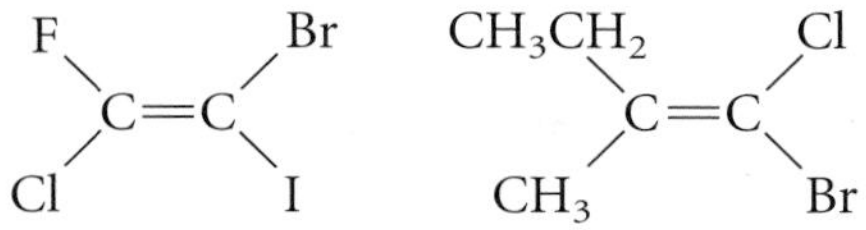

cis or *trans?*   *cis* or *trans?*

The system we have just discussed for stereogenic centers has been extended to double-bond isomers. We use exactly the same priority rules. *The two groups attached to each carbon of the double bond are assigned priorities.* If the two higher priority groups are on *opposite* sides of the double bond, the prefix E (from the German *entgegen*, opposite) is used. If the two higher priority groups are on the *same* side of the double bond, the prefix is Z (from the German *zusammen*, together). The higher priority groups for the above examples are shown here in color, and the correct names are given below the structures.

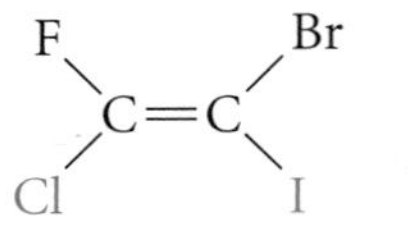

$$\underset{Cl}{\overset{F}{\diagdown}} C = C \underset{I}{\overset{Br}{\diagup}}$$

$$\underset{CH_3}{\overset{CH_3CH_2}{\diagdown}} C = C \underset{Br}{\overset{Cl}{\diagup}}$$

(Z)-1-bromo-2-chloro-
2-fluoro-1-iodoethene

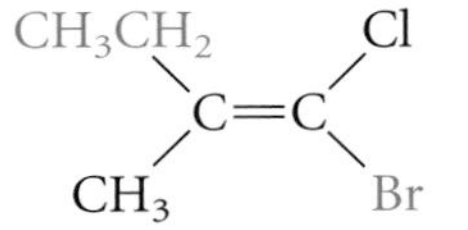

(E)-1-bromo-1-chloro-
2-methyl-1-butene

PROBLEM 5.12   Name each compound by the E-Z system.

a. 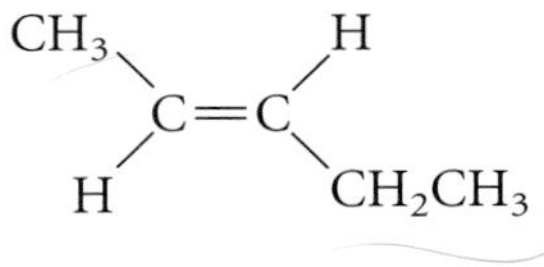 $$\underset{H}{\overset{CH_3}{\diagdown}} C = C \underset{CH_2CH_3}{\overset{H}{\diagup}}$$

b. 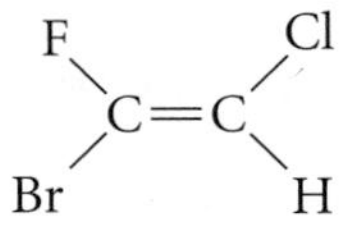 $$\underset{Br}{\overset{F}{\diagdown}} C = C \underset{H}{\overset{Cl}{\diagup}}$$

PROBLEM 5.13   Write the structure for

a. (Z)-2-pentene.   b. (E)-1,3-pentadiene.

5.5   Polarized Light and Optical Activity

The concept of molecular chirality follows logically from the tetrahedral geometry of carbon, as developed in Sections 5.1 and 5.2. Historically, however, these concepts were developed in the reverse order; how this happened is one of the most elegant and logically beautiful stories in the history of science. The story began in the early eighteenth century with the discovery of polarized light and with studies on how molecules placed in the path of such a light beam affect it.

An ordinary light beam consists of waves that vibrate in all possible planes perpendicular to its path. However, if this light beam is passed through certain types of substances, the waves of the transmitted beam will all vibrate in parallel planes. Such a light beam, said to be **plane polarized,** is illustrated in Figure 5.7. One convenient way to polarize light is to pass it through a device composed of Iceland spar (crystalline calcium carbonate) called a **Nicol prism** (invented in 1828 by the British physicist William Nicol). A more recently developed polarizing material is Polaroid, invented by the American

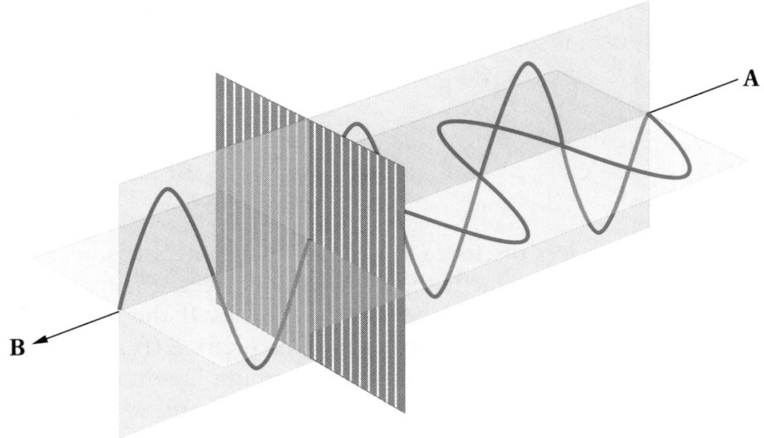

Plane-polarized light is a light beam consisting of waves that vibrate in parallel planes.

Figure 5.7
A beam of light, AB, initially vibrating in all directions, passes through a polarizing substance that "strains" the light so that only the vertical component emerges.

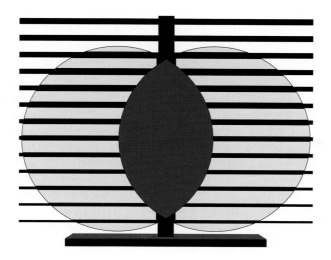

Figure 5.8
The two sheets of polarizing material shown have their axes aligned perpendicularly. Although each disk alone is almost transparent, the area where they overlap is opaque. You can duplicate this effect using two pairs of Polaroid sunglasses. Try it! (Courtesy of the Polaroid Corporation.)

E. H. Land. It contains a crystalline organic compound properly oriented and embedded in a transparent plastic. Sunglasses, for example, are often made from Polaroid.

A light beam will pass through *two* samples of polarizing material only if their polarizing axes are aligned. If the axes are perpendicular, no light will pass through. This result, illustrated in Figure 5.8, is the basis of an instrument used to study the effect of various substances on plane-polarized light.

A **polarimeter is** shown schematically in Figure 5.9. Here is how it works. With the light on and the sample tube empty, the analyzer prism is rotated so that the light beam that has been polarized by the polarizing prism is completely blocked and the field of view is dark. At this point, the prism axes of the polarizer prism and the analyzer prism are perpendicular to one another. Now the sample is placed in the sample tube. If the substance is **optically inactive,** nothing changes. The field of view remains dark. But if an **optically active** substance is placed in the tube, it rotates the plane of polarization, and some light passes through the analyzer to the observer. By turning the analyzer prism clockwise or counterclockwise, the observer can again block the light beam and restore the dark field.

The angle through which the analyzer prism must be rotated in this experiment is called α, the **observed rotation.** It is equal to the number of degrees that the optically active substance rotated the beam of plane-polarized light. If the analyzer must be rotated to the *right* (clockwise), the optically active substance is said to

A **polarimeter,** or **spectropolarimeter,** is an instrument used to detect optical activity. An **optically active** substance rotates plane-polarized light, whereas an **optically inactive** substance does not.

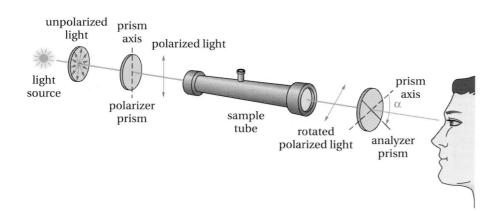

Figure 5.9
Diagram of a polarimeter.

A WORD ABOUT ...

Pasteur's Experiments and the van't Hoff–LeBel Explanation

The great French scientist Louis Pasteur (1822–1895) was the first to recognize that optical activity is related to what we now call chirality. He realized that similar molecules that rotate plane-polarized light through equal angles but in opposite directions must be related to one another as an object and its nonsuperimposable mirror image (that is, as a pair of enantiomers). Here is how he came to that conclusion.

Working in the mid-nineteenth century in a country famous for its wine industry, Pasteur was aware of two *isomeric* acids that deposit in wine casks during fermentation. One of these, called **tartaric acid,** was optically active and dextrorotatory. The other, at the time called **racemic acid,** was optically inactive.

Pasteur prepared various salts of these acids. He noticed that *crystals* of the sodium ammonium salt of *tartaric acid* were *not* symmetric (that is, they were not identical to their mirror images). In other words, they exhibited the property of handedness (chirality). Let us say that the crystals were all left-handed.

Louis Pasteur in his laboratory.

When Pasteur next examined crystals of the same salt of *racemic acid,* he found that they, too, were chiral but that some of the crystals were left-handed and others were right-handed. The crystals were related to one another as an object and its nonsuperimposable mirror image, and they were present in equal amounts. With a magnifying lens and a pair of tweezers, Pasteur carefully separated these crystals into two piles: the left-handed ones and the right-handed ones.

Then Pasteur made a crucial observation. When he *dissolved* the two types of crystals *separately* in water and placed the solutions in a polarimeter, he found that each solution was optically active (remember, he obtained these crystals from racemic acid, which was optically inactive). One solution had a specific rotation identical to that of the sodium ammonium salt of tartaric acid! The other had an equal but *opposite* specific rotation. This meant that it must be the mirror image, or levorotatory tartaric acid. Pasteur correctly concluded that racemic acid was not really a single substance, but a 50:50 mixture of (+) and (−)-tartaric acids. Racemic acid was optically inactive because it contained equal amounts of two enantiomers. We now define a **racemic mixture** as a 50:50 mixture of enantiomers and, of course, such a mixture is optically inactive because the rotations of the two enantiomers cancel out.

Pasteur recognized that optical activity must be due to some property of tartaric acid molecules themselves, *not* to some property of the crystals, because the crystalline shape was lost when the crystals were dissolved in

be **dextrorotatory** (+); if rotated to the *left* (counterclockwise), the substance is **levorotatory** (−).*

The observed rotation, α, of a sample of an optically active substance depends on its molecular structure and also on the number of molecules in the sample tube, the length of the tube, the wavelength of the polarized light, and the temperature. All of these have to be standardized if we want to compare the optical activity of different substances. This is done using the **specific rotation** $[\alpha]$, defined as follows:

The **specific rotation** of an optically active substance (a standardized version of its **observed rotation**) is a characteristic physical property of the substance.

$$\text{Specific rotation} = [\alpha]_{\lambda}^{t} = \frac{\alpha}{l \times c}\,(\text{solvent})$$

where l is the length of the sample tube in *decimeters,* c is the concentration in *grams per milliliter,* t is the temperature of the solution, and λ is the wavelength of light. The

---

*It is not possible to tell from a single measurement whether a rotation is + or −. For example, is a reading +10° or −350°? We can distinguish between these alternatives by, for example, increasing the sample concentration by 10%. Then a +10° reading would change to +11°, and a −350° reading would change to −385° (that is, −25°).

water in order to measure the specific rotation. However, the precise explanation in terms of molecular structure eluded Pasteur and was not to come for another 25 years.

Pasteur's experiments were performed at about the same time that Kekulé in Germany was developing his theories about organic structures. Kekulé recognized that carbon is tetravalent, and there is even a hint in some of his writings (about 1867) and also in the writings of the Russian chemist A. M. Butlerov (1862) and the Italian E. Paterno (1869) that carbon might be tetrahedral. But it was not until 1874 that the Dutch physical chemist J. H. van't Hoff (1852–1911) and the Frenchman J. A. LeBel (1847–1930) simultaneously but quite independently made a bold hypothesis about carbon that would explain the optical activity of some organic molecules and the optical inactivity of others.

These scientists knew their solid geometry. They knew that four different objects can be arranged in two different ways at the corners of a tetrahedron, and that these two ways are related to one another as an object and its nonsuperimposable mirror image. They also knew that this arrangement resulted in right- and left-handedness, as shown in the drawing at the top of the next column.

They made the bold hypothesis that the four valences of carbon were directed toward the corners of a tetrahedron, and that optically active molecules would contain at least one carbon atom with four different groups attached to it. This idea would explain why Pasteur's (+) and (−)-tartaric acids rotated plane-polarized light to

equal extents but in opposite (right- and left-handed) directions. Optically inactive organic substances would either contain no asymmetric carbon atom (stereogenic carbon) or be a 50:50 mixture of enantiomers.

This, then, is how the tetrahedral geometry of carbon was first recognized.* The boldness and brilliance of this proposal is remarkable when one realizes that neither the electron nor the nucleus of an atom had yet been discovered and that almost nothing was then known about the physical nature of chemical bonds. At the time they made their proposal, van't Hoff and LeBel were relatively unknown chemists. Their hypothesis was ridiculed by at least one establishment chemist at the time, but it was soon generally accepted, has survived many tests, and can now be regarded as fact. In 1901, van't Hoff received the first Nobel Prize in chemistry.

---

*Actually, LeBel developed his ideas based on symmetry considerations, whereas van't Hoff based his (at least at first) on the idea of an asymmetric carbon atom. For a stimulating article on these differences, see R. B. Grossman, *J. Chem. Educ.* **1989**, *66*, 30–33.

solvent used is indicated in parentheses. Measurements are usually made at room temperature, and the most common light source is the D-line of a sodium vapor lamp ($\lambda = 589.3$ nm), although modern instruments called **spectropolarimeters** allow the light wavelength to be varied at will. The specific rotation of an optically active substance at a particular wavelength is as definite a property of the substance as its melting point, boiling point, or density.

PROBLEM 5.14 Camphor is optically active. A camphor sample (1.5 g) dissolved in ethanol (optically inactive) to a total volume of 50 mL, placed in a 5-cm polarimeter sample tube, gives an observed rotation of +0.66° at 20°C (using the sodium D-line). Calculate and express the specific rotation of camphor.

In the early nineteenth century, the French physicist Jean Baptiste Biot (1774–1862) studied the behavior of a great many substances in a polarimeter. Some, such as turpentine, lemon oil, solutions of camphor in alcohol, and solutions of cane sugar in water, were optically active. Others, such as water, alcohol, and solutions of salt in water, were optically inactive. Later, many natural products (carbohydrates, proteins, and steroids, to name just a few) were added to the list of optically active compounds. What is it about the structure of molecules that causes some to be optically active and others inactive?

When plane-polarized light passes through a single molecule, the light and the electrons in the molecule interact. This interaction causes the plane of polarization to rotate slightly.* But when we place a substance in a polarimeter, *we do not place a single molecule there, we place a large collection of molecules there* (recall that even as little as a thousandth of a mole contains 6×10^{20} molecules).

Now, if the substance is achiral, then for every single molecule in one orientation that rotates the plane of polarization in one direction, there will be another molecule with the mirror-image orientation that will rotate the plane of polarization an equal amount in the opposite direction. The result is that the light beam passes through a sample of achiral molecules without any net change in the plane of polarization. *Achiral molecules are optically inactive.*

But for chiral molecules the situation is different. Consider a sample containing one enantiomer (say, *R*) of a chiral molecule. For any molecule with a given orientation in the sample, *there can be no mirror-image orientation* (because the mirror image gives a different molecule, the *S* enantiomer). Therefore, the rotation in the polarization plane caused by one molecule is *not* cancelled by any other molecule, and the light beam passes through the sample with a net change in the plane of polarization. *Chiral molecules are optically active.*

5.6 Properties of Enantiomers

In what properties do enantiomers differ from one another? Enantiomers differ only with respect to chirality. In all other respects they are identical. For this reason, they differ from one another *only in properties that are also chiral.* Let us illustrate this idea first with familiar objects.

A left-handed baseball player (chiral) can use the same ball or bat (achiral) as can a right-handed player. But, of course, a left-handed player (chiral) can use only a left-handed baseball glove (chiral). A bolt with a right-handed thread (chiral) can use the same washer (achiral) as a bolt with a left-handed thread, but it can only fit into a nut (chiral) with a right-handed thread. To generalize, *the chirality of an object is most significant when the object interacts with another chiral object.*

Enantiomers have identical achiral properties, such as melting point, boiling point, density, and various types of spectra. Their solubilities in an ordinary, achiral solvent are also identical. However, *enantiomers have different chiral properties,* one of which is the *direction* in which they rotate plane-polarized light (clockwise or counterclockwise). Although enantiomers rotate plane-polarized light in opposite directions, they have specific rotations of the same magnitude (but with opposite signs), because the *number of degrees* is not a chiral property. Only the *direction* of rotation is a chiral property. Here is a specific example.

Lactic acid is an optically active hydroxyacid that is important in several biological processes. It has one stereogenic center. Its structure and some of its properties are shown in Figure 5.10. Note that both enantiomers have identical melting points and, *except for sign,* identical specific rotations.

There is no obvious relationship between configuration (*R* or *S*) and sign of rotation (+ or −). For example, (*R*)-lactic acid is levorotatory. When (*R*)-lactic acid

---

*This is because the electric and magnetic fields that result from electronic motions in the molecule affect the electric and magnetic fields of the light.

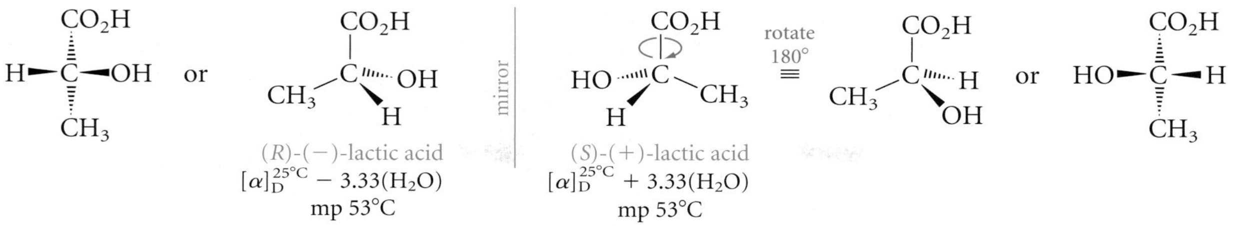

Figure 5.10
The structures and properties of the lactic acid enantiomers.

is converted to its methyl ester (eq. 5.1), the configuration is unchanged because none of the bonds to the stereogenic carbon is involved in the reaction. Yet the sign of rotation of the product, a physical property, changes from − to +.

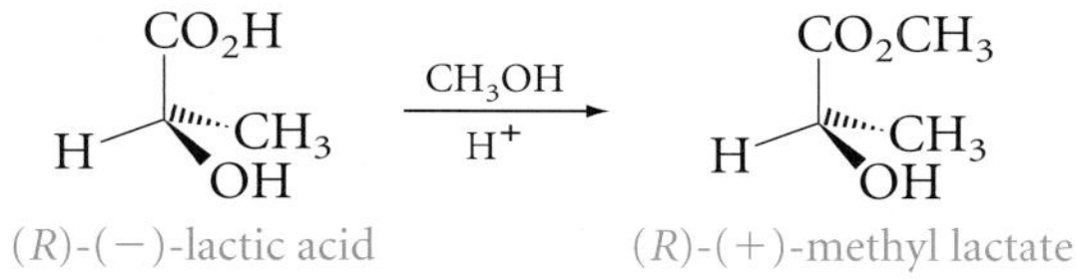

(5.1)

Enantiomers often behave differently biologically because these properties usually involve a reaction with another chiral molecule. For example, the enzyme *lactic acid dehydrogenase* will oxidize (+)-lactic acid to pyruvic acid, but it will *not* oxidize (−)-lactic acid (eq. 5.2).

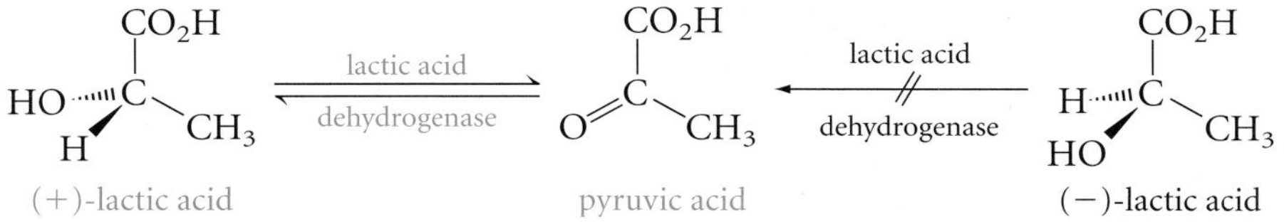

(5.2)

Why? The enzyme itself is chiral and can distinguish between right- and left-handed lactic acid molecules, just as a right hand distinguishes between left-handed and right-handed gloves.

Enantiomers differ in many types of biological activity. One enantiomer may be a drug, whereas its enantiomer may be ineffective. For example, only (−)-adrenalin is a cardiac stimulant; (+)-adrenalin is ineffective. One enantiomer may be toxic, another harmless. One may be an antibiotic, the other useless. One may be an insect sex attractant, the other without effect or perhaps a repellant. Chirality is of paramount importance in the biological world.

5.7　Fischer Projection Formulas

Instead of using dashed and solid wedges to show the three-dimensional arrangements of groups in a chiral molecule, it is sometimes convenient to have a two-dimensional way of doing so. A useful way to do this was devised many years ago by Emil Fischer*; the formulas are called **Fischer projections.**

A **Fischer projection** is a type of two-dimensional formula of a molecule used to represent the three-dimensional configurations of stereogenic centers.

---

*Emil Fischer (1852–1919), who devised these formulas, was one of the early giants in the field of organic chemistry. He did much to elucidate the structures of carbohydrates, proteins, and other natural products and received the 1902 Nobel Prize in chemistry.

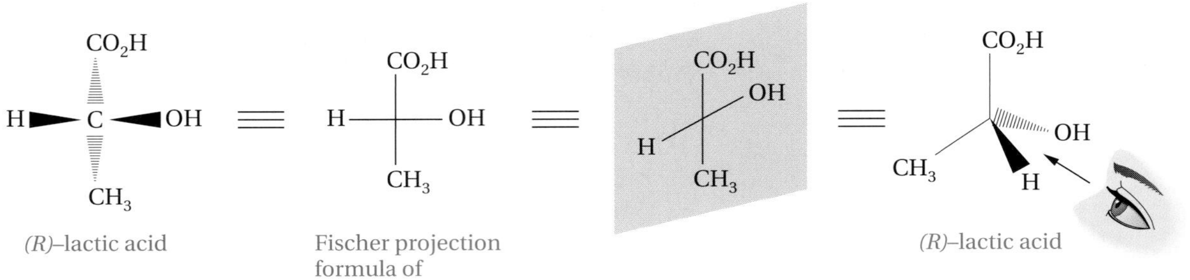

(*R*)–lactic acid   Fischer projection formula of (*R*)–lactic acid   (*R*)–lactic acid

Figure 5.11
Projecting the model at the right onto a plane gives the Fischer projection formula.

Consider the formula for (*R*)-lactic acid, to the left of the mirror in Figure 5.10. If we project that three-dimensional formula onto a plane, as illustrated in Figure 5.11, we obtain the flattened Fischer projection formula.

There are two important things to notice about Fischer projection formulas. First, the C for the stereogenic carbon atom is omitted and is represented simply as the crossing point of the horizontal and vertical lines. Second, horizontal lines connect the stereogenic center to groups that project *above* the plane of the page, *toward* the viewer; vertical lines lead to groups that project *below* the plane of the page, *away* from the viewer. As with other stereorepresentations, interchange of any two groups always gives the enantiomer.

PROBLEM 5.15 Draw a Fischer projection formula for (*S*)-lactic acid.

The following example demonstrates how the absolute configuration of a stereocenter is determined from its Fischer projection.

EXAMPLE 5.7

Determine the absolute (*R* or *S*) configuration of the stereoisomer of 2-chlorobutane shown in the Fischer projection below.

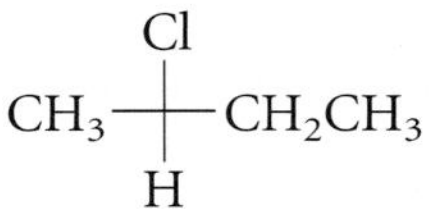

Solution First prioritize the four groups attached to the central carbon according to the Cahn–Ingold–Prelog rules:

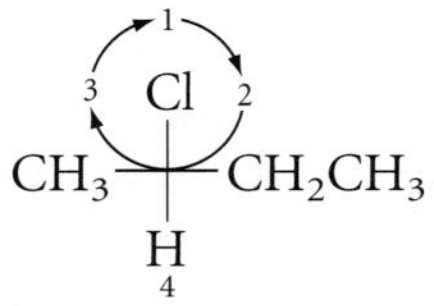

clockwise = *R*

Since atoms attached to vertical arms project away from the viewer, you can now determine whether the sequence 1→2→3 is clockwise or counterclockwise. In this case, it is clockwise, so the absolute configuration of this enantiomer of 2-chlorobutane is *R*.

Note that if the lowest priority group is on a horizontal group, you can still determine the absolute configuration of the stereocenter. This can be done by *rotating three* of the four groups so that the lowest priority group is located on a vertical arm and then proceding as above:

$$
\begin{array}{c}
\overset{3}{CH_3} \\
\overset{4}{H}\!\!-\!\!\overset{2}{CH_2CH_3} \\
\underset{1}{Cl}
\end{array}
\longrightarrow
\begin{array}{c}
\overset{1}{Cl} \\
\overset{3}{CH_3}\!\!-\!\!\overset{2}{CH_2CH_3} \\
\underset{4}{H}
\end{array}
$$

It can also be done by remembering that a substituent on a horizontal arm is pointing out at the viewer. If the sequence of the three priority groups 1→2→3 is counterclockwise, as in this case, the absolute configuration is *R*. (If the direction were clockwise, the absolute configuration would be *S*.)

PROBLEM 5.16   Determine the absolute configuration of the following enantiomer of 2-butanol from its Fischer projection:

$$
\begin{array}{c}
H \\
CH_3\!\!-\!\!CH_2CH_3 \\
OH
\end{array}
$$

Fischer projections are used extensively in biochemistry and in carbohydrate chemistry, where compounds frequently contain more than one stereocenter. In the next section, you will see how useful Fischer projections are in dealing with compounds containing more than one stereogenic center.

5.8 Compounds with More Than One Stereogenic Center; Diastereomers

Compounds may have more than one stereogenic center, so it is important to be able to determine how many isomers exist in such cases and how they are related to one another. Consider the molecule 2-bromo-3-chlorobutane.

$$
\overset{1}{CH_3}\!-\!\overset{2*}{CH}\!-\!\overset{3*}{CH}\!-\!\overset{4}{CH_3}
$$
$$
\qquad\quad \underset{Br}{|} \quad \underset{Cl}{|}
$$

2-bromo-3-chlorobutane

As indicated by the asterisks, the molecule has two stereogenic centers. Each of these could have the configuration *R* or *S*. Thus, four isomers in all are possible: (2*R*,3*R*), (2*S*,3*S*), (2*R*,3*S*), and (2*S*,3*R*). We can draw these four isomers as shown in Figure 5.12. Note that there are two pairs of enantiomers. The (2*R*,3*R*) and (2*S*,3*S*) forms are nonsuperimposable mirror images, and the (2*R*,3*S*) and (2*S*,3*R*) forms are another such pair.

Let us see how to use Fischer projection formulas for these molecules. Consider the (2*R*,3*R*) isomer, the one at the left in Figure 5.12. The solid-dashed wedge drawing has horizontal groups projecting out of the plane of the paper toward us and vertical

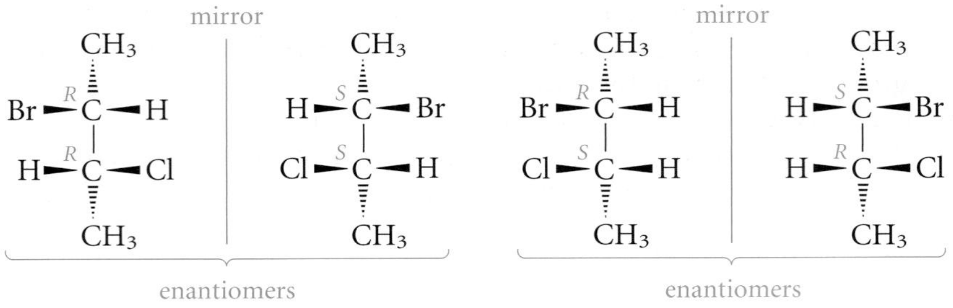

Figure 5.12
The four stereoisomers of 2-bromo-3-chlorobutane, a compound with two stereogenic centers.

groups going away from us, behind the paper. These facts are expressed in the equivalent Fischer projection formula as shown.*

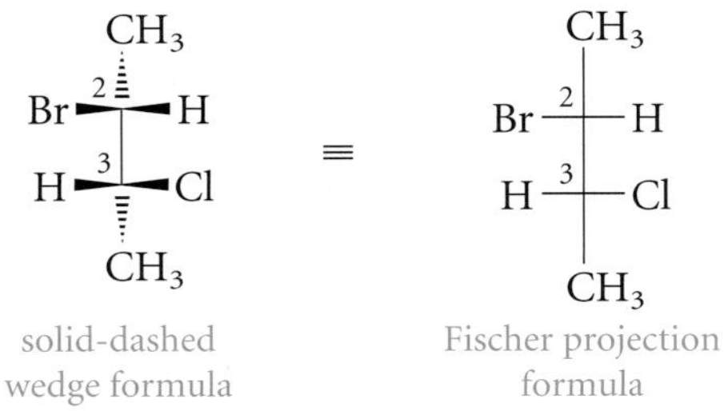

solid-dashed wedge formula   Fischer projection formula

PROBLEM 5.17 Draw the Fischer projection formulas for the remaining stereoisomers of 2-bromo-3-chlorobutane shown in Figure 5.12.

Now we come to an extremely important new idea. Consider the relationship between, for example, the (2R,3R) and (2R,3S) forms of the isomers in Figure 5.12. These forms are *not* mirror images because they have the *same* configuration at carbon-2, though they have opposite configurations at carbon-3. They are certainly stereoisomers, but they are not enantiomers. For such pairs of stereoisomers we use the term **diastereomers**. Diastereomers are stereoisomers that are not mirror images of one another.

Diastereomers are stereoisomers that are not mirror images of each other.

---

*Make 3D models of the molecules in Figure 5.12 to help you visualize them. Notice that these structures are derived from an eclipsed conformation of the molecule, viewed from above so that horizontal groups project toward the viewer. The actual molecule is an equilibrium mixture of several staggered conformations, one of which is shown below. Fischer formulas are used to represent the correct *configurations*, but not necessarily the lowest energy *conformations* of a molecule.

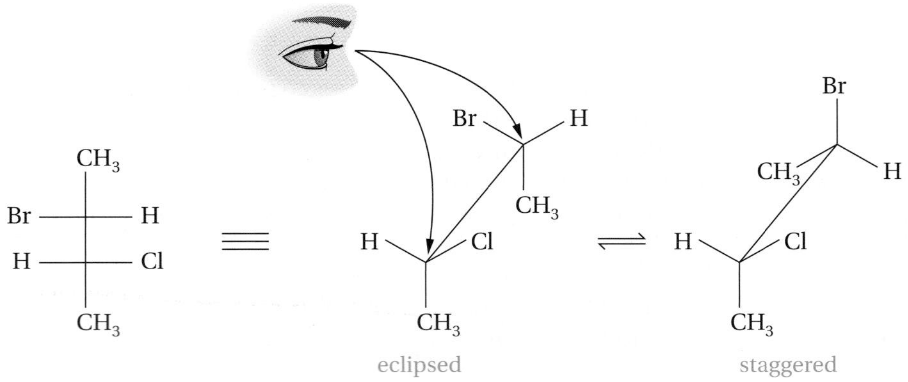

eclipsed   staggered

There is an important, fundamental difference between enantiomers and diastereomers. Because they are mirror images, enantiomers differ *only* in mirror-image (chiral) properties. They have the same achiral properties, such as melting point, boiling point, and solubility in ordinary solvents. Enantiomers cannot be separated from one another by methods that depend on achiral properties, such as recrystallization or distillation. On the other hand, diastereomers are *not* mirror images. They may differ in *all* properties, whether chiral or achiral. As a consequence, diastereomers may differ in melting point, boiling point, solubility, and not only direction but also the number of degrees that they rotate plane-polarized light—in short, they behave as two different chemical substances.

PROBLEM 5.18 How do you expect the specific rotations of the $(2R,3R)$ and $(2S,3S)$ forms of 2-bromo-3-chlorobutane to be related? Answer the same question for the $(2R,3R)$ and $(2S,3R)$ forms.

Can we generalize about the number of stereoisomers possible when a larger number of stereogenic centers is present? Suppose, for example, that we add a third stereogenic center to the compounds shown in Figure 5.12 (say, 2-bromo-3-chloro-4-iodopentane). The new stereogenic center added to each of the four structures can once again have either an *R* or an *S* configuration, so that with three different stereogenic centers, eight stereoisomers are possible. The situation is summed up in a single rule: If a molecule has *n* different stereogenic centers, it may exist in a maximum of 2^n stereoisomeric forms. There will be a maximum of $2^n/2$ pairs of enantiomers.

PROBLEM 5.19 The Fischer projection formula for glucose (blood sugar, Sec. 16.4, p. 454) is

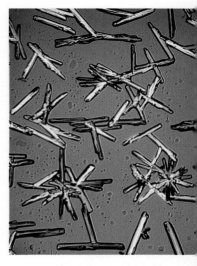

Polarized light micrograph of glucose crystals.

Altogether, how many stereoisomers of this sugar are possible?

Actually, the number of isomers predicted by this rule is the *maximum* number possible. Sometimes certain structural features reduce the actual number of isomers. In the next section, we examine a case of this type.

5.9 *Meso* Compounds; the Stereoisomers of Tartaric Acid

Consider the stereoisomers of 2,3-dichlorobutane. There are two stereogenic centers.

2,3-dichlorobutane

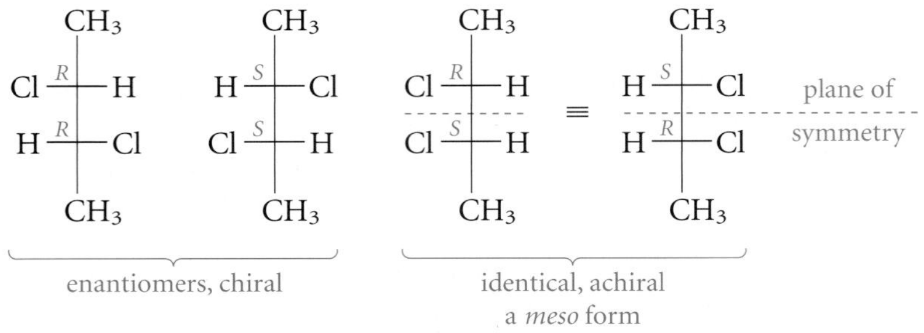

Figure 5.13
Fischer projections of the stereoisomers of 2,3-dichlorobutane.

We can write out the stereoisomers just as we did in Figure 5.12; they are shown in Figure 5.13. Once again, the (R,R) and (S,S) isomers constitute a pair of nonsuperimposable mirror images, or enantiomers. *However, the other "two" structures, (R,S) and (S,R), in fact, now represent a single compound.*

Look more closely at the structures to the right in Figure 5.13. Notice that they have a plane of symmetry that is perpendicular to the plane of the paper and bisects the central C—C bond. The reason is that each stereogenic center has the *same* four groups attached. The structures are identical, superimposable mirror images and therefore *achiral*. We call such a structure a *meso* **compound.** A *meso* compound is an achiral diastereomer of a compound with stereogenic centers. Its stereogenic centers have opposite configurations. Being achiral, *meso* compounds are optically inactive.*

A *meso* **compound** is an achiral diastereomer of a compound with stereogenic centers.

Now let us take a look at tartaric acid, the compound whose optical activity was so carefully studied by Louis Pasteur (see A Word About Pasteur's Experiments on pages 160–161). It has two identical stereogenic centers.

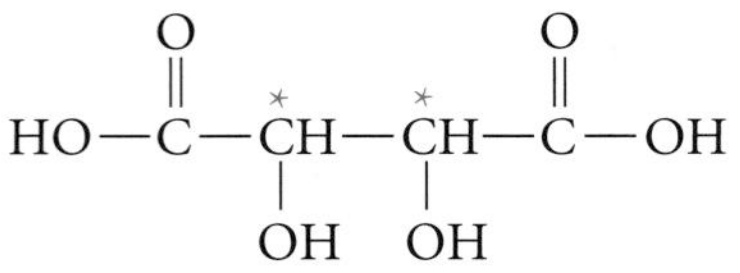

tartaric acid

The structures of these three stereoisomers and two of their properties are shown in Figure 5.14. Note that the enantiomers have identical properties except for the *sign* of the specific rotation, whereas the *meso* form, being a diastereomer of each enantiomer, differs from them in both properties.

For about 100 years after Pasteur's research, it was still not possible to determine the configuration associated with a particular enantiomer of tartaric acid. For example, it

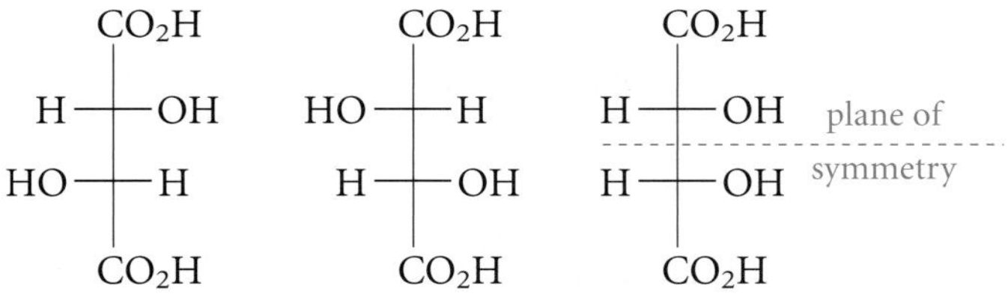

Figure 5.14
The stereoisomers of tartaric acid.

Configuration	(R,R)	(S,S)	meso (R,S)
$[\alpha]_D^{20°}$ (H_2O)	+12	−12	0
Melting point, °C	170	170	140

---

*Fischer projections may be turned 180° in the plane of the paper without changing configuration. You can see that such an operation on the enantiomeric pair in Figure 5.13 does *not* interconvert them, but when performed on the *meso* form, it does.

was not known whether (+)-tartaric acid had the (R,R) or the (S,S) configuration. It was known that (+)-tartaric acid had to have one of these two configurations and that (−)-tartaric acid had to have the opposite configuration, but which isomer had which?

In 1951, the Dutch scientist J. M. Bijvoet developed a special x-ray technique that solved the problem. Using this technique on crystals of the sodium rubidium salt of (+)-tartaric acid, Bijvoet showed that it had the (R,R) configuration. So this was the tartaric acid studied by Pasteur, and racemic acid was a 50:50 mixture of the (R,R) and (S,S) isomers. The *meso* form was not studied until later.

Since tartaric acid had been converted chemically into other chiral compounds and these in turn into still others, it became possible as a result of Bijvoet's work to assign **absolute configurations** (that is, the correct *R* or *S* configuration for each stereocenter) to many pairs of enantiomers.

The correctly assigned (*R* or *S*) configuration of a stereocenter in a molecule is called the **absolute configuration** of the stereocenter.

PROBLEM 5.20 Show that *trans*-1,2-dimethylcyclopentane can exist in chiral, enantiomeric forms.

PROBLEM 5.21 Is *cis*-1,2-dimethylcyclopentane chiral or achiral? What stereochemical term can we give to it?

5.10 Stereochemistry: A Recap of Definitions

We have seen here and in Section 2.11 that *stereoisomers* can be classified in three different ways. They may be either *conformers* or *configurational isomers*; they may be *chiral* or *achiral*; and they may be *enantiomers* or *diastereomers*.

A	*Conformers:*	interconvertible by rotation about single bonds
	Configurational Isomers:	not interconvertible by rotation, only by breaking and making bonds
B	*Chiral:*	mirror image not superimposable on itself
	Achiral:	molecule and mirror image are identical
C	*Enantiomers:*	mirror images; have opposite configurations at all stereogenic centers
	Diastereomers:	stereoisomers but not mirror images; have same configuration at one or more centers, but differ at the remaining stereogenic centers

Various combinations of these three sets of terms can be applied to any pair of stereoisomers. Here are a few examples:

1. *Cis-* and *trans*-2-butene (*Z*- and *E*-2-butene).

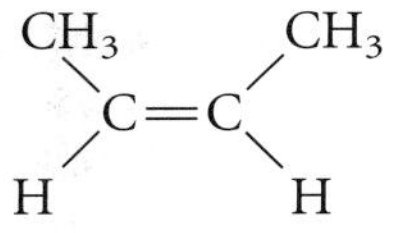

$$CH_3 \quad CH_3 \qquad\qquad CH_3 \quad H$$
$$\diagdown\diagup\qquad\qquad\diagdown\diagup$$
$$C{=}C \qquad \text{and} \qquad C{=}C$$
$$\diagup\diagdown\qquad\qquad\qquad\diagup\diagdown$$
$$H \qquad\quad H \qquad\qquad\quad H \qquad\quad CH_3$$

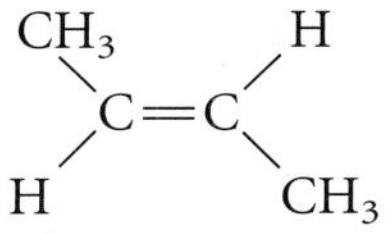

These isomers are *configurational* (not interconverted by rotation about single bonds), *achiral* (the mirror image of each is superimposable on the original), and *diastereomers* (although they are stereoisomers, they are *not* mirror images of one another; hence they must be diastereomers).

2. Staggered and eclipsed ethane.

and

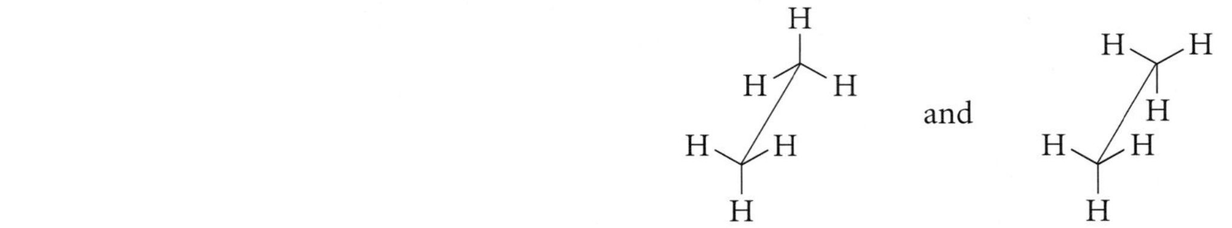

These are *achiral conformers*. They are *diastereomeric conformers* (but without stereogenic centers) because they are not mirror images.

3. (*R*)- and (*S*)-lactic acid.

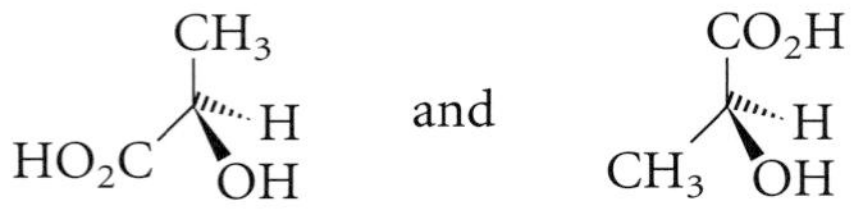

and

These isomers are *configurational*, each is *chiral*, and they constitute a pair of *enantiomers*.

4. *Meso*- and (*R,R*)-tartaric acids.

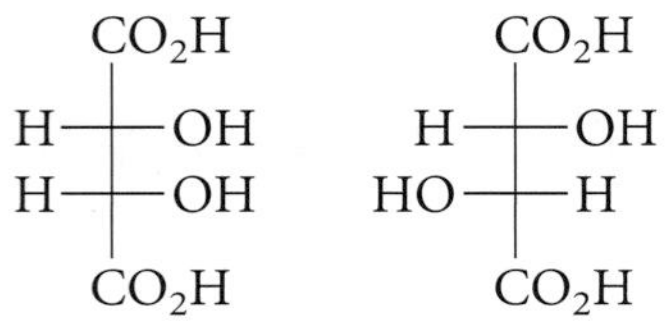

These isomers are *configurational* and *diastereomers*. One is *achiral*, and the other is *chiral*.

Enantiomers, such as (*R*)- and (*S*)-lactic acid, differ only in chiral properties and therefore cannot be separated by ordinary achiral methods such as distillation or recrystallization. Diastereomers differ in all properties, chiral or achiral. *If* they are also configurational isomers (such as *cis*- and *trans*-2-butene, or *meso*- and *R,R*-tartaric acid), they can be separated by ordinary achiral methods, such as distillation or recrystallization. *If*, on the other hand, they are conformers (such as staggered and eclipsed ethane), they may interconvert so readily by bond rotation as to not be separable.

PROBLEM 5.22 Draw the two stereoisomers of 1,3-dimethylcyclobutane, and classify the pair according to the categories listed in A, B, and C above.

5.11 Stereochemistry and Chemical Reactions

How important is stereochemistry in chemical reactions? The answer depends on the nature of the reactants. First, consider the formation of a chiral product from achiral reactants; for example, the addition of hydrogen bromide to 1-butene to give 2-bromobutane in accord with Markovnikov's rule.

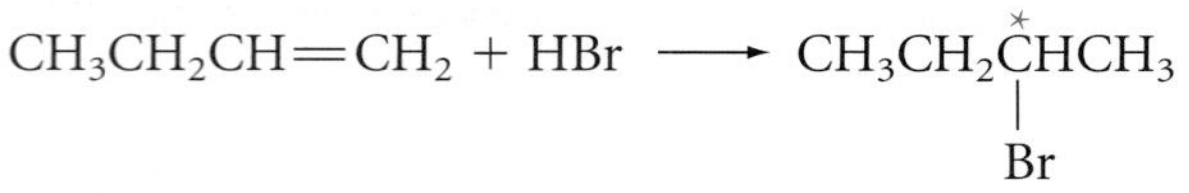

$$CH_3CH_2CH=CH_2 + HBr \longrightarrow CH_3CH_2\overset{*}{C}HCH_3 \qquad \textbf{(5.3)}$$

$$|$$
$$Br$$

1-butene   2-bromobutane

The product has one stereogenic center, marked with an asterisk, but both enantiomers are formed in exactly equal amounts. The product is a racemic mixture. Why? Although this result will be obtained *regardless* of the reaction mechanism, let us consider the generally accepted mechanism.

$$CH_3CH_2CH\!=\!CH_2 + H^+ \longrightarrow CH_3CH_2\overset{+}{C}HCH_3 \xrightarrow{\ Br^-\ } CH_3CH_2CHCH_3 \qquad (5.4)$$
$$\text{2-butyl cation} \qquad\qquad\qquad\qquad \underset{Br}{|}$$

The intermediate 2-butyl cation obtained by adding a proton to the end carbon is planar, and bromide ion can combine with it from the "top" or "bottom" side with exactly equal probability.

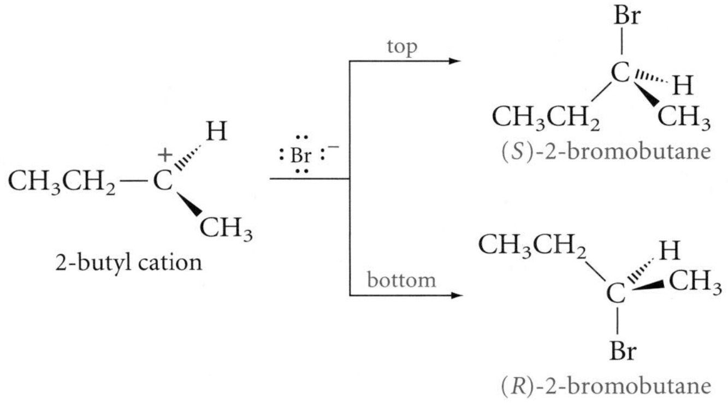

(5.5)

(S)-2-bromobutane

(R)-2-bromobutane

The product is therefore a racemic mixture, an optically inactive 50 : 50 mixture of the two enantiomers.

We can generalize this result. *When chiral products are obtained from achiral reactants, both enantiomers are formed at the same rates, in equal amounts.*

PROBLEM 5.23 Show that, if the mechanism of addition of HBr to 1-butene involved *no* intermediates, but *simultaneous one-step* addition (in the Markovnikov sense), the product would still be racemic 2-bromobutane.

PROBLEM 5.24 Show that the chlorination of butane at carbon-2 will give a 50 : 50 mixture of enantiomers.

Now consider a different situation, the reaction of a *chiral* molecule with an achiral reagent to create a second stereogenic center. Consider, for example, the addition of HBr to 3-chloro-1-butene.

$$CH_3\overset{*}{C}HCH\!=\!CH_2 + HBr \longrightarrow CH_3\overset{*}{C}H\!-\!\overset{*}{C}HCH_3 \qquad (5.6)$$
$$\underset{Cl}{|} \qquad\qquad\qquad\qquad \underset{Cl}{|}\ \ \underset{Br}{|}$$
$$\text{3-chloro-1-butene} \qquad\qquad \text{2-bromo-3-chlorobutane}$$

A WORD ABOUT ...

Enantiomers and Biological Activity

Enantiomers of chiral molecules can elicit vastly different biological responses when ingested by living organisms. The tastes, odors, medicinal properties, toxicity, bactericidal, fungicidal, insecticidal, and other properties of enantiomers often differ widely. Here are some examples: The amino acid (R)-asparagine tastes sweet, while (S)-asparagine tastes bitter; (R)-carvone has the odor of spearmint, whereas (S)-carvone is responsible for the smell of caraway; (S)-naproxen is an important anti-inflammatory drug, while its enantiomer is a liver toxin; (R,R)-chloramphenicol is a useful antibiotic but its enantiomer is harmless to bacteria; (R,R)-paclobutrazol is a fungicide, while its enantiomer is a plant growth regulator; and (R)-thalidomide is a sedative and hypnotic, while its enantiomer is a potent teratogen.

How can it be that two molecules as similar in structure as enantiomers have such different biological activities? The reason is that biological activity is initiated by binding of the small molecule to a receptor molecule in the living organism to form a small molecule–receptor complex. The receptors are usually chiral, nonracemic

molecules such as proteins, complex carbohydrates, or nucleic acids that bind well to only one of a pair of enantiomers. Because of their different three-dimensional shapes, (R)-asparagine binds to a receptor in humans that triggers a sensation of sweetness, whereas (S)-asparagine does not bind to that receptor. Instead, it binds to a differently shaped receptor that results in a bitter sensation.

A useful analogy for this selective binding is the manner in which your left shoe (receptor) interacts with your left and right feet (a pair of enantiomers). The left foot fits (binds) into the shoe comfortably (one response) whereas the right foot does not, or only uncomfortably (a different response).

There are now a number of examples where one enantiomer of a chiral molecule elicits a desirable biological response, while the other enantiomer results in a detrimental response (naproxen, for example). Therefore, it has become increasingly important for chiral pharmaceutical and agrochemical compounds to be marketed as single enantiomers rather than as racemic mixtures. This has recently stimulated the development of new synthetic

Suppose we start with one pure enantiomer of 3-chloro-1-butene, say, the *R* isomer. What can we say about the stereochemistry of the products? One way to see the answer quickly is to draw Fischer projections.

(R)-3-chloro-1-butene $(2R,3R)$-2-bromo-3-chlorobutane $(2S,3R)$-2-bromo-3-chlorobutane   (5.7)

The configuration where the chloro substituent is located remains unchanged and *R*, but the new stereogenic center can be either *R* or *S*. Therefore, the products are *diastereomers*. Are they formed in equal amounts? No. Looking at the starting material in eq. 5.7, we can see that it has no plane of symmetry. Approach of the bromine to the double bond from the H side or from the Cl side of the stereogenic center should not occur with equal ease.

We can generalize this result. *Reaction of a chiral reagent with an achiral reagent, when it creates a new stereogenic center, leads to diastereomeric products at different rates and in unequal amounts.*

methods that provide only a single enantiomer of a chiral molecule, a process called *asymmetric synthesis*. In fact, the 2001 Nobel Prize in chemistry was awarded to William S. Knowles (Monsanto), K. Barry Sharpless (Scripps Institute), and Ryoji Noyori (Nagoya University) for their seminal and practical contributions to this field of research. The need for enantiomerically pure compounds has also led to the development of new methods for separating racemic mixtures into their enantiomeric components, a process called resolution.

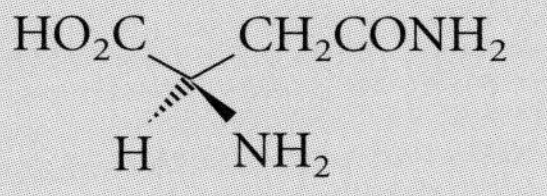

(R)-asparagine 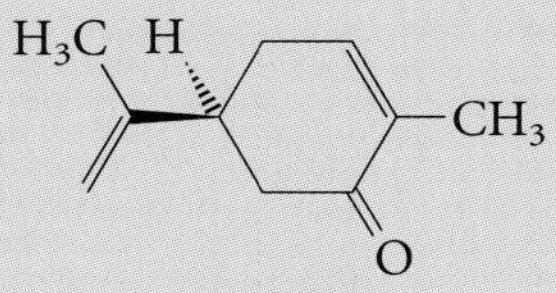 (S)-carvone 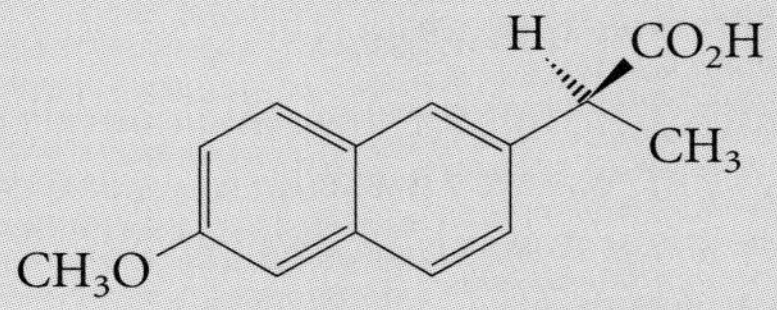 (S)-naproxen

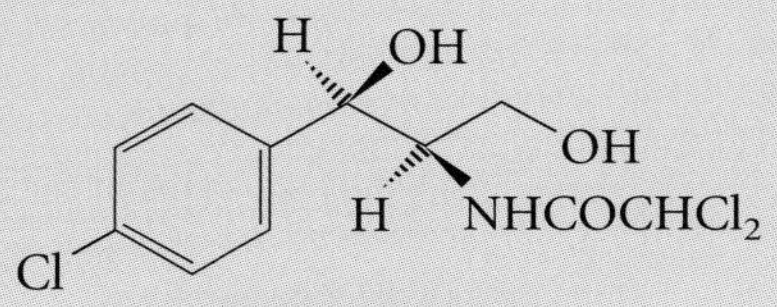

(R,R)-chloramphenicol 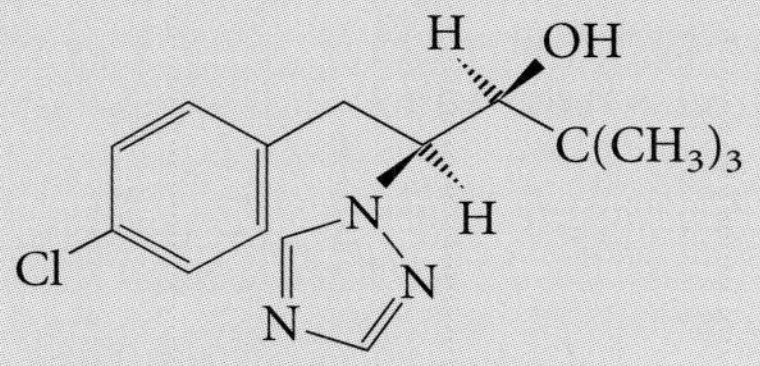 (S,S)-paclobutrazol 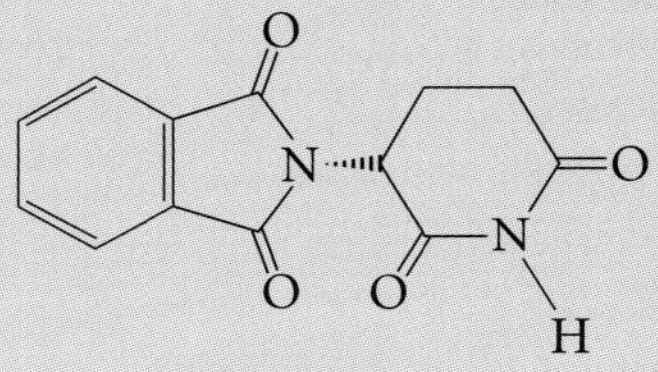 (R)-thalidomide

Enantiomers of Some Biologically Active Compounds

PROBLEM 5.25 Let us say that the (2R,3R) and (2S,3R) products in eq. 5.7 are formed in a 60:40 ratio. What products would be formed and in what ratio by adding HBr to pure (S)-3-chloro-1-butene? by adding HBr to a racemic mixture of (R)- and (S)-3-chloro-1-butene?

5.12 Resolution of a Racemic Mixture

We have just seen (eq. 5.5) that, when reaction between two achiral reagents leads to a chiral product, it always gives a **racemic (50:50) mixture** of enantiomers. Suppose we want to obtain each enantiomer pure and free of the other. The process of separating a racemic mixture into its enantiomers is called **resolution.** Since enantiomers have identical achiral properties, how can we resolve a racemic mixture into its components? The answer is to convert them to diastereomers, separate the *diastereomers,* and then reconvert the now-separated diastereomers back to enantiomers.

To separate two enantiomers, we first let them react with a chiral reagent. The product will be a pair of *diastereomers.* These, as we have seen, differ in all types of prop-

The process of separating the two enantiomers of a **racemic (50:50) mixture** is called **resolution.**

A CLOSER LOOK AT ...

Thalidomide

L og on to http://www.college.hmco.com and follow the links to the student text web site. Use the Web Link section for Chapter 5 to answer the following questions and find more information on the topics below.

Historical Information About Thalidomide

1. When and where was the drug thalidomide first used?
2. For what purpose was it used?
3. What are the unintended consequences of its use by pregnant women?

Patients Taking Thalidomide

1. During what period of pregnancy does this drug cause damage? How much of the drug is required to produce this effect?

2. In addition to its teratogenic effects, thalidomide also can give rise to other side effects. What are some of these side effects?

Why We Still Use Thalidomide

1. What are some diseases for which thalidomide is proving to be an effective medication?
2. What are other potential uses of this drug?

Lots More Information

For more information on a number of related topics, click this link!

Tartaric acid crystals under polarized light.

erties and can be separated by ordinary methods. This principle is illustrated in the following equation:

$$\left\{\begin{array}{c} R \\ S \end{array}\right\} \quad + \quad R \quad \longrightarrow \quad \left\{\begin{array}{c} R-R \\ S-R \end{array}\right\} \tag{5.8}$$

pair of enantiomers (not separable)   chiral reagent   diastereomeric products (separable)

After the diastereomers are separated, we then carry out reactions that regenerate the chiral reagent and the separated enantiomers.

$$R-R \longrightarrow R + R$$

and $\qquad\qquad\qquad\qquad\qquad\qquad\qquad$ (5.9)

$$S-R \longrightarrow S + R$$

Louis Pasteur was the first to resolve a racemic mixture when he separated the sodium ammonium salts of $(+)$- and $(-)$-tartaric acid. In a sense, he was the chiral reagent, since he could distinguish between the right- and left-handed crystals. In Chapter 11, we will see a specific example of how this is done chemically.

The principle behind the resolution of racemic mixtures is the same as the principle involved in the specificity of many biological reactions. That is, a chiral reagent (in a cell, usually an enzyme) can discriminate between enantiomers.

ADDITIONAL PROBLEMS

Stereochemistry: Definitions and Stereogenic Centers

5.26 Define or describe the following terms.

a. stereogenic center
b. chiral molecule
c. enantiomers
d. plane-polarized light
e. specific rotation
f. diastereomers
g. plane of symmetry
h. *meso* form
i. racemic mixture
j. resolution

5.27 Which of the following substances contain stereogenic centers? (Drawing the structures will help you answer this question.)

a. 2,2-dichloropropane
b. 1,2-dichlorobutane
c. 2-methylhexane
d. 2,3-dimethylhexane
e. 1-deuterioethanol (CH_3CHDOH)
f. methylcyclopropane

5.28 Locate with an asterisk the stereogenic centers (if any) in the following structures.

a. $CH_3CHClCF_3$
b. $CH_2(OH)CH(OH)CH(OH)CHO$
c. $C_6H_5CH(OH)CO_2H$

d.  —CH(OH)CH_3
e. 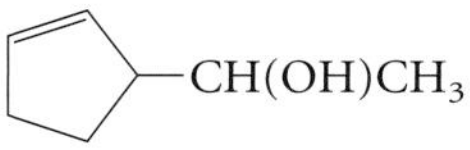 ⬠—CH(OH)CH_3
f. 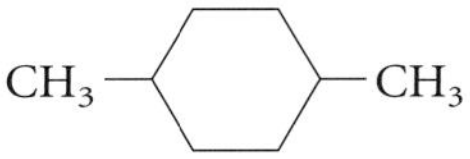 CH_3—⬡—CH_3

Optical Activity

5.29 What would happen to the observed and to the *specific* rotation if, in measuring the optical activity of a solution of sugar in water, we

a. doubled the concentration of the solution?
b. doubled the length of the sample tube?

5.30 The observed rotation for 100 mL of an aqueous solution containing 1 g of sucrose (ordinary sugar), placed in a 2-decimeter sample tube, is +1.33° at 25°C (using a sodium lamp). Calculate and express the specific rotation of sucrose.

Relationships Between Stereoisomers

5.31 Tell whether the following structures are identical or enantiomers:

5.32 Draw a structural formula for an optically active compound with the molecular formula

a. $C_4H_{10}O$.
b. $C_5H_{11}Cl$.
c. $C_4H_8(OH)_2$.
d. C_6H_{12}.

5.33 Draw the formula of an unsaturated bromide, C_5H_9Cl, that can show

a. neither *cis–trans* isomerism nor optical activity.
b. *cis–trans* isomerism but no optical activity.
c. no *cis–trans* isomerism but optical activity.
d. *cis–trans* isomerism and optical activity.

The *R-S* and *E-Z* Conventions

5.34 Place the members of the following groups in order of decreasing priority according to the *R-S* convention:

    a. CH_3—, HS—, CH_2=CH—, H—

    b. H—, CH_3—, C_6H_5—, I—   (see p. 123 for structure of C_6H_5— group)

    c. CH_3—, HO—, CH_3CH_2—, $HOCH_2$—

    d. CH_3CH_2—, $CH_3CH_2CH_2$—, CH_2=CH—, —CH=O

5.35 Assume that the four groups in each part of Problem 5.34 are attached to one carbon atom.

    a. Draw a three-dimensional formula for the *R* configuration of the molecule in 5.34a and 5.34b.

    b. Draw a three-dimensional formula for the *S* configuration of the molecule 5.34c and 5.34d.

5.36 Tell whether the stereogenic centers marked with an asterisk in the following structures have the *R* or the *S* configuration:

a.

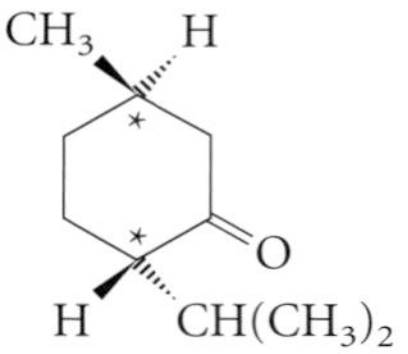

(−)-menthone
(found in peppermint)

b. 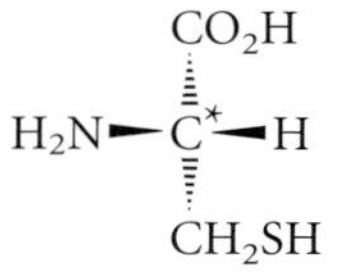 H_2N—$\overset{*}{C}$—H

CO$_2$H

CH$_2$SH

(−)-cysteine
(an amino acid
found in proteins)

c. 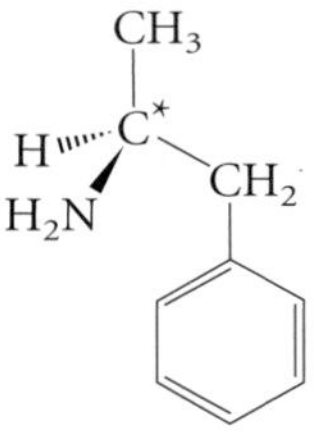 H...$\overset{*}{C}$

CH$_3$

CH$_2$

H_2N

(+)-amphetamine
(central nervous
system stimulant)

5.37 Determine the configuration, *R* or *S*, of (+)-carvone, the compound responsible for the odor of caraway seeds.

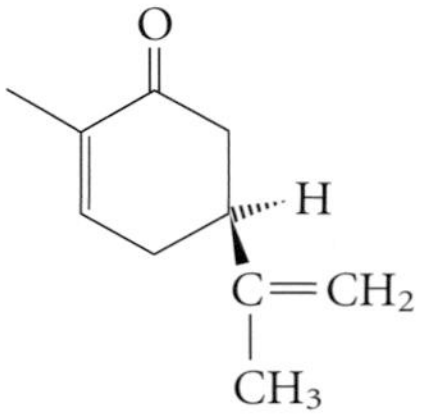

..H

C=CH$_2$

CH$_3$

(+)-carvone

5.38 In a recent collaboration, French and American chemists found that (−)-bromochlorofluoromethane (CHBrClF), one of the simplest chiral molecules, has the *R* configuration. Draw a three-dimensional structural formula for (*R*)-(−)-bromochlorofluoromethane.

5.39 Name the following compounds, using *E-Z* notation:

    a. 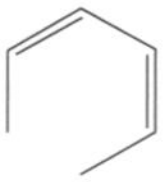   **b.** 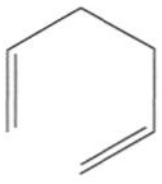   **c.** 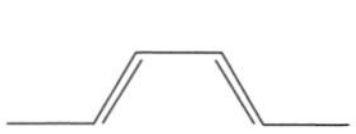   **d.**

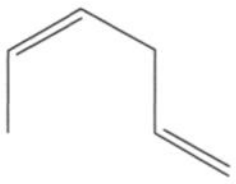

5.40 Two possible isomers of 1,2-dichloroethene are:

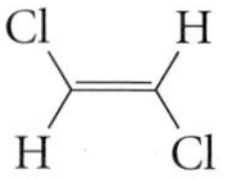

Cl   H

H   Cl

and

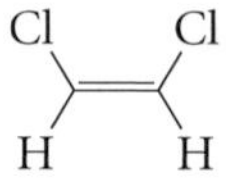

Cl   Cl

H   H

Classify them fully, according to the discussion in Section 5.10.

─────────────

5.41 4-Bromo-2-pentene has a double bond that can have either the *E* or the *Z* configuration and a stereogenic center that can have either the *R* or the *S* configuration. How many stereoisomers are possible altogether? Draw the structure of each, and group the pairs of enantiomers.

5.42 How many stereoisomers are possible for each of the following structures? Draw them, and name each by the *R-S* and *E-Z* conventions. (See Problem 5.41.)

    a. 3-chloro-1,4-pentadiene   **b.** 3-methyl-1,4-hexadiene

    c. 2-bromo-5-chloro-3-hexene   **d.** 2,5-dibromo-3-hexene

Fischer and Newman Projections

5.43 Which of the following Fischer projection formulas have the same configuration as **A**,

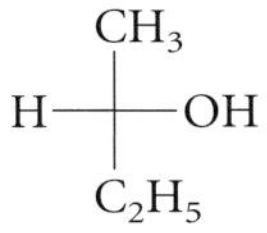

and which are its enantiomer?

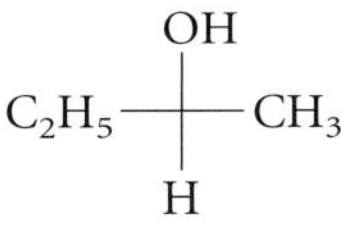

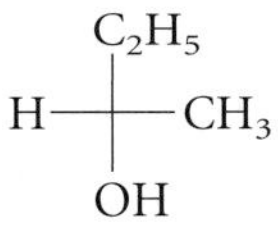

5.44 Below are Newman projections for the three tartaric acids (*R,R*), (*S,S*), and *meso*. Which is which?

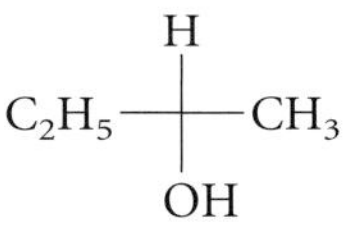

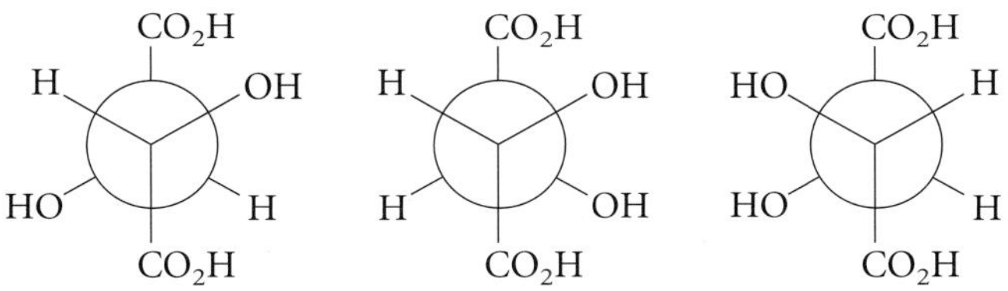

5.45 Convert the sawhorse formula below for one isomer of tartaric acid to a Fischer projection formula. Which isomer of tartaric acid is it?

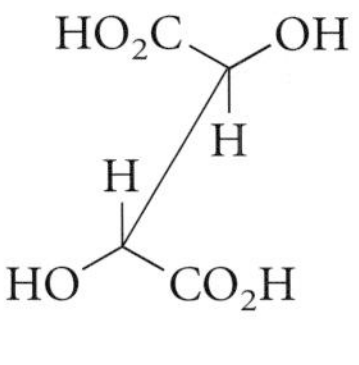

5.46 Two possible isomeric structures of 1,2-dichloroethane are

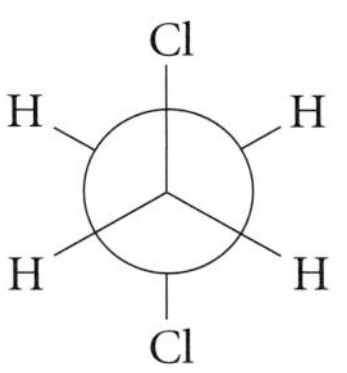

and

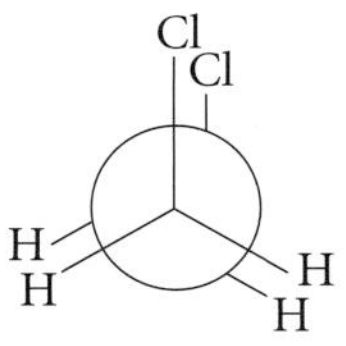

Classify them fully, according to the discussion in Section 5.10.

5.47 Two possible configurations for a molecule with three different stereogenic centers are (R,R,R) and its mirror image (S,S,S). What are all the remaining possibilities?

5.48 When (R)-2-chlorobutane is chlorinated, we obtain some 2,3-dichlorobutane. It consists of 71% *meso* isomer and 29% racemic isomers. Explain why the mixture need not be 50:50 *meso* and $(2R,3R)$-2,3-dichlorobutane.

Stereochemistry: Natural and Synthetic Applications

5.49 The formula for muscarine, a toxic constituent of poisonous mushrooms, is

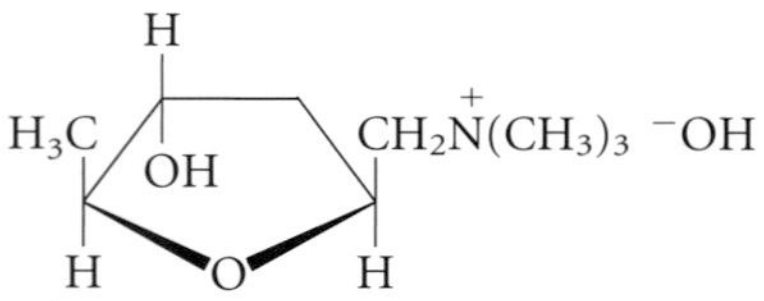

$$H_3C \overset{H}{\underset{H}{\bigvee}} \overset{OH}{\underset{O}{\diagup}} CH_2\overset{+}{N}(CH_3)_3 \ ^-OH$$

Is it chiral? How many stereoisomers of this structure are possible? An interesting murder mystery, which you might enjoy reading and which depends for its solution on the distinction between optically active and racemic forms of this poison, is Dorothy L. Sayers's *The Documents in the Case*, published in paperback by Avon Books. (See an article by H. Hart, "Accident, Suicide, or Murder? A Question of Stereochemistry," *J. Chem. Educ.*, **1975**, *52*, 444.)

5.50 Chloramphenicol is an antibiotic that is particularly effective against typhoid fever. Its structure is

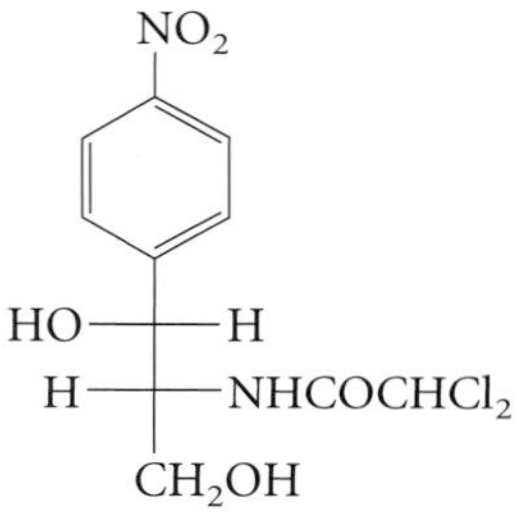

$$NO_2$$

$$HO \text{---} H$$
$$H \text{---} NHCOCHCl_2$$
$$CH_2OH$$

What is the configuration (R,S) at each stereogenic center?

5.51 Methoprene (marketed as Precor), an insect juvenile hormone mimic used in flea control products for pets, works by preventing the development of flea eggs and larvae. The effective form of methoprene, shown below, is optically active. Locate the stereogenic center and determine its configuration (R,S).

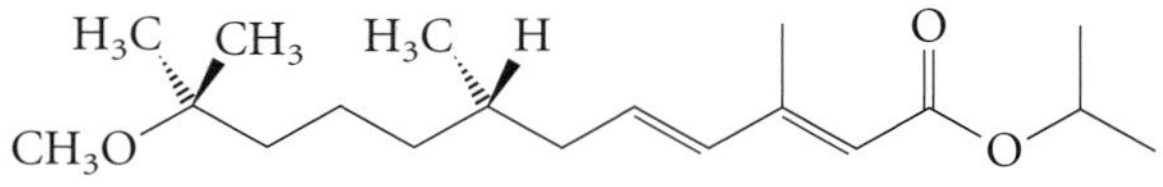

methoprene

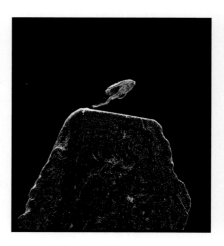

Adult cat flea jumping.

5.52 Mature crocodiles secrete from their skin glands the compound with the structure shown below. This compound is thought to be a communication pheromone for nesting or mating.

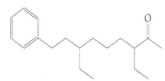

Adult crocodiles.

a. How many stereogenic centers are there in this compoud? Mark them with an asterisk.

b. Two stereoisomers of this compound have been isolated from crocodile skin gland secretions. How many possible stereoisomers of this compound are there?

5.53 Extract of *Ephedra sinica*, a Chinese herbal treatment for asthma, contains the compound ephedrine, which dilates the air passages of the lungs. The naturally occurring stereoisomer is levorotatory and has the structure shown below. (a) What is the configuration (R,S) at each stereogenic center? (b) How many stereoisomers of ephedrine are possible altogether? (c) Compare the structure of $(-)$-ephedrine to that of $(-)$-epinephrine. How are they similar and how do they differ?

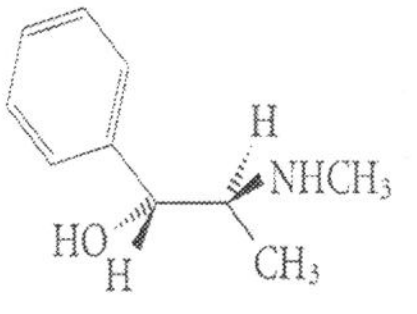

$(-)$-ephedrine 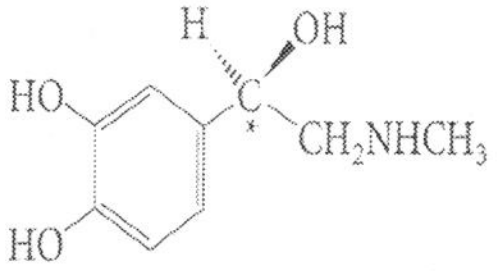 $(-)$-epinephrine
(adrenalin)

Stereochemistry and Chemical Reactions

5.54 What can you say about the stereochemistry of the products in the following reactions? (See Sec. 5.11 and eq. 3.28.)

a. 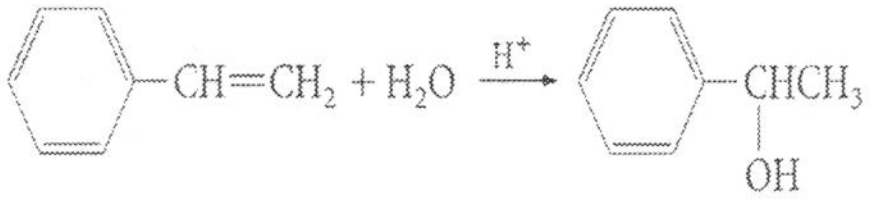 $\text{—CH}{=}\text{CH}_2 + \text{H}_2\text{O} \xrightarrow{\text{H}^+}$ —CHCH_3 | OH

b. 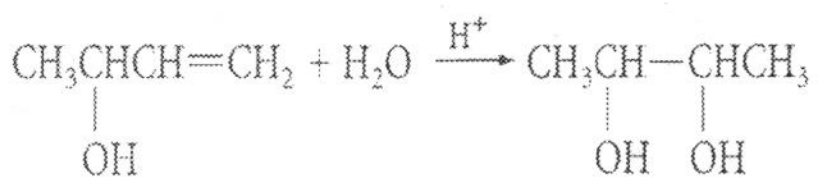 $\text{CH}_3\text{CHCH}{=}\text{CH}_2 + \text{H}_2\text{O} \xrightarrow{\text{H}^+} \text{CH}_3\text{CH—CHCH}_3$ | OH | $\text{OH} \ \text{OH}$
(*R*-enantiomer)

5.55 $(+)$- and $(-)$-Carvone (see Problem 5.37 for the structure) are enantiomers that have very different odors and are responsible for the odors of caraway seeds and spearmint, respectively. Suggest a possible explanation.

CH₃CH₂OH
ethanol

ALCOHOLS, PHENOLS, AND THIOLS

The word *alcohol* immediately brings to mind ethanol, the intoxicating compound in wine and beer. But ethanol is just one member of a family of organic compounds called alcohols that abound in nature. Naturally occuring alcohols include 2-phenylethanol, the compound responsible for the intoxicating smell of a rose; cholesterol, a tasty alcohol with which many of us have developed a love–hate relationship; sucrose, a sugar we use to satisfy our sweet tooth; and many others. In this chapter we will discuss the structural and physical properties and main chemical reactions of alcohols and their structural relatives, phenols and thiols.

$$H-\overset{..}{\underset{..}{O}}-H \qquad R-\overset{..}{\underset{..}{O}}-H \qquad Ar-\overset{..}{\underset{..}{O}}-H \qquad R-\overset{..}{\underset{..}{S}}-H \qquad Ar-\overset{..}{\underset{..}{S}}-H$$

water   an alcohol   a phenol   a thiol   a thiophenol

Alcohols have the general formula **R—OH** and are characterized by the presence of a **hydroxyl group, —OH.** They are structurally similar to water, but with one of the hydrogens replaced by an alkyl group. **Phenols** have a hydroxyl group attached directly to an aromatic ring. **Thiols** and thiophenols are similar to alcohols and phenols, except the oxygen is replaced by sulfur.

7.1 Nomenclature of Alcohols

In the IUPAC system, the hydroxyl group in alcohols is indicated by the ending **-ol.** In common names the separate word *alcohol* is placed after the name of the

▲ The alcohol ethanol is obtained from the fermentation of carbohydrates contained in fruits and grains.

alkyl group. The following examples illustrate the use of IUPAC rules, with common names given in parentheses.

$$CH_3OH \qquad CH_3CH_2OH \qquad \overset{3}{C}H_3\overset{2}{C}H_2\overset{1}{C}H_2OH \qquad \overset{1}{C}H_3\overset{2}{\underset{\underset{\displaystyle OH}{|}}{C}}H\overset{3}{C}H_3$$

methanol   ethanol   1-propanol   2-propanol
(methyl alcohol)   (ethyl alcohol)   (*n*-propyl alcohol)   (isopropyl alcohol)

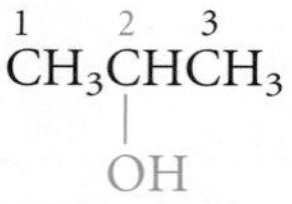

$$CH_3CH_2CH_2CH_2OH \qquad CH_3\underset{\underset{\displaystyle OH}{|}}{C}HCH_2CH_3 \qquad CH_3\underset{\underset{\displaystyle CH_3}{|}}{C}HCH_2OH \qquad CH_3-\underset{\underset{\displaystyle CH_3}{|}}{\overset{\overset{\displaystyle CH_3}{|}}{C}}-OH$$

1-butanol   2-butanol   2-methyl-1-propanol   2-methyl-2-propanol
(*n*-butyl alcohol)   (*sec*-butyl alcohol)   (isobutyl alcohol)   (*tert*-butyl alcohol)

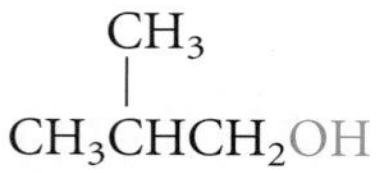

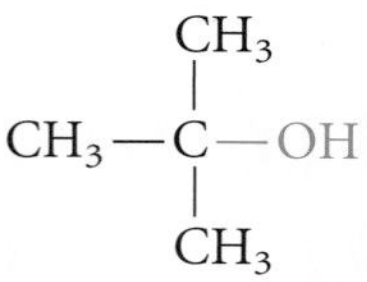

$$CH_2{=}CHCH_2OH$$

2-propen-1-ol   cyclohexanol   phenylmethanol
(allyl alcohol)   (cyclohexyl alcohol)   (benzyl alcohol)

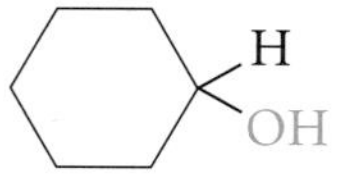

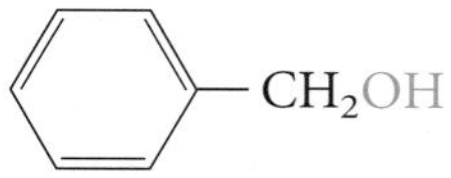

With unsaturated alcohols, two endings are needed: one for the double or triple bond and one for the hydroxyl group (see the IUPAC name for allyl alcohol). In these cases, the *-ol* suffix comes last and takes precedence in numbering.

Alcohols contain the **hydroxyl (—OH) group.** In **phenols** the hydroxyl group is attached to an aromatic ring, and in **thiols** oxygen is replaced by sulfur.

EXAMPLE 7.1

Name the following alcohols by the IUPAC system:

a. $ClCH_2CH_2OH$ b. 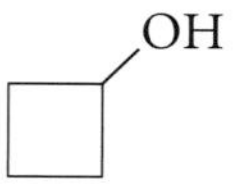 [cyclobutane with OH]   c. $CH_3C{\equiv}CCH_2CH_2OH$

Solution

a. 2-chloroethanol (number from the hydroxyl-bearing carbon)
b. cyclobutanol
c. 3-pentyne-1-ol (*not* 2-pentyne-5-ol)

PROBLEM 7.1   Name these alcohols by the IUPAC system:

a. $BrCH_2CH_2CH_2OH$ b. 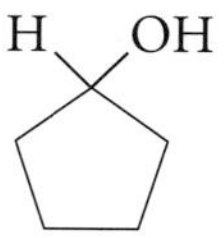 [cyclopentane with H, OH]   c. $CH_2{=}CHCH_2CH_2OH$

PROBLEM 7.2   Write a structural formula for

a. 2-pentanol.   b. 2-phenylethanol.   c. 3-pentyn-2-ol.

A   W O R D   A B O U T . . .

Industrial Alcohols

The lower alcohols (those with up to four carbon atoms) are manufactured on a large scale. They are used as raw materials for other valuable chemicals and also have important uses in their own right.

Methanol was at one time produced from wood by distillation and is still sometimes called wood alcohol. At present, however, methanol is manufactured from carbon monoxide and hydrogen.

$$CO + 2H_2 \xrightarrow[\text{400°C, 150 atm}]{ZnO-Cr_2O_3} CH_3OH$$

The U.S. production of methanol is 73 million tons per year. Most of it is used to produce formaldehyde and other chemicals, but some is used as a solvent and an antifreeze. Methanol is highly toxic and can cause permanent blindness because when taken internally it is oxidized to formaldehyde ($CH_2=O$), which binds to opsin, preventing formation of rhodopsin, the light-sensitive pigment needed for vision (see A Word About the Chemistry of Vision p. 80).

Ethanol is prepared by the fermentation of blackstrap molasses, the residue that results from the purification of cane sugar.

$$\underset{\text{cane sugar}}{C_{12}H_{22}O_{11}} + H_2O \xrightarrow{\text{yeast}} 4 \underset{\text{ethanol}}{CH_3CH_2OH} + 4 CO_2$$

The starch in grain, potatoes, and rice can be fermented similarly to produce ethanol, sometimes called grain alcohol.

Besides fermentation, ethanol is also manufactured by the acid-catalyzed hydration of ethylene (eq. 3.7). This method, using sulfuric acid or other acid catalysts, results in an annual U.S. production of 8.3 million tons.

Commercial alcohol is a constant-boiling mixture containing 95% ethanol and 5% water and cannot be further purified by distillation. To remove the remaining water to obtain **absolute alcohol,** one adds quicklime (CaO), which reacts with water to form calcium hydroxide but does not react with ethanol.

Since earlier times, ethanol has been known as an ingredient in fermented beverages (beer, wine, whiskey). The term *proof,* as used in the United States in reference to alcoholic beverages, is approximately twice the volume percentage of alcohol present. For example 100-proof whiskey contains 50% ethanol.

Ethanol is used as a solvent, as a topical antiseptic (for example, when drawing blood), and as a starting material for the manufacture of ether and ethyl esters. It also can be used as a fuel (gasohol).

2-Propanol (isopropyl alcohol) is manufactured commercially by the acid-catalyzed hydration of propene (eq. 3.13). It is the main component of rubbing alcohol. More than half the isopropyl alcohol produced (more than 1 million tons annually) is used to make acetone, by oxidation.

7.2 Classification of Alcohols

Alcohols are classified as primary (1°), secondary (2°), or tertiary (3°), depending on whether one, two, or three organic groups are connected to the hydroxyl-bearing carbon atom.

$$R-CH_2OH \qquad R-\overset{\displaystyle R}{\underset{}{C}}HOH \qquad R-\overset{\displaystyle R}{\underset{\displaystyle R}{C}}-OH$$

primary (1°)   secondary (2°)   tertiary (3°)

Methyl alcohol, which is not strictly covered by this classification, is usually grouped with the primary alcohols. This classification is similar to that for carbocations

(Sec. 3.10). We will see that the chemistry of an alcohol sometimes depends on its class.

PROBLEM 7.3   Classify as 1°, 2°, or 3° the eleven alcohols listed in Sec. 7.1.

7.3   Nomenclature of Phenols

Phenols are usually named as derivatives of the parent compounds.

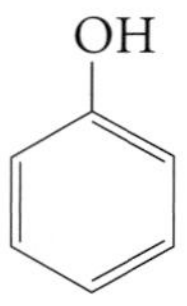

phenol    *p*-chlorophenol 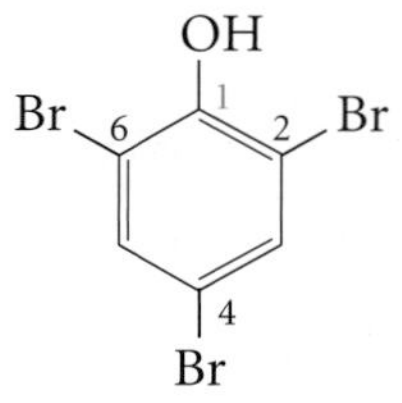   2,4,6-tribromophenol

The hydroxyl group is named as a substituent when it occurs in the same molecule with carboxylic acid, aldehyde, or ketone functionalities, which have priority in naming. Examples are

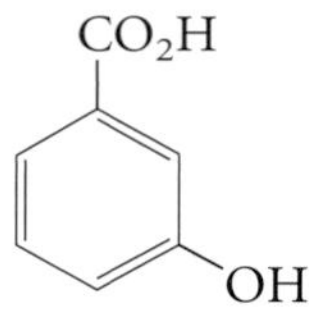

m-hydroxy   *p*-hydroxybenzaldehyde   *p*-nitrophenol
benzoic acid   (*not p*-hydroxynitrobenzene)

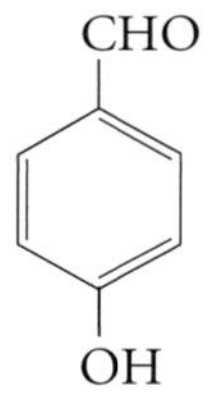

but

PROBLEM 7.4   Write the structure for

a. *p*-ethylphenol.
b. pentachlorophenol (an insecticide for termite control, and a fungicide).
c. *o*-hydroxyacetophenone (for the structure of acetophenone see Sec. 4.6).

7.4   Hydrogen Bonding in Alcohols and Phenols

The boiling points of alcohols are much higher than those of ethers or hydrocarbons with similar molecular weights.

	CH_3CH_2OH	CH_3OCH_3	$CH_3CH_2CH_3$
mol wt	46	46	44
bp	$+78.5°C$	$-24°C$	$-42°C$

Why? Because alcohols form *hydrogen bonds* with one another (see Sec. 2.7). The O—H bond is polarized by the high electronegativity of the oxygen atom. This

Table 7.1	Boiling point and water solubility of some alcohols		
Name	**Formula**	**bp, °C**	**Solubility in H_2O g/100 g at 20°C**
methanol	CH_3OH	65	completely miscible
ethanol	CH_3CH_2OH	78.5	completely miscible
1-propanol	$CH_3CH_2CH_2OH$	97	completely miscible
1-butanol	$CH_3CH_2CH_2CH_2OH$	117.7	7.9
1-pentanol	$CH_3CH_2CH_2CH_2CH_2OH$	137.9	2.7
1-hexanol	$CH_3CH_2CH_2CH_2CH_2CH_2OH$	155.8	0.59

polarization places a partial positive charge on the hydrogen atom and a partial negative charge on the oxygen atom. Because of its small size and partial positive charge, the hydrogen atom can link two electronegative atoms such as oxygen.

$$
\underset{\text{two separate alcohol molecules}}{\overset{R}{\underset{\delta-}{O}}\overset{}{-}\overset{\delta+}{H} \;+\; \overset{R}{\underset{\delta-}{O}}\overset{}{-}\overset{\delta+}{H}} \;\rightleftharpoons\; \underset{\text{a hydrogen bond}}{\overset{R}{\underset{\delta-}{O}}\overset{}{-}\overset{\delta+}{H} \text{---} \overset{R}{\underset{\delta-}{O}}\overset{}{-}\overset{\delta+}{H}} \tag{7.1}
$$

Two or more alcohol molecules thus become loosely bonded to one another through hydrogen bonds.

Hydrogen bonds are weaker than ordinary covalent bonds.* Nevertheless, their strength is significant, about 5 to 10 kcal/mol (20 to 40 kJ/mol). Consequently, alcohols and phenols have relatively high boiling points because we must not only supply enough heat (energy) to vaporize each molecule but must also supply enough heat to break the hydrogen bonds before each molecule can be vaporized.

Water, of course, is also a hydrogen-bonded liquid (see Figure 2.2). The lower-molecular-weight alcohols can readily replace water molecules in the hydrogen-bonded network.

This accounts for the complete miscibility of the lower alcohols with water. However, as the organic chain lengthens and the alcohol becomes relatively more hydrocarbon-like, its water solubility decreases. Table 7.1 illustrates these properties.

7.5 Acidity and Basicity Reviewed

The acid–base behavior of organic compounds often helps to explain their chemistry; this is certainly true of alcohols. It is a good idea, therefore, to review the fundamental concepts of acidity and basicity.

---

*Covalent O—H bond strengths are about 120 kcal/mol (480 kJ/mol).

A **Brønsted–Lowry acid** is a proton donor, whereas a **Brønsted–Lowry base** is a proton acceptor.

Acids and **bases** are defined in two ways. According to the **Brønsted–Lowry definition**, an acid is a proton donor, and a base is a proton acceptor. For example, in eq. 7.2, which represents what occurs when hydrogen chloride dissolves in water, the water accepts a proton from the hydrogen chloride.

$$\overset{\cdot\cdot}{H-O:} + H-\overset{\cdot\cdot}{\underset{\cdot\cdot}{Cl}}: \; \rightleftharpoons \; H-\overset{\cdot\cdot}{\overset{\pm}{O}}-H \; + \; :\overset{\cdot\cdot}{\underset{\cdot\cdot}{Cl}}:^- \qquad (7.2)$$

$$\underset{\text{base}}{\underset{|}{H}} \qquad \text{acid} \qquad \underset{\substack{\text{conjugate} \\ \text{acid of} \\ \text{water}}}{\underset{|}{H}} \qquad \underset{\substack{\text{conjugate} \\ \text{base of} \\ \text{hydrogen} \\ \text{chloride}}}{}$$

Here water acts as a base or proton acceptor, and hydrogen chloride acts as an acid or proton donor. The products of this proton exchange are called the *conjugate acid* and the *conjugate base*.

The **acidity** (or **ionization**) **constant, K_a,** of an acid is a quantitative measure of its strength.

The strength of an acid is measured quantitatively by its **acidity constant,** or **ionization constant, K_a.** For example, any acid dissolved in water is in equilibrium with hydronium ions and its conjugate base A^-:

$$HA + H_2O \rightleftharpoons H_3O^+ + A^- \qquad (7.3)$$

K_a is related to the equilibrium constant (see p. 90) for this reaction and is defined as follows*:

$$K_a = \frac{[H_3O^+][A^-]}{[HA]} \qquad (7.4)$$

The stronger the acid, the more this equilibrium is shifted to the right, thus increasing the concentration of H_3O^+ and the value of K_a.

For water, these expressions are

$$H_2O + H_2O \rightleftharpoons H_3O^+ + HO^- \qquad (7.5)$$

and

$$K_a = \frac{[H_3O^+][HO^-]}{[H_2O]} = 1.8 \times 10^{-16} \qquad (7.6)$$

PROBLEM 7.5 Verify from eq. 7.6 and from the molarity of water (55.5 *M*) that the concentrations of both H_3O^+ and HO^- in water are 10^{-7} moles per liter.

The **pK_a** of an acid is the negative logarithm of the acidity constant.

To avoid using numbers with negative exponents, such as those we have just seen for the acidity constant K_a for water, we often express acidity as **pK_a,** the negative logarithm of the acidity constant.

$$pK_a = -\log K_a \qquad (7.7)$$

---

*The square brackets used in the expression for K_a indicate concentration, at equilibrium, of the enclosed species in moles per liter. The acidity constant K_a is related to the equilibrium constant for the reaction shown in eq. 7.3; only the concentration of water $[H_2O]$ is omitted from the denominator of the expression since it remains nearly constant at 55.5 *M*, very large compared to the concentrations of the other three species. For a discussion of reaction equilibria and equilibrium constants, see Section 3.11.

The pK_a of water is

$$-\log(1.8 \times 10^{-16}) = -\log 1.8 - \log 10^{-16} = -0.26 + 16 = +15.74$$

The mathematical relationship between the values for K_a and pK_a means that *the smaller K_a or the larger pK_a, the weaker the acid.*

It is useful to keep in mind that there is an inverse relationship between the strength of an acid and the strength of its conjugate base. In eq. 7.2, for example, hydrogen chloride is a *strong* acid since the equilibrium is shifted largely to the right. It follows that the chloride ion must be a *weak* base, since it has relatively little affinity for a proton. Similarly, since water is a *weak* acid, its conjugate base, hydroxide ion, must be a *strong* base.

Another way to define acids and bases was first proposed by G. N. Lewis. A **Lewis acid** is a substance that can accept an electron pair, and a **Lewis base** is a substance that can donate an electron pair. According to this definition, a proton is considered a Lewis acid because it can accept an electron pair from a donor (a Lewis base) to fill its 1s shell.

A **Lewis acid** is an electron pair acceptor; a **Lewis base** is an electron pair donor.

$$H^+ + :\overset{\displaystyle ..}{\underset{\displaystyle H}{O}}{-}H \rightleftharpoons H{-}\overset{+ ..}{\underset{\displaystyle H}{O}}{-}H \qquad (7.8)$$

Lewis acid   Lewis base

Any atom with an unshared electron pair can act as a Lewis base.

Compounds with an element whose valence shell is incomplete also act as Lewis acids. For example,

$$\underset{\displaystyle F}{\overset{\displaystyle F}{F{-}B}} + :\!\overset{..}{\underset{..}{F}}\!:^- \rightleftharpoons \underset{\displaystyle F}{\overset{\displaystyle F}{F{-}B{-}}}\overset{..}{\underset{..}{F}}: \qquad (7.9)$$

Lewis acid   Lewis base

Similarly, when $FeCl_3$ or $AlCl_3$ acts as a catalyst for electrophilic aromatic chlorination (eqs. 4.13 and 4.14) or the Friedel–Crafts reaction (eqs. 4.20 and 4.22), they are acting as Lewis acids; the metal atom accepts an electron pair from chlorine or from an alkyl or acyl chloride to complete its valence shell of electrons.

Finally, some substances can act as either an acid or a base, depending on the other reactant. For example, in eq. 7.2, water acts as a base (a proton acceptor). However, in its reaction with ammonia, water acts as an acid (a proton donor).

$$:\overset{\displaystyle ..}{\underset{\displaystyle H}{O}}{-}H + :NH_3 \rightleftharpoons H{-}\overset{..}{\underset{..}{O}}:^- + H{-}\overset{+}{N}H_3 \qquad (7.10)$$

water   ammonia   hydroxide ion   ammonium ion
(acid)   (base)   (conjugate base)   (conjugate acid)

Water acts as a base toward acids that are stronger than itself (HCl) and as an acid toward bases that are stronger than itself (NH_3). Substances that can act as either an acid or a base are said to be **amphoteric.**

An **amphoteric** substance can act as an acid or as a base.

PROBLEM 7.6   The K_a for ethanol is 1.0×10^{-16}. What is its pK_a?

PROBLEM 7.7   The pK_a's of hydrogen cyanide and acetic acid are 9.2 and 4.7, respectively. Which is the stronger acid?

PROBLEM 7.8   Which of the following are Lewis acids and which are Lewis bases?

a. $(CH_3)_3C:^-$ b. $(CH_3)_3B$ c. Zn^{2+}
d. CH_3OCH_3 e. $(CH_3)_3C^+$ f. CH_3NH_2
g. $(CH_3)_3N$ h. $H:^-$ i. Mg^{2+}

PROBLEM 7.9   In eq. 3.53, is the amide ion, NH_2^-, functioning as an acid or as a base?

7.6   The Acidity of Alcohols and Phenols

Like water, alcohols and phenols are weak acids. The hydroxyl group can act as a proton donor, and dissociation occurs in a manner similar to that for water:

$$\overset{..}{\underset{..}{RO}}-H \;\rightleftharpoons\; \overset{..}{\underset{..}{RO}}:^- + H^+ \qquad\qquad \textbf{(7.11)}$$

alcohol   alkoxide
ion

The conjugate base of an alcohol is an alkoxide ion.

The conjugate base of an alcohol is an **alkoxide ion** (for example, *meth*oxide ion from *meth*anol, *eth*oxide ion from *eth*anol, and so on).

Table 7.2 lists pK_a values for selected alcohols and phenols.* Methanol and ethanol have nearly the same acid strength as water; bulky alcohols such as *t*-butyl alcohol are somewhat weaker because their bulk makes it difficult to solvate the corresponding alkoxide ion.

Phenol is a much stronger acid than ethanol. How can we explain this acidity difference between alcohols and phenols, since in both types of compounds the proton donor is a hydroxyl group?

Phenols are stronger acids than alcohols mainly because the corresponding phenoxide ions are stabilized by resonance. The negative charge of an alkoxide ion is concentrated on the oxygen atom, but the negative charge on a phenoxide ion can be delocalized to the *ortho* and *para* ring positions through resonance.

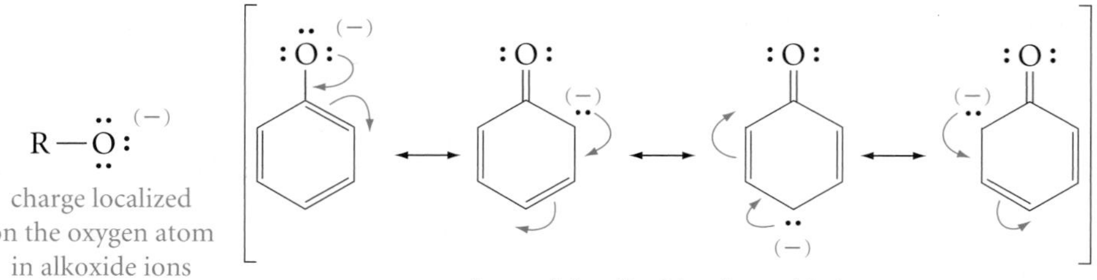

charge delocalized in phenoxide ion

R—$\overset{..}{\underset{..}{O}}$: $^{(-)}$
charge localized
on the oxygen atom
in alkoxide ions

Since phenoxide ions are stabilized in this way, the equilibrium for their formation is more favorable than that for alkoxide ions. Thus, phenols are stronger acids than alcohols.

*To compare the acidity of alcohols and phenols with that of other organic compounds, see Table C in the Appendix.

Table 7.2	pKₐ's of selected alcohols and phenols in aqueous solution	

Name	Formula	pK_a
water	HO—H	15.7
methanol	CH_3O—H	15.5
ethanol	CH_3CH_2O—H	15.9
t-butyl alcohol	$(CH_3)_3CO$—H	18
2,2,2-trifluoroethanol	CF_3CH_2O—H	12.4
phenol	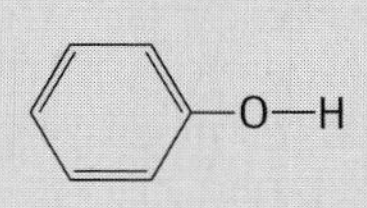 ⟨benzene ring⟩—O—H	10.0
p-nitrophenol	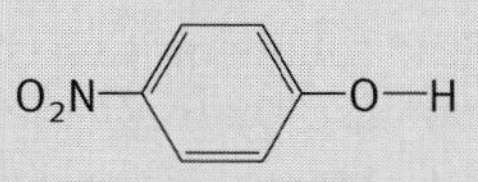 O_2N—⟨benzene ring⟩—O—H	7.2
picric acid	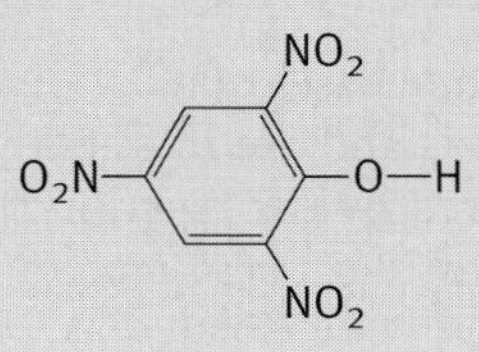 O_2N—⟨benzene ring with NO_2 groups⟩—O—H	0.25

We see in Table 7.2 that 2,2,2-trifluoroethanol is a much stronger acid than ethanol. How can we explain this effect of fluorine substitution? Again, think about the stabilities of the respective anions. Fluorine is a strongly electronegative element, so each C—F bond is polarized, with the fluorine partially negative and the carbon partially positive.

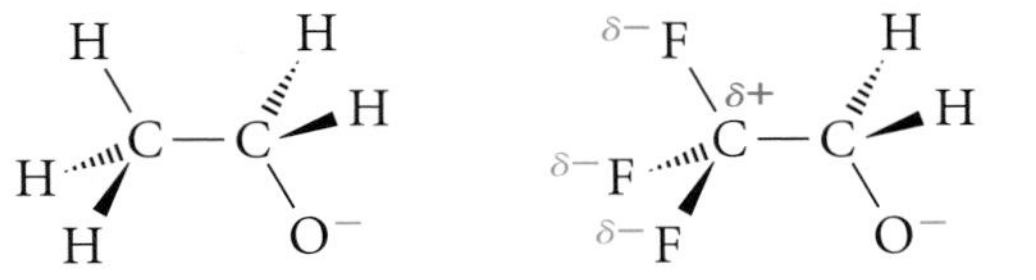

ethoxide ion   2,2,2-trifluoroethoxide ion

The positive charge on the carbon is located near the negative charge on the nearby oxygen atom, where it can partially neutralize and hence stabilize it. This **inductive effect,** as it is called, is absent in ethoxide ion.

The acidity-increasing effect of fluorine seen here is not a special case, but a general phenomenon. *All electron-withdrawing groups increase acidity* by stabilizing the conjugate base. *Electron-donating groups decrease acidity* because they destabilize the conjugate base.

Here is another example. *p*-Nitrophenol (Table 7.2) is a much stronger acid than phenol. In this case, the nitro group acts in two ways to stabilize the *p*-nitrophenoxide ion.

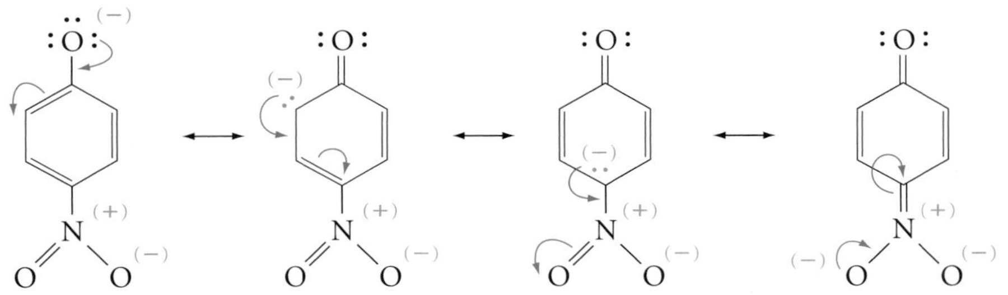

Polar bonds that place a partial positive charge near the negative charge on an alkoxide ion stabilize the ion by an **inductive effect.**

I   II   III   IV

p-nitrophenoxide ion resonance contributors

First, the nitrogen atom has a formal positive charge and is therefore strongly electron withdrawing. It therefore increases the acidity of *p*-nitrophenol through the inductive effect. Second, the negative charge on the oxygen of the hydroxyl group can be delocalized through resonance, not only to the *ortho* and *para* ring carbons, as in phenoxide itself, but to the oxygen atoms of the nitro group as well (structure IV). Both the inductive and the resonance effects of the nitro group are acid-strengthening.

Additional nitro groups on the benzene ring further increase phenolic acidity. Picric acid (2,4,6-trinitrophenol) is an even stronger acid than *p*-nitrophenol.

PROBLEM 7.10   Draw the resonance contributors for the 2,4,6-trinitrophenoxide (picrate) ion, and show that the negative charge can be delocalized to every oxygen atom.

PROBLEM 7.11   Rank the following five compounds in order of increasing acid strength: 2-chloroethanol, *p*-chlorophenol, *p*-methylphenol, ethanol, phenol.

Alkoxides, the conjugate bases of alcohols, are strong bases just like hydroxide ion. They are ionic compounds and are frequently used as strong bases in organic chemistry. They can be prepared by the reaction of an alcohol with sodium or potassium metal or with a metal hydride. These reactions proceed irreversibly to give the metal alkoxides that can frequently be isolated as white solids.

$$2\ \text{R\ddot{O}-H} + 2\ \text{K} \longrightarrow 2\ \text{R\ddot{O}:}^- \text{K}^+ + \text{H}_2 \tag{7.12}$$

$$\underset{\text{alcohol}}{} \qquad\qquad \underset{\substack{\text{potassium} \\ \text{alkoxide}}}{}$$

$$\text{R\ddot{O}-H} + \text{NaH} \longrightarrow \text{R\ddot{O}:}^- \text{Na}^+ + \text{H-H} \tag{7.13}$$

$$\qquad\quad \underset{\substack{\text{sodium} \\ \text{hydride}}}{} \qquad \underset{\substack{\text{sodium} \\ \text{alkoxide}}}{}$$

PROBLEM 7.12   Write the equation for the reaction of *t*-butyl alcohol with potassium metal. Name the product.

Ordinarily, treatment of alcohols with sodium hydroxide does not convert them to their alkoxides. This is because alkoxides are stronger bases than hydroxide ion, so the reaction goes in the reverse direction. Phenols, however, can be converted to phenoxide ions in this way.

$$\text{ROH} + \text{Na}^+\text{HO}^- \;\rlap{\,/}{\rightleftharpoons}\; \text{RO}^-\text{Na}^+ + \text{H}_2\text{O} \tag{7.14}$$

$$\underset{\text{phenol}}{\text{\hexagon}-\text{OH}} + \text{Na}^+\text{HO}^- \longrightarrow \underset{\text{sodium phenoxide}}{\text{\hexagon}-\text{O}^-\text{Na}^+} + \text{HOH} \tag{7.15}$$

PROBLEM 7.13   Write an equation for the reaction, if any, between

a. *p*-nitrophenol and aqueous potassium hydroxide.
b. cyclohexanol and aqueous potassium hydroxide.

7.7 The Basicity of Alcohols and Phenols

Alcohols (and phenols) function not only as weak acids but also as weak bases. They have unshared electron pairs on the oxygen and are therefore Lewis bases. They can be protonated by strong acids. The product, analogous to the oxonium ion H_3O^+, is an alkyloxonium ion.

$$R-\overset{..}{\underset{..}{O}}-H \;+\; H^+ \;\rightleftharpoons\; R-\overset{\overset{\displaystyle H}{|}}{\underset{\underset{\displaystyle +}{..}}{O}}-H \qquad (7.16)$$

alcohol acting   alkyloxonium ion
as a base

This protonation is the first step in two important reactions of alcohols that are discussed in the following two sections: their dehydration to alkenes and their conversion to alkyl halides.

7.8 Dehydration of Alcohols to Alkenes

Alcohols can be dehydrated by heating them with a strong acid. For example, when ethanol is heated at 180°C with a small amount of concentrated sulfuric acid, a good yield of ethylene is obtained.

$$H-CH_2CH_2-OH \xrightarrow{\;H^+,\,180°C\;} CH_2{=}CH_2 \;+\; H-OH \qquad (7.17)$$

ethanol   ethylene

This type of reaction, which can be used to prepare alkenes, is the reverse of hydration (Sec. 3.7.b). It is an *elimination reaction* and can occur by either an E1 or an E2 mechanism, depending on the class of the alcohol.

Tertiary alcohols dehydrate by the E1 mechanism. *t*-Butyl alcohol is a typical example. The first step involves rapid and reversible protonation of the hydroxyl group.

$$(CH_3)_3C-\overset{..}{\underset{..}{O}}H \;+\; H^+ \;\rightleftharpoons\; (CH_3)_3C-\overset{\overset{\displaystyle +}{\overset{\displaystyle ..}{}}}{\underset{\underset{\displaystyle H}{|}}{O}}-H \qquad (7.18)$$

Ionization (the rate-determining step), with water as the leaving group, occurs readily because the resulting carbocation is tertiary.

$$(CH_3)_3C\overset{\overset{\displaystyle +}{\overset{\displaystyle ..}{}}}{\underset{\underset{\displaystyle H}{|}}{O}}-H \;\rightleftharpoons\; (CH_3)_3C^+ \;+\; H_2O \qquad (7.19)$$

t-butyl cation

Proton loss from a carbon atom adjacent to the positive carbon completes the reaction.

$$\underset{\underset{\displaystyle CH_3}{|}}{\overset{\overset{\displaystyle H \qquad CH_3}{|\qquad\;\; |}}{CH_2-C^+}} \longrightarrow CH_2{=}C\overset{\diagup CH_3}{\diagdown CH_3} \;+\; H^+ \qquad (7.20)$$

The overall dehydration reaction is the sum of all three steps.

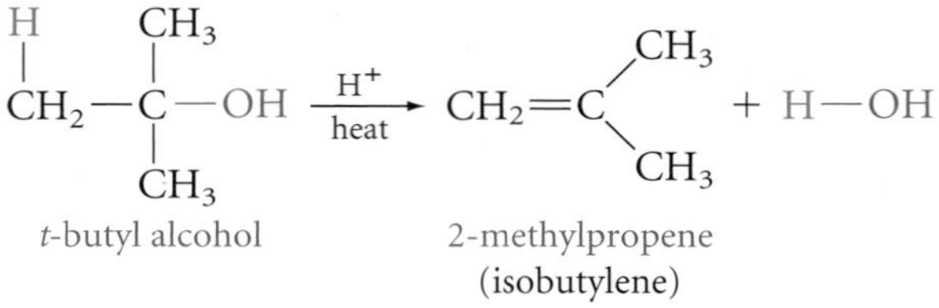

$$\underset{\substack{| \\ CH_3}}{\overset{\substack{H \quad CH_3 \\ | \quad\;\; |}}{CH_2-C-OH}} \xrightarrow[\text{heat}]{H^+} CH_2{=}C\underset{CH_3}{\overset{CH_3}{<}} + H-OH \qquad (7.21)$$

t-butyl alcohol   2-methylpropene
(isobutylene)

With a primary alcohol, a primary carbocation intermediate is avoided by combining the last two steps of the mechanism. The loss of water and an adjacent proton occur simultaneously in an E2 mechanism.

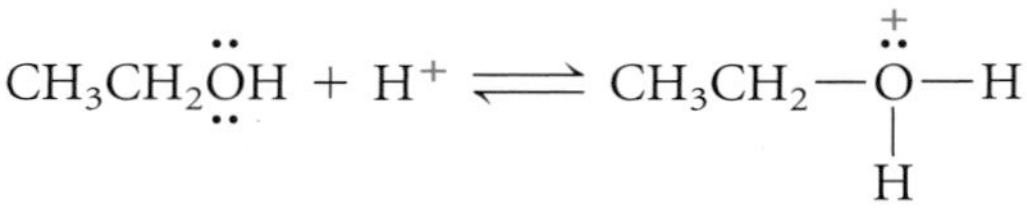

$$CH_3CH_2\overset{..}{\underset{..}{O}}H + H^+ \;\rightleftharpoons\; CH_3CH_2-\overset{+}{\underset{|}{\overset{..}{O}}}-H \qquad (7.22)$$
$$\phantom{CH_3CH_2\overset{..}{\underset{..}{O}}H + H^+ \;\rightleftharpoons\; CH_3CH_2-}H$$

$$\underset{\substack{| \\ H}}{\overset{\substack{H \\ |}}{CH_2-CH_2-\overset{+}{\underset{|}{\overset{..}{O}}}-H}} \longrightarrow CH_2{=}CH_2 + H^+ + H_2O \qquad (7.23)$$

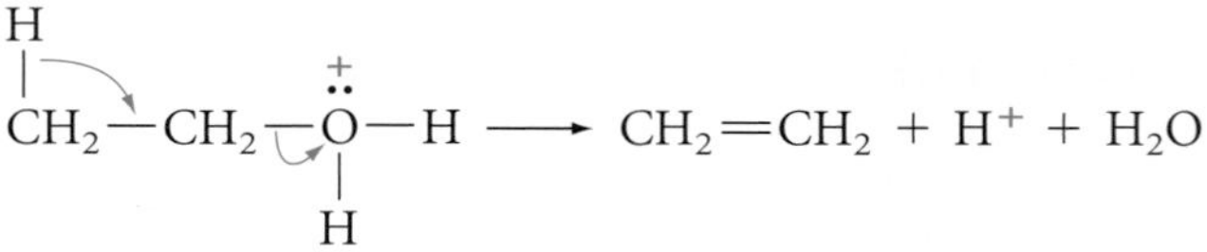

The important things to remember about alcohol dehydrations are that (1) they all begin by protonation of the hydroxyl group (that is, the alcohol acts as a base) and (2) the ease of alcohol dehydration is 3° > 2° > 1° (the same as the order of carbocation stability).

Sometimes a single alcohol gives two or more alkenes because the proton lost during dehydration can come from any carbon atom that is directly attached to the hydroxyl-bearing carbon. For example, 2-methyl-2-butanol can give two alkenes.

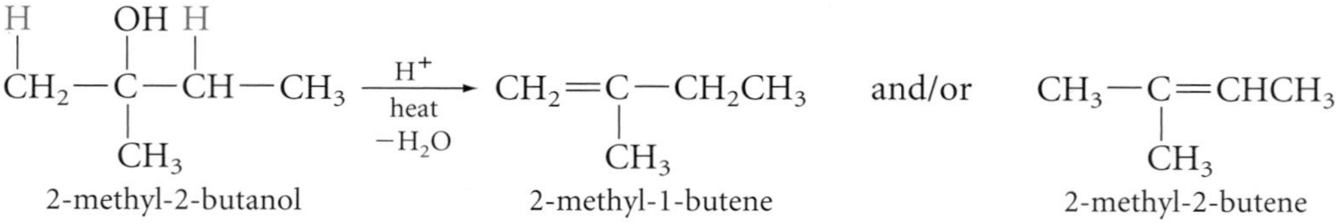

$$\underset{\substack{| \\ CH_3}}{\overset{\substack{H \quad OH\; H \\ | \quad\;\; |\;\;\; |}}{CH_2-C-CH-CH_3}} \xrightarrow[\substack{\text{heat} \\ -H_2O}]{H^+} \underset{\substack{| \\ CH_3}}{CH_2{=}C-CH_2CH_3} \quad \text{and/or} \quad \underset{\substack{| \\ CH_3}}{CH_3-C{=}CHCH_3} \qquad (7.24)$$

2-methyl-2-butanol   2-methyl-1-butene   2-methyl-2-butene

In these cases, *the alkene with the most substituted double bond usually predominates.* By "most substituted" we mean the alkene with the greatest number of alkyl groups on the doubly bonded carbons. Thus, in the example shown, the major product is 2-methyl-2-butene.

PROBLEM 7.14 Write the structure for all possible dehydration products of

   a. 3-methyl-3-pentanol.   b. 1-methylcyclohexanol.

In each case, which product do you expect to predominate?

7.9 The Reaction of Alcohols with Hydrogen Halides

Alcohols react with hydrogen halides (HCl, HBr, and HI) to give alkyl halides (chlorides, bromides, and iodides).

$$R\text{—OH} + H\text{—X} \longrightarrow \underset{\text{alkyl halide}}{R\text{—X}} + H\text{—OH} \qquad (7.25)$$
$$\underset{\text{alcohol}}{}$$

This substitution reaction provides a useful general route to alkyl halides. Because halide ions are good nucleophiles, we obtain mainly substitution products instead of dehydration. Once again, the reaction rate and mechanism depend on the class of alcohol (tertiary, secondary, or primary).

Tertiary alcohols react the fastest. For example, we can convert *t*-butyl alcohol to *t*-butyl chloride simply by shaking it for a few minutes at room temperature (rt) with concentrated hydrochloric acid.

$$\underset{\text{t-butyl alcohol}}{(CH_3)_3COH} + H\text{—Cl} \xrightarrow[\text{15 min}]{\text{rt}} \underset{\text{t-butyl chloride}}{(CH_3)_3C\text{—Cl}} + H\text{—OH} \qquad (7.26)$$

The reaction occurs by an S_N1 mechanism and involves a carbocation intermediate. The first two steps in the mechanism are identical to those shown in eqs. 7.18 and 7.19. The final step involves capture of the *t*-butyl carbocation by chloride ion.

$$(CH_3)_3C^+ + Cl^- \xrightarrow{\text{fast}} (CH_3)_3CCl \qquad (7.27)$$

On the other hand, 1-butanol, a primary alcohol, reacts slowly and must be heated for several hours with a mixture of concentrated hydrochloric acid and a Lewis acid catalyst such as zinc chloride to accomplish the same type of reaction.

$$\underset{\text{1-butanol}}{CH_3CH_2CH_2CH_2OH} + H\text{—Cl} \xrightarrow[\text{several hours}]{\text{heat, ZnCl}_2} \underset{\text{1-chlorobutane}}{CH_3CH_2CH_2CH_2\text{—Cl}} + H\text{—OH} \qquad (7.28)$$

The reaction occurs by an S_N2 mechanism. In the first step, the alcohol is protonated by the acid.

$$CH_3CH_2CH_2CH_2\text{—}\overset{..}{\underset{..}{O}}H + H^+ \rightleftharpoons CH_3CH_2CH_2CH_2\text{—}\overset{\overset{+}{..}}{O}\text{—}H \qquad (7.29)$$
$$\underset{\phantom{CH_3CH_2CH_2CH_2\text{—}}}{\underset{H}{|}}$$

In the second step, chloride ion displaces water in a typical S_N2 process. The zinc chloride is a good Lewis acid and can serve the same role as a proton in sharing an electron pair of the hydroxyl oxygen. It also increases the chloride ion concentration, thus speeding up the S_N2 displacement.

$$CH_3CH_2CH_2 \atop Cl^- \quad C\text{—}\overset{+}{\overset{..}{O}}\text{—}H \longrightarrow CH_3CH_2CH_2CH_2Cl + H_2O \qquad (7.30)$$

Secondary alcohols react at intermediate rates by both S_N1 and S_N2 mechanisms.

EXAMPLE 7.2

Explain why *t*-butyl alcohol reacts at equal rates with HCl, HBr, and HI (to form, in each case, the corresponding *t*-butyl halide).

Solution   *t*-Butyl alcohol is a tertiary alcohol; thus it reacts by an S_N1 mechanism. As in all S_N1 reactions, the rate-determining step involves formation of a carbocation, in this case the *t*-butyl carbocation. The rate of this step does not depend on which acid is used, so all of the reactions proceed at equal rates.

$$(CH_3)_3COH + H^+ \rightleftharpoons (CH_3)_3C\overset{\cdot\cdot}{\underset{\underset{H}{|^+}}{O}}{-}H \xrightarrow[\text{slow step}]{S_N1} (CH_3)_3C^+ + H_2O$$
$$\text{\small{\textit{t}-butyl cation}}$$

The reaction of the carbocation with Cl^-, Br^-, or I^- is then fast.

PROBLEM 7.15   Explain why 1-butanol reacts with hydrogen halides in the rate order HI > HBr > HCl (to form, in each case, the corresponding butyl halide).

PROBLEM 7.16   Write equations for reactions of the following alcohols with concentrated HBr.

a. [structure: cyclohexane ring with CH₃ and OH substituents]   b. [structure: branched chain with OH]

7.10   Other Ways to Prepare Alkyl Halides from Alcohols

Since alkyl halides are extremely useful in synthesis, it is not surprising that chemists have devised several ways to prepare them from alcohols. For example, **thionyl chloride** (eq. 7.31) reacts with alcohols to give alkyl chlorides. The alcohol is first converted to a chlorosulfite ester intermediate, a step that converts the hydroxyl group into a good leaving group. This is followed by a nucleophilic substitution whose mechanism (S_N1 or S_N2) depends on whether the alcohol is primary, secondary, or tertiary.

$$R{-}OH + Cl{-}\overset{O}{\overset{||}{S}}{-}Cl \xrightarrow{\text{heat}} \left[R{-}O{-}\overset{O}{\overset{||}{S}}{-}Cl \right] \longrightarrow R{-}Cl + \overset{O}{\underset{\underset{O}{||}}{S}}\uparrow + HCl\uparrow \qquad \textbf{(7.31)}$$
$$\text{\small thionyl chloride} \qquad \text{\small chlorosulfite ester} \qquad \qquad \text{\small intermediate}$$

One advantage of this method is that two of the reaction products, hydrogen chloride and sulfur dioxide, are gases and evolve from the reaction mixture (indicated by the upward pointing arrows), leaving behind only the desired alkyl chloride. The method is not effective, however, for preparing low-boiling alkyl chlorides (in which R has only a few carbon atoms), because they easily boil out of the reaction mixture with the gaseous products.

Phosphorus halides (eq. 7.32) also convert alcohols to alkyl halides.

$$3\,ROH + PX_3 \longrightarrow 3\,RX + H_3PO_3\,(X = Cl\ or\ Br) \qquad \textbf{(7.32)}$$
$$\text{\small phosphorus halide}$$

In this case, the other reaction product, phosphorous acid, has a rather high boiling point. Thus, the alkyl halide is usually the lowest boiling component of the reaction mixture and can be isolated by distillation.

Both of these methods are used mainly with primary and secondary alcohols, whose reaction with hydrogen halides is slow.

PROBLEM 7.17 Write balanced equations for the preparation of the following alkyl halides from the corresponding alcohol and either $SOCl_2$, PCl_3 or PBr_3.

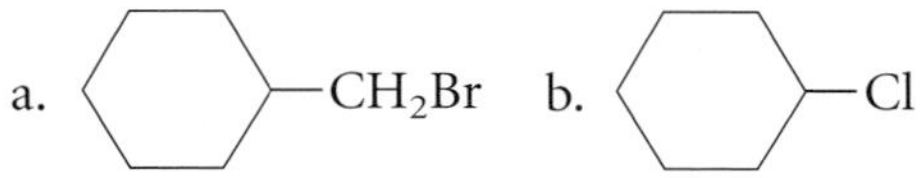

a. ⟨benzene ring⟩—CH_2Br b. ⟨benzene ring⟩—Cl

7.11 A Comparison of Alcohols and Phenols

Because they have the same functional group, alcohols and phenols have many similar properties. But whereas it is relatively easy, with acid catalysis, to break the C—OH bond of alcohols, this bond is difficult to break in phenols. Protonation of the phenolic hydroxyl group can occur, but loss of a water molecule would give a phenyl cation.

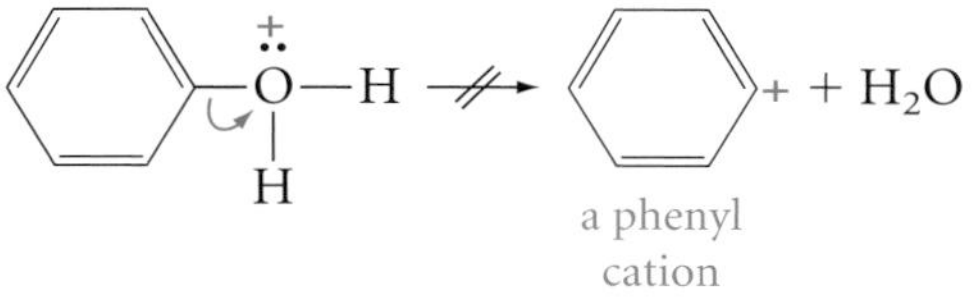

$$\text{⟨benzene ring⟩}-\overset{\overset{+}{\cdot\cdot}}{O}-H \xrightarrow{\;/\!\!/\;} \text{⟨benzene ring⟩}^+ + H_2O \qquad (7.33)$$

a phenyl
cation

With only two attached groups, the positive carbon in a phenyl cation should be *sp*-hybridized and linear. But this geometry is prevented by the structure of the benzene ring, so *phenyl cations are exceedingly difficult to form.* Consequently, phenols cannot undergo replacement of the hydroxyl group by an S_N1 mechanism. Neither can phenols undergo displacement by the S_N2 mechanism. (The geometry of the ring makes the usual inversion mechanism impossible.) Therefore, hydrogen halides, phosphorus halides, or thionyl halides cannot readily cause replacement of the hydroxyl group by halogens in phenols.

PROBLEM 7.18 Compare the reactions of cyclopentanol and phenol with

a. HBr.   b. H_2SO_4, heat.

7.12 Oxidation of Alcohols to Aldehydes, Ketones, and Carboxylic Acids

Alcohols with at least one hydrogen attached to the hydroxyl-bearing carbon can be oxidized to carbonyl compounds. Primary alcohols give aldehydes, which may be further oxidized to carboxylic acids. Secondary alcohols give ketones. Notice that as an alcohol is oxidized to an aldehyde or ketone and then to a carboxylic acid, the number of bonds between the reactive carbon atom and oxygen atoms increases from one to two to three. In other words, we say that the oxidation state of that carbon increases as we go from an alcohol to an aldehyde or ketone to a carboxylic acid.

$$
\underset{\text{primary alcohol}}{\overset{\displaystyle \text{OH}}{\underset{\displaystyle \text{H}}{R-C-H}}} \xrightarrow[\text{agent}]{\text{oxidizing}} \underset{\text{aldehyde}}{\overset{\displaystyle O}{R-C-H}} \xrightarrow[\text{agent}]{\text{oxidizing}} \underset{\text{acid}}{\overset{\displaystyle O}{R-C-OH}} \tag{7.34}
$$

$$
\underset{\text{secondary alcohol}}{\overset{\displaystyle \text{OH}}{\underset{\displaystyle \text{H}}{R-C-R'}}} \xrightarrow[\text{agent}]{\text{oxidizing}} \underset{\text{ketone}}{\overset{\displaystyle O}{R-C-R'}} \tag{7.35}
$$

Tertiary alcohols, having no hydrogen atom on the hydroxyl-bearing carbon, do not undergo this type of oxidation.

A common laboratory oxidizing agent for alcohols is chromic anhydride, CrO_3, dissolved in aqueous sulfuric acid (**Jones' reagent**). Acetone is used as a solvent in such oxidations. Typical examples are

Jones' reagent is an oxidizing agent composed of CrO_3 dissolved in aqueous H_2SO_4.

$$
\underset{\text{cyclohexanol}}{\text{(structure)}\overset{\text{H}}{\underset{\text{OH}}{}}} \xrightarrow[\substack{\text{H}^+,\,\text{acetone} \\ \text{(Jones' reagent)}}]{CrO_3} \underset{\text{cyclohexanone}}{\text{(structure)}=O} \tag{7.36}
$$

$$
\underset{\text{1-octanol}}{CH_3(CH_2)_6CH_2OH} \xrightarrow[\text{reagent}]{\text{Jones'}} \underset{\text{octanoic acid}}{CH_3(CH_2)_6CO_2H} \tag{7.37}
$$

With primary alcohols, oxidation can be stopped at the aldehyde stage by special reagents, such as **pyridinium chlorochromate (PCC),** shown in eq. 7.38.

$$
\underset{\text{1-octanol}}{CH_3(CH_2)_6CH_2OH} \xrightarrow[CH_2Cl_2,\,25^\circ C]{PCC} \underset{\text{octanal}}{\overset{\displaystyle O}{CH_3(CH_2)_6C-H}} \tag{7.38}
$$

PCC is prepared by dissolving CrO_3 in hydrochloric acid and then adding pyridine:

$$
CrO_3 + HCl + \underset{\text{pyridine}}{\text{(structure)}N:} \longrightarrow \underset{\substack{\text{pyridinium chlorochromate} \\ \text{(PCC)}}}{\text{(structure)}N^+-H \;\; CrO_3Cl^-} \tag{7.39}
$$

PROBLEM 7.19   Write an equation for the oxidation of

a. 1-hexanol with Jones' reagent.
b. 1-hexanol with PCC.
c. 4-phenyl-2-butanol with Jones' reagent.
d. 4-phenyl-2-butanol with PCC.

---

*In the oxidation reactions shown in eqs. 7.36–7.38, the chromium is reduced from Cr^{6+} to Cr^{3+}. Aqueous solutions of Cr^{6+} are orange, whereas aqueous solutions of Cr^{3+} are green. This color change has been used as the basis for detecting ethanol in Breathalyzer tests.

Biologically Important Alcohols and Phenols

The hydroxyl group appears in many biologically important molecules.

Four metabolically important unsaturated primary alcohols are 3-methyl-2-buten-1-ol, 3-methyl-3-buten-1-ol, geraniol, and farnesol.

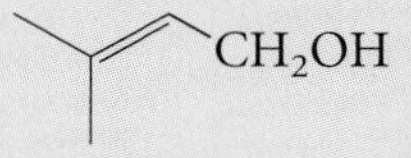

3-methyl-2-buten-1-ol 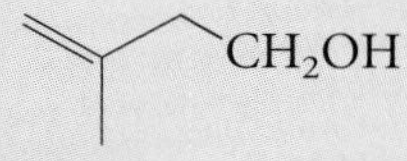   3-methyl-3-buten-1-ol

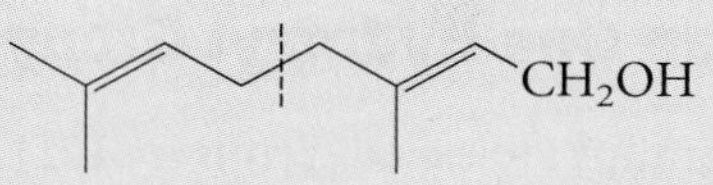

geraniol

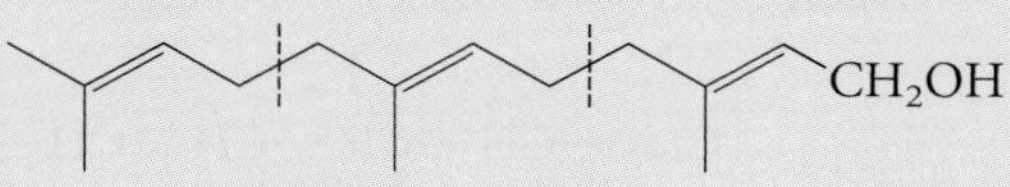

farnesol

The two smaller alcohols contain a five-carbon unit, called an **isoprene unit,** that is present in many natural products (see p. 411). This unit consists of a four-carbon chain with a one-carbon branch at carbon-2. These five-carbon alcohols can combine to give geraniol (10 carbons), which then can add yet another five-carbon unit to give farnesol (15 carbons). Note the isoprene units, marked off by dotted lines, in the structures of geraniol and farnesol.

Compounds of this type are called **terpenes.** Terpenes occur in many plants and flowers. They have 10, 15, 20, or more carbon atoms and are formed by linking isoprene units in various ways.

Geraniol, as its name implies, occurs in oil of geranium but also constitutes about 50% of rose oil, the extract of rose petals. **Farnesol,** which occurs in the essential oils of rose and cyclamen, has a pleasing lily-of-the-valley odor. Both geraniol and farnesol are used in making perfumes.

Combination of two farnesol units leads to **squalene,** a 30-carbon hydrocarbon present in small amounts in the livers of most higher animals. Squalene is the biological precursor of steroids.

Cholesterol, a typical steroidal alcohol, has the following structure:

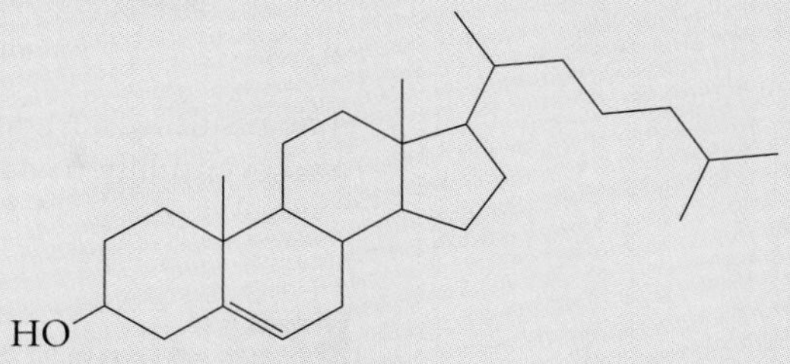

cholesterol

Although it has 27 carbon atoms (instead of 30) and is therefore not strictly a terpene, cholesterol is synthesized in the body from the terpene squalene through a complex process that, in its final stages, involves the loss of 3 carbon atoms.

Phenols are less involved than alcohols in fundamental metabolic processes. Three phenolic alcohols do, however, form the basic building blocks of **lignins,** complex polymeric substances that, together with cellulose, form the woody parts of trees and shrubs. They have very similar structures.

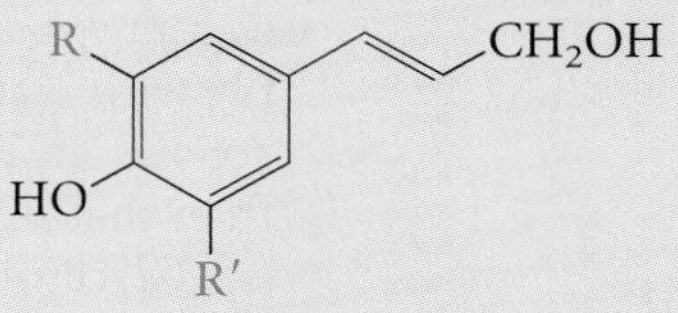

coniferyl alcohol (R = OCH₃, R′ = H)
sinapyl alcohol (R = R′ = OCH₃)
p-coumaryl alcohol (R = R′ = H)

Some phenolic natural products to be avoided are **urushiols,** the active allergenic ingredients in poison ivy and poison oak.

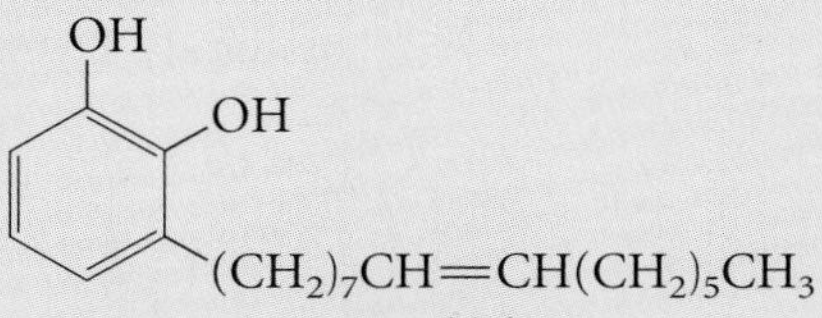

an urushiol

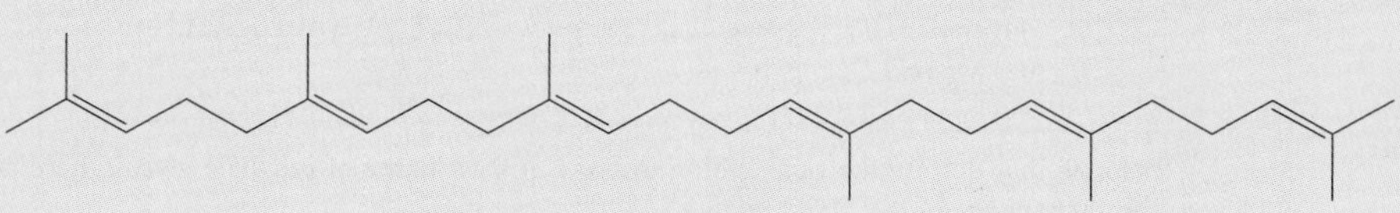

squalene

In the body, similar oxidations are accomplished by enzymes, together with a rather complex coenzyme called nicotinamide adenine dinucleotide, NAD^+ (for its structure, see p. 535). Oxidation occurs in the liver and is a key step in the body's attempt to rid itself of imbibed alcohol.

$$CH_3CH_2OH + NAD^+ \underset{\text{dehydrogenase}}{\overset{\text{alcohol}}{\rightleftarrows}} CH_3\overset{\overset{\text{O}}{\|}}{C}{-}H + NADH \qquad (7.40)$$

$$\text{ethanol} \qquad\qquad\qquad\qquad\qquad \text{acetaldehyde}$$

The resulting acetaldehyde—also toxic—is further oxidized in the body to acetic acid and eventually to carbon dioxide and water.

7.13   Alcohols with More Than One Hydroxyl Group

Some important polyols are **ethylene glycol, glycerol,** and **sorbitol.**

Compounds with two adjacent alcohol groups are called *glycols* (p. 101). The most important example is **ethylene glycol.*** Compounds with more than two hydroxyl groups are also known, and several, such as **glycerol** and **sorbitol,** are important commercial chemicals.

$$\underset{\substack{\text{ethylene glycol} \\ \text{(1,2-ethanediol)} \\ \text{bp } 198°C}}{\overset{\displaystyle CH_2{-}CH_2}{\underset{\displaystyle OH \quad OH}{|\qquad|}}} \qquad \underset{\substack{\text{glycerol (glycerine)} \\ \text{(1,2,3-propanetriol)} \\ \text{bp } 290°C \text{ (decomposes)}}}{\overset{\displaystyle CH_2{-}CH{-}CH_2}{\underset{\displaystyle OH \quad OH \quad OH}{|\qquad|\qquad|}}} \qquad \underset{\substack{\text{sorbitol} \\ \text{(1,2,3,4,5,6-hexanehexaol)} \\ \text{mp } 110–112°C}}{\overset{\displaystyle CH_2{-}CH{-}CH{-}CH{-}CH{-}CH_2}{\underset{\displaystyle OH \quad OH \quad OH \quad OH \quad OH \quad OH}{|\qquad|\qquad|\qquad|\qquad|\qquad|}}}$$

Ethylene glycol is used as the "permanent" antifreeze in automobile radiators and as a raw material in the manufacture of Dacron. Ethylene glycol is completely miscible with water. Because of its increased capacity for hydrogen bonding, ethylene glycol has an exceptionally high boiling point for its molecular weight—much higher than that of ethanol.

Glycerol is a syrupy, colorless, water-soluble high-boiling liquid with a distinctly sweet taste. Its soothing qualities make it useful in shaving and toilet soaps and in cough drops and syrups. Triesters of glycerol are fats and oils, whose chemistry is discussed in Chapter 15.

Nitration of glycerol gives **glyceryl trinitrate** (nitroglycerine), a powerful and shock-sensitive explosive. Alfred Nobel, who invented dynamite in 1866, found that glyceryl trinitrate could be controlled by absorbing it on an inert porous material. Dynamite contains about 15% glyceryl (and glycol) nitrate along with other explosive materials. Dynamite is used mainly in mining and construction.

$$\underset{\substack{\text{glycerol}}}{\overset{\displaystyle CH_2OH}{\underset{\displaystyle CH_2OH}{\underset{\displaystyle |}{\overset{\displaystyle |}{CHOH}}}}} + 3\ \underset{\substack{\text{nitric} \\ \text{acid}}}{HONO_2} \xrightarrow{\ H_2SO_4\ } \underset{\substack{\text{glyceryl trinitrate} \\ \text{(nitroglycerine)}}}{\overset{\displaystyle CH_2ONO_2}{\underset{\displaystyle CH_2ONO_2}{\underset{\displaystyle |}{\overset{\displaystyle |}{CHONO_2}}}}} + 3\ H_2O \qquad (7.41)$$

*Notice that despite the *-ene* ending in the common name of ethylene glycol, there is *no double bond* between the carbons.

Nitroglycerine is also used in medicine as a vasodilator, to prevent heart attacks in patients who suffer with angina.

Sorbitol, with its many hydroxyl groups, is water soluble. It is almost as sweet as cane sugar and is used in candy making and as a sugar substitute for diabetics. In Chapter 16 we will see that carbohydrates, for example, sucrose (table sugar), starch, and cellulose, have many hydroxyl groups.

7.14 Aromatic Substitution in Phenols

Now we will examine some reactions that occur with phenols, but not with alcohols. Phenols undergo electrophilic aromatic substitution under very mild conditions because the hydroxyl group is strongly ring activating. For example, phenol can be nitrated with *dilute aqueous* nitric acid.

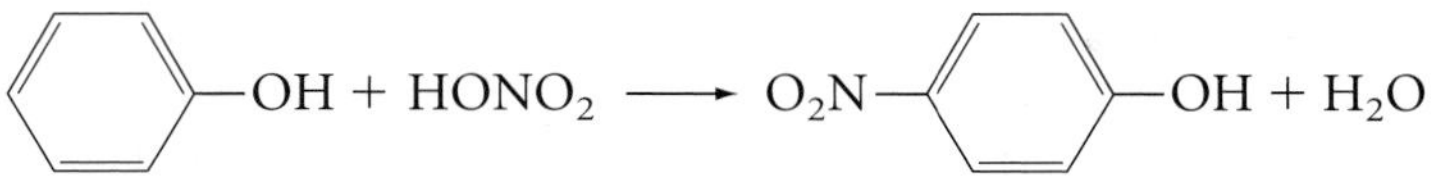

$$\text{phenol} \qquad \text{dilute nitric acid} \qquad \qquad p\text{-nitrophenol} \qquad \textbf{(7.42)}$$

Phenol is also brominated rapidly with *bromine* water, to produce 2,4,6-tribromophenol.

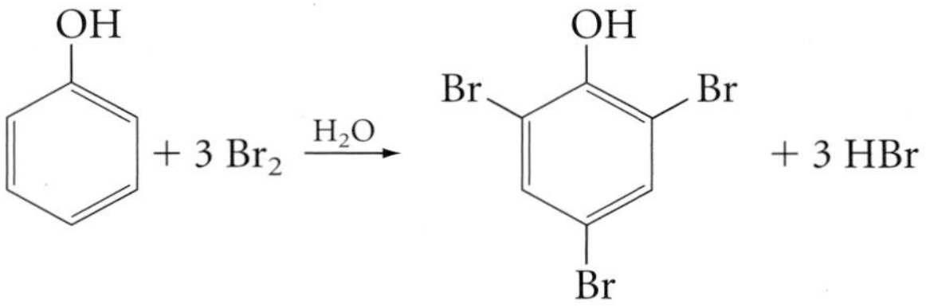

$$\text{phenol} \qquad \qquad \text{2,4,6-tribromophenol} \qquad \textbf{(7.43)}$$

EXAMPLE 7.3

Draw the intermediate in electrophilic aromatic substitution *para* to a hydroxyl group, and show how the intermediate benzenonium ion is stabilized by the hydroxyl group.

Solution

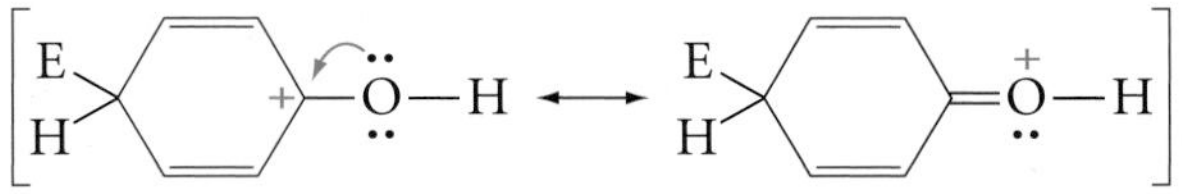

An unshared electron pair on the oxygen atom helps to delocalize the positive charge.

PROBLEM 7.20 Explain why phenoxide ion undergoes electrophilic aromatic substitution even more easily than does phenol.

PROBLEM 7.21 Write an equation for the reaction of

a. *p*-methylphenol + $HONO_2$ (1 mole).
b. *o*-chlorophenol + Br_2 (1 mole).

A WORD ABOUT ...

Quinones and the Bombardier Beetle

The bombardier beetle has a curious way of defending itself from attack by a predator. It sprays the attacker with a hot mixture of noxious chemicals. These are discharged with remarkable accuracy and an audible pop from a pair of glands at the tip of its abdomen. The components of the spray are 1,4-benzoquinone, 2-methyl-1,4-benzoquinone, and hot water. The propellant is oxygen. This surprising and unpleasant event deters predators and provides the beetle with a chance to retreat to cover. How does the bombardier beetle accomplish this feat?

Bombardier beetle spraying.

Each abdominal gland of the bombardier beetle consists of two connected chambers. The inner chamber contains a solution of 1,4-hydroquinone, 2-methyl-1,4-hydroquinone, and hydrogen peroxide dissolved in water. This is connected by a tube to an outer chamber. A one-way valve in the tube prevents the contents of the two chambers from mixing when closed, and allows flow only from the inner chamber to the outer chamber when opened. The outer chamber contains a water solution of enzymes called **peroxidases** and **catalases.** The peroxidase promotes a reaction between the hydrogen peroxide and hydroquinones to produce **quinones,** while the catalase converts hydrogen peroxide to water and oxygen. These reactions are exothermic enough to rapidly bring water to its boiling point. When under attack, the bombardier beetle opens the valve and the contents of the inner compartment flow into the outer compartment. The aforementioned chemical reactions occur and the quinones and hot water are, by pressure produced by the oxygen, ejected as a pulsating spray through a vent in the outer chamber, accompanied by the audible pop.

$$R = H \quad \text{1,4-hydroquinone}$$
$$R = CH_3 \quad \text{2-methyl-1,4-hydroquinone}$$

$$R = H \quad \text{1,4-benzoquinone}$$
$$R = CH_3 \quad \text{2-methyl-1,4-benzoquinone}$$

$$2 H_2O_2 \xrightarrow{\text{catalase}} 2 H_2O + O_2$$

Although the delivery system used by the bombardier beetle is particularly impressive, quinones are also used by other beetles, and certain earwigs, cockroaches, and spiders, as components of defensive secretions. Furthermore, two-chambered glands related to that of the bombardier beetle are used by other organisms to deliver defensive substances (Sec. 9.10). For some fascinating reading, see W. Agosta, *Bombardier Beetles and Fever Trees: A Close-Up Look at Chemical Warfare and Signals in Animals and Plants,* Addison-Wesley (1996). Also see T. Eisner and J. Meinwald, *Science,* **1966,** *153,* 1341.

7.15 Oxidation of Phenols

As indicated in the A Word About on page 222, phenols are easily oxidized. Samples that stand exposed to air for some time often become highly colored due to the formation of oxidation products. With **hydroquinone** (1,4-dihydroxybenzene), the reaction is easily controlled to give **1,4-benzoquinone** (commonly called *quinone*).

$$
\underset{\substack{\text{hydroquinone}\\\text{colorless, mp 171°C}}}{\text{OH}\text{—}\bigcirc\text{—}\text{OH}}
\xrightarrow[\text{H}_2\text{SO}_4,\ 30°C]{\text{Na}_2\text{Cr}_2\text{O}_7}
\underset{\substack{\text{1,4-benzoquinone}\\\text{yellow, mp 116°C}}}{\text{O}=\bigcirc=\text{O}}
\qquad (7.44)
$$

Hydroquinone and related compounds are used in photographic developers. They reduce silver ion that has not been exposed to light to metallic silver (and, in turn, they are oxidized to quinones). The oxidation of hydroquinones to quinones is reversible; this interconversion plays an important role in several biological oxidation–reduction reactions.

7.16 Phenols as Antioxidants

Substances that are sensitive to air oxidation, such as foods and lubricating oils, can be protected by phenolic additives. Phenols function as **antioxidants.** They react with and destroy peroxy (ROO•) and hydroxy (HO•) radicals, which otherwise react with the alkenes present in foods and oils to cause their degradation. The peroxy and **hydroxy radicals** abstract the phenolic hydrogen atom to produce more stable **phenoxy radicals** that cause less damage to the alkenes (eq. 7.45).

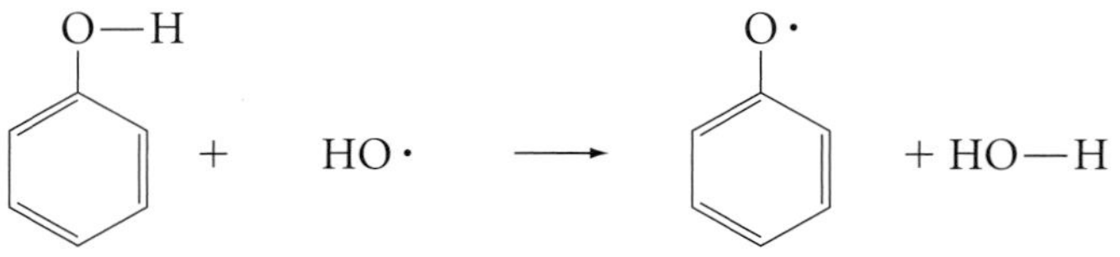

Phenols are **antioxidants,** preventing oxidation of substances sensitive to air oxidation.

$$
\underset{\text{hydroxy radical}}{\text{O—H}\ \bigcirc\ +\ \text{HO}\cdot}
\longrightarrow
\underset{\text{phenoxy radical}}{\text{O}\cdot\ \bigcirc\ +\ \text{HO—H}}
\qquad (7.45)
$$

PROBLEM 7.22   Write resonance structures that indicate how the unpaired electron in the phenoxy radical can be delocalized to the *ortho* and *para* positions (reason by analogy with the resonance structures for the phenoxide anion on p. 210).

Two commercial phenolic antioxidants are **BHA** (butylated hydroxyanisole) and **BHT** (butylated hydroxytoluene). BHA is used as an antioxidant in foods, especially meat products. BHT is used not only in foods, animal feeds, and vegetable oils, but also in lubricating oils, synthetic rubber, and various plastics.

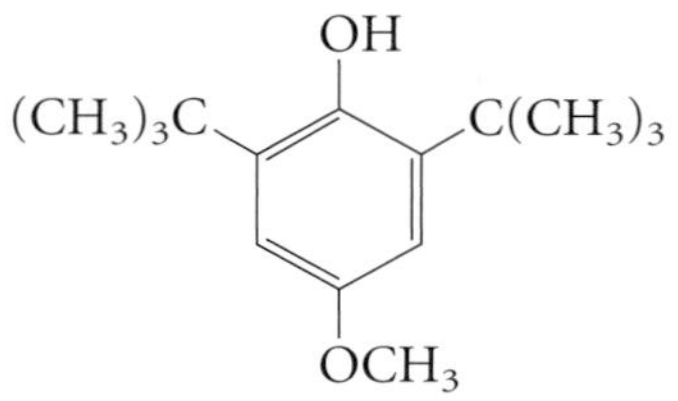

BHA 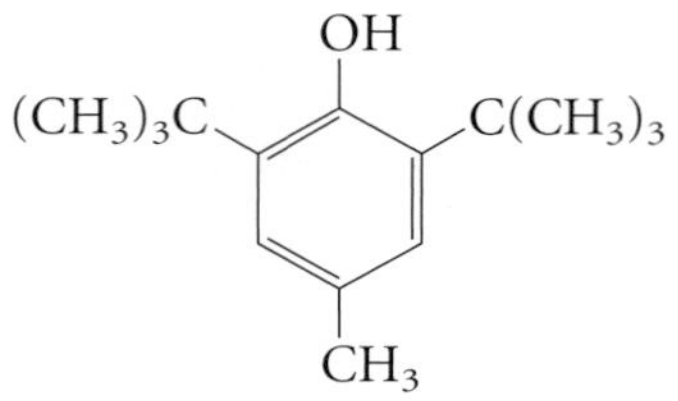   BHT

PROBLEM 7.23   Write an equation for the reaction of BHT with hydroxy radical.

Human beings and other animals also use antioxidants for protection against biological sources of peroxy and hydroxy radicals. **Vitamin E** (α-tocopherol) is a natural phenolic antioxidant that protects the body against free radicals. It is obtained largely through dietary sources such as leafy vegetables, egg yolks, wheat germ, vegetable oil, and legumes. Deficiencies of vitamin E can cause eye problems in premature infants and nerve damage in older children. **Resveratrol** is another phenolic natural product that is a common constituent of the human diet. It is found in a number of foods including peanuts and grapes. Resveratrol is also an antioxidant and has been studied as a possible cancer chemopreventive agent.

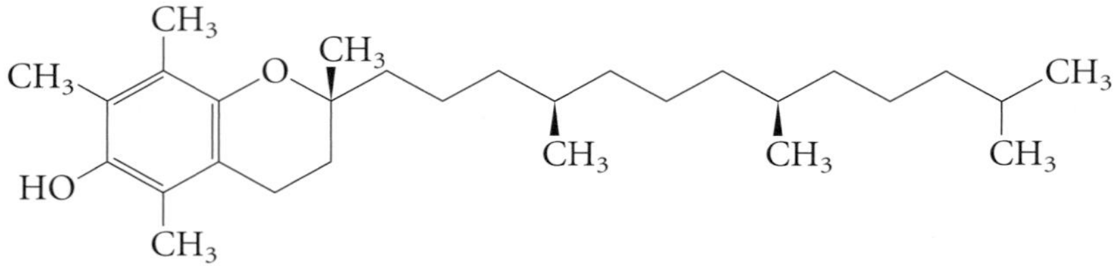

vitamin E (α-tocopherol) 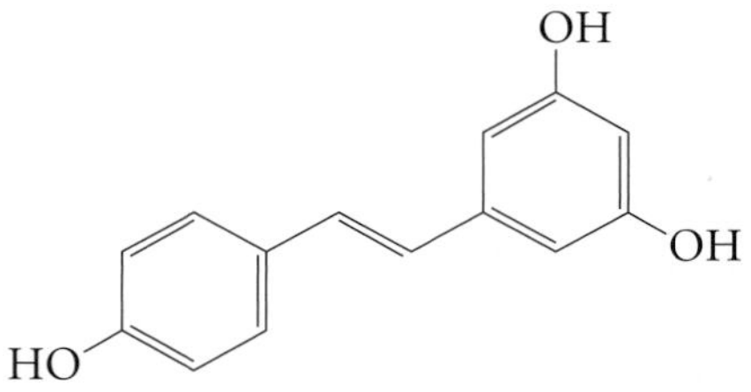   resveratrol

7.17   Thiols, the Sulfur Analogs of Alcohols and Phenols

The **sulfhydryl group, —SH** is the functional group of thiols.

Sulfur is immediately beneath oxygen in the periodic table and can often take its place in organic structures. The **—SH** group, called the **sulfhydryl group,** is the functional group of thiols (p. 203). Thiols are named as follows:

$$CH_3SH \qquad CH_3CH_2CH_2CH_2SH \qquad \text{⟨ ⟩—SH}$$

methanethiol   1-butanethiol   thiophenol
(methyl mercaptan)   (*n*-butyl mercaptan)   (phenyl mercaptan)

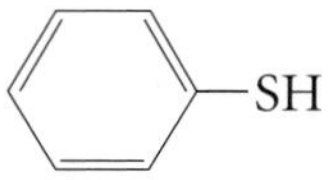

Thiols are also called **mercaptans;** their mercury salts are called **mercaptides.**

Thiols are sometimes called **mercaptans,** a name that refers to their reaction with mercuric ion to form mercury salts, called **mercaptides.**

$$2\,RSH \;+\; HgCl_2 \;\longrightarrow\; (RS)_2Hg \;+\; 2\,HCl \qquad\qquad \textbf{(7.46)}$$

a mercaptide

A W O R D A B O U T . . .

Hair, Curly or Straight

Hair consists of a fibrous protein called **keratin,** which, as proteins go, contains an unusually large percentage of the sulfur-containing amino acid **cystine.** Horse hair, for example, contains about 8% cystine:

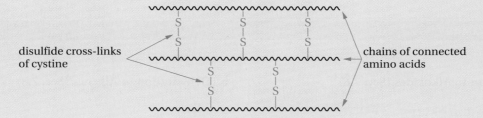

cystine (CyS—SCy)

The disulfide link in cystine serves to cross-link the chains of amino acids that make up the protein (Figure 7.1).

The chemistry used in waving or straightening hair involves the oxidation–reduction chemistry of the disulfide bond (eq. 7.49). First, the hair is treated with a reducing agent, which breaks the S—S bonds, converting each sulfur to an —SH group. This breaks the cross-links between the long protein chains. The reduced hair can now be shaped as desired, either waved or straightened. Finally, the reduced and rearranged hair is treated with an oxidizing agent to reform the disulfide cross-links. The new disulfide bonds, no longer in their original positions, hold the hair in its new shape.

disulfide cross-links of cystine

chains of connected amino acids

Figure 7.1
Schematic structure of hair.

PROBLEM 7.24   Draw the structure for

a. 2-butanethiol.   b. isopropyl mercaptan.

Alkyl thiols can be made from alkyl halides by nucleophilic displacement with sulfhydryl ion (Table 6.1, entry 10).

$$R-X + {}^-SH \longrightarrow R-SH + X^-$$ **(7.47)**

Perhaps the most distinctive feature of thiols is their intense and disagreeable odor. The thiols $CH_3CH=CHCH_2SH$ and $(CH_3)_2CHCH_2CH_2SH$, for example, are responsible for the odor of a skunk. The structurally related thiol $(CH_3)_2C=CHCH_2SH$ has recently been shown to be responsible for the skunky odor and taste of beer that has been exposed to light.

Thiols are more acidic than alcohols. The pK_a of ethanethiol, for example, is 10.6 whereas that of ethanol is 15.9. Hence, thiols react with aqueous base to give **thiolates.**

$$RSH + Na^+OH^- \longrightarrow RS^-Na^+ + HOH$$ **(7.48)**
a sodium thiolate

The striped skunk (*Mephitis mephitis*) sprays a foul mixture of thiols at its enemies.

PROBLEM 7.25   Write an equation for the reaction of ethanethiol (CH_3CH_2SH) with

a. KOH.   b. $CH_3CH_2O^-Na^+$.   c. $HgCl_2$.

Disulfides are compounds containing an S—S bond.

Thiols are easily oxidized to **disulfides,** compounds containing an S—S bond, by mild oxidizing agents such as hydrogen peroxide or iodine. A naturally occurring disulfide whose smell you are probably familiar with is diallyl disulfide ($CH_2{=}CHCH_2S{-}SCH_2CH{=}CH_2$), which is responsible for the odor of fresh garlic.*

$$\underset{\text{thiol}}{2\ RS{-}H} \underset{\text{reduction}}{\overset{\text{oxidation}}{\rightleftharpoons}} \underset{\text{disulfide}}{RS{-}SR} \quad \textbf{(7.49)}$$

The reaction shown in eq. 7.49 can be reversed with a variety of reducing agents. Since proteins contain disulfide links, these reversible oxidation–reduction reactions can be used to manipulate protein structures.

REACTION SUMMARY

1. Alcohols

a. Conversion to Alkoxides (Sec. 7.6)

$$2\ RO{-}H + 2\ Na \longrightarrow 2\ RO^- Na^+ + H_2$$

$$RO{-}H + NaH \longrightarrow RO^- Na^+ + H_2$$

b. Dehydration to Alkenes (Sec. 7.8)

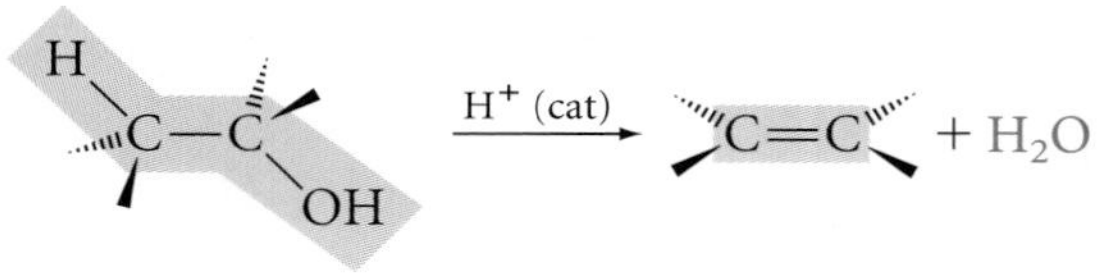

c. Conversion to Alkyl Halides (Secs. 7.9–7.10)

$$R{-}OH + HX \longrightarrow R{-}X + H_2O\ (X{=}Cl, Br, I)$$

$$R{-}OH + SOCl_2 \longrightarrow R{-}Cl + HCl + SO_2$$

$$R{-}OH + PX_3 \longrightarrow R{-}X + H_3PO_3\ (X{=}Cl, Br)$$

d. Oxidation (Sec. 7.12)

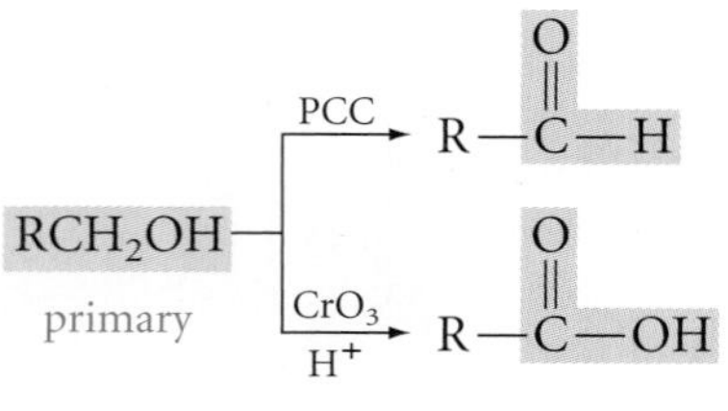

(aldehyde)

(carboxylic acid)

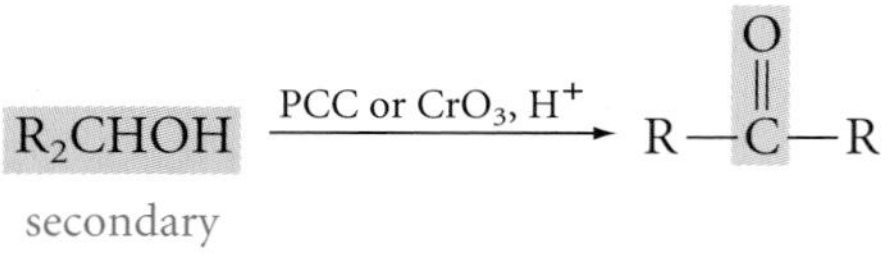

(ketone)

*Garlic belongs to the plant family *Allium,* from which the *allyl* group gets its name.

2. Phenols

a. Preparation of Phenoxides (Sec. 7.6)

$$ArO\!-\!H + NaOH \longrightarrow ArO^- Na^+ + H_2O$$

b. Electrophilic Aromatic Substitution (Sec. 7.14)

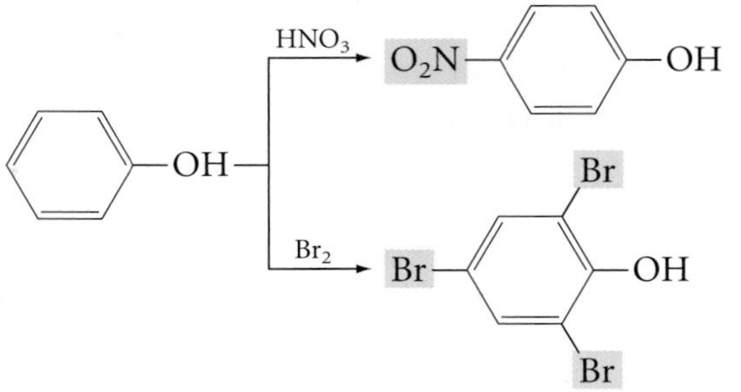

c. Oxidation to Quinones (Sec. 7.15)

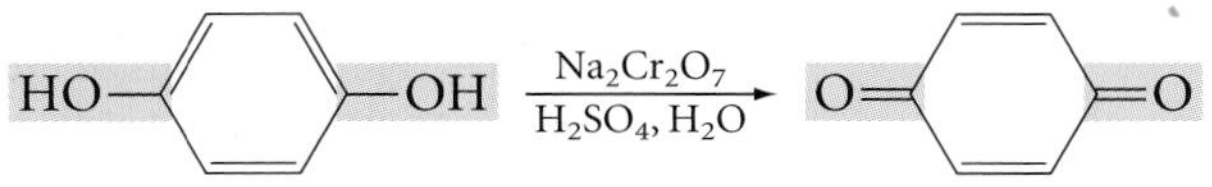

quinone

3. Thiols

a. Conversion to Thiolates (Sec. 7.17)

$$RS\!-\!H + NaOH \longrightarrow RS^- Na^+ + H_2O$$

thiol   thiolate

b. Oxidation to Disulfides (Sec. 7.17)

$$2\,RSH \xrightarrow{\text{oxidation}} RS\!-\!SR$$

thiol   disulfide

ADDITIONAL PROBLEMS

Nomenclature and Structure of Alcohols

7.26 Name each of the following alcohols:
  a. $CH_3CH_2CH(OH)CH_2CH_3$
  b. $(CH_3)_2CHCH(OH)CH_2CH_3$
  c. $CH_3CH(Cl)CH(OH)CH_2CH_3$
  d. $CH_3CH(Cl)CH_2CH(OH)CH_3$

7.27 Write a structural formula for each of the following compounds:
  a. 2,2-dimethyl-1-butanol
  b. *m*-bromophenol
  c. 2,3-pentanediol
  d. 2-phenylethanol
  e. sodium ethoxide
  f. 1-methylcyclopentanol
  g. *cis*-2-methylcyclopentanol
  h. (*S*)-2-butanol
  i. 2-methyl-2-propen-1-ol
  j. 2-cyclohexenol

7.28 Name each of the following compounds:
  a. $CH_3C(CH_3)_2CH(OH)CH_3$
  b. $CH_3CHBrC(CH_3)_2OH$

  c. 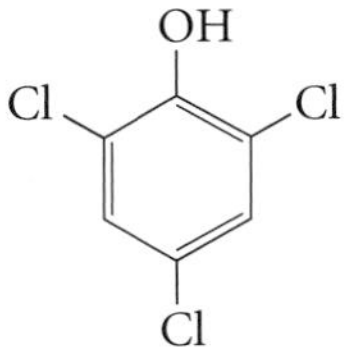
  d.

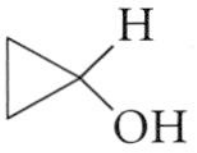

OH

Br

e.

f.

OH

CH$_3$

g. CH$_3$CH=CHCH$_2$OH
i. HOCH$_2$CH(OH)CH(OH)CH$_2$OH

h. CH$_3$CH(SH)CH$_3$
j. (CH$_3$)$_2$CHO$^-$K$^+$

7.29 Explain why each of the following names is unsatisfactory, and give a correct name:

a. 2-ethyl-1-propanol
c. 1-propene-3-ol
e. 3,6-dibromophenol

b. 2,2-dimethyl-3-butanol
d. 2-chloro-4-pentanol

7.30 Thymol is an antibacterial oil obtained from thyme (*Thymus vulgaris*). The IUPAC name of this compound is 2-isopropyl-5-methylphenol. Draw the structure of thymol.

Thyme (*Thymus vulgaris*), source of the antibacterial oil thymol.

Properties of Alcohols

7.31 Classify the alcohols in parts a, d, f, g, h, i, and j of Problem 7.27 as primary, secondary, or tertiary.

7.32 Arrange the compounds in each of the following groups in order of increasing solubility in water, and briefly explain your answers:

a. ethanol; ethyl chloride; 1-hexanol

b. 1-pentanol; 1,5-pentanediol; HOCH$_2$(CHOH)$_3$CH$_2$OH

Acid–Base Reactions of Alcohols and Thiols

7.33 The following classes of organic compounds are Lewis bases. Write an equation that shows how each class might react with H$^+$.

a. ether, RÖR

b. amine, R$_3$N:

c. ketone, R$_2$C=Ö

7.34 Arrange the following compounds in order of increasing acidity, and explain the reasons for your choice of order: cyclohexanol, phenol, *p*-nitrophenol, 2-chlorocyclohexanol.

7.35 Which is the stronger base, potassium *t*-butoxide or potassium ethoxide? (*Hint:* Use the data in Table 7.2.)

7.36 Complete each of the following equations:

a. CH$_3$CH(OH)CH$_2$CH$_3$ + K $\longrightarrow$

b. (CH$_3$)$_2$CHOH + NaH $\longrightarrow$

c. Cl—⟨ ⟩—OH + NaOH $\longrightarrow$

d.

H

OH

+ NaOH $\longrightarrow$

e. CH$_3$CH=CHCH$_2$SH + NaOH $\longrightarrow$

7.37 Explain why your answers to parts c, d, and e of Problem 7.36 are consistent with the pK_a's of the starting acids and product acids (see eqs. 7.14, 7.15, and 7.48).

🔗 = concept connections

Acid-Catalyzed Dehydration of Alcohols

7.38 Show the structures of all possible acid-catalyzed dehydration products of the following. If more than one alkene is possible, predict which one will be formed in the largest amount.

    a. cyclopentanol   **b.** 1-methylcyclopentanol

    c. 2-butanol   **d.** 2-phenylethanol

7.39 Explain why the reaction shown in eq. 7.19 occurs much more easily than the reaction $(CH_3)_3C-OH \rightleftharpoons (CH_3)_3C^+ + HO^-$. (That is, why is it necessary to protonate the alcohol before ionization can occur?)

7.40 Draw a reaction energy diagram for the dehydration of *tert*-butyl alcohol (eq. 7.21). Include the steps shown in eqs. 7.18–7.20 in your diagram.

7.41 Write out all the steps in the mechanism for eq. 7.24, showing how each product is formed.

Alkyl Halides from Alcohols

7.42 Although the reaction shown in eq. 7.26 occurs faster than that shown in eq. 7.28, the yield of product is lower. The yield of *t*-butyl chloride is only 80%, whereas the yield of *n*-butyl chloride is nearly 100%. What by-product is formed in eq. 7.26, and by what mechanism is it formed? Why is a similar by-product *not* formed in eq. 7.28?

7.43 Treatment of 3-buten-2-ol with concentrated hydrochloric acid gives a mixture of two products, 3-chloro-1-butene and 1-chloro-2-butene. Write a reaction mechanism that explains how both products are formed.

Synthesis and Reactions of Alcohols

7.44 Write an equation for each of the following reactions:

    a. 2-methyl-2-butanol + HCl   **b.** 1-pentanol + Na   **c.** cyclopentanol + PBr₃

    d. 2-phenylethanol + SOCl₂   **e.** 1-methylcyclopentanol + H₂SO₄, heat   **f.** ethylene glycol + HONO₂

    g. 1-octanol + HBr + ZnBr₂   **h.** 1-pentanol + aqueous NaOH   **i.** 1-pentanol + CrO₃, H⁺

    j. 2-cyclohexylethanol + PCC

7.45 Write an equation for each of the following two-step syntheses:

    a. cyclohexene to cyclohexanone   **b.** 1-chlorobutane to butanal   **c.** 1-butanol to 1-butanethiol

Oxidation Reactions of Alcohols, Phenols and Thiols

7.46 The alcohol citronellol is a terpene (p. 219) found in rose oil. The product formed when citronellol is oxidized with pyridinium chlorochromate (PCC) is a constituent of lemon oil. Draw the structure of the product when citronellol is oxidized with PCC.

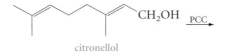

citronellol

7.47 What product do you expect from the oxidation of cholesterol with CrO₃ and H⁺? (See p. 219 for the formula of cholesterol.)

7.48 Draw the structure of the quinone expected from the oxidation of

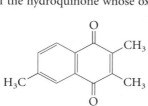

a.   b.   c.

7.49 Dimethyl disulfide, $CH_3S—SCH_3$, found in the vaginal secretions of female hamsters, acts as a sexual attractant for the male hamster. Write an equation for its synthesis from methanethiol.

7.50 The disulfide $[(CH_3)_2CHCH_2CH_2S]_2$ is a component of the odorous secretion of mink. Describe a synthesis of this disulfide, starting with 3-methyl-1-butanol.

7.51 2,3,6-Trimethyl-1,4-naphthoquinone (TMNQ) is a quinone that was recently isolated from tobacco leaves and was shown to slow the metabolism of dopamine, a neurotransmitter whose depletion can lead to Parkinson's disease. What is the structure of the hydroquinone whose oxidation gives TMNQ?

TMNQ

Diethyl ether (CH₃CH₂OCH₂CH₃)

8

ETHERS AND EPOXIDES

For many people the word *ether* is associated with the well-known general anesthetic (see A Word About Ether and Anesthesia later in this chapter). *Ether* is also a common ingredient in starter fluid for car engines. However, this is only one member of a class of compounds known as **ethers.** The natural antibiotic monensin and the pheromone disparlure (see A Word About the Gypsy Moth's Epoxide later in this chapter) both contain the ether functional group. The drug tetrahydrocannabinol (THC), the active ingredient in marijuana, contains a cyclic ether. Synthetic ethers include the gasoline additive MTBE (p. 237) and ethylene oxide, the industrial precursor to ethylene glycol, which is used as antifreeze and as a starting material in the manufacture of polyesters (Chapter 14).

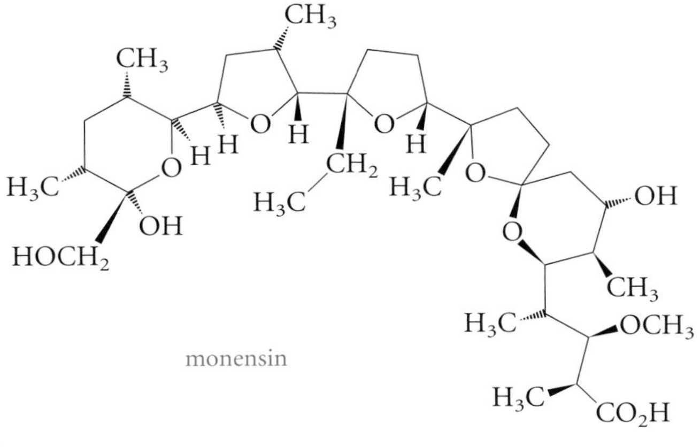

monensin

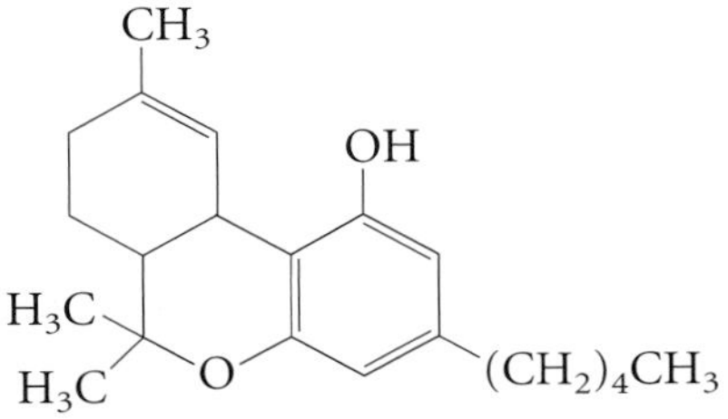

tetrahydrocannabinol (THC)

All ethers are compounds in which two organic groups are connected to a single oxygen atom. The general formula for ethers is R—O—R′, where R and

▲ Start your engines! Diethyl ether is the major flammable compound in starting fluid.

Ethers are compounds that have two organic groups connected to a single oxygen atom. **Epoxides** are cyclic three-membered ring ethers.

R' may be identical or different, and they may be alkyl or aryl groups. In the common anesthetic, both R and R' are ethyl groups, $CH_3CH_2—O—CH_2CH_3$. In this chapter, we will describe the physical and chemical properties of ethers. Their excellent solvent properties are applied in the preparation of Grignard reagents, organometallic compounds with a carbon–magnesium bond. We will give special attention to **epoxides,** cyclic three-membered ring ethers that have important industrial utility.

8.1   Nomenclature of Ethers

Ethers are usually named by giving the name of each alkyl or aryl group, in alphabetical order, followed by the word *ether.*

$$CH_3CH_2—O—CH_3 \qquad CH_3CH_2—O—CH_2CH_3$$

ethyl methyl ether   diethyl ether (the prefix   diphenyl ether
                            di- is sometimes omitted)

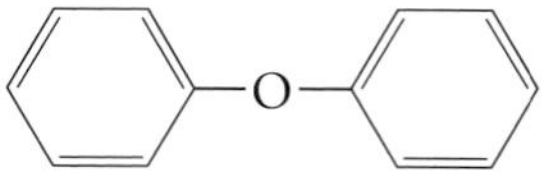

For ethers with more complex structures, it may be necessary to name the —OR group as an **alkoxy group.** In the IUPAC system, the smaller alkoxy group is named as a substituent.

$$CH_3CHCH_2CH_2CH_3$$
$$|$$
$$OCH_3$$

2-methoxypentane   *trans*-2-methoxycyclohexanol 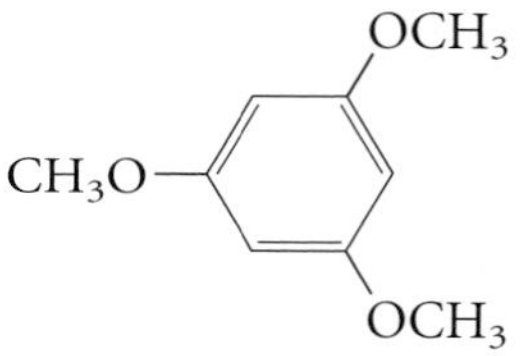   1,3,5-trimethoxybenzene

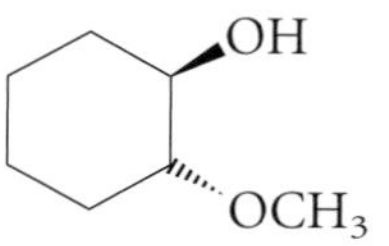

EXAMPLE 8.1

Give a correct name for $CH_3CHCH(CH_3)_2$.
$\qquad\qquad\qquad\qquad\qquad\qquad |$
$\qquad\qquad\qquad\qquad\qquad OCH_2CH_3$

Solution

$$\overset{\displaystyle CH_3}{\underset{\displaystyle OCH_2CH_3}{\overset{1\quad2\quad|\,3\;4}{CH_3CHCHCH_3}}}$$

2-ethoxy-3-methylbutane

PROBLEM 8.1   Give a correct name for

a. $(CH_3)_2CHOCH_3$.   b. 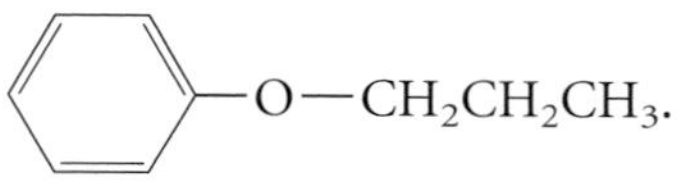 —O—$CH_2CH_2CH_3$.

c.

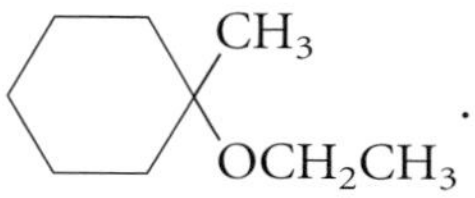

PROBLEM 8.2   Write the structural formula for

a. dicyclopropyl ether.   b. 2-methoxyoctane.

8.2 Physical Properties of Ethers

Ethers are colorless compounds with characteristic, relatively pleasant odors. They have lower boiling points than alcohols with an equal number of carbon atoms. In fact, an ether has nearly the same boiling point as the corresponding hydrocarbon in which a $-CH_2-$ group replaces the ether's oxygen. The following data illustrate these facts:

Compound	Formula	bp	mol wt	Water solubility (g/100 mL, 20°C)
1-butanol	$CH_3CH_2CH_2CH_2OH$	118°C	74	7.9
diethyl ether	$CH_3CH_2-O-CH_2CH_3$	35°C	74	7.5
pentane	$CH_3CH_2-CH_2-CH_2CH_3$	36°C	72	0.03

Because of their structures (no O—H bonds), ether molecules cannot form hydrogen bonds with one another. This is why they boil at much lower temperatures than their isomeric alcohols.*

> **PROBLEM 8.3** Write structures for each of the following *isomers*, and arrange them in order of decreasing boiling point: 3-methoxy-1-propanol, 1,2-dimethoxyethane, 1,4-butanediol.

Although ethers cannot form hydrogen bonds with one another, they do form hydrogen bonds with alcohols:

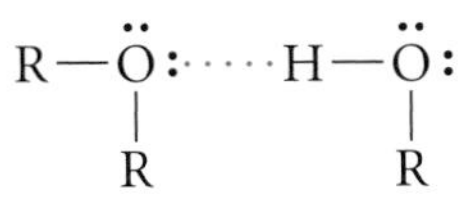

$$R-\overset{..}{\underset{\underset{R}{|}}{O}}:\cdots\cdots H-\overset{..}{\underset{\underset{R}{|}}{O}}:$$

For this reason, alcohols and ethers are usually mutually soluble. Low-molecular-weight ethers, such as dimethyl ether, are quite soluble in water. Likewise, the modest solubility of diethyl ether in water is similar to that of its isomer l-butanol (see data tabulated above) because each can form a hydrogen bond to water. Ethers are less dense than water.

8.3 Ethers as Solvents

Ethers are relatively inert compounds. They do not usually react with dilute acids, with dilute bases, or with common oxidizing and reducing agents. They do not react with metallic sodium—a property that distinguishes them from alcohols. This general inertness, coupled with the fact that most organic compounds are ether-soluble, makes ethers excellent solvents in which to carry out organic reactions.

---

*Although ethers are slightly more polar than alkanes, the major attractive forces between ether molecules are van der Waals attractions (see Sec. 2.7).

Ethers are also used frequently to extract organic compounds from their natural sources. Diethyl ether is particularly good for this purpose. Its low boiling point makes it easy to remove from an extract and easy to recover by distillation. It is highly flammable, however, and must not be used if there are any flames in the same laboratory.

Ethers that have been in a laboratory for a long time, exposed to air, may contain organic peroxides as a result of oxidation.

$$CH_3CH_2OCH_2CH_3 + O_2 \longrightarrow CH_3CH_2OCHCH_3 \qquad (8.1)$$
$$| \atop OOH$$

an ether hydroperoxide

These peroxides are explosive and must be removed before the ether can be used safely. Test papers are used to detect peroxides, and shaking with aqueous ferrous sulfate ($FeSO_4$) destroys these peroxides by reduction.

8.4 The Grignard Reagent; an Organometallic Compound

Grignard reagents are alkyl- or arylmagnesium halides.

One of the most striking examples of the solvating power of ethers is in the preparation of **Grignard reagents.** These reagents, which are exceedingly useful in organic synthesis, were discovered by the French organic chemist Victor Grignard [pronounced "greenyar(d)"]. In 1912 he received a Nobel Prize for this contribution to organic synthesis.*

Grignard found that when magnesium turnings are stirred with an ether solution of an alkyl or aryl halide, an exothermic reaction occurs. The magnesium, which is insoluble in ether, disappears as it reacts with the halide to give solutions of ether-soluble Grignard reagents.

$$R-X + Mg \xrightarrow{\text{dry ether}} R-MgX \qquad (8.2)$$

a Grignard reagent

The carbon–halogen bond is broken and both the alkyl group and halogen become bonded to the magnesium.

Although the ether used as a solvent for this reaction is normally not shown as part of the Grignard reagent structure, it does play an important role. The unshared electron pairs on the ether oxygen help to stabilize the magnesium through coordination.

R   R
  \ ‥ /
    O
    ‥
R′—Mg—X   Acting as a Lewis base, ether
stabilizes a Grignard reagent.
    ‥
    O
  / ‥ \
R   R

---

*For a brief account of how Grignard discovered these reagents, see D. Hodson, *Chemistry in Britain* **1987,** 141–142.

The two ethers most commonly used in Grignard preparations are diethyl ether and the cyclic ether tetrahydrofuran, abbreviated THF (p. 244). The Grignard reagent will not form unless the ether is scrupulously dry, free of traces of water or alcohols.

Grignard reagents are named as shown in the following equations:

$$CH_3{-}I + Mg \xrightarrow{\text{ether}} CH_3MgI \qquad (8.3)$$

methyl
iodide
methylmagnesium
iodide

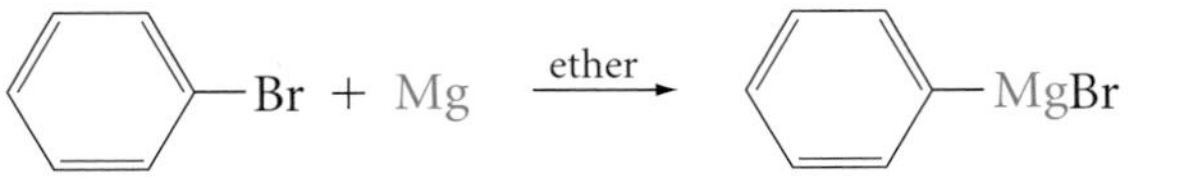

$$\text{bromobenzene} \qquad \text{phenylmagnesium bromide}$$ (8.4)

Notice that there is no space between the name of the organic group and magnesium, but that there is a space before the halide name.

Grignard reagents usually react as if the alkyl or aryl group is negatively charged (a carbanion) and the magnesium atom is positively charged.

$$\overset{\delta-}{R}{-}\overset{\delta+}{MgX}$$

Carbanions *are strong bases.* They are the conjugate bases of hydrocarbons, which are very weak acids. It is not surprising, then, that Grignard reagents react vigorously with even such a weak acid as water, or with any other compound with an OH, SH, or NH bond.

A **carbanion** is an alkyl or aryl group with a negatively charged carbon atom. Carbanions are strong bases.

$$\overset{\delta-}{R}{-}MgX + \overset{\delta+}{H}{-}OH \longrightarrow R{-}H + Mg^{2+}(OH)^-X^- \qquad (8.5)$$

stronger base   stronger acid   weaker acid   weaker base

This is why the ether used as a solvent for the Grignard reagent *must be scrupulously free of water or alcohol.*

PROBLEM 8.4   Write an equation for the reaction between

a. methylmagnesium iodide and water.
b. phenylmagnesium bromide and methanol.

EXAMPLE 8.2

Is it possible to prepare a Grignard reagent from $HOCH_2CH_2CH_2Br$ and magnesium?

Solution   No! Any Grignard reagent that forms is immediately destroyed by protons from the OH group. Grignard and hydroxyl groups in the same molecule are incompatible.

PROBLEM 8.5   Is it possible or impossible to prepare a Grignard reagent from $CH_3OCH_2CH_2CH_2Br$? Explain.

The reaction of a Grignard reagent with water can be put to useful purpose. For example, if heavy water (D_2O) is used, deuterium (an isotope of hydrogen) can be substituted for a halogen.

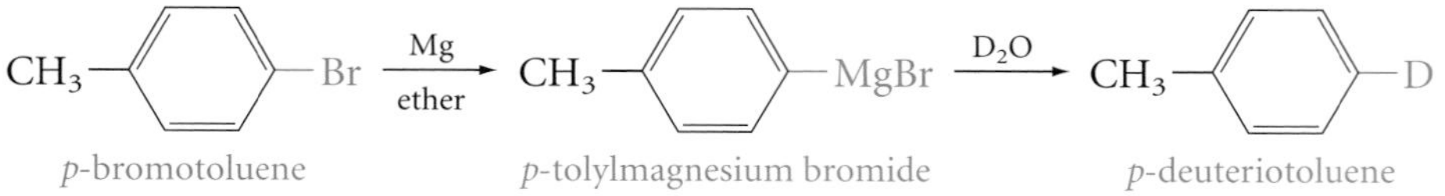

$$CH_3-\!\!\!\left\langle\!\!\!\bigcirc\!\!\!\right\rangle\!\!\!-Br \xrightarrow[\text{ether}]{Mg} CH_3-\!\!\!\left\langle\!\!\!\bigcirc\!\!\!\right\rangle\!\!\!-MgBr \xrightarrow{D_2O} CH_3-\!\!\!\left\langle\!\!\!\bigcirc\!\!\!\right\rangle\!\!\!-D \qquad \textbf{(8.6)}$$

p-bromotoluene   p-tolylmagnesium bromide   p-deuteriotoluene

This is a useful way to introduce an isotopic label into an organic compound.*

EXAMPLE 8.3

Show how to prepare CH_3CHDCH_3 from $CH_2{=}CHCH_3$.

Solution

$$CH_2{=}CHCH_3 \xrightarrow{HBr} \underset{\underset{Br}{|}}{CH_3CHCH_3} \xrightarrow[\text{ether}]{Mg} \underset{\underset{MgBr}{|}}{CH_3CHCH_3} \xrightarrow{D_2O} CH_3CHDCH_3$$

PROBLEM 8.6   Show how to prepare CH_3CHDCH_3 from $(CH_3)_2CHOH$.

Organometallic compounds are organic compounds that contain a carbon–metal bond.

Grignard reagents are **organometallic compounds;** they contain a carbon–metal bond. Many other types of organometallic compounds are known. Acetylides (eq. 3.53), for example, are similar to Grignard reagents in their reactions. **Organolithium compounds,** which can be prepared in a manner similar to that for Grignard reagents, are also useful in synthesis.

$$R{-}X + 2\,Li \xrightarrow{\text{ether}} R{-}Li + Li^+\,X^- \qquad \textbf{(8.7)}$$

an alkyllithium

PROBLEM 8.7   Write an equation for the preparation of

a. propyllithium   b. the acetylide of 1-butyne

and for the reaction of each with D_2O.

Later in this chapter and elsewhere in this book, we will see examples of the synthetic utility of organometallic reagents.

---

*An isotope label serves as a "tag" that allows chemists to obtain information about such things as reaction mechanisms and rates. For a detailed example of how deuterium labels have been used to determine reaction mechanisms, see Section 9.16.

8.5 Preparation of Ethers

The most important commercial ether is diethyl ether. It is prepared from ethanol and sulfuric acid.

$$CH_3CH_2OH + HOCH_2CH_3 \xrightarrow[140°C]{H_2SO_4} CH_3CH_2OCH_2CH_3 + H_2O \qquad (8.8)$$

ethanol   diethyl ether

Note that ethanol can be dehydrated by sulfuric acid to give either ethylene (eq. 7.17) or diethyl ether (eq. 8.8). Of course, the reaction conditions are different in each case. These reactions provide a good example of how important it is to control reaction conditions and to specify them in equations.

> **PROBLEM 8.8** The reaction in eq. 7.17 occurs by an E2 mechanism (review eqs. 7.22 and 7.23). By what mechanism does the reaction in eq. 8.8 occur?

Although it can be adapted to other ethers, the alcohol–sulfuric acid method is most commonly used to make symmetric ethers from primary alcohols.

> **PROBLEM 8.9** Write an equation for the synthesis of propyl ether from 1-propanol.

The commercial production of *t*-butyl methyl ether has become important in recent years. In 1998, worldwide consumption of MTBE was 6.6 billion gallons. With an octane value of 110, it is used as an octane number enhancer in unleaded gasolines.* It is prepared by the acid-catalyzed addition of methanol to 2-methyl-propene. The reaction is related to the hydration of alkenes (Sec. 3.7.b). The only difference is that an alcohol, methanol, is used as the nucleophile instead of water.

$$CH_3OH + CH_2{=}C(CH_3)_2 \xrightarrow{H^+} CH_3O{-}\underset{\underset{CH_3}{|}}{\overset{\overset{CH_3}{|}}{C}}{-}CH_3 \qquad (8.9)$$

methanol   2-methylpropene   *t*-butyl methyl ether

> **PROBLEM 8.10** Write out the steps in the mechanism for eq. 8.9 (see eqs. 3.18 and 3.20).

Most important for the laboratory synthesis of unsymmetric ethers is the **Williamson synthesis,** named after the British chemist who devised it. This method

---

*See A Word About Petroleum, Gasoline, and Octane Number, p. 106.

MTBE

Log on to http://www.college.hmco.com and follow the links to the student text web site. Use the Web Link section for Chapter 8 to answer the following questions and find more information on the topics below.

FAQs about MTBE from the EPA

1. What is MTBE and how is it made?
2. Why is MTBE used in gasoline? What is RFG? To what extent is MTBE used in RFG?
3. According to the EPA, does the use of MTBE improve air quality? Is MTBE hazardous to humans when inhaled? When ingested in drinking water?
4. How does MTBE get into the environment? What properties of MTBE make it likely to be found far from the original site of a spill or leak?
5. How can MTBE be removed from the environment?

FAQs about MTBE from the ATSDR

1. What are some possible effects of inhaling or drinking MTBE?
2. How likely is MTBE to cause cancer?

European Fuel Oxygenates Association

1. What are some specific ways in which use of MTBE improves air quality?
2. What evidence is there with regard to the health dangers presented by MTBE?
3. How does MTBE affect water quality, especially with respect to taste and odor?

You now have reviewed several different points of view concerning the use of MTBE. What do you think?

---

has two steps, both of which we have already discussed. In the first step, an alcohol is converted to its alkoxide by treatment with a reactive metal (sodium or potassium) or metal hydride (review eqs. 7.12 and 7.13). In the second step, an S_N2 displacement is carried out between the alkoxide and an alkyl halide (see Table 6.1, entry 2). The Williamson synthesis is summarized by the general equations

$$2\,ROH + 2\,Na \longrightarrow 2\,RO^-Na^+ + H_2 \qquad \textbf{(8.10)}$$

$$RO^-Na^+ + R'{-}X \longrightarrow ROR' + Na^+X^- \qquad \textbf{(8.11)}$$

Since the second step is an S_N2 reaction, it works best if R′ in the alkyl halide is primary and not well at all if R′ is tertiary.

EXAMPLE 8.4

Write an equation for the synthesis of $CH_3OCH_2CH_2CH_3$ using the Williamson method.

Solution There are two possibilities, depending on which alcohol and which alkyl halide are used:

$$\begin{array}{ccc} CH_3OCH_2CH_2CH_3 & \text{or} & CH_3OCH_2CH_2CH_3 \\ \text{from} & & \text{from} \\ CH_3O^-Na^+ + XCH_2CH_2CH_3 & & CH_3X + Na^{+\,-}OCH_2CH_2CH_3 \end{array}$$

The equations are

$$2\ CH_3OH + 2\ Na \longrightarrow 2\ CH_3O^-Na^+ + H_2$$

$$CH_3O^-Na^+ + CH_3CH_2CH_2X \longrightarrow CH_3OCH_2CH_2CH_3 + Na^+X^-$$

or

$$2\ CH_3CH_2CH_2OH + 2\ Na \longrightarrow 2\ CH_3CH_2CH_2O^-Na^+ + H_2$$

$$CH_3CH_2CH_2O^-Na^+ + CH_3X \longrightarrow CH_3CH_2CH_2OCH_3 + Na^+X^-$$

X is usually Cl, Br, or I.

PROBLEM 8.11 Write equations for the synthesis of the following ethers by the Williamson method:

a. 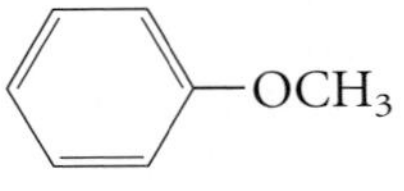 (benzene ring)—OCH_3 b. $(CH_3)_3COCH_3$ (*Reminder:* The second step proceeds by the S_N2 mechanism.)

8.6 Cleavage of Ethers

Ethers have unshared electron pairs on the oxygen atom and are therefore Lewis bases. They react with strong proton acids and with Lewis acids such as the boron halides.

$$R-\overset{..}{\underset{..}{O}}-R' + H^+ \rightleftharpoons R-\overset{H}{\overset{|}{\underset{..}{O}^+}}-R' \qquad (8.12)$$

$$R-\overset{..}{\underset{..}{O}}-R' + Br-\overset{|}{\underset{Br}{B}}-Br \rightleftharpoons R-\overset{..}{\overset{+}{O}}-R' \qquad (8.13)$$
$$\qquad\qquad\qquad\qquad\qquad\qquad Br-\overset{|}{\underset{Br}{B}^-}-Br$$

These reactions are similar to the reaction of alcohols with strong acids (eq. 7.18). If the alkyl groups R and/or R' are primary or secondary, the bond to oxygen can be broken by reaction with a strong nucleophile such as I^- or Br^- (by an S_N2 process). For example,

$$CH_3CH_2OCH(CH_3)_2 + HI \xrightarrow{\text{heat}} CH_3CH_2I + HOCH(CH_3)_2 \qquad (8.14)$$ 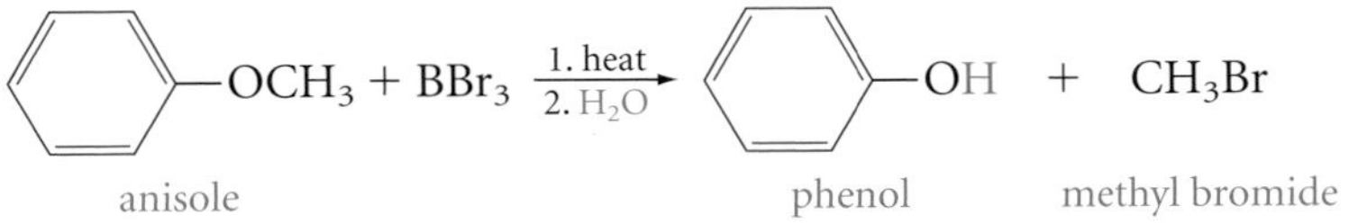
$$\text{ethyl isopropyl ether} \qquad\qquad \text{ethyl iodide} \quad \text{isopropyl alcohol}$$

$$\text{(benzene ring)}-OCH_3 + BBr_3 \xrightarrow[\text{2. }H_2O]{\text{1. heat}} \text{(benzene ring)}-OH + CH_3Br \qquad (8.15)$$
$$\text{anisole} \qquad\qquad\qquad \text{phenol} \qquad \text{methyl bromide}$$

A WORD ABOUT ...

Ether and Anesthesia

Prior to the 1840s, pain during surgery was relieved by various methods (asphyxiation, pressure on nerves, administration of narcotics or alcohol), but on the whole it was almost worse torture to undergo an operation than to endure the disease. Modern use of anesthesia during surgery has changed all that. Anesthesia stems from the work of several physicians in the mid-nineteenth century. The earliest experiments used nitrous oxide, ether, or chloroform. Perhaps the best known of these experiments was the removal of a tumor from the jaw of a patient anesthetized by ether, performed by Boston dentist William T. G. Morton in 1846.

Dr. William T. G. Morton making the first public demonstration of anesthetization with ether.

Anesthetics fall into two major categories, general and local. **General anesthetics** are usually administered to accomplish three ends: insensitivity to pain (analgesia), loss of consciousness, and muscle relaxation. Gases such as nitrous oxide and cyclopropane and volatile liquids such as ether are administered by inhalation, but other general anesthetics such as barbiturates are injected intravenously.

The exact mechanism by which anesthetics affect the central nervous system is not completely known. Unconsciousness may result from several factors: changes in the properties of nerve cell membranes, suppression of certain enzymatic reactions, and solubility of the anesthetic in lipid membranes.

A good inhalation anesthetic should vaporize readily and have appropriate solubility in the blood and tissues. It should also be stable, inert, nonflammable, potent, and minimally toxic. It should have an acceptable odor and cause minimal side effects such as nausea or vomiting. No anesthetic that meets *all* of these specifications has yet been developed. Although **diethyl ether** is perhaps the best known general anesthetic to the layperson, it fails on several counts (flammability, side effects of nausea or vomiting, and relatively slow action). It is quite potent, however, and produces good analgesia and muscle relaxation. The use of ether at present is rather limited, mainly because of its undesirable side effects. **Halothane,** $CF_3CHBrCl$, comes closest to an ideal inhalation anesthetic at present, but halogenated ethers such as **enflurane,** $CF_2H—O—CF_2CHClF$, are also used.

Local anesthetics are either applied to body surfaces or injected near nerves to desensitize a particular region of the body to pain. The best known of these anesthetics is procaine (Novocain), an aromatic amino-ester (see A Word About Morphine and Other Nitrogen-Containing Drugs, p. 395).

The discovery of anesthetics enabled physicians to perform surgery with deliberation and care, leading to many of the advances of modern medicine.

If R or R′ is tertiary, a strong nucleophile is not required since reaction will occur by an S_N1 (or E1) mechanism.

$$\text{(t-butyl phenyl ether)} \quad OC(CH_3)_3 \xrightarrow[\text{H}_2\text{O}]{\text{H}^+} \quad \text{(phenol)}\quad OH \; + \; (CH_3)_3COH \qquad \textbf{(8.16)}$$

t-butyl phenyl ether   phenol   t-butyl alcohol
(and $(CH_3)_2C{=}CH_2$)

The net result of these reactions is **cleavage** of the ether at one of the C—O bonds. Ether cleavage is a useful reaction for determining the structure of a complex, naturally occurring ether because it allows one to break the large molecule into more easily handled, smaller fragments.

EXAMPLE 8.5

Write out the steps in the mechanism for eq. 8.14.

Solution   The ether is first protonated by the acid.

$$\overset{..}{\underset{..}{CH_3CH_2\overset{..}{\underset{..}{O}}CH(CH_3)_2}} \xrightleftharpoons{H^+} CH_3CH_2\overset{\overset{\overset{H}{|}}{\underset{..}{O}}+}{CH(CH_3)_2}$$

oxonium ion

The resulting **oxonium ion** is then cleaved by S_N2 attack of iodide ion at the primary carbon (recall that $1° > 2°$ in S_N2 reactions).

$$I^- + CH_3CH_2 \overset{\overset{H}{\underset{|}{+}}}{O}CH(CH_3)_2 \longrightarrow CH_3CH_2I + HOCH(CH_3)_2$$

PROBLEM 8.12   Write out the steps in the mechanism for formation of *t*-butyl alcohol in eq. 8.16. Which C—O bond cleaves, the one to the phenyl or the one to the *t*-butyl group?

8.7   Epoxides (Oxiranes)

Epoxides (or oxiranes) are cyclic ethers with a three-membered ring containing one oxygen atom.

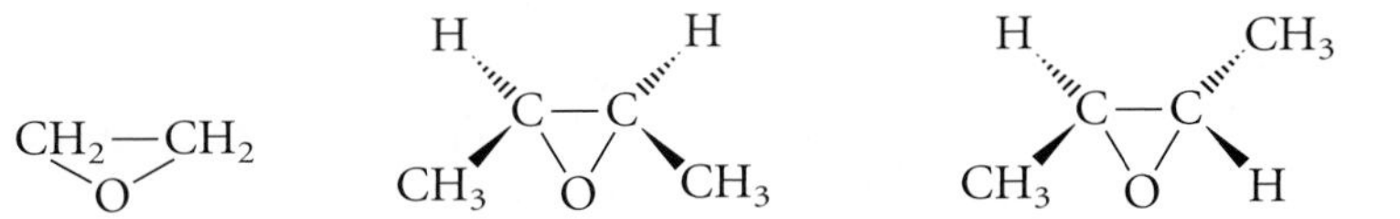

CH₂—CH₂ with O bridge	*cis*-2-butene oxide structure	*trans*-2-butene oxide structure
ethylene oxide	*cis*-2-butene oxide	*trans*-2-butene oxide
(oxirane)	(*cis*-2,3-dimethyloxirane)	(*trans*-2,3-dimethyloxirane)
bp 13.5°C	bp 60°C	bp 54°C

The most important commercial epoxide is ethylene oxide, produced by the silver-catalyzed air oxidation of ethylene.

$$CH_2{=}CH_2 + O_2 \xrightarrow[\text{250°C, pressure}]{\text{silver catalyst}} \underset{\text{ethylene oxide}}{CH_2{-}CH_2} \qquad \textbf{(8.17)}$$

Annual U.S. production of ethylene oxide exceeds 9.2 billion pounds. Only rather small amounts are used directly (for example, as a fumigant in grain storage). Most of the ethylene oxide constitutes a versatile raw material for the manufacture of other products, the main one being ethylene glycol.

A WORD ABOUT ...

The Gypsy Moth's Epoxide

The main mode of communication among insects is via the emission and detection of specific chemical substances. These substances are called **pheromones.** The word is from the Greek (*pherein,* to carry, and *horman,* to excite). Even though they are emitted and detected in exceedingly small amounts, pheromones have profound biological effects. One of their main effects is sexual attraction and stimulation, but they are also used as alarm substances to alert members of the same species to danger, as aggregation substances to call together both sexes of a species, and as trail substances to lead members of a species to food.

Male (top) and female (bottom) gypsy moths.

Often pheromones are chemically simple compounds—alcohols, esters, aldehydes, ketones, ethers, epoxides, or even hydrocarbons. Two examples are **muscalure** and **bombykol,** the sex attractants of the common housefly and the silkworm moth, respectively.

$$CH_3(CH_2)_7 \qquad (CH_2)_{12}CH_3$$
$$C=C$$
$$H \qquad\qquad H$$

muscalure

$$CH_3(CH_2)_2 \qquad H \qquad (CH_2)_8CH_2OH$$
$$C=C$$
$$C=C \qquad\qquad H$$
$$H \qquad H$$

bombykol

Their molecular weights are low enough that the substances are volatile, yet not so low that they disperse too rapidly. Also, their molecular structures must be distinctive to make them species-specific; survival of the species would not be served by attracting another species. Often this specificity is attained through stereoisomerism (at double bonds and/or at chiral centers), but it can also be achieved by using specific ratios of two or more pheromones for a particular communication purpose.

Let us consider a specific pheromone, **disparlure,** the sex attractant of the gypsy moth (*Lymantria dispar*). The gypsy moth is a serious despoiler of forest and shade trees as well as fruit orchards. Gypsy moth larvae, which hatch each spring, are voracious eaters and can strip a tree bare of leaves in just a few weeks.

The abdominal tips (last two segments) of the virgin female moth contain the sex attractant. Extraction of 78,000 tips led to isolation of the main sex attractant, which was the following *cis*-epoxide.

$$(CH_3)_2CH(CH_2)_4 \qquad\qquad (CH_2)_9CH_3$$
$$7 \quad 8$$
$$H \quad O \quad H$$

(7R,8S)-(+)-7,8-epoxy-2-methyloctadecane
(disparlure)

The active isomer has the *R* configuration at carbon-7 and the *S* configuration at carbon-8. This isomer can be detected by the male gypsy moth at a concentration as low as 10^{-10} g/mL; its enantiomer is inactive in solutions a million times more concentrated.

Disparlure has been synthesized in the laboratory. The synthetic material can be used to lure the male to traps and in that way to control the insect population. This form of insect control sometimes has advantages over spraying with insecticides.

For more on pheromones, chemical ecology, and organic synthesis, see K. Mori, *Acc. Chem. Res.,* **2000,** *33,* 102–110.

The reaction in eq. 8.17 is suitable only for ethylene oxide. Other epoxides are usually prepared by the reaction of an alkene with an organic peroxyacid (often called simply a peracid).

$$\text{(cyclohexene)} + R-\overset{O}{\underset{\|}{C}}-O-O-H \longrightarrow \text{(cyclohexene oxide)} O + R-\overset{O}{\underset{\|}{C}}-OH \qquad \textbf{(8.18)}$$

cyclohexene   organic peroxy acid   cyclohexene oxide   organic acid

Peroxy acids, like hydrogen peroxide H—O—O—H, to which they are structurally related, are good oxidizing agents. On a large scale, peracetic acid (R = CH_3) is used, whereas in the laboratory organic peroxy acids such as *m*-chloroperbenzoic acid (R = *m*-chlorophenyl) are frequently used.

PROBLEM 8.13 Write an equation for the reaction of cyclopentene with *m*-chloroperbenzoic acid,

---

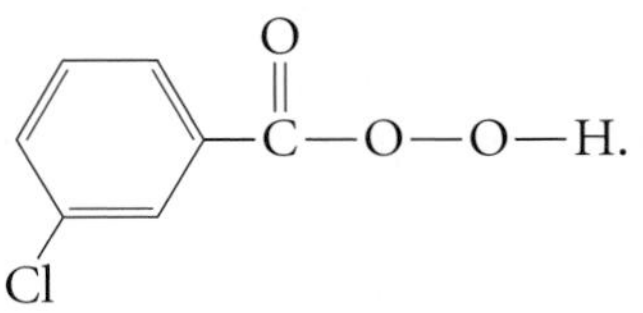

8.8 Reactions of Epoxides

Because of the strain in the three-membered ring, epoxides are much more reactive than ordinary ethers and give products in which the ring has opened. For example, with water they undergo acid-catalyzed ring opening to give glycols. [Remember that treating alkenes with alkaline $KMnO_4$ (Sec. 3.17.a) also produces glycols.]

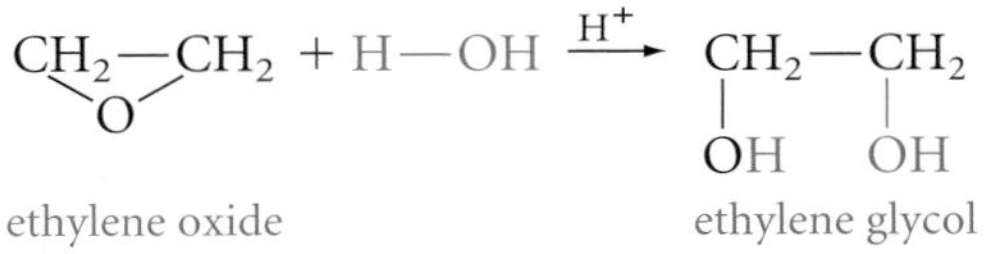

$$CH_2—CH_2 + H—OH \xrightarrow{H^+} CH_2—CH_2 \qquad (8.19)$$
$$\underset{O}{\diagdown\diagup} \qquad\qquad\qquad \underset{OH\ \ OH}{|\ \ \ \ |}$$

ethylene oxide   ethylene glycol

In this way, about 8.2 billion pounds of ethylene glycol are produced annually in the United States alone. Approximately half of it is used in automobile cooling systems as antifreeze. Most of the rest is used to prepare polyester fibers.

EXAMPLE 8.6

Write equations that show the mechanism for eq. 8.19.

Solution The first step is reversible protonation of the epoxide oxygen, as in eq. 8.12.

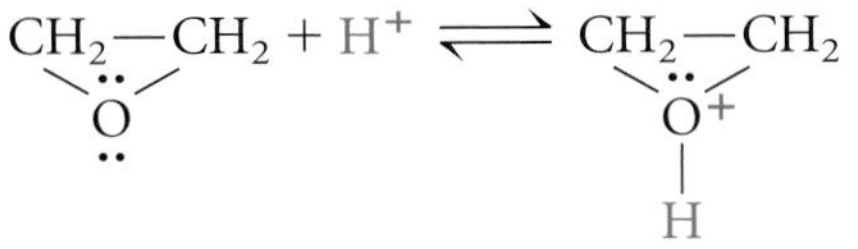

$$CH_2—CH_2 + H^+ \rightleftharpoons CH_2—CH_2$$

The second step is a nucleophilic S_N2 displacement on the primary carbon, with water as the nucleophile. Then proton loss yields the glycol (see Table 6.1, entry 3).

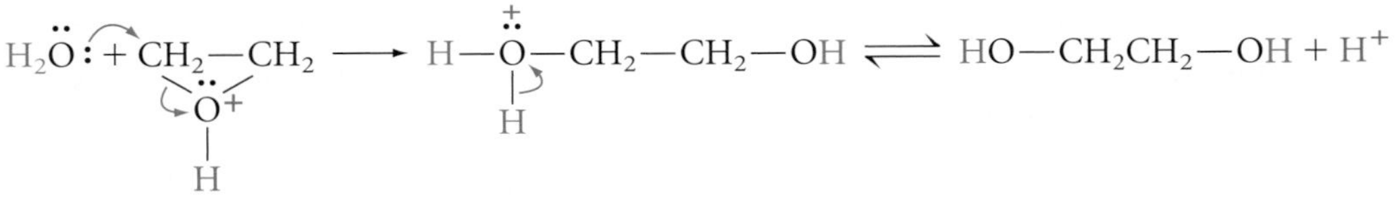

$$H_2\ddot{O}:+ CH_2—CH_2 \longrightarrow H—\overset{+}{\underset{H}{\ddot{O}}}—CH_2—CH_2—OH \rightleftharpoons HO—CH_2CH_2—OH + H^+$$

PROBLEM 8.14 Write an equation for the acid-catalyzed reaction of cyclohexene oxide with water. Predict the stereochemistry of the product.

Other nucleophiles add to epoxides in a similar way.

$$
\underset{\overset{\displaystyle |}{\text{O}}}{\text{CH}_2\text{—CH}_2} \xrightarrow{\text{H}^+}
\begin{cases}
\xrightarrow{\text{CH}_3\text{OH}} \text{HOCH}_2\text{CH}_2\text{OCH}_3 \\
\quad\quad\quad\quad \text{2-methoxyethanol} \\[1em]
\xrightarrow{\text{HOCH}_2\text{CH}_2\text{OH}} \text{HOCH}_2\text{CH}_2\text{OCH}_2\text{CH}_2\text{OH} \\
\quad\quad\quad\quad\quad\quad \text{diethylene glycol}
\end{cases}
\tag{8.20}
$$

2-Methoxyethanol is an additive for jet fuels, used to prevent water from freezing in fuel lines. Being both an alcohol and an ether, it is soluble in both water and organic solvents. *Diethylene glycol* is useful as a plasticizer (softener) in cork gaskets and tiles.

Grignard reagents and organolithium compounds are strong nucleophiles capable of opening the ethylene oxide ring. The initial product is a magnesium alkoxide or lithium alkoxide, but after hydrolysis we obtain a primary alcohol with two more carbon atoms than the organometallic reagent.

$$
\overset{\delta-\ \ \ \delta+}{\text{R—MgX}} + \overset{\delta+\ \ \ \ \delta+}{\text{H}_2\text{C—CH}_2} \longrightarrow \underset{\substack{\text{a magnesium} \\ \text{alkoxide}}}{\text{RCH}_2\text{CH}_2\text{OMgX}} \xrightarrow{\text{H—OH}} \text{RCH}_2\text{CH}_2\text{OH} + \text{Mg(OH)X}
\tag{8.21}
$$

$$
\overset{\delta-\ \ \ \delta+}{\text{R—Li}} + \overset{\delta+\ \ \ \ \delta+}{\text{H}_2\text{C—CH}_2} \longrightarrow \underset{\substack{\text{a lithium} \\ \text{alkoxide}}}{\text{RCH}_2\text{CH}_2\text{OLi}} \xrightarrow{\text{H—OH}} \text{RCH}_2\text{CH}_2\text{OH} + \text{LiOH}
\tag{8.22}
$$

PROBLEM 8.15   Write an equation for the reaction between ethylene oxide and

a. $CH_3CH_2CH_2MgCl$ followed by hydrolysis.
b. $H_2C{=}CHLi$ followed by hydrolysis.
c. $CH_3C{\equiv}C^-Na^+$ followed by hydrolysis.

8.9 Cyclic Ethers

Cyclic ethers whose rings are larger than the three-membered epoxides are known. The most common are five- or six-membered rings. Some examples include

tetrahydrofuran	tetrahydropyran	1,4-dioxane
(oxolane)	(oxane)	bp 101°C
bp 67°C	bp 88°C	

Tetrahydrofuran (THF) is a particularly useful solvent that not only dissolves many organic compounds but is miscible with water. THF is an excellent solvent—often superior to diethyl ether—in which to prepare Grignard reagents. Although it has the same number of carbon atoms as diethyl ether, they are "pinned back" in a ring. The

Tetrahydrofuran (THF), tetrahydropyran, and 1,4-dioxane are important cyclic ethers.

oxygen in THF is therefore less hindered and better at coordinating with the magnesium in a Grignard reagent. **Tetrahydropyran** and **1,4-dioxane** are also soluble in both water and organic solvents.

In recent years, there has been much interest in macrocyclic (large-ring) polyethers. Some examples are

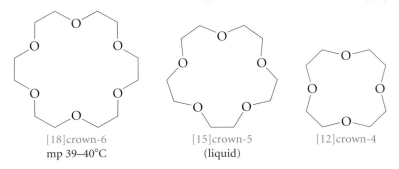

| [18]crown-6 | [15]crown-5 | [12]crown-4 |
| mp 39–40°C | (liquid) | |

These compounds are called **crown ethers** because their molecules have a crownlike shape. The bracketed number in their common names gives the ring size, and the terminal number gives the number of oxygens. The oxygens are usually separated from one another by two carbon atoms.

Crown ethers have the unique property of forming complexes with positive ions (Na^+, K^+, and so on). The positive ions fit within the macrocyclic rings selectively, depending on the sizes of the particular ring and ion. For example, [18]crown-6 binds K^+ more tightly than it does the smaller Na^+ (too loose a fit) or the larger Cs^+ (too large to fit in the hole). Similarly, [15]crown-5 binds Na^+, and [12]crown-4 binds Li^+. The crown ethers act as hosts for their ionic guests.

Crown ethers are crown-shaped macrocyclic polyethers that form complexes with positive ions.

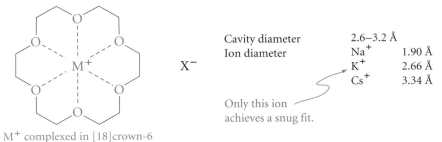

M^+ complexed in [18]crown-6

Cavity diameter	2.6–3.2 Å
Ion diameter	
Na^+	1.90 Å
K^+	2.66 Å
Cs^+	3.34 Å

Only this ion achieves a snug fit.

This complexing ability is so strong that ionic compounds can be dissolved in organic solvents that contain a crown ether. For example, potassium permanganate ($KMnO_4$) is soluble in water but insoluble in benzene. However, if some dicyclohexyl[18]crown-6 is dissolved in the benzene, it is possible to extract the potassium permanganate from the water into the benzene! The resulting "purple benzene," containing free, essentially unsolvated permanganate ions, is a powerful oxidizing agent.*

The selective binding of metallic ions by macrocyclic compounds is important in nature. Several antibiotics, such as **nonactin,** have large rings that contain

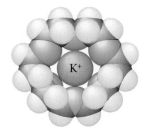

Model of [18]crown-6 complex with K⁺.

---

*Crown ethers were discovered by Charles J. Pedersen, working at Du Pont Company. This discovery had broad implications for a field now known as molecular recognition, or host–guest chemistry. Pedersen, Donald J. Cram (U.S.), and Jean-Marie Lehn (France) shared the 1987 Nobel Prize in chemistry for their imaginative development of this field. You might enjoy Pedersen's personal account of this discovery (*Journal of Inclusion Phenomena*, **1988**, *6*, 337–350); the same journal contains the Nobel lectures by Cram and Lehn on their work.

regularly spaced oxygen atoms. Nonactin (which contains four tetrahydrofuran rings joined by four ester links) selectively binds K$^+$ (in the presence of Na$^+$) in aqueous media, thus allowing the selective transport of K$^+$ (but not Na$^+$) through cell membranes.

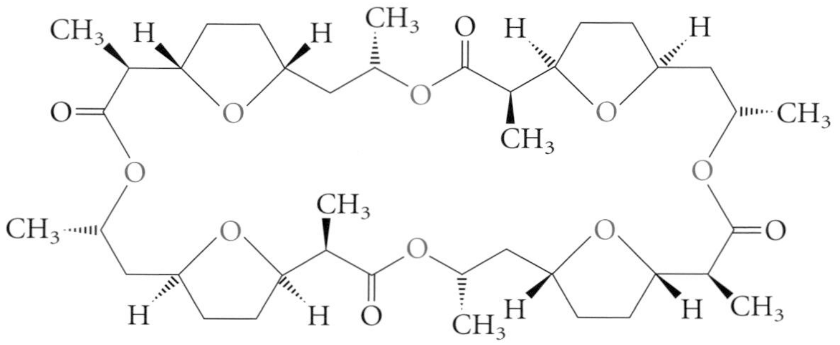

nonactin

REACTION SUMMARY

1. Organometallic Compounds

a. Preparation of Grignard Reagents (Sec. 8.4)

$$R\text{—}X + Mg \xrightarrow{\text{ether or THF}} R\text{—}MgX$$
X = Cl, Br, I

b. Preparation of Organolithium Reagents (Sec. 8.4)

$$R\text{—}X + 2 Li \longrightarrow R\text{—}Li + LiX$$
X = Cl, Br, I

c. Hydrolysis of Organometallics to Alkanes (Sec. 8.4)

$$R\text{—}MgX + H_2O \longrightarrow R\text{—}H + Mg(OH)X$$
$$R\text{—}Li + H_2O \longrightarrow R\text{—}H + LiOH$$

---

2. Ethers

a. Preparation by Dehydration of Alcohols (Sec. 8.5)

$$2 R\text{—}OH \xrightarrow[\Delta]{H_2SO_4} R\text{—}O\text{—}R + H_2O$$

b. Preparation from Alkenes and Alcohols (Sec. 8.5)

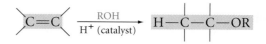

c. Preparation from Alcohols and Alkyl Halides (Sec. 8.5)

$$2 ROH + 2 Na \longrightarrow 2 RO^- + H_2$$
$$\text{or } ROH + NaH \longrightarrow RO^-Na^+ + H_2$$
$$RO^-Na^+ + R'\text{—}X \longrightarrow RO\text{—}R' + Na^+X^-$$
(best for R′ = primary)

d. Cleavage by Hydrogen Halides (Sec. 8.6)

$$R\text{—}O\text{—}R + HX \longrightarrow R\text{—}X + R\text{—}OH$$

e. Cleavage by Boron Tribromide (Sec. 8.6)

$$R\text{—}O\text{—}R \xrightarrow[\text{2. H}_2\text{O}]{\text{1. BBr}_3} RBr + ROH$$

3. Epoxides

a. Preparation from Alkenes (Sec. 8.7)

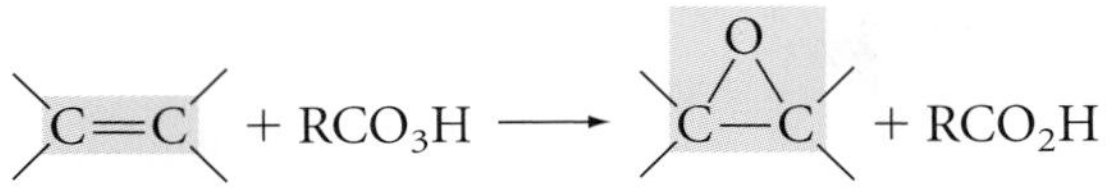

$$\text{C=C} + RCO_3H \longrightarrow \text{C}-\text{C} \text{ (epoxide)} + RCO_2H$$

b. Reaction with Water and Alcohols (Sec. 8.8)

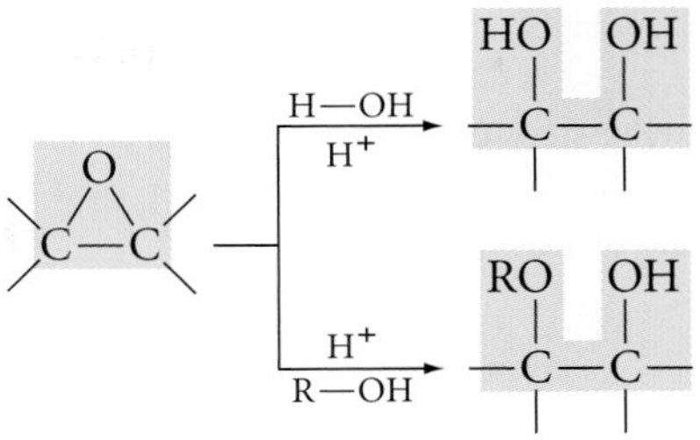

$$\xrightarrow[H^+]{H-OH} \begin{array}{cc} HO & OH \\ | & | \\ -C-C- \\ | & | \end{array}$$

$$\xrightarrow[R-OH]{H^+} \begin{array}{cc} RO & OH \\ | & | \\ -C-C- \\ | & | \end{array}$$

c. Reaction with Organometallic Reagents (Sec. 8.8)

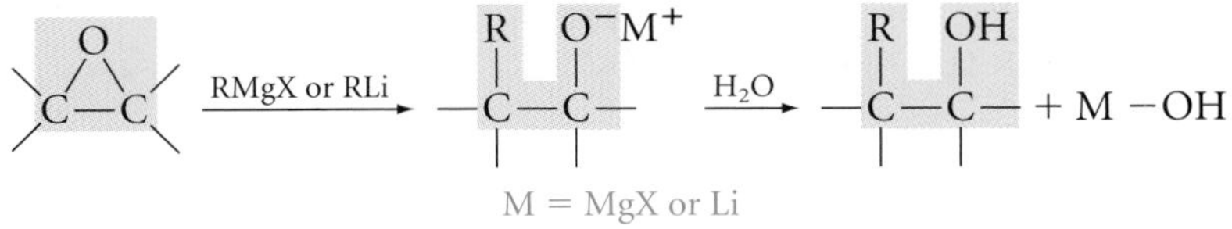

$$\xrightarrow{RMgX \text{ or } RLi} \begin{array}{cc} R & O^-M^+ \\ | & | \\ -C-C- \\ | & | \end{array} \xrightarrow{H_2O} \begin{array}{cc} R & OH \\ | & | \\ -C-C- \\ | & | \end{array} + M-OH$$

M = MgX or Li

ADDITIONAL PROBLEMS

Ethers and Epoxides: Structure, Nomenclature, and Properties

8.16 Write a structural formula for each of the following compounds:

a. dimethyl ether
b. ethyl isopropyl ether
c. 2-methoxyhexane
d. allyl propyl ether
e. *p*-chlorophenyl methyl ether
f. *trans*-2-ethoxycyclopentanol
g. ethylene glycol diethyl ether
h. 1-methoxypropene
i. propylene oxide
j. *p*-ethoxyanisole

8.17 Name each of the following compounds:

a. $CH_3CH_2CH_2OCH(CH_3)_2$

b. $CH_3OCH_2CH_2CH_3$

c. 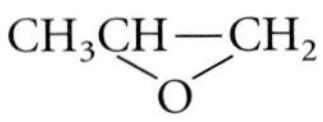 CH_3CH-CH_2 (epoxide, O bridging the two carbons)

d. 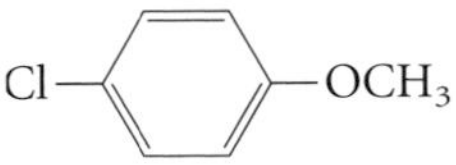 Cl—⟨benzene ring⟩—OCH_3

e. $CH_3CH_2CH(OCH_2CH_3)CH_2CH_3$

f. $CH_3CH_2OCH_2CH_2OH$

g. 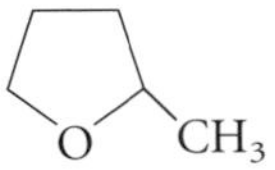 ⟨tetrahydrofuran ring with O and CH_3 substituent⟩

h. $CH_3OCH_2CH_2C\equiv CH$

8.18 Ethers and alcohols can be isomeric (see Sec. 1.8). Write the structures and give the names for all possible isomers with the molecular formula $C_4H_{10}O$.

8.19 Consider four compounds that have nearly the same molecular weights: 1,2-dimethoxyethane, ethyl propyl ether, hexane, and 1-pentanol.

a. Draw the structural formulas of these four molecules.
b. Which would you expect to have the highest boiling point? Which would be most soluble in water? Explain the reasons for your choices.

= concept connections

Preparation and Reaction of Grignard Reagents

8.20 Write equations for the reaction of each of the following with (1) Mg in ether followed by (2) addition of D_2O to the resulting solution:

  a. $CH_3CH_2CH_2Br$ **b.** $CH_3CH_2OCH_2CH_2CH_2Br$

8.21 The following steps can be used to convert anisole to *o-t*-butylanisole. Give the reagent for each step. Explain why the overall result cannot be achieved in one step by a Friedel–Crafts alkylation. (*Hint*: Sections 4.10 and 4.11 may be helpful for the second part of this question.)

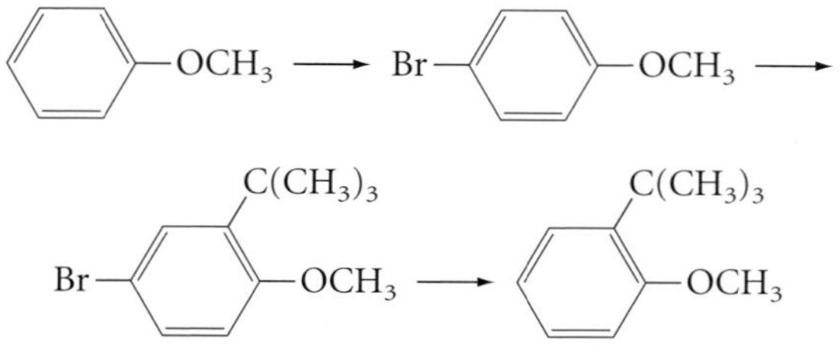

Preparation of Ethers

8.22 Write equations for the best method to prepare each of the following ethers:

  a. $(CH_3CH_2CH_2)_2O$ **b.** 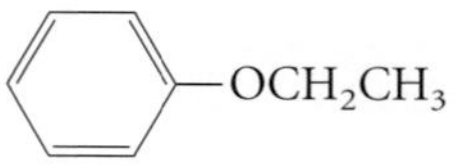 $—OCH_2CH_3$ **c.** $CH_3CH_2OC(CH_3)_3$

8.23 Explain why the Williamson synthesis cannot be used to prepare diphenyl ether.

8.24 Several companies are developing the manufacture and use of dimethyl ether, CH_3OCH_3, as an efficient and clean alternative to diesel fuel [*Chemical and Engineering News*, **1995** (May 29), 37–39]. Show how dimethyl ether could be synthesized from methanol.

Behavior of Ethers in Acids and Bases

8.25 Write an equation for each of the following reactions. If no reaction occurs, say so.

  a. methyl propyl ether + excess HBr (hot) $\longrightarrow$ **b.** dibutyl ether + boiling aqueous NaOH $\longrightarrow$
  c. ethyl ether + cold concentrated H_2SO_4 $\longrightarrow$ **d.** dipropyl ether + Na $\longrightarrow$
  e. ethyl phenyl ether $\xrightarrow[\text{2. H}_2\text{O}]{\text{1. BBr}_3}$

8.26 Ethers are soluble in cold, concentrated sulfuric acid, but alkanes are not. This difference can be used as a simple chemical test to distinguish between these two classes of compounds. What chemistry (show an equation) is the basis for this difference? (*Hint*: See Sec. 8.6.)

8.27 When heated with excess HBr, a cyclic ether gave 1,5-dibromopentane as the only organic product. Write a structure for the ether and an equation for the reaction. (*Hint*: See Secs. 8.6 and 7.9.)

Preparation and Reactions of Epoxides

8.28 Using the peroxyacid epoxidation of an alkene and the ring opening of an epoxide, devise a two-step synthesis of 1,2-butanediol from 1-butene.

8.29 Write an equation for the reaction of ethylene oxide with

  a. 1 mole of HCl.   **b.** excess HCl.   **c.** phenol + H^+.

8.30 $CH_3CH_2OCH_2CH_2OH$ (ethyl cellosolve) and $CH_3CH_2OCH_2CH_2OCH_2CH_2OH$ (ethyl carbitol) are solvents used in the formulation of lacquers. They are produced commercially from ethylene oxide and certain other reagents. Show with equations how this might be done.

8.31 2-Phenylethanol, which has the aroma of oil of roses, is used in perfumes. Write equations to show how 2-phenylethanol can be synthesized from bromobenzene and ethylene oxide, using a Grignard reagent.

8.32 2,2-Dimethyloxirane dissolved in excess methanol and treated with a little acid yields the product 2-methoxy-2-methyl-1-propanol (and no 1-methoxy-2-methyl-2-propanol). What reaction mechanism explains this result? (*Hint:* See Secs. 6.5 and 3.10.)

8.33 Write an equation for the reaction of ammonia with ethylene oxide. The product is a water-soluble organic base used to absorb and concentrate CO_2 in the manufacture of dry ice.

8.34 Design a synthesis of 3-pentyn-1-ol using propyne and ethylene oxide as the only sources of carbon atoms.

8.35 Write out the steps in the reaction mechanisms for the reactions given in eq. 8.20.

Cyclic Ethers

8.36 The cyclic polyether *nonactin* selectively binds and transports K^+ ions, while the acyclic polyether *monensin* (see p. 231 for the structure) selectively binds and transports Na^+ ions. Use information in Section 8.9 to help you answer the following questions.

   a. Which atoms in these two compounds are likely to be involved in binding the metal cations? Explain.
   b. Suggest a reason why *nonactin* is selective for K^+ while *monensin* is selective for Na^+.

Puzzles

8.37 What chemical test will distinguish between the compounds in each of the following pairs? Indicate what is visually observed with each test. (*Hint:* Make a list of tests you have learned from this chapter and previous chapters for identifying the functional groups in these compounds.)

   a. dipropyl ether and hexane   **b.** ethyl phenyl ether and allyl phenyl ether
   c. phenol and anisole   **d.** 1-butanol and methyl propyl ether

8.38 An organic compound with the molecular formula $C_4H_{10}O_3$ shows properties of both an alcohol and an ether. When treated with an excess of hydrogen bromide, it yields only one organic compound, 1,2-dibromoethane. Draw a structural formula for the original compound.

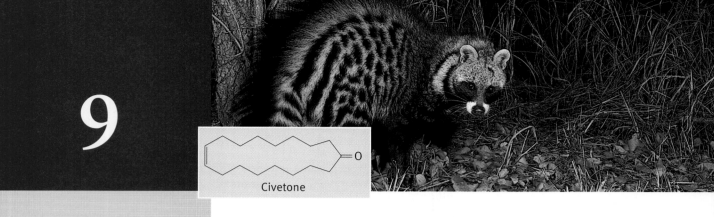

Civetone

9

ALDEHYDES AND KETONES

Aldehydes and ketones are a large family of organic compounds that permeate our everyday lives. They are responsible for the fragrant odors of many fruits and fine perfumes. For example, cinnamaldehyde (an aldehyde) provides the smell we associate with cinnamon, and civetone (a ketone) is used to provide the musky odor of many perfumes. Formaldehyde is a component of many building materials we use to construct our houses. The ketones testosterone and estrone are known to many as hormones responsible for our sexual characteristics. And the chemistry of aldehydes and ketones plays a role in how we digest food and even in how we can see the words on this page (chemistry of vision). So what are aldehydes and ketones?

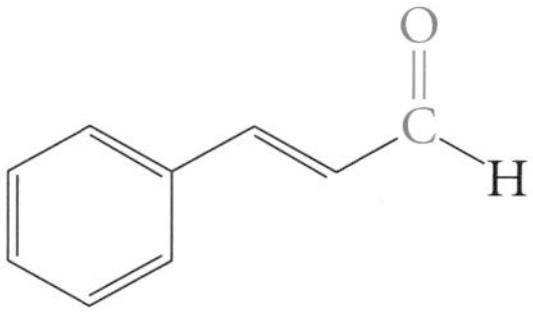

cinnamaldehyde

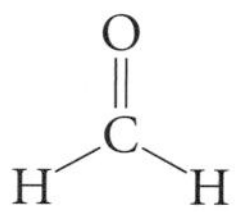

formaldehyde

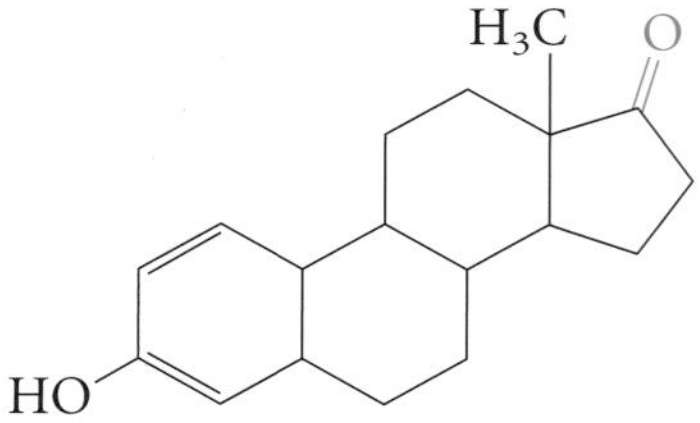

estrone

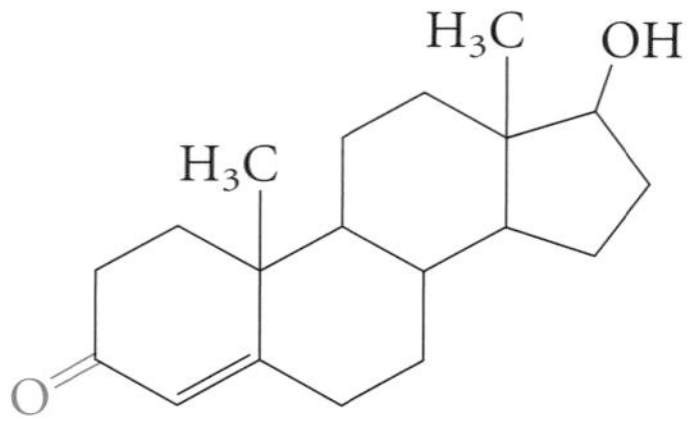

testosterone

▲ The civet cat (*Viverra civetta*) is the original source of civetone, a sweet and pungent ketone, now produced synthetically and used as a fixative in perfumery.

Aldehydes and ketones are characterized by the presence of the **carbonyl group,** perhaps the most important functional group in organic chemistry. Aldehydes have at least one hydrogen atom attached to the carbonyl carbon atom. The remaining group may be another hydrogen atom or any aliphatic or aromatic organic group. The —CH=O group characteristic of aldehydes is often called a **formyl group.** In ketones, the carbonyl carbon atom is connected to two other carbon atoms.

Aldehydes have at least one hydrogen atom attached to the carbonyl carbon atom. In **ketones,** the carbonyl carbon atom is connected to two other carbon atoms.

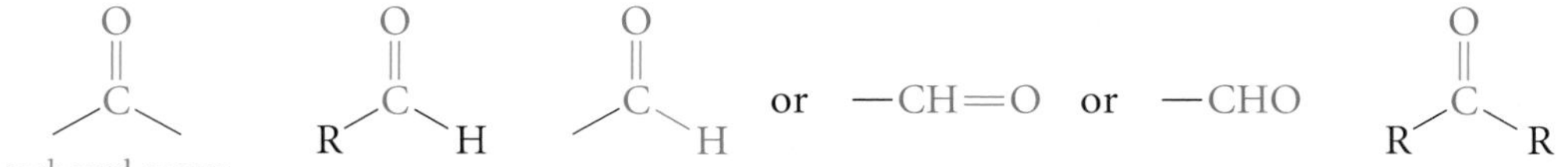

carbonyl group   aldehyde   aldehyde group **or** formyl group   ketone

We will see that the carbonyl group appears in many other organic compounds including carboxylic acids and their derivatives (Chapter 10). This chapter, however, will focus only on aldehydes and ketones.

9.1   Nomenclature of Aldehydes and Ketones

In the IUPAC system, the characteristic ending for aldehydes is *-al* (from the first syllable of aldehyde). The following examples illustrate the system:

methanal   ethanal   propanal   butanal
(formaldehyde)   (acetaldehyde)   (propionaldehyde)   (*n*-butyraldehyde)

The common names shown below the IUPAC names are frequently used, so you should learn them.

For substituted aldehydes, we number the chain starting with the aldehyde carbon, as the following examples illustrate:

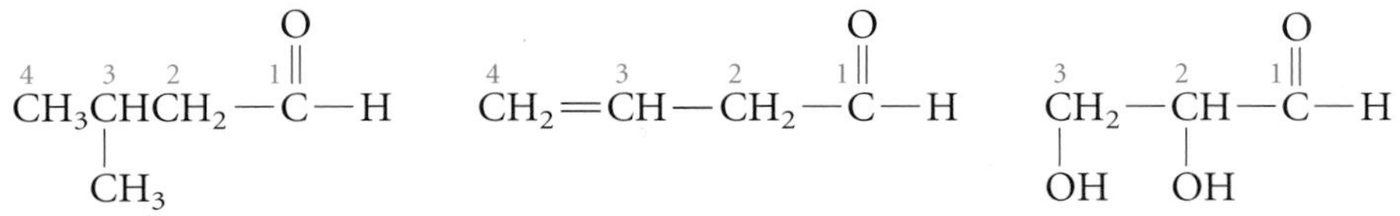

3-methylbutanal   3-butenal   2,3-dihydroxypropanal
(glyceraldehyde)

Notice from the last two examples that an aldehyde group has priority over a double bond or a hydroxyl group, not only in numbering, but also as the suffix. For cyclic aldehydes, the suffix *-carbaldehyde* is used. Aromatic aldehydes often have common names:

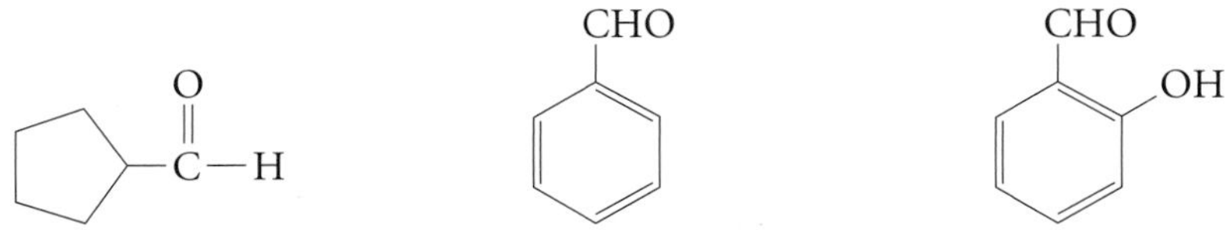

cyclopentanecarbaldehyde   benzaldehyde   salicylaldehyde
(formylcyclopentane)   (benzenecarbaldehyde)   (2-hydroxybenzenecarbaldehyde)

In the IUPAC system, the ending for ketones is *-one* (from the last syllable of ketone). The chain is numbered so that the carbonyl carbon has the lowest possible number. Common names of ketones are formed by adding the word *ketone* to the names of the alkyl or aryl groups attached to the carbonyl carbon. In still other cases, traditional names are used. The following examples illustrate these methods:

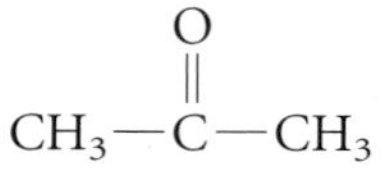

$$CH_3-\overset{\overset{\displaystyle O}{\|}}{C}-CH_3$$

propanone
(acetone)

$$\overset{1}{C}H_3-\overset{2}{\overset{\overset{\displaystyle O}{\|}}{C}}-\overset{3}{C}H_2\overset{4}{C}H_3$$

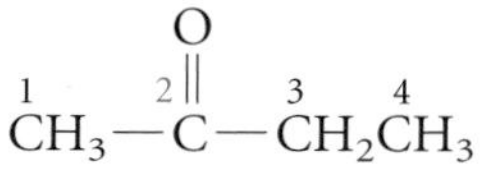

2-butanone
(ethyl methyl ketone)

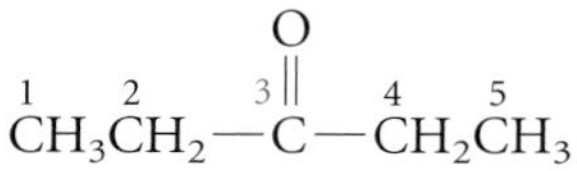

$$\overset{1}{C}H_3\overset{2}{C}H_2-\overset{3}{\overset{\overset{\displaystyle O}{\|}}{C}}-\overset{4}{C}H_2\overset{5}{C}H_3$$

3-pentanone
(diethyl ketone)

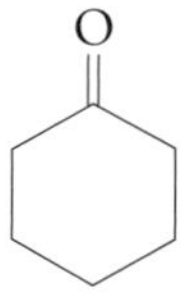

cyclohexanone

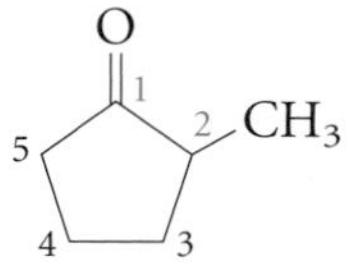

2-methylcyclopentanone

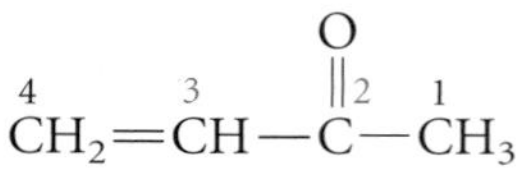

$$\overset{4}{C}H_2=\overset{3}{C}H-\overset{2}{\overset{\overset{\displaystyle O}{\|}}{C}}-\overset{1}{C}H_3$$

3-buten-2-one
(methyl vinyl ketone)

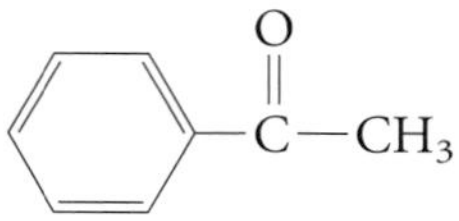

acetophenone
(methyl phenyl ketone)

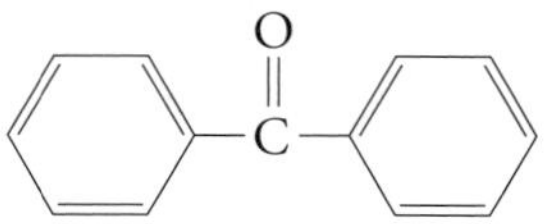

benzophenone
(diphenyl ketone)

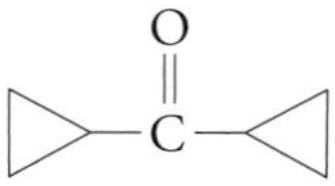

dicyclopropyl ketone

PROBLEM 9.1   Using the examples as a guide, write a structure for

a. pentanal.
c. 2-pentanone.
e. cyclohexanecarbaldehyde.

b. *m*-bromobenzaldehyde.
d. isopropyl methyl ketone.
f. 3-pentyn-2-one.

PROBLEM 9.2   Using the examples as a guide, write a correct name for

a. $(CH_3)_2CHCH_2CH{=}O$ b. $CH_3CH{=}CHCH{=}O$

c.

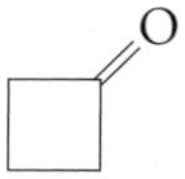

d. $CH_3(CH_2)_3\overset{\overset{\displaystyle O}{\|}}{C}CH_2CH_3$

Some Common Aldehydes and Ketones

Formaldehyde, the simplest aldehyde, is manufactured on a very large scale by the oxidation of methanol.

$$CH_3OH \xrightarrow[\text{600–700°C}]{\text{Ag catalyst}} CH_2{=}O \ + H_2 \tag{9.1}$$

formaldehyde

Annual world production is more than 8 billion pounds. Formaldehyde is a gas (bp -21°C), but it cannot be stored in a free state because it polymerizes readily.*

---

*The polymer derived from formaldehyde is a long chain of alternating CH_2 and oxygen units, which can be described by the structure $(CH_2O)_n$. See Chapter 14 for a discussion of polymers.

Normally it is supplied as a 37% aqueous solution called **formalin.** In this form it is used as a disinfectant and preservative, but most formaldehyde is used in the manufacture of plastics, building insulation, particle board, and plywood.

Acetaldehyde boils close to room temperature (bp 20°C). It is manufactured mainly by the oxidation of ethylene over a palladium–copper catalyst.

$$2 \ CH_2{=}CH_2 + O_2 \xrightarrow[\text{100–300°C}]{\text{Pd–Cu}} 2 \ CH_3CH{=}O \qquad \textbf{(9.2)}$$

About half the acetaldehyde produced annually is oxidized to acetic acid. The rest is used for the production of 1-butanol and other commercial chemicals.

Acetone, the simplest ketone, is also produced on a large scale—about 4 billion pounds annually. The most common methods for its commercial synthesis are the oxidation of propene (analogous to eq. 9.2), the oxidation of isopropyl alcohol (eq. 7.35, R = R′ = CH₃), and the oxidation of isopropylbenzene (eq. 9.3).

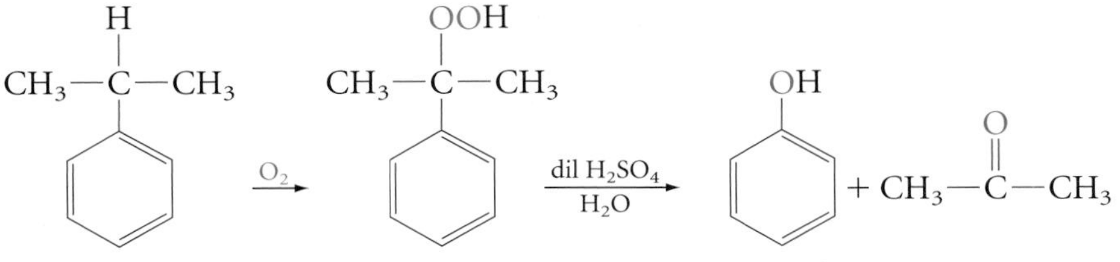

phenol   acetone

$$\textbf{(9.3)}$$

About 30% of the acetone is used directly, because it is not only completely miscible with water but is also an excellent solvent for many organic substances (resins, paints, dyes, and nail polish). The rest is used to manufacture other commercial chemicals, including bisphenol-A for epoxy resins (Sec. 14.9).

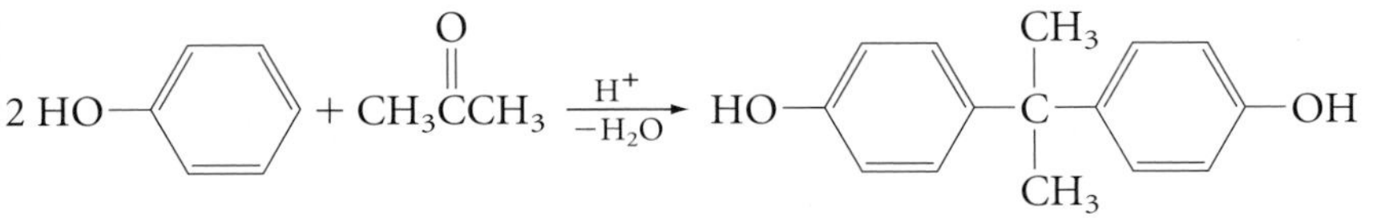

phenol   acetone   bisphenol-A

$$\textbf{(9.4)}$$

Quinones constitute a unique class of carbonyl compounds. They are cyclic conjugated diketones. The simplest example is **1,4-benzoquinone** (eq. 7.44). All quinones are colored and many are naturally occurring pigments that are used as dyes. **Alizarin** is an orange-red quinone that was used to dye the red coats of the British army during the American Revolution. **Vitamin K** is a quinone that is required for the normal clotting of blood.

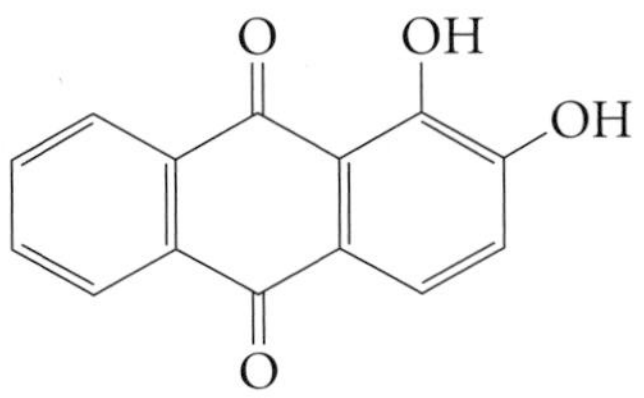

alizarin
mp 290°C

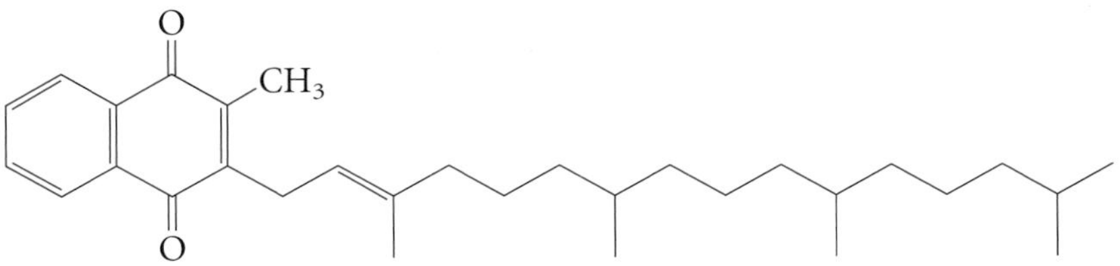

vitamin K
mp −20°C

9.3 Synthesis of Aldehydes and Ketones

We have already seen, in previous chapters, several ways to prepare aldehydes and ketones. One of the most useful is the oxidation of alcohols.

$$\underset{\text{OH}}{\overset{\text{H}}{\underset{|}{\overset{|}{\text{C}}}}} \xrightarrow[\text{agent}]{\text{oxidizing}} \text{C}=\text{O} \tag{9.5}$$

Oxidation of a primary alcohol gives an aldehyde, and oxidation of a secondary alcohol gives a ketone. Chromium reagents such as pyridinium chlorochromate (PCC) are commonly used in the laboratory for this purpose (review Sec. 7.12).

PROBLEM 9.3   Give the product expected from treatment of

a. cyclopentanol with Jones' reagent (see Sec. 7.12).
b. 5-methyl-1-hexanol with pyridinium chlorochromate.

EXAMPLE 9.1

Write an equation for the oxidation of an appropriate alcohol to $(CH_3)_2CHCH_2CHO$ (3-methylbutanal).

Solution   Aldehydes are prepared by oxidation of 1° alcohols (RCH_2OH) with pyridinium chlorochromate (PCC). First find the carbonyl carbon in 3-methyl-butanal (marked with an asterisk). Convert this carbon to a primary alcohol. A proper equation is:

$$(CH_3)_2CHCH_2{-}CH_2{-}OH \xrightarrow{\text{PCC}} (CH_3)_2CHCH_2{-}\overset{\overset{\textstyle O}{\|}}{\underset{*}{C}}{-}H$$

PROBLEM 9.4   Write an equation for the oxidation of an appropriate alcohol to

a. ⬠—CHO.   b. ⬠=O.

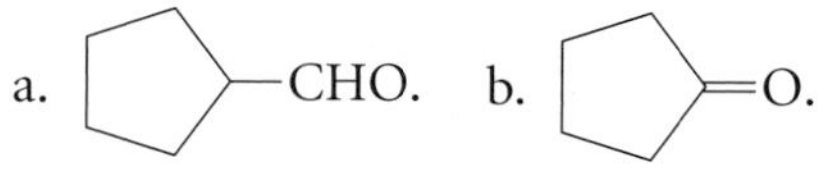

Aromatic ketones can be made by Friedel–Crafts acylation of an aromatic ring (review eq. 4.12 and Sec. 4.9d). For example,

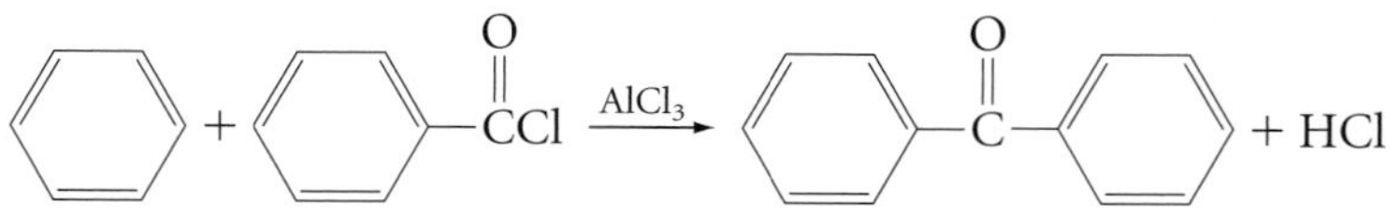

$$\text{benzene} + \text{benzoyl chloride} \xrightarrow{\text{AlCl}_3} \text{benzophenone} + \text{HCl} \tag{9.6}$$

PROBLEM 9.5   Complete the following equation and name the product:

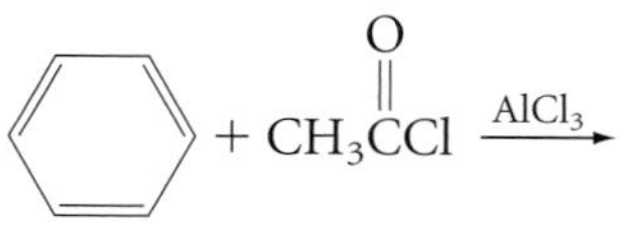

$$\text{⬡} + CH_3\overset{\overset{\textstyle O}{\|}}{C}Cl \xrightarrow{\text{AlCl}_3}$$

Methyl ketones can be prepared by hydration of terminal alkynes, catalyzed by acid and mercuric ion (review eq. 3.52). For example,

$$CH_3(CH_2)_5C\equiv CH \xrightarrow[Hg^{2+}]{H^+, H_2O} CH_3(CH_2)_5\overset{\displaystyle O}{\overset{\|}{C}}CH_3 \qquad (9.7)$$

1-octyne   2-octanone

> **PROBLEM 9.6** What alkyne would be useful for the synthesis of 2-heptanone (oil of cloves)?

Cloves, source of 2-heptanone.

9.4 Aldehydes and Ketones in Nature

Aldehydes and ketones occur very widely in nature. Figures 1.11 and 1.12 show three examples, and Figure 9.1 gives several more. Many aldehydes and ketones have pleasant odors and flavors and are used for these properties in perfumes and other consumer products (soaps, bleaches, and air fresheners, for example). The gathering and extraction of these fragrant substances from flowers, plants, and animal glands is extremely expensive, however. Chanel No. 5, introduced to the perfume market in 1921, was the first fine fragrance to use *synthetic* organic chemicals. Today most fragrances do.

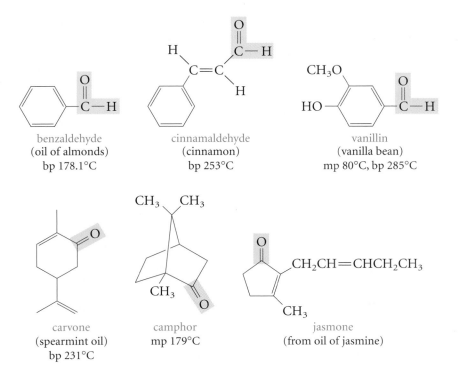

benzaldehyde
(oil of almonds)
bp 178.1°C

cinnamaldehyde
(cinnamon)
bp 253°C

vanillin
(vanilla bean)
mp 80°C, bp 285°C

carvone
(spearmint oil)
bp 231°C

camphor
mp 179°C

jasmone
(from oil of jasmine)

Vanilla beans, source of vanillin.

Figure 9.1
Some naturally occurring aldehydes and ketones.

9.5 The Carbonyl Group

To best understand the reactions of aldehydes, ketones, and other carbonyl compounds, we must first appreciate the structure and properties of the carbonyl group.

The carbon–oxygen double bond consists of a sigma bond and a pi bond (Figure 9.2). The carbon atom is sp^2-hybridized. *The three atoms attached to the carbonyl carbon lie in a plane with bond angles of 120°.* The pi bond is formed by overlap of a *p* orbital on carbon with an oxygen *p* orbital. There are also two unshared electron pairs on the oxygen atom. The $C{=}O$ bond distance is 1.24 Å, shorter than the $C{-}O$ distance in alcohols and ethers (1.43 Å).

Oxygen is much more electronegative than carbon. Therefore the electrons in the $C{=}O$ bond are attracted to the oxygen, producing a highly polarized bond. This effect is especially pronounced for the pi electrons and can be expressed in the following ways:

resonance contributors
to the carbonyl group

polarization of the
carbonyl group

As a consequence of this polarization, *most carbonyl reactions involve* **nucleophilic attack** *at the carbonyl carbon,* often accompanied by addition of a proton to the oxygen.

attack here by a ⟶   ⟵ may react
nucleophile   with a proton

$C{=}O$ bonds are quite different, then, from $C{=}C$ bonds, which are not polarized and where attack at carbon is usually by an electrophile (Sec. 3.9).

In addition to its effect on reactivity, polarization of the $C{=}O$ bond influences the physical properties of carbonyl compounds. For example, carbonyl compounds boil at higher temperatures than hydrocarbons, but at lower temperatures than alcohols of comparable molecular weight.

$$CH_3(CH_2)_4CH_3 \qquad CH_3(CH_2)_3CH{=}O \qquad CH_3(CH_2)_3CH_2OH$$

hexane (bp 69°C)   pentanal (bp 102°C)   pentanol (bp 118°C)

Why is this so? Unlike hydrocarbon molecules, which can be temporarily polarized, molecules of carbonyl compounds have *permanently polar* $C{=}O$ *bonds* and thus have

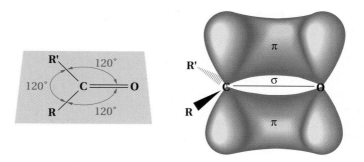

Figure 9.2
Bonding in the carbonyl group (see the text for a description of this bonding).

a stronger tendency to associate. *The positive part of one molecule is attracted to the negative part of another molecule.* These intermolecular forces of attraction, called **dipole–dipole interactions,** are generally stronger than van der Waals attractions (Sec. 2.7) but not as strong as hydrogen bonds (Sec. 7.4). Carbonyl compounds such as aldehydes and ketones that have a $C=O$ bond, but no $O-H$ bond, cannot form hydrogen bonds with one another, as can alcohols. Consequently, carbonyl compounds require more energy (heat) than hydrocarbons of comparable molecular weight to overcome intermolecular attractive forces when converted from liquid to vapor, but less than alcohols.

The polarity of the carbonyl group also affects the solubility properties of aldehydes and ketones. For example, carbonyl compounds with low molecular weights are soluble in water. Although they cannot form hydrogen bonds with themselves, they can form hydrogen bonds with $O-H$ or $N-H$ compounds.

Dipole–dipole interactions are opposite pole attractions between polar molecules.

$$\overset{\delta+}{\underset{}{}}C=\overset{\delta-}{O}: \cdots H-\overset{\delta-}{O}\diagup^{H}$$

PROBLEM 9.7 Arrange benzaldehyde (mol. wt. 106), benzyl alcohol (mol. wt. 108), and *p*-xylene (mol. wt. 106) in order of

a. increasing boiling point.   b. increasing water solubility.

9.6 Nucleophilic Addition to Carbonyl Groups: An Overview

Nucleophiles attack the carbon atom of a carbon–oxygen double bond because that carbon has a partial positive charge. The pi electrons of the $C=O$ bond move to the oxygen atom, which, because of its electronegativity, can easily accommodate the negative charge that it acquires. When these reactions are carried out in a hydroxylic solvent such as alcohol or water, the reaction is usually completed by addition of a proton to the negative oxygen. The overall reaction involves addition of a nucleophile and a proton across the pi bond of the carbonyl group.

$$\text{Nu}:^- + \ \ C=\overset{..}{\underset{..}{O}}: \ \rightleftharpoons \ \ \underset{\text{tetrahedral}}{\overset{\text{Nu}}{\underset{}{C}-\overset{..}{\underset{..}{O}}:^-}} \ \xrightarrow{\text{H}_2\text{O}} \ \overset{\text{Nu}}{\underset{}{C-\overset{..}{\underset{..}{O}}H}} \qquad \textbf{(9.8)}$$

trigonal reactant   tetrahedral intermediate   tetrahedral product

The carbonyl carbon, which is trigonal and sp^2-hybridized in the starting aldehyde or ketone, becomes tetrahedral and sp^3-hybridized in the reaction product.

Because of the unshared electron pairs on the oxygen atom, carbonyl compounds are weak Lewis bases and can be protonated. *Acids can catalyze the addition of weak nucleophiles to carbonyl compounds* by protonating the carbonyl oxygen atom.

$$C=\overset{..}{\underset{..}{O}}: + H^+ \longrightarrow \left[C=\overset{+}{\underset{..}{O}}H \longleftrightarrow \overset{+}{C}-\overset{..}{\underset{..}{O}}H \right] \xrightarrow{\text{Nu}:^-} \overset{\text{Nu}}{\underset{}{C-\overset{..}{\underset{..}{O}}H}} \qquad \textbf{(9.9)}$$

a resonance-stabilized carbocation

This converts the carbonyl carbon to a carbocation and enhances its susceptibility to attack by nucleophiles.

Nucleophiles can be classified as those that add reversibly to the carbonyl carbon and those that add irreversibly. Nucleophiles that add reversibly are also good leaving groups. In other words, they are the conjugate bases of relatively strong acids. Nucleophiles that add irreversibly are poor leaving groups, the conjugate bases of weak acids. This classification will be useful when we consider the mechanism of carbonyl additions.

In general, *ketones are somewhat less reactive than aldehydes toward nucleophiles.* There are two main reasons for this reactivity difference. The first reason is *steric.* The carbonyl carbon atom is more crowded in ketones (two organic groups) than in aldehydes (one organic group and one hydrogen atom). In nucleophilic addition, we bring these attached groups closer together because the hybridization changes from sp^2 to sp^3 and the bond angles decrease from 120° to 109.5° (eq. 9.8). Less strain is involved in additions to aldehydes than in additions to ketones because one of the groups (H) is small. The second reason is *electronic.* As we have already seen in connection with carbocation stability, alkyl groups are usually electron-donating compared to hydrogen (Sec. 3.10). They therefore tend to neutralize the partial positive charge on the carbonyl carbon, decreasing its reactivity toward nucleophiles. Ketones have two such alkyl groups; aldehydes have only one. If, however, the attached groups are strongly electron-withdrawing (contain halogens, for example), they can have the opposite effect and increase carbonyl reactivity toward nucleophiles.

In the following discussion, we will classify nucleophilic additions to aldehydes and ketones according to the type of new bond formed to the carbonyl carbon. We will consider oxygen, carbon, and nitrogen nucleophiles, in that sequence.

9.7   Addition of Alcohols: Formation of Hemiacetals and Acetals

The reactions discussed in this section are extremely important because they are crucial to understanding the chemistry of carbohydrates, which we will discuss later in Chapter 16.

Alcohols are oxygen nucleophiles. They add to the C=O bond, the OR group becoming attached to the carbon, and the proton becoming attached to the oxygen:

$$ ROH \ + \ \underset{\text{alcohol}}{} \ \underset{\substack{R' \\ | \\ H}}{\overset{}{C}}=O \ \underset{}{\overset{H^+}{\rightleftharpoons}} \ \underset{\substack{R' \\ | \\ H}}{\overset{RO}{C}}-OH \qquad \textbf{(9.10)} $$

alcohol   aldehyde   hemiacetal

Because alcohols are *weak* nucleophiles, an acid catalyst is required.* The product is a **hemiacetal;** it contains both alcohol and ether functional groups on the same carbon atom. The addition is reversible.

A **hemiacetal** contains both alcohol and ether functional groups on the same carbon atom.

---

*Many acid catalysts can be used. Sulfuric acid and *p*-toluenesulfonic acid are commonly used in the laboratory.

The mechanism of hemiacetal formation involves three steps. First, the carbonyl oxygen is protonated by the acid catalyst. The alcohol oxygen then attacks the carbonyl carbon, and a proton is lost from the resulting positive oxygen. *Each step is reversible.* In terms of acid–base reactions, the starting acid in each step is converted to a product acid of similar strength (Sec. 7.5).

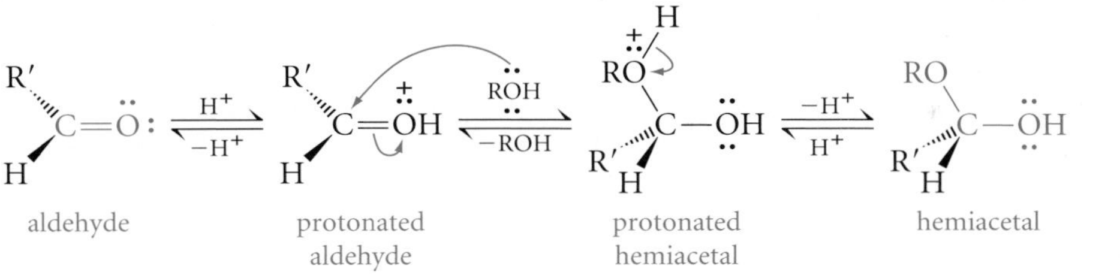

(9.11)

aldehyde   protonated aldehyde   protonated hemiacetal   hemiacetal

Ch. 9, Ref. 1. *View animation of acid-catalyzed nucleophilic addition of methanol to acetaldehyde.*

PROBLEM 9.8 Write an equation for the formation of a hemiacetal from acetaldehyde (CH_3CHO), ethanol (CH_3CH_2OH), and H^+. Show each step in the reaction mechanism.

In the presence of *excess alcohol,* hemiacetals react further to form **acetals.**

An **acetal** has two ether functional groups on the same carbon atom.

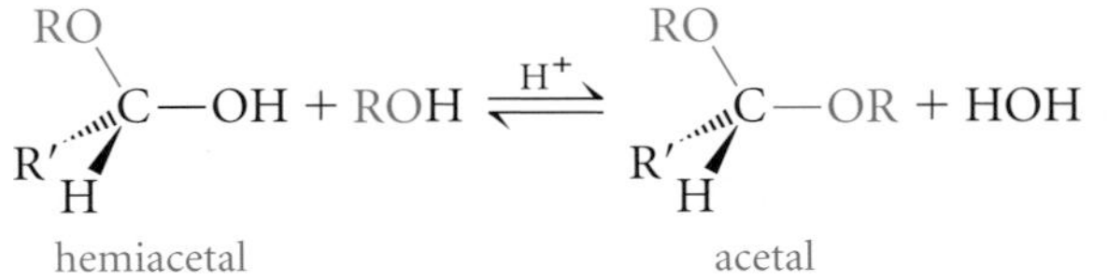

(9.12)

hemiacetal   acetal

The hydroxyl group of the hemiacetal is replaced by an alkoxyl group. Acetals have two ether functions at the same carbon atom.

The mechanism of acetal formation involves the following steps.

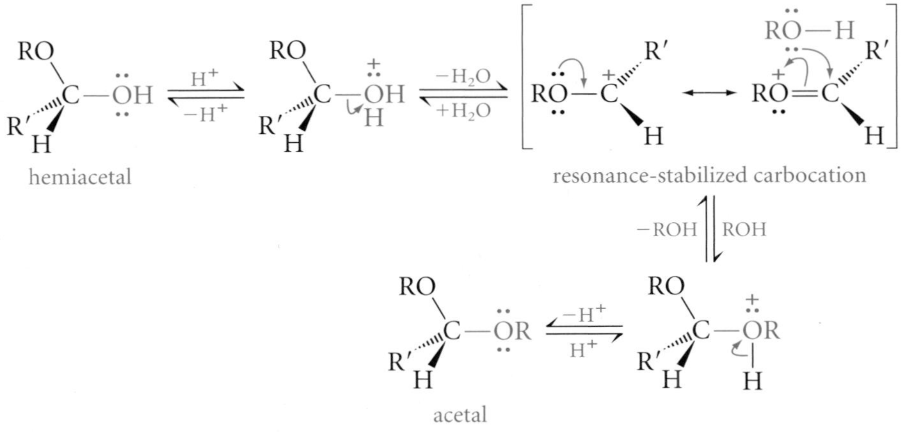

(9.13)

hemiacetal   resonance-stabilized carbocation

acetal

Either oxygen of the hemiacetal can be protonated. When the hydroxyl oxygen is protonated, loss of water leads to a resonance-stabilized carbocation. Reaction of this carbocation with the alcohol, which is usually the solvent and present in large excess, gives (after proton loss) the acetal. The mechanism is like an S_N1 reaction. *Each step is reversible.*

> **PROBLEM 9.9** Write an equation for the reaction of the hemiacetal
>
> $$\underset{\displaystyle CH_3\overset{\displaystyle |}{\underset{}{C}}HOCH_2CH_3}{\overset{\displaystyle OH}{}}$$
>
> with excess ethanol and H⁺. Show each step in the mechanism.

In a cyclic hemiacetal, the ether functional group is cyclic.

Aldehydes that have an appropriately located hydroxyl group *in the same molecule* may exist in equilibrium with a **cyclic hemiacetal,** formed by *intramolecular* nucleophilic addition. For example, 5-hydroxypentanal exists mainly in the cyclic hemiacetal form:

(9.14)

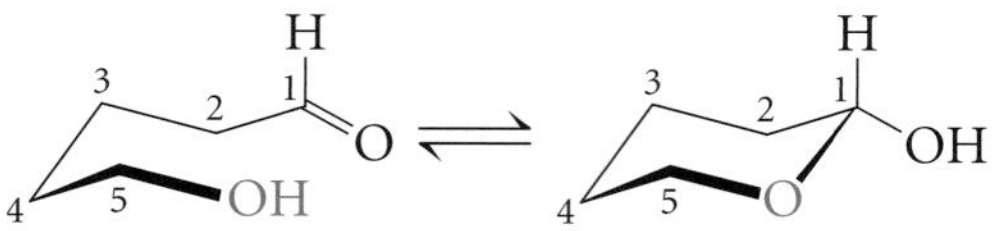

5-hydroxypentanal   hemiacetal form
of 5-hydroxypentanal

The hydroxyl group is favorably located to act as a nucleophile toward the carbonyl carbon, and cyclization occurs by the following mechanism:

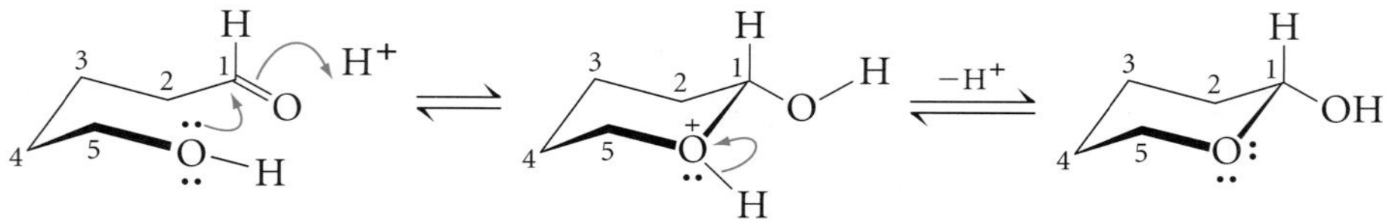

(9.15)

Compounds with a hydroxyl group four or five carbons from the aldehyde group tend to form cyclic hemiacetals and acetals because the ring size (five- or six-membered) is relatively strain free. As we will see in Chapter 16, these structures are crucial to the chemistry of carbohydrates. For example, glucose is an important carbohydrate that exists as a cyclic hemiacetal.

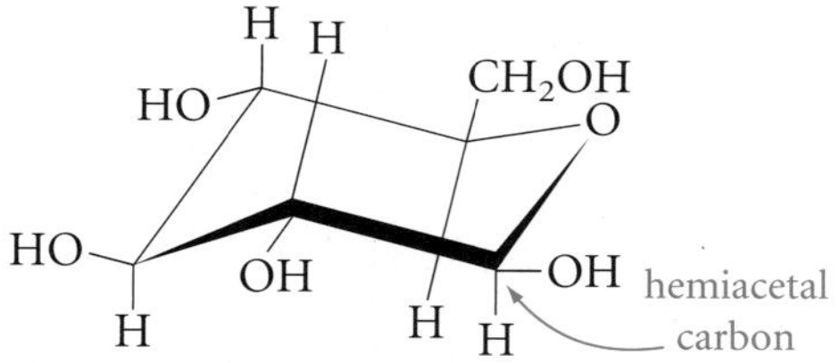

glucose

Ketones also form acetals. If, as in the following example, a glycol is used as the alcohol, the product will be cyclic.

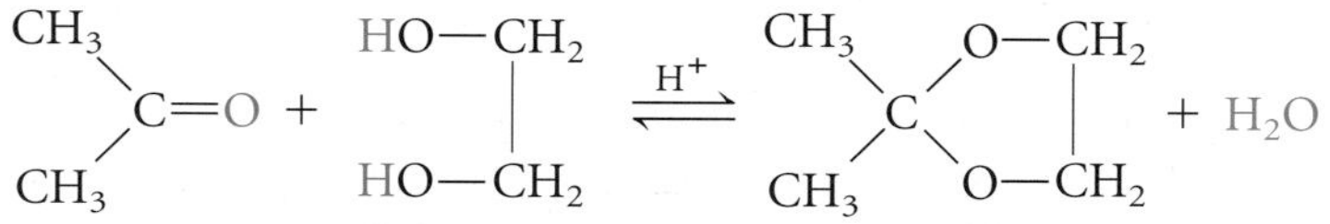

(9.16)

acetone   ethylene glycol   acetone–ethylene
glycol acetal

To summarize, aldehydes and ketones react with alcohols to form, first, hemiacetals and then, if excess alcohol is present, acetals.

$$R'-\underset{\substack{\| \\ O}}{C}-R'' \underset{H^+}{\overset{RO-H}{\rightleftharpoons}} R'-\underset{\substack{| \\ R''}}{\overset{| \\ OH}}{C}-OR \underset{H^+}{\overset{RO-H}{\rightleftharpoons}} R'-\underset{\substack{| \\ R''}}{\overset{| \\ OR}}{C}-OR + HOH \qquad (9.17)$$

aldehyde or ketone   hemiacetal   acetal

EXAMPLE 9.2

Write an equation for the reaction of benzaldehyde with excess methanol and an acid catalyst.

Solution

$$\text{C}_6\text{H}_5-\text{CHO} \xrightarrow[\text{H}^+ \text{(catalyst)}]{\text{CH}_3\text{OH (excess)}} \text{C}_6\text{H}_5-\underset{\substack{| \\ \text{OCH}_3}}{\overset{\substack{\text{OCH}_3 \\ |}}{\text{C}}}-\text{H} + \text{H}_2\text{O} \qquad (9.18)$$

PROBLEM 9.10 Show the steps in the mechanism for eq. 9.18.

PROBLEM 9.11 Write an equation for the acid catalyzed reactions between cyclohexanone and

a. excess ethanol.
b. excess ethylene glycol ($HOCH_2CH_2OH$).

Notice that acetal formation is a reversible process that involves a series of equilibria (eq. 9.17). How can these reactions be driven in the forward direction? One way is to use a large excess of alcohol. Another way is to remove water, a product of the forward reaction, as it is formed.* The reverse of acetal formation, called **acetal hydrolysis,** cannot proceed without water. On the other hand, an acetal can be hydrolyzed to its aldehyde or ketone and alcohol components by treatment with *excess water* in the presence of an acid catalyst. The hemiacetal intermediate in both the forward and reverse processes usually cannot be isolated when R′ and R″ are simple alkyl or aryl groups.

Acetal hydrolysis is the reverse of acetal formation.

EXAMPLE 9.3

Write an equation for the reaction of benzaldehyde dimethylacetal with aqueous acid.

Solution

$$\text{C}_6\text{H}_5-\underset{\substack{\diagdown \\ \text{OCH}_3}}{\overset{\diagup \text{OCH}_3}{\text{CH}}} \xrightarrow[\text{H}^+]{\text{H}_2\text{O}} \text{C}_6\text{H}_5-\text{CH}{=}\text{O} + 2\,\text{CH}_3\text{OH} \qquad (9.19)$$

PROBLEM 9.12 Show the steps in the mechanism for eq. 9.19.

---

*In the laboratory this can be accomplished in several ways. One method involves distilling the water from the reaction mixture. Another method involves trapping the water with molecular sieves, inorganic materials with cavities of the size and shape required to hold water molecules.

The acid-catalyzed cleavage of acetals occurs much more readily than the acid-catalyzed cleavage of simple ethers (Sec. 8.6) because the intermediate carbocation is resonance stabilized. However, acetals, like ordinary ethers, are stable toward bases.

9.8 Addition of Water; Hydration of Aldehydes and Ketones

Water, like alcohols, is an oxygen nucleophile and can add reversibly to aldehydes and ketones. For example, formaldehyde in aqueous solution exists mainly as its hydrate.

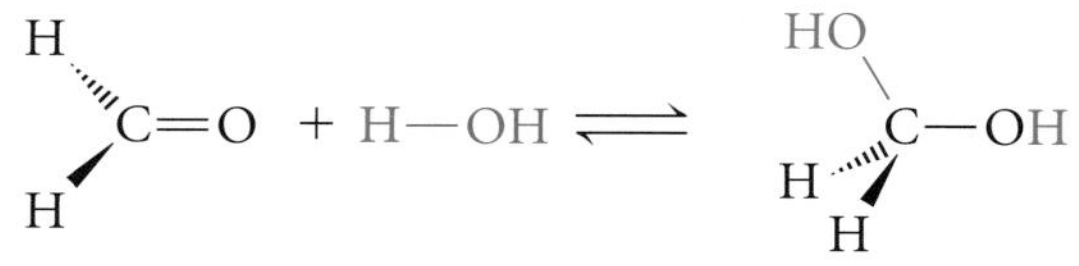

$$\begin{array}{c} \text{H} \\ \diagdown \\ \diagup \\ \text{H} \end{array} \!\!\! C{=}O \; + \; H{-}OH \; \rightleftharpoons \; \begin{array}{c} \text{HO} \\ \diagdown \\ \text{H} \diagup \\ \quad \text{H} \end{array} \!\!\! C{-}OH \qquad \textbf{(9.20)}$$

formaldehyde   formaldehyde hydrate

With most other aldehydes or ketones, however, the hydrates cannot be isolated because they readily lose water to reform the carbonyl compound. An exception is trichloroacetaldehyde (chloral), which forms a stable crystalline hydrate, $CCl_3CH(OH)_2$. **Chloral hydrate** is used in medicine as a sedative and in veterinary medicine as a narcotic and anesthetic for horses, cattle, swine, and poultry. The potent drink known as a Mickey Finn is a combination of alcohol and chloral hydrate.

> **PROBLEM 9.13** Hydrolysis of $CH_3CBr_2CH_3$ with sodium hydroxide does *not* give $CH_3C(OH)_2CH_3$. Instead, it gives acetone. Explain.

9.9 Addition of Grignard Reagents and Acetylides

Grignard reagents act as carbon nucleophiles toward carbonyl compounds. The R group of the Grignard reagent adds irreversibly to the carbonyl carbon, forming a new carbon–carbon bond. In terms of acid–base reactions, the addition is favorable because the product (an alkoxide) is a much weaker base than the starting carbanion (Grignard reagent). The alkoxide can be protonated to give an alcohol.

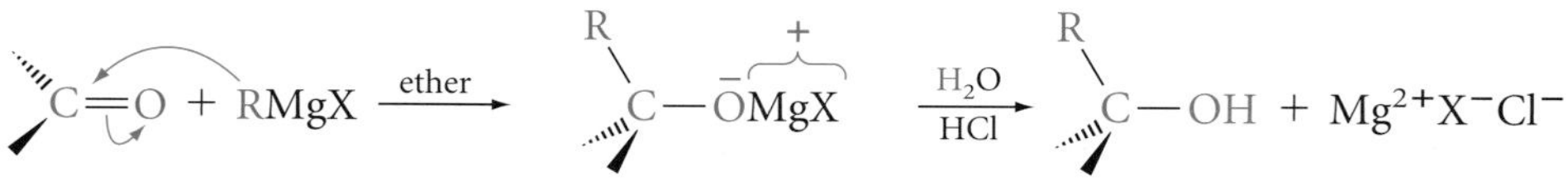

$$\diagup\!\!\! C{=}O \; + \; RMgX \; \xrightarrow{\text{ether}} \; \begin{array}{c} R \\ \diagdown \\ C{-}\overset{+}{O}MgX \\ \diagup \end{array} \; \xrightarrow[\text{HCl}]{H_2O} \; \begin{array}{c} R \\ \diagdown \\ C{-}OH \\ \diagup \end{array} \; + \; Mg^{2+}X^-Cl^- \qquad \textbf{(9.21)}$$

intermediate addition   an alcohol
product (a magnesium
alkoxide)

The reaction is normally carried out by slowly adding an ether solution of the aldehyde or ketone to an ether solution of the Grignard reagent. After all the carbonyl compound is added and the reaction is complete, the resulting magnesium alkoxide is hydrolyzed with aqueous acid.

The reaction of a Grignard reagent with a carbonyl compound provides a useful route to alcohols. The type of carbonyl compound chosen determines the class of alcohol produced. *Formaldehyde gives primary alcohols.*

$$R\!-\!MgX + H\!-\!\overset{\overset{\displaystyle O}{\|}}{C}\!-\!H \longrightarrow R\!-\!\overset{\overset{\displaystyle H}{|}}{\underset{\underset{\displaystyle H}{|}}{C}}\!-\!OMgX \xrightarrow[\text{H}^+]{\text{H}_2\text{O}} R\!-\!\overset{\overset{\displaystyle H}{|}}{\underset{\underset{\displaystyle H}{|}}{C}}\!-\!OH \qquad \textbf{(9.22)}$$

formaldehyde   a primary alcohol

Other aldehydes give secondary alcohols.

$$R\!-\!MgX + R'\!-\!\overset{\overset{\displaystyle O}{\|}}{C}\!-\!H \longrightarrow R\!-\!\overset{\overset{\displaystyle R'}{|}}{\underset{\underset{\displaystyle H}{|}}{C}}\!-\!OMgX \xrightarrow[\text{H}^+]{\text{H}_2\text{O}} R\!-\!\overset{\overset{\displaystyle R'}{|}}{\underset{\underset{\displaystyle H}{|}}{C}}\!-\!OH \qquad \textbf{(9.23)}$$

aldehyde   a secondary alcohol

Ketones give tertiary alcohols.

$$R\!-\!MgX + R'\!-\!\overset{\overset{\displaystyle O}{\|}}{C}\!-\!R'' \longrightarrow R\!-\!\overset{\overset{\displaystyle R'}{|}}{\underset{\underset{\displaystyle R''}{|}}{C}}\!-\!OMgX \xrightarrow[\text{H}^+]{\text{H}_2\text{O}} R\!-\!\overset{\overset{\displaystyle R'}{|}}{\underset{\underset{\displaystyle R''}{|}}{C}}\!-\!OH \qquad \textbf{(9.24)}$$

ketone   a tertiary alcohol

Note that only *one* of the R groups (shown in black) attached to the hydroxyl-bearing carbon of the alcohol comes from the Grignard reagent. The rest of the alcohol's carbon skeleton comes from the carbonyl compound.

EXAMPLE 9.4

What is the product expected from the reaction between ethylmagnesium bromide and 3-pentanone followed by hydrolysis?

Solution   3-Pentanone is a ketone. Following eq. 9.24 as an example, the product is 3-ethyl-3-pentanol.

$$\begin{array}{c} \overset{\overset{\displaystyle O}{\|}}{CH_3CH_2\!-\!C\!-\!CH_2CH_3} \\ + \\ CH_3CH_2MgBr \end{array} \longrightarrow \overset{\overset{\displaystyle OMgBr}{|}}{\underset{\underset{\displaystyle CH_2CH_3}{|}}{CH_3CH_2\!-\!C\!-\!CH_2CH_3}} \xrightarrow[\text{H}^+]{\text{H}_2\text{O}} \overset{\overset{\displaystyle OH}{|}}{\underset{\underset{\displaystyle CH_2CH_3}{|}}{CH_3CH_2\!-\!C\!-\!CH_2CH_3}}$$

PROBLEM 9.14   Provide the products expected from the reaction of

a. formaldehyde with propylmagnesium bromide followed by hydrolysis.
b. pentanal with ethylmagnesium bromide followed by hydrolysis.

EXAMPLE 9.5

Show how the following alcohol can be synthesized from a Grignard reagent and a carbonyl compound:

$$
\begin{array}{c}
\text{OH} \\
| \\
\text{C}_6\text{H}_5-\text{CHCH}_3
\end{array}
$$

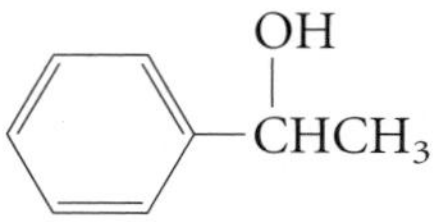

Solution   The alcohol is secondary, so the carbonyl compound must be an aldehyde. We can use either a methyl or a phenyl Grignard reagent.

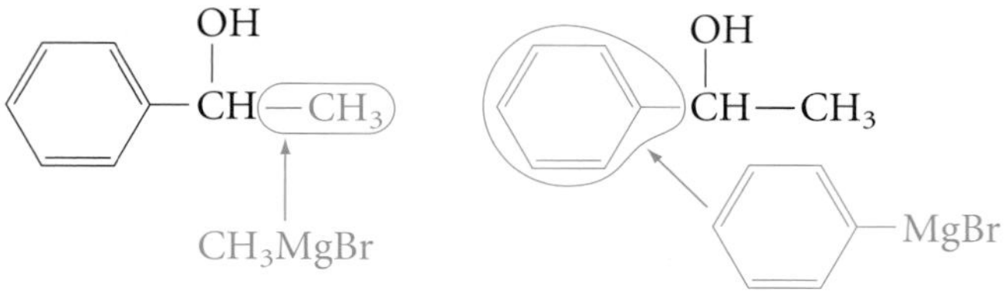

The equations are

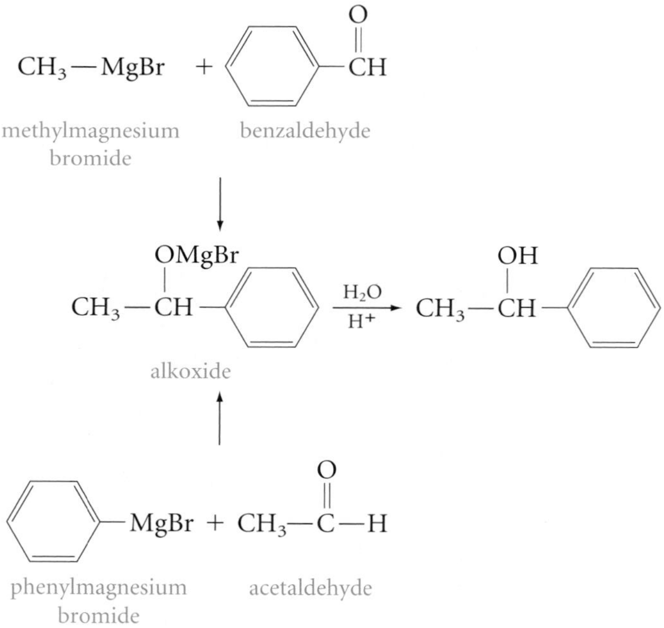

$$
\text{CH}_3-\text{MgBr} \; + \; \text{benzaldehyde}
$$

methylmagnesium bromide   benzaldehyde

$$
\begin{array}{c}
\text{OMgBr} \\
| \\
\text{CH}_3-\text{CH}-\text{C}_6\text{H}_5
\end{array}
\xrightarrow[\text{H}^+]{\text{H}_2\text{O}}
\begin{array}{c}
\text{OH} \\
| \\
\text{CH}_3-\text{CH}-\text{C}_6\text{H}_5
\end{array}
\qquad \textbf{(9.25)}
$$

alkoxide

$$
\text{C}_6\text{H}_5-\text{MgBr} \; + \; \text{CH}_3-\overset{\text{O}}{\overset{\|}{\text{C}}}-\text{H}
$$

phenylmagnesium bromide   acetaldehyde

The choice between the possible sets of reactants may be made by availability or cost of reactants, or for chemical reasons (for example, the more reactive aldehyde or ketone might be selected).

PROBLEM 9.15   Show how each of the following alcohols can be made from a Grignard reagent and a carbonyl compound:

a. 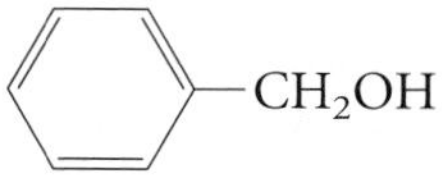 $\text{C}_6\text{H}_5-\text{CH}_2\text{OH}$ b. 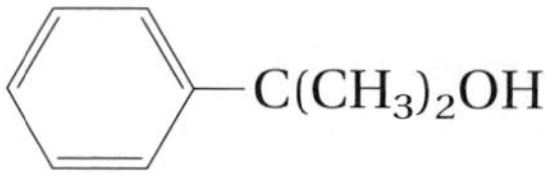 $\text{C}_6\text{H}_5-\text{C}(\text{CH}_3)_2\text{OH}$

Other organometallic reagents, such as organolithium compounds and acetylides, react with carbonyl compounds similarly to Grignard reagents. For example,

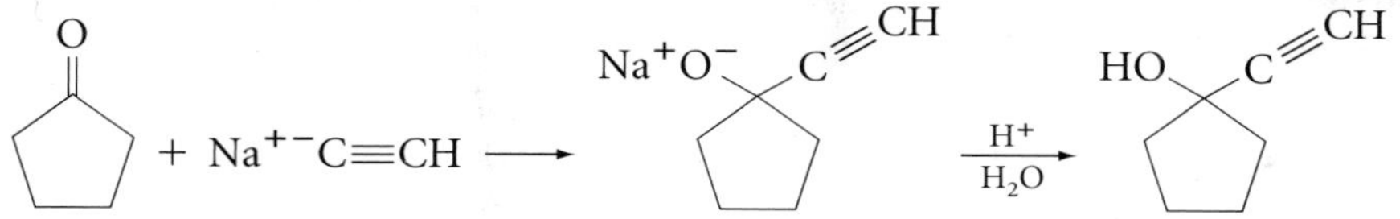

$$
\underset{\text{a ketone}}{\ce{O}} + \underset{\substack{\text{sodium} \\ \text{acetylide}}}{\ce{Na+ ^- C#CH}} \longrightarrow \underset{}{\ce{Na+ O^- \cdots C#CH}} \xrightarrow[\ce{H2O}]{\ce{H+}} \underset{\substack{\text{a tertiary} \\ \text{acetylenic alcohol}}}{\ce{HO \cdots C#CH}} \tag{9.26}
$$

PROBLEM 9.16   Write the structural formula of the product expected from the reaction of $\ce{CH3C#C^- Na+}$ with cyclohexanone followed by $\ce{H3O+}$.

9.10 Addition of Hydrogen Cyanide; Cyanohydrins

Hydrogen cyanide adds reversibly to the carbonyl group of aldehydes and ketones to form **cyanohydrins**, compounds with a hydroxyl and a cyano group attached to the same carbon. A basic catalyst is required.

Cyanohydrins are compounds with a hydroxyl and a cyano group attached to the same carbon.

$$
\ce{C=O} + \ce{HCN} \xrightarrow{\ce{KOH}} \underset{\text{a cyanohydrin}}{\overset{\ce{CN}}{\ce{C-OH}}} \tag{9.27}
$$

Acetone, for example, reacts as follows:

$$
\underset{\text{acetone}}{\ce{CH3-\overset{O}{\underset{||}{C}}-CH3}} + \ce{HCN} \xrightarrow{\ce{KOH}} \underset{\text{acetone cyanohydrin}}{\ce{CH3-\overset{OH}{\underset{CN}{C}}-CH3}} \tag{9.28}
$$

Hydrogen cyanide has no unshared electron pair on its carbon, so it cannot function as a carbon nucleophile. The base converts some of the hydrogen cyanide to cyanide ion, however, which then acts as a carbon nucleophile.

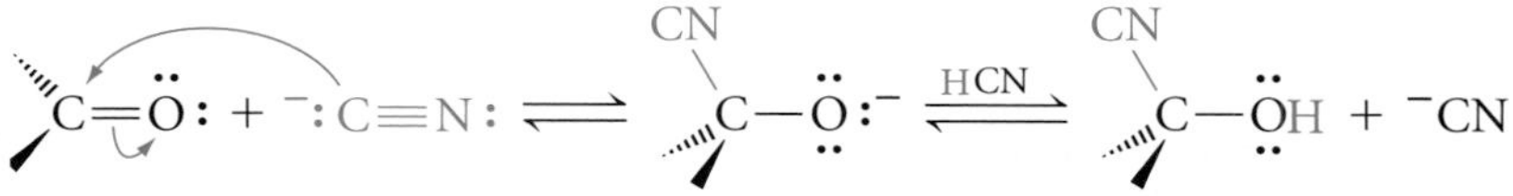

$$
\ce{C=\overset{..}{O}:} + {}^-\!:\!\ce{C#N:} \rightleftharpoons \overset{\ce{CN}}{\ce{C-\overset{..}{\underset{..}{O}}:{}^-}} \overset{\ce{HCN}}{\rightleftharpoons} \underset{\text{cyanohydrin}}{\overset{\ce{CN}}{\ce{C-\overset{..}{\underset{..}{O}}H}}} + {}^-\ce{CN} \tag{9.29}
$$

PROBLEM 9.17   Write an equation for the addition of HCN to

a. acetaldehyde.   b. benzaldehyde.

Cyanohydrin chemistry plays a central role in the defense system of *Apheloria corrugata.* This millipede uses a two-chamber gland much like that used by the bombardier beetle (see A Word About Quinones and the Bombardier Beetle in

Chapter 7) to deliver a secretion that contains hydrogen cyanide. *Apheloria* stores benzaldehyde cyanohydrin and, when threatened, converts it to a mixture of benzaldehyde and hydrogen cyanide which is then secreted. The hydrogen cyanide gas that emanates from the secretion is an effective deterrent of predators.

$$\underset{\text{benzaldehyde cyanohydrin}}{\text{HO}\diagdown\!\diagup\text{CN}} \xrightarrow{\text{enzyme catalyst}} \underset{\text{benzaldehyde}}{\text{O}\diagdown\!\diagup\text{H}} + \text{HCN} \qquad \textbf{(9.30)}$$

9.11 Addition of Nitrogen Nucleophiles

Ammonia, amines, and certain related compounds have an unshared electron pair on the nitrogen atom and act as nitrogen nucleophiles toward the carbonyl carbon atom. For example, primary amines react as follows:

$$\underset{\substack{\text{primary}\\\text{amine}}}{\searrow\!\!C\!=\!O + \ddot{N}H_2\!-\!R} \rightleftharpoons \underset{\substack{\text{tetrahedral}\\\text{addition product}}}{\left[\begin{matrix}\text{OH}\\|\\ C\!-\!\text{NHR}\end{matrix}\right]} \xrightarrow{-\text{HOH}} \underset{\text{imine}}{\searrow\!\!C\!=\!\text{NR}} \qquad \textbf{(9.31)}$$

The tetrahedral addition product that is formed first is similar to a hemiacetal, but with an NH group in place of one of the oxygens. These addition products are normally not stable. They eliminate water to form a product with a carbon–nitrogen double bond. With primary amines, the products are called **imines.** Imines are like carbonyl compounds, except that the O is replaced by NR. They are important intermediates in some biochemical reactions, particularly in binding carbonyl compounds to the free amino groups that are present in most enzymes.

Imines are compounds containing a carbon–nitrogen double bond.

$$\underset{\substack{\text{enzyme}\\\\\text{substrate}}}{\overset{\text{NH}_2}{\big|}} \underset{\overset{\text{O}}{\underset{\|}{\text{C}}}}{} \longrightarrow \underset{\substack{\text{enzyme-substrate}\\\text{compound}}}{\overset{\text{N}}{\underset{\overset{\|}{\text{C}}}{}}} + \text{H}_2\text{O} \qquad \textbf{(9.32)}$$

For example, retinal (see A Word About the Chemistry of Vision in Chapter 3) binds to the protein opsin in this way, to form rhodopsin.

> **PROBLEM 9.18** Write an equation for the reaction of benzaldehyde with aniline (the formula of which is $C_6H_5NH_2$).

Other ammonia derivatives containing an —NH$_2$ group react with carbonyl compounds similarly to primary amines. Table 9.1 lists some specific examples. Notice that in each of these reactions the two hydrogens attached to nitrogen and the oxygen of the carbonyl group are eliminated as water.

Table 9.1	Nitrogen derivatives of carbonyl compounds			
Formula of ammonia derivative	**Name**	**Formula of carbonyl derivative**	**Name**	
RNH_2 or $ArNH_2$	primary amine	$C{=}NR$ or $C{=}NAr$	imine	
NH_2OH	hydroxylamine	$C{=}NOH$	oxime	
NH_2NH_2	hydrazine	$C{=}NNH_2$	hydrazone	
$NH_2NHC_6H_5$	phenylhydrazine	$C{=}NNHC_6H_5$	phenylhydrazone	

EXAMPLE 9.6

Using Table 9.1 as a guide, write an equation for the reaction of hydrazine with cyclohexanone.

Solution

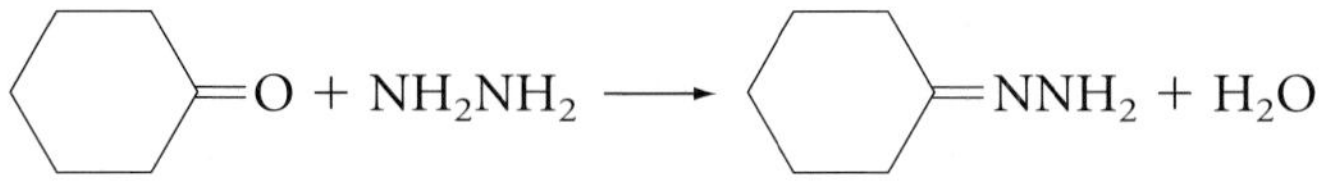

$$\text{cyclohexanone}{=}O + NH_2NH_2 \longrightarrow {=}NNH_2 + H_2O$$

The product is a hydrazone.

PROBLEM 9.19 Using Table 9.1 as a guide, write an equation for the reaction of propanal ($CH_3CH_2CH{=}O$) with

a. hydroxylamine.
b. phenylhydrazine.

9.12 Reduction of Carbonyl Compounds

Aldehydes and ketones are easily reduced to primary and secondary alcohols, respectively. Reduction can be accomplished in many ways, most commonly by metal hydrides.

The most common metal hydrides used to reduce carbonyl compounds are **lithium aluminum hydride** ($LiAlH_4$) and **sodium borohydride** ($NaBH_4$). The metal–hydride bond is polarized, with the metal positive and the hydrogen negative. The reaction therefore involves irreversible nucleophilic attack of the hydride (H^-) at the carbonyl carbon:

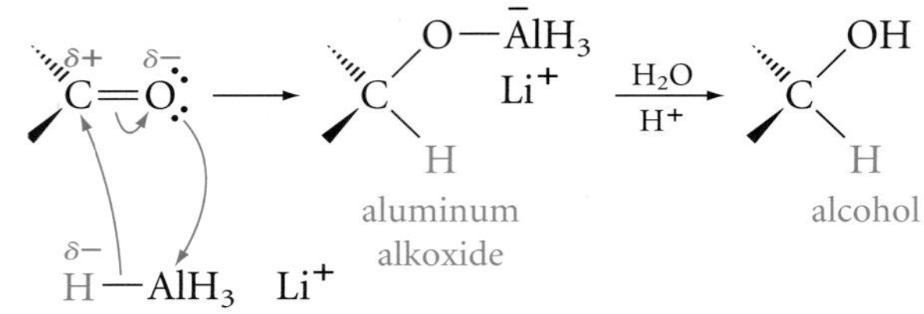

$$\underset{\substack{\delta+ \quad \delta- \\[-2pt]}}{C=O} \quad \longrightarrow \quad \underset{\substack{\text{aluminum} \\ \text{alkoxide}}}{\overset{O-\bar{A}lH_3 \; Li^+}{\underset{H}{C}}} \quad \xrightarrow[H^+]{H_2O} \quad \underset{\substack{\text{alcohol}}}{\overset{OH}{\underset{H}{C}}} \tag{9.33}$$

$$\underset{\delta-}{H-\bar{A}lH_3} \;\; Li^+$$

The initial product is an aluminum alkoxide, which is subsequently hydrolyzed by water and acid to give the alcohol. The net result is addition of hydrogen across the carbon–oxygen double bond. A specific example is

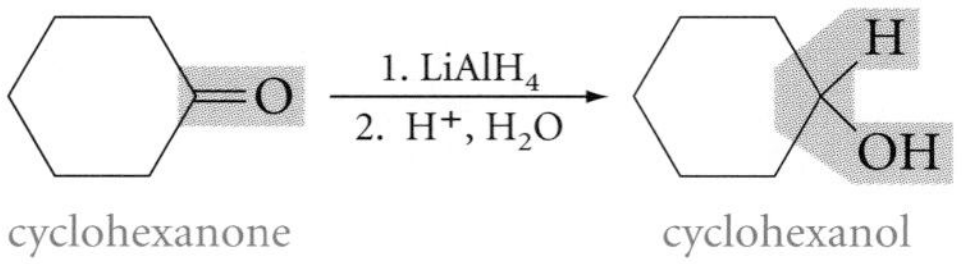

$$\text{cyclohexanone} \xrightarrow[\substack{\text{2. } H^+, H_2O}]{\text{1. LiAlH}_4} \text{cyclohexanol} \tag{9.34}$$

PROBLEM 9.20   Show how the following alcohols can be made from lithium aluminum hydride and a carbonyl compound:

a. 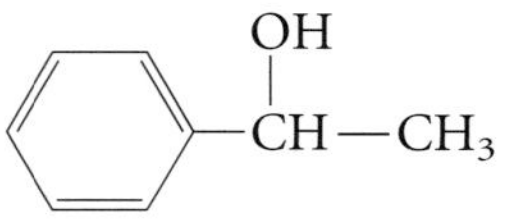 (structure: phenyl ring)—$\underset{\substack{| \\ OH}}{CH}-CH_3$ b. $CH_3CH_2CH_2CH_2CH_2OH$

Since a carbon–carbon double bond is not readily attacked by nucleophiles, metal hydrides can be used to reduce a carbon–oxygen double bond to the corresponding alcohol without reducing a carbon–carbon double bond present in the same compound.

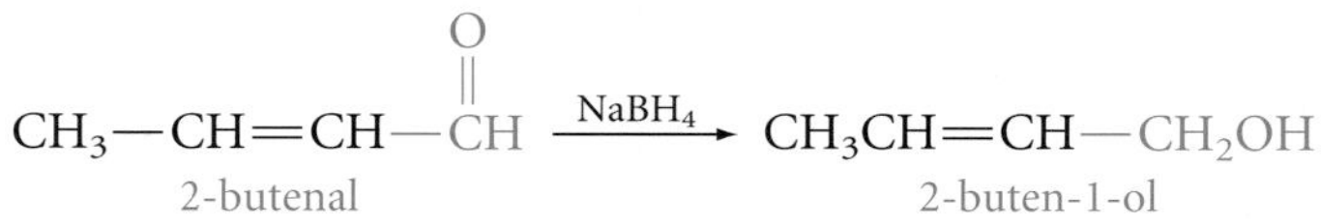

$$\underset{\text{2-butenal}}{CH_3-CH=CH-\overset{O}{\overset{\|}{CH}}} \xrightarrow{\text{NaBH}_4} \underset{\text{2-buten-1-ol}}{CH_3CH=CH-CH_2OH} \tag{9.35}$$

PROBLEM 9.21   Show how

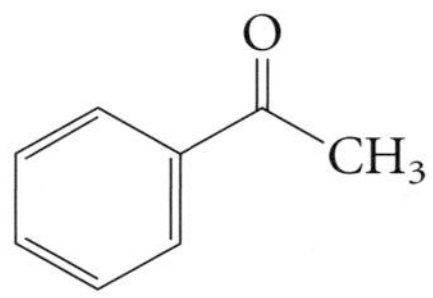

can be reduced to

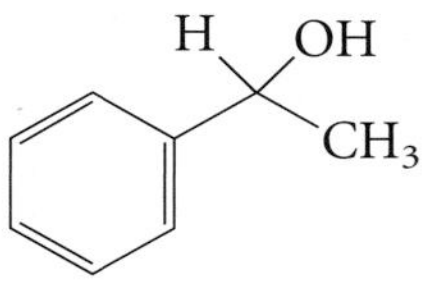

9.13   Oxidation of Carbonyl Compounds

Aldehydes are more easily oxidized than ketones. Oxidation of an aldehyde gives an acid with the same number of carbon atoms.

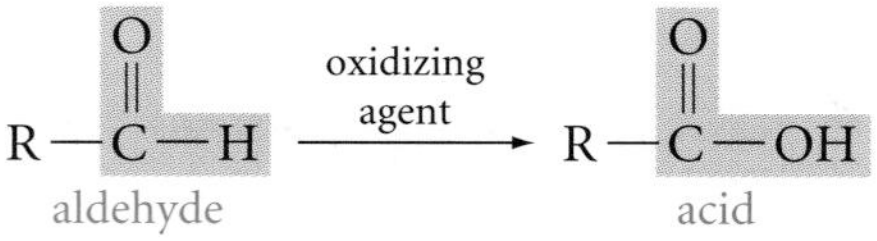

$$\underset{\text{aldehyde}}{R-\overset{O}{\overset{\|}{C}}-H} \xrightarrow[\substack{\text{agent}}]{\text{oxidizing}} \underset{\text{acid}}{R-\overset{O}{\overset{\|}{C}}-OH} \tag{9.36}$$

Since the reaction occurs easily, many oxidizing agents, such as $KMnO_4$, CrO_3, Ag_2O, and peracids (see eq. 8.18), will work. Specific examples are

$$CH_3(CH_2)_5CH{=}O \xrightarrow[\text{(Jones' reagent)}]{CrO_3, \, H^+} CH_3(CH_2)_5CO_2H \qquad \textbf{(9.37)}$$

$$\text{(cyclohexane-CHO)} \xrightarrow{Ag_2O} \text{(cyclohexane-CO}_2\text{H)} \qquad \textbf{(9.38)}$$

Silver ion as an oxidant is expensive but has the virtue that it selectively oxidizes aldehydes to carboxylic acids in the presence of alkenes (eq. 9.38).

A laboratory test that distinguishes aldehydes from ketones takes advantage of their different ease of oxidation. In the **Tollens' silver mirror test,** the silver–ammonia complex ion is reduced by aldehydes (but not by ketones) to metallic silver.* The equation for the reaction may be written as follows:

$$\underset{\substack{\text{aldehyde}}}{\overset{\overset{\textstyle O}{\|}}{R\text{C}H}} + \underset{\substack{\text{silver–ammonia}\\\text{complex ion}\\\text{(colorless)}}}{2\,Ag(NH_3)_2{}^+} + 3\,HO^- \longrightarrow \underset{\substack{\text{acid}\\\text{anion}}}{\overset{\overset{\textstyle O}{\|}}{R\text{C}{-}O^-}} + \underset{\substack{\text{silver}\\\text{mirror}}}{2\,Ag{\downarrow}} + 4\,NH_3{\uparrow} + 2\,H_2O \qquad \textbf{(9.39)}$$

The symbol ↓ indicates formation of a precipitate; the symbol ↑ indicates the formation of a gas.

If the glass vessel in which the test is performed is thoroughly clean, the silver deposits as a mirror on the glass surface. This reaction is also employed to silver glass, using the inexpensive aldehyde formaldehyde.

> **PROBLEM 9.22** Write an equation for the formation of a silver mirror from formaldehyde and Tollens' reagent.

Aldehydes are so easily oxidized that stored samples usually contain some of the corresponding acid. This contamination is caused by air oxidation.

$$2\,RCHO + O_2 \longrightarrow 2\,RCO_2H \qquad \textbf{(9.40)}$$

Ketones also can be oxidized, but require special oxidizing conditions. For example, cyclohexanone is oxidized commercially to **adipic acid,** an important industrial chemical used to manufacture nylon.

Tollens' silver mirror test (Eq. 9.39).

one of
these C—C
bonds is
cleaved in
the oxidation

$$\text{cyclohexanone} + HNO_3 \xrightarrow{V_2O_5} \underset{\text{adipic acid}}{HO{-}\overset{\overset{\textstyle O}{\|}}{C}{-}CH_2CH_2CH_2CH_2{-}\overset{\overset{\textstyle O}{\|}}{C}{-}OH} \qquad \textbf{(9.41)}$$

---

*Silver hydroxide is insoluble in water, so the silver ion must be complexed with ammonia to keep it in solution in a basic medium.

9.14 Keto–Enol Tautomerism

Aldehydes and ketones may exist as an equilibrium mixture of two forms, called the **keto form** and the **enol form.** The two forms differ in the location of a proton and a double bond.

$$
\underset{\text{keto form}}{\overset{\overset{\displaystyle H}{\underset{\displaystyle |}{}}\overset{\displaystyle O}{\underset{\displaystyle \|}{}}}{-C-C-}} \;\rightleftharpoons\; \underset{\text{enol form}}{C=C\overset{OH}{\diagup}}
$$
(9.42)

Tautomers are structural isomers that differ in the location of a proton and a double bond.

This type of structural isomerism is called **tautomerism** (from the Greek *tauto,* the same, and *meros,* part). The two forms of the aldehyde or ketone are called **tautomers.**

EXAMPLE 9.7

Write formulas for the keto and enol forms of acetone.

Solution

$$
\underset{\text{keto form}}{\overset{\displaystyle O}{\overset{\displaystyle \|}{CH_3-C-CH_3}}} \qquad \underset{\text{enol form}}{\overset{\displaystyle OH}{\overset{\displaystyle |}{CH_2=C-CH_3}}}
$$

PROBLEM 9.23 Draw the structural formula for the enol form of

   a. cyclohexanone.   b. acetaldehyde (CH_3CHO).

Tautomers are structural isomers, *not* contributors to a resonance hybrid. They readily equilibrate, and we indicate that fact by using the equilibrium symbol $\rightleftharpoons$ between their structures.

To be capable of existing in an enol form, a carbonyl compound must have a hydrogen atom attached to the carbon atom adjacent to the carbonyl group. This hydrogen is called an **α-hydrogen** and is attached to the **α-carbon atom** (from the first letter of the Greek alphabet, α, or alpha).

An **α-hydrogen** is attached to the **α-carbon atom,** the carbon atom adjacent to a carbonyl group.

$$
\alpha\text{-hydrogen} \quad \alpha\text{-carbon} \qquad \overset{\overset{\displaystyle H}{\underset{\displaystyle |}{}}\overset{\displaystyle O}{\underset{\displaystyle \|}{}}}{-C-C-}
$$

Most simple aldehydes and ketones exist mainly in the keto form. Acetone, for example, is 99.9997% in the keto form, with only 0.0003% of the enol present. The main reason for the greater stability of the keto form is that the $C=O$ plus $C-H$ bond energy present in the keto form is greater than the $C=C$ plus $O-H$ bond energy of the enol form. We have already encountered some molecules, however, that have mainly the enol structure—the *phenols.* In this case, the resonance stabilization of the aromatic ring is greater than the usual energy difference that favors the keto over the enol form. Aromaticity would be destroyed if the molecule existed in the keto form; therefore, the enol form is preferred.

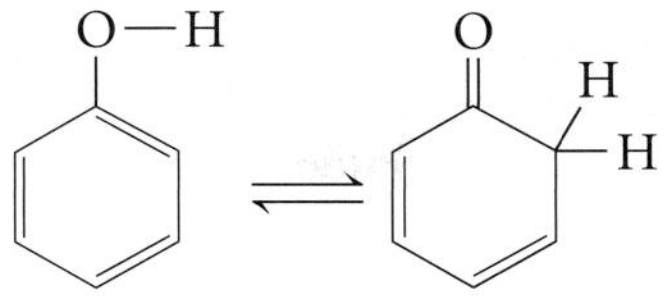

| | | | | | (9.43) |

enol form
of phenol

keto form
of phenol

Carbonyl compounds that do not have an α-hydrogen cannot form enols and exist only in the keto form. Examples include

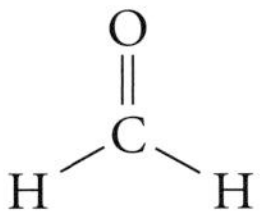

formaldehyde 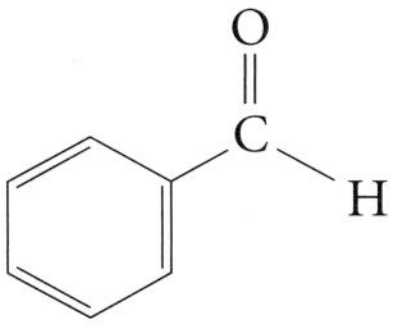   benzaldehyde 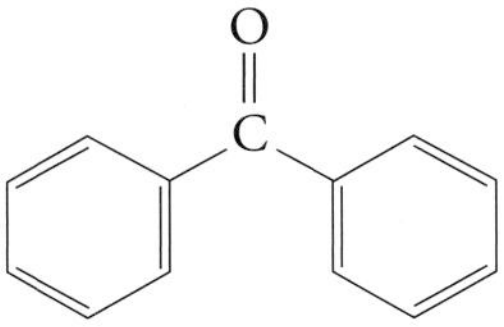   benzophenone

9.15 Acidity of α-Hydrogens; the Enolate Anion

The α-hydrogen in a carbonyl compound is more acidic than a normal hydrogen bonded to a carbon atom. Table 9.2 shows the pK_a values for a typical aldehyde and ketone, as well as for reference compounds. The result of placing a carbonyl group adjacent to methyl protons is truly striking, an increase in their acidity of over 30 powers of 10! (Compare acetaldehyde or acetone with propane.) Indeed, these compounds are almost as acidic as the O—H protons in alcohols. Why is this?

There are two reasons. First, the carbonyl carbon carries a partial positive charge. Bonding electrons are displaced toward the carbonyl carbon and away from the α-hydrogen (shown by the blue arrows below), making it easy for a base to remove the α-hydrogen as a proton (that is, without its bonding electrons).

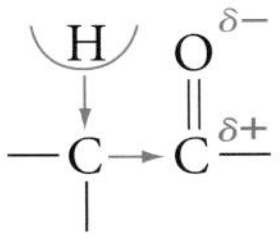

Table 9.2	Acidity of α-hydrogens	
Compound	Name	pK_a
$CH_3CH_2CH_3$	propane	~50
CH_3CCH_3 (with O double bond)	acetone	19
CH_3CH (with O double bond)	acetaldehyde	17
CH_3CH_2OH	ethanol	16

Second, the resulting anion is stabilized by resonance.

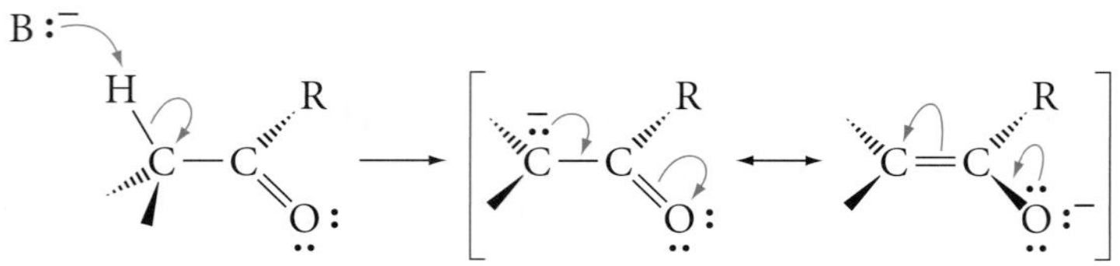

resonance structures of an enolate anion   (9.44)

An **enolate anion** is formed by removal of the α-hydrogen of a ketone or aldehyde.

The anion is called an **enolate anion.** Its negative charge is distributed between the α-carbon and the carbonyl oxygen atom.

EXAMPLE 9.8

Draw the formula for the enolate anion of acetone.

Solution

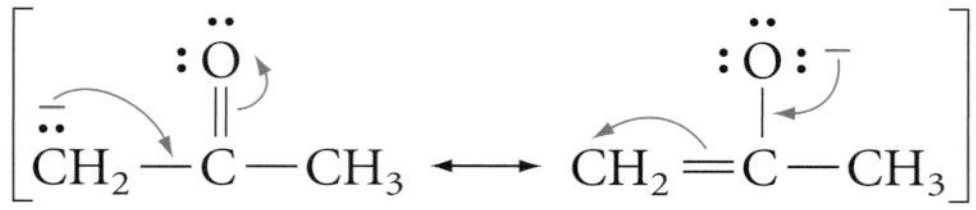

An enolate anion is a resonance hybrid of two contributing structures that differ *only* in the arrangement of the electrons.

PROBLEM 9.24   Draw the resonance contributors to the enolate anion of

   a. cyclohexanone.   b. acetaldehyde.

9.16 Deuterium Exchange in Carbonyl Compounds

Even though its concentration is very low, the presence of the enol form of ordinary aldehydes and ketones can be demonstrated experimentally. For example, the α-hydrogens can be exchanged for deuterium by placing the carbonyl compound in a solvent such as D_2O or CH_3OD. The exchange is catalyzed by acid or base. *Only the α-hydrogens exchange,* as illustrated by the following examples:

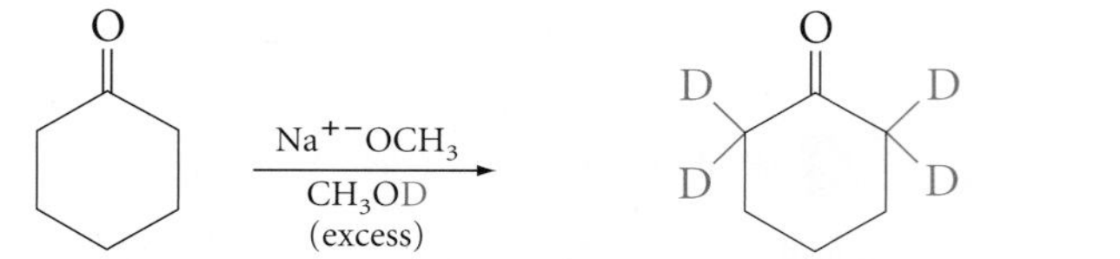

cyclohexanone   2,2,6,6-tetradeuteriocyclohexanone   (9.45)

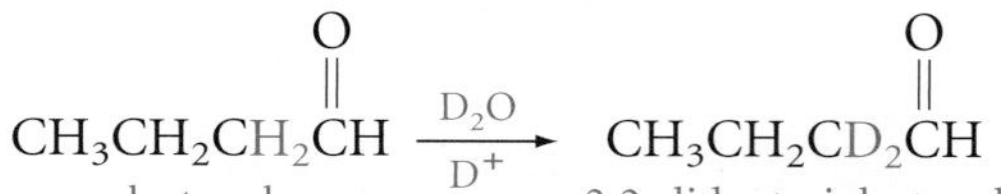

$$CH_3CH_2CH_2\overset{\overset{\textstyle O}{\|}}{C}H \xrightarrow[\text{D}^+]{\text{D}_2\text{O}} CH_3CH_2CD_2\overset{\overset{\textstyle O}{\|}}{C}H$$

butanal   2,2-dideuteriobutanal   (9.46)

The mechanism of the base-catalyzed exchange of the α-hydrogens (eq. 9.45) involves two steps.

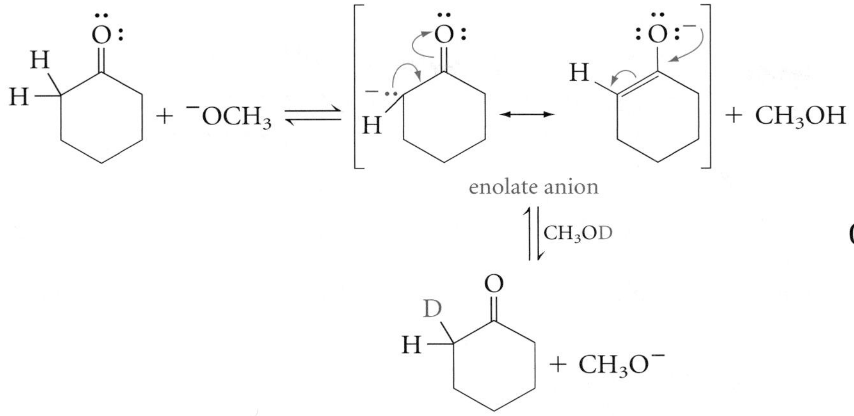

$$\text{(9.47)}$$

enolate anion

The base (methoxide ion) removes an α-proton to form the enolate anion. Reprotonation, but with CH_3OD, replaces the α-hydrogen with deuterium. With excess CH_3OD, all four α-hydrogens are eventually exchanged.

The mechanism of the acid-catalyzed exchange of the α-hydrogens (eq. 9.46) also involves several steps. The keto form is first protonated and, by loss of an α-hydrogen, converted to its enol.

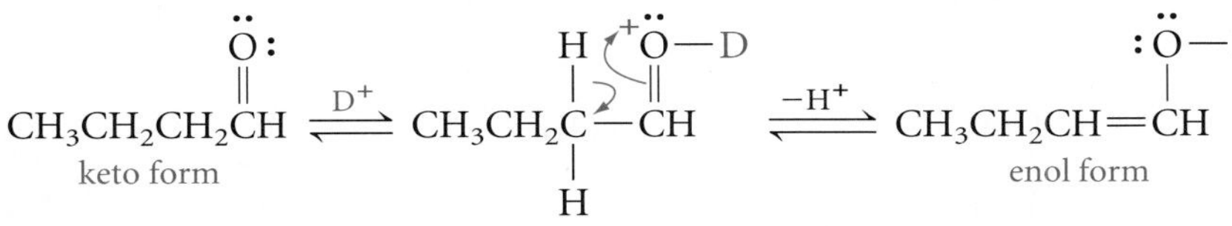

$$\text{(9.48)}$$

keto form   enol form

In the reversal of these equilibria, the enol then adds D^+ at the α-carbon.

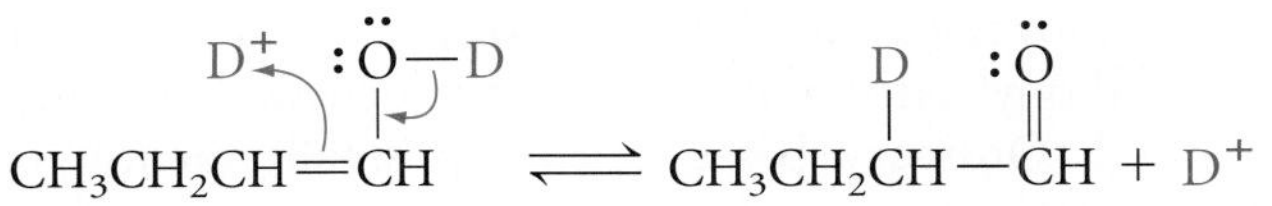

$$\text{(9.49)}$$

Repetition of this sequence results in exchange of the other α-hydrogen.

PROBLEM 9.25 Identify the hydrogens that are readily exchanged for deuterium in

a. 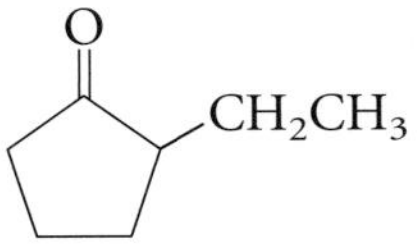   b. $(CH_3)_3CCCH_3$

9.17 ## The Aldol Condensation

Enolate anions may act as carbon nucleophiles. They add reversibly to the carbonyl group of another aldehyde or ketone molecule in a reaction called the **aldol condensation,** an extremely useful carbon–carbon bond-forming reaction.

An enolate anion adds to the carbonyl group of an aldehyde or ketone in an **aldol condensation.** An **aldol** is a 3-hydroxyaldehyde or 3-hydroxyketone.

The simplest example of an aldol condensation is the combination of two acetaldehyde molecules, which occurs when a solution of acetaldehyde is treated with catalytic amounts of aqueous base.

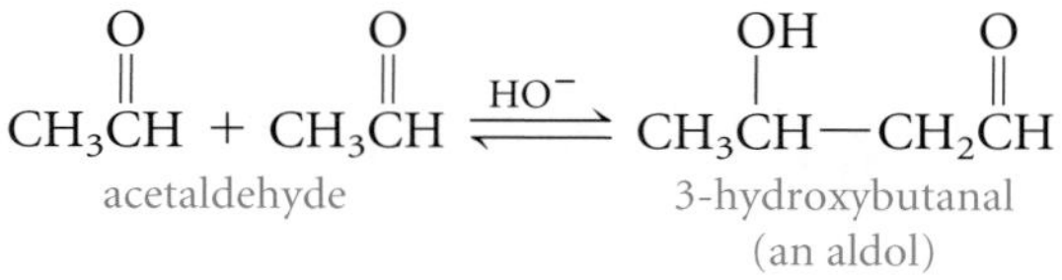

$$\underset{\text{acetaldehyde}}{CH_3\overset{\displaystyle O}{\overset{\|}{C}}H} + CH_3\overset{\displaystyle O}{\overset{\|}{C}}H \;\underset{}{\overset{HO^-}{\rightleftharpoons}}\; \underset{\substack{\text{3-hydroxybutanal}\\ \text{(an aldol)}}}{CH_3\overset{\displaystyle OH}{\overset{\|}{C}}H-CH_2\overset{\displaystyle O}{\overset{\|}{C}}H} \qquad (9.50)$$

The product is called an **aldol** (so named because the product is both an *alde*hyde and an alcoh*ol*).

The aldol condensation of acetaldehyde occurs according to the following three-step mechanism:

Step 1. $\underset{\alpha}{CH_3}-\overset{\displaystyle \overset{..}{\underset{..}{O}}:}{\overset{\|}{C}}-H + HO^- \;\rightleftharpoons\; \underset{\text{enolate anion}}{\overset{=}{CH_2}-\overset{\displaystyle \overset{..}{\underset{..}{O}}:}{\overset{\|}{C}}-H} + HOH \qquad (9.51)$

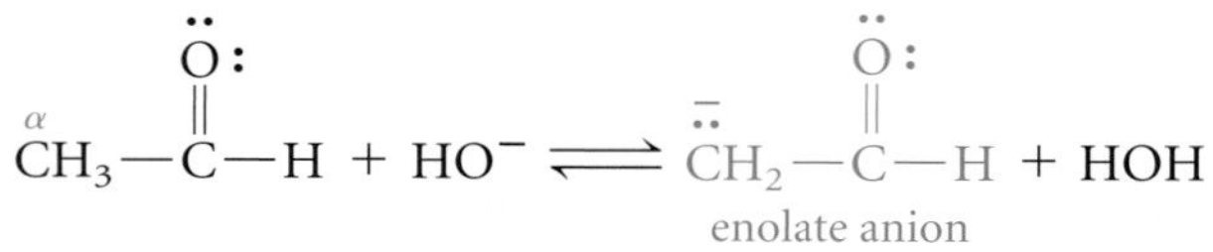

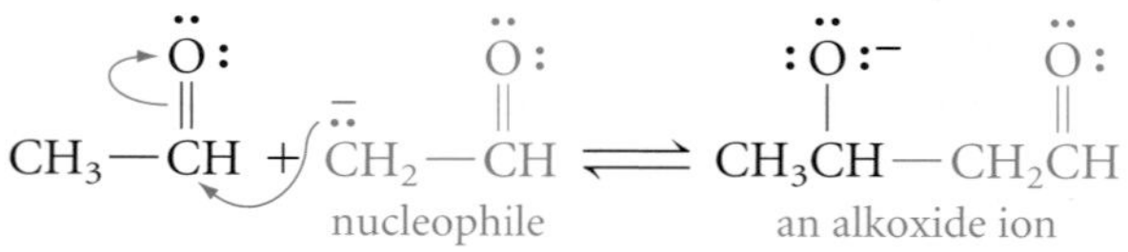

Step 2. $CH_3-\overset{\displaystyle \overset{..}{\underset{..}{O}}:}{\overset{\|}{C}}H + \underset{\text{nucleophile}}{\overset{=}{CH_2}-\overset{\displaystyle \overset{..}{\underset{..}{O}}:}{\overset{\|}{C}}H} \;\rightleftharpoons\; \underset{\text{an alkoxide ion}}{CH_3\overset{\displaystyle :\overset{..}{O}:^-}{\overset{|}{C}}H-CH_2\overset{\displaystyle \overset{..}{\underset{..}{O}}:}{\overset{\|}{C}}H} \qquad (9.52)$

Step 3. $CH_3\overset{\displaystyle :\overset{..}{O}:^-}{\overset{|}{C}}H-CH_2\overset{\displaystyle \overset{..}{O}:}{\overset{\|}{C}}H + HOH \;\rightleftharpoons\; \underset{\text{aldol}}{CH_3\overset{\displaystyle :\overset{..}{O}H}{\overset{|}{C}}H-\underset{\alpha}{CH_2}\overset{\displaystyle \overset{..}{O}:}{\overset{\|}{C}}H} + HO^- \qquad (9.53)$

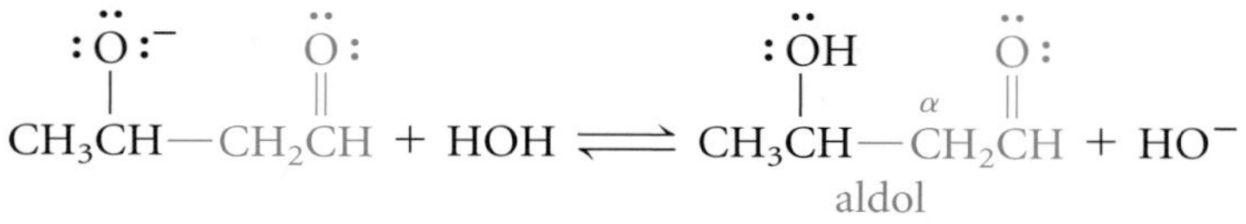

In Step 1, the base removes an α-hydrogen to form the enolate anion. In Step 2, this anion adds to the carbonyl carbon of *another* acetaldehyde molecule, forming a new carbon–carbon bond. Ordinary bases convert a small fraction of the carbonyl compound to the enolate anion, so that a substantial fraction of the aldehyde is still present in the un-ionized carbonyl form needed for this step. In Step 3, the alkoxide ion formed in Step 2 accepts a proton from the solvent, thus regenerating the hydroxide ion needed for the first step.

In the aldol condensation, the α-carbon of one aldehyde molecule becomes connected to the carbonyl carbon of another aldehyde molecule.

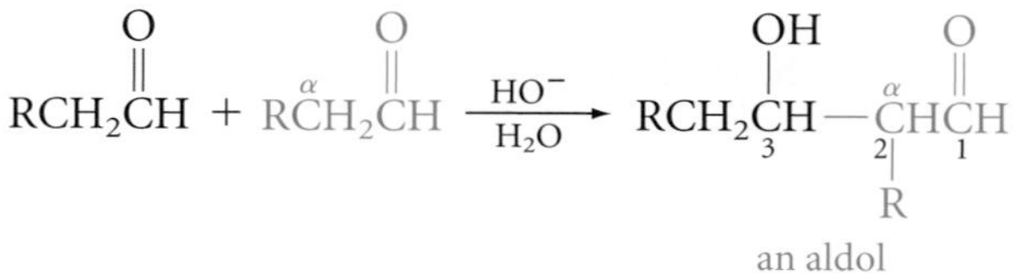

$$RCH_2\overset{\displaystyle O}{\overset{\|}{C}}H + R\underset{\alpha}{CH_2}\overset{\displaystyle O}{\overset{\|}{C}}H \;\overset{HO^-}{\underset{H_2O}{\longrightarrow}}\; \underset{\text{an aldol}}{RCH_2\underset{3}{\overset{\displaystyle OH}{\overset{|}{C}}}H-\underset{\substack{2\\ \underset{\displaystyle R}{|}}}{\overset{\alpha}{C}}H\underset{1}{\overset{\displaystyle O}{\overset{\|}{C}}}H} \qquad (9.54)$$

Aldols are therefore 3-hydroxyaldehydes. *Since it is always the α-carbon that acts as a nucleophile, the product always has just one carbon atom between the aldehyde and alcohol carbons,* regardless of how long the carbon chain is in the starting aldehyde.

EXAMPLE 9.9

Give the structure of the aldol that is obtained by treating propanal ($CH_3CH_2CH=O$) with base.

Solution   Rewriting eq. 9.54 with R = CH_3, the product is

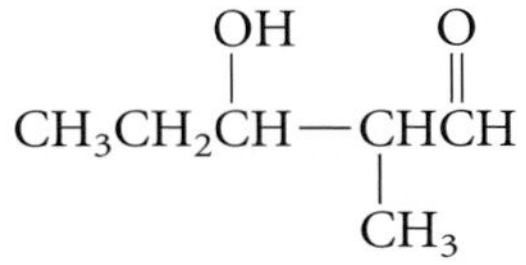

$$\underset{\underset{\displaystyle CH_3}{|}}{CH_3CH_2\underset{\overset{\displaystyle |}{\overset{\displaystyle OH}{}}}{CH}-\underset{\overset{\displaystyle \|}{\overset{\displaystyle O}{}}}{CH}CH}$$

PROBLEM 9.26   Write out the steps in the mechanism for formation of the product in Example 9.9.

9.18   The Mixed Aldol Condensation

The aldol condensation is very versatile in that the enolate anion of *one* carbonyl compound can be made to add to the carbonyl carbon of *another*, provided that the reaction partners are carefully selected. Consider, for example, the reaction between acetaldehyde and benzaldehyde, when treated with base. Only acetaldehyde can form an enolate anion (benzaldehyde has no α-hydrogen). If the enolate ion of acetaldehyde adds to the benzaldehyde carbonyl group, a mixed aldol condensation occurs.

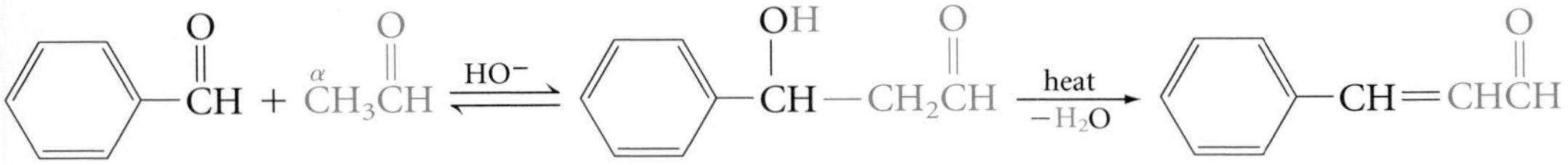

a mixed aldol   cinnamaldehyde

(9.55)

In this particular example, the resulting mixed aldol eliminates water on heating to give **cinnamaldehyde** (the flavor constituent of cinnamon).

EXAMPLE 9.10

Write the structure of the mixed aldol obtained from acetone and formaldehyde.

Solution   Of the two reactants, only acetone has α-hydrogens.

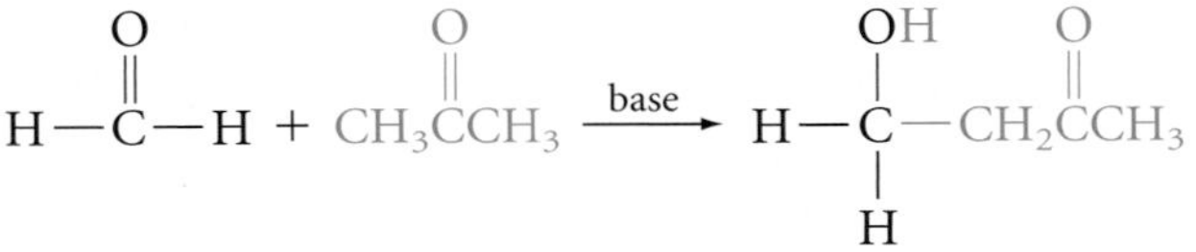

PROBLEM 9.27   Using eqs. 9.51 through 9.53 as a guide, write out the steps in the mechanism for eq. 9.55.

PROBLEM 9.28   Write the structure of the mixed aldol obtained from propanal and benzaldehyde. What structure is obtained from dehydration of this mixed aldol?

Water Treatment and the Chemistry of Enols/Enolates

In a city with a population of about 35,000, around 1.2 billion gallons of drinking water is delivered annually to homes. This water is usually purified before use to remove contaminants such as viruses and bacteria, organic chemicals including pesticides and herbicides, and inorganic chemicals including radioactive substances. These contaminants come from a variety of industrial and natural sources. The cleanup process involves several steps, including removal of water-insoluble material by sedimentation, coagulation, and filtration; softening of the water using lime (calcium oxide) and soda ash (sodium carbonate); adjustment of the pH to 7.5–8.0 with carbon dioxide; and disinfection with chlorine (Cl_2). A typical water quality report will show that while the water delivered to homes is quite clean, α-halogenated ketones, halogenated carboxylic acids, and trihalomethanes (such as chloroform) are sometimes present at very low levels due to chemistry that occurs during the treatment process. What are these compounds and how do they get into the water during the treatment process?

When ketones are treated with halogens (Cl_2, Br_2, I_2), they undergo α-halogenation. This is illustrated

below for acetone (2-propanone) and chlorine. Enolization of the acetone is followed by reaction of the enol with chlorine to provide α-chloroacetone. The enol (or enolate) of acetone behaves as a nucleophile and the chlorine behaves as an electrophile in this reaction. Repetition of this process several times provides polychlorinated compounds such as 1,1,1-trichloroacetone. Polychlorinated ketones can then undergo further reactions that provide chlorinated carboxylic acids and chloroform. Because acetone is a common water-soluble organic solvent and is also an end product of fat metabolism in humans and other animals, it is not surprising that it shows up in water treatment plants. The normal chemistry of ketones (via their enols/enolates) that takes place during the water treatment process, therefore, is responsible in part for the presence of low levels of chlorinated organic compounds in our drinking water. Whereas one could argue that water treatment introduces potentially toxic compounds into the drinking water supply, this is balanced by the removal of far more dangerous materials by the treatment process.

9.19 Commercial Syntheses via the Aldol Condensation

Aldols are useful in synthesis. For example, acetaldehyde is converted commercially to crotonaldehyde, 1-butanol, and butanal using the aldol condensation.

The particular product obtained in the hydrogenation step depends on the catalyst and reaction conditions.

Butanal is the starting material for the synthesis of the mosquito repellent "6-12" (2-ethylhexane-1,3-diol). The first step is an aldol condensation, and the second step is reduction of the aldehyde group to a primary alcohol.

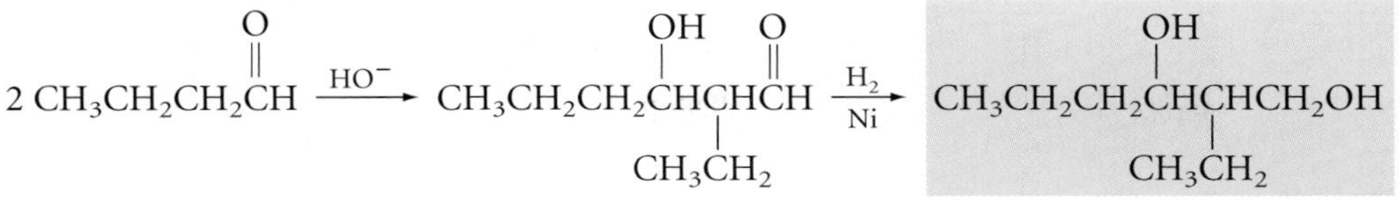

$$2 \; CH_3CH_2CH_2CH \xrightarrow{HO^-} CH_3CH_2CH_2\underset{\underset{CH_3CH_2}{|}}{CH}CHCH \xrightarrow[Ni]{H_2} CH_3CH_2CH_2\underset{\underset{CH_3CH_2}{|}}{CH}CHCH_2OH \qquad (9.57)$$

butanal   butanal aldol   2-ethylhexane-1,3-diol
                                                                                           ("6-12")

The aldol condensation is also used in nature to build up (and, in the case of *reverse* aldol condensations, to break down) carbon chains.

PROBLEM 9.29 2-Ethylhexanol, used commercially in the manufacture of plasticizers and synthetic lubricants, is synthesized from butanal via its aldol. Devise a route to it.

REACTION SUMMARY

1. Preparation of Aldehydes and Ketones

a. Oxidation of Alcohols (Sec. 9.3)

$$*R-CH_2-OH \xrightarrow{PCC} R-\overset{\overset{O}{\|}}{C}-H$$

1° alcohol   aldehyde

$$R-\underset{\underset{|}{\overset{\overset{OH}{|}}{CH}}}{}-R \xrightarrow{PCC} R-\overset{\overset{O}{\|}}{C}-R$$

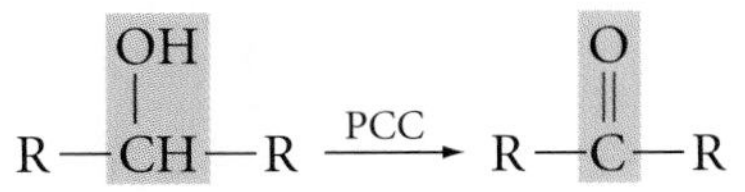

2° alcohol   ketone

b. Friedel–Crafts Acylation (Sec. 9.3)

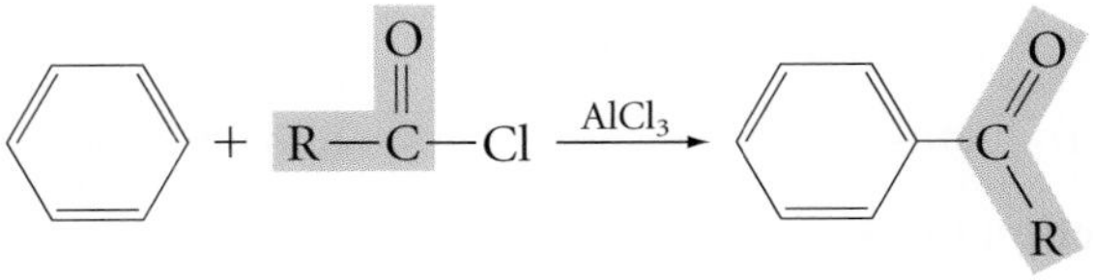

$$\bigcirc + R-\overset{\overset{O}{\|}}{C}-Cl \xrightarrow{AlCl_3} \bigcirc\overset{\overset{O}{\|}}{C}\diagdown R$$

c. Hydration of Alkynes (Sec. 9.3)

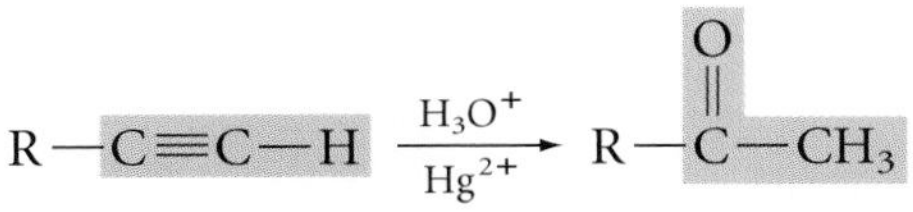

$$R-C\equiv C-H \xrightarrow[Hg^{2+}]{H_3O^+} R-\overset{\overset{O}{\|}}{C}-CH_3$$

---

*For all syntheses and reactions of carbonyl compounds in this summary, R can be alkyl or aryl.

2. Reactions of Aldehydes and Ketones

a. Formation and Hydrolysis of Acetals (Sec. 9.7)

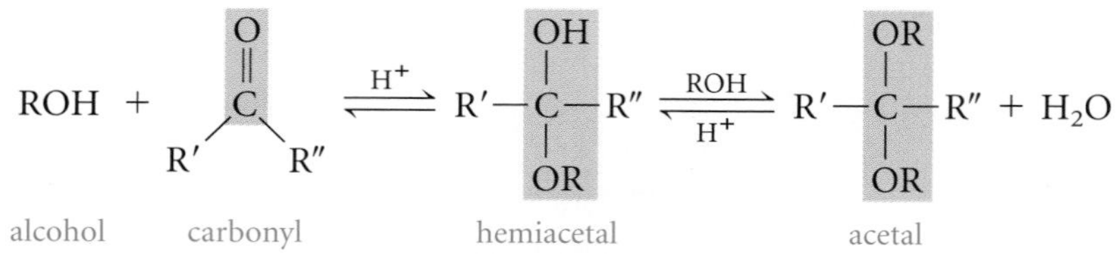

$$ROH + \underset{\underset{\text{carbonyl}}{R' \quad R''}}{\overset{O}{\underset{\|}{C}}} \xrightleftharpoons{H^+} \underset{\underset{\text{hemiacetal}}{OR}}{R'\overset{OH}{\underset{|}{\underset{|}{C}}}R''} \xrightarrow[H^+]{ROH} \underset{\underset{\text{acetal}}{OR}}{R'\overset{OR}{\underset{|}{\underset{|}{C}}}R''} + H_2O$$

alcohol   carbonyl   hemiacetal   acetal

b. Addition of Grignard Reagents (Sec. 9.9)

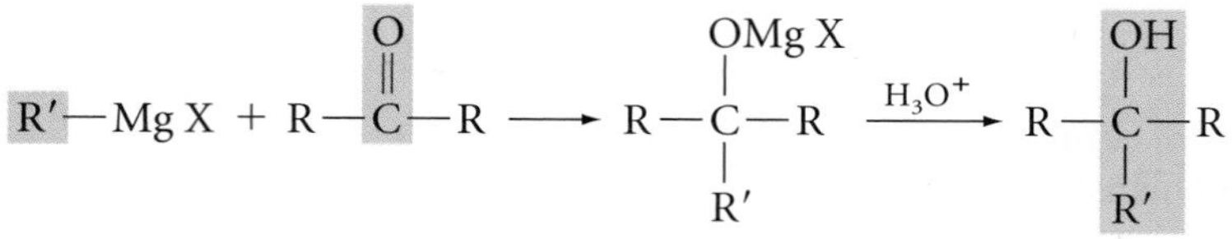

$$R'-MgX + R-\overset{O}{\underset{\|}{C}}-R \longrightarrow R-\underset{R'}{\overset{OMgX}{\underset{|}{\underset{|}{C}}}}-R \xrightarrow{H_3O^+} R-\underset{R'}{\overset{OH}{\underset{|}{\underset{|}{C}}}}-R$$

Formaldehyde gives 1° alcohols; other aldehydes give 2° alcohols; ketones give 3° alcohols.

c. Formation of Cyanohydrins (Sec. 9.10)

$$HC\equiv N + R-\overset{O}{\underset{\|}{C}}-R \xrightleftharpoons{NaOH \ catalyst} R-\underset{CN}{\overset{OH}{\underset{|}{\underset{|}{C}}}}-R$$

R=H, alkyl   cyanohydrin

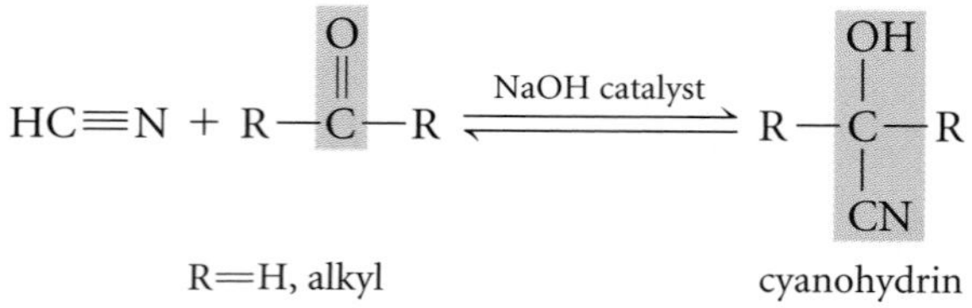

d. Addition of Nitrogen Nucleophiles (Sec. 9.11 and Table 9.1)

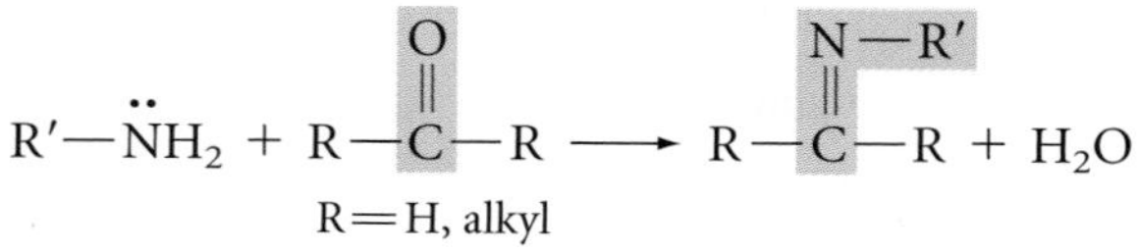

$$R'-\overset{..}{N}H_2 + R-\overset{O}{\underset{\|}{C}}-R \longrightarrow R-\overset{N-R'}{\underset{\|}{C}}-R + H_2O$$

R=H, alkyl

The product is an imine when R′ is an alkyl group.

e. Reduction to Alcohols (Sec. 9.12)

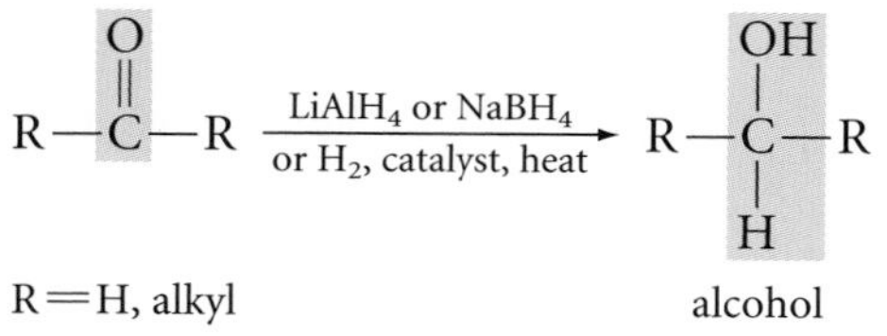

$$R-\overset{O}{\underset{\|}{C}}-R \xrightarrow[\text{or } H_2, \text{ catalyst, heat}]{LiAlH_4 \text{ or } NaBH_4} R-\underset{H}{\overset{OH}{\underset{|}{\underset{|}{C}}}}-R$$

R=H, alkyl   alcohol

f. Oxidation to Carboxylic Acids (Sec. 9.13)

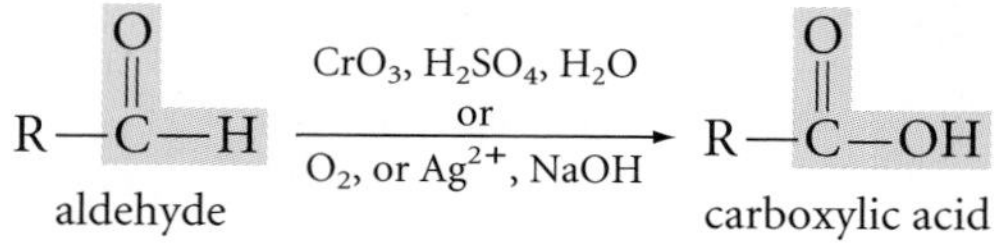

$$R-\overset{O}{\underset{\|}{C}}-H \xrightarrow[\text{O}_2, \text{ or } Ag^{2+}, NaOH]{CrO_3, H_2SO_4, H_2O \atop \text{or}} R-\overset{O}{\underset{\|}{C}}-OH$$

aldehyde   carboxylic acid

g. Aldol condensation (Sec. 9.17)

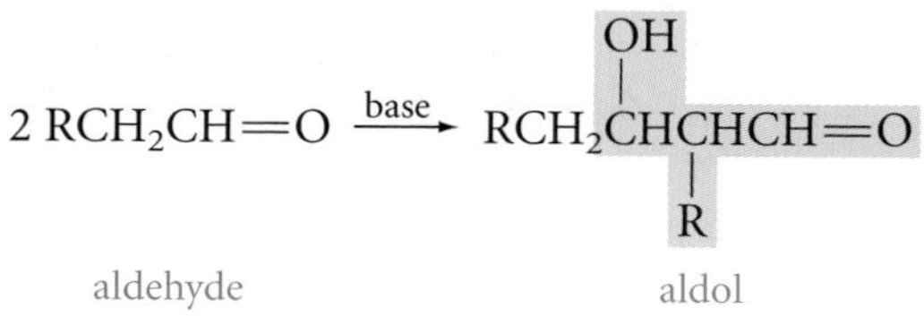

$$2 \ RCH_2CH{=}O \xrightarrow{base} RCH_2\underset{R}{\overset{OH}{\underset{|}{\underset{|}{C}H}}}CHCH{=}O$$

aldehyde   aldol

MECHANISM SUMMARY

Nucleophilic Addition (Sec. 9.6)

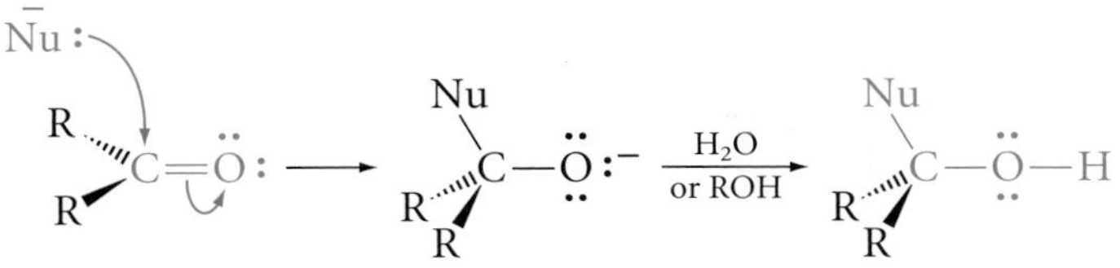

ADDITIONAL PROBLEMS

Nomenclature, Structure, and Properties of Aldehydes and Ketones

9.30 Name each of the following compounds.

a. $CH_3CH_2\overset{\displaystyle O}{\overset{\|}{C}}CH_2CH_3$ **b.** $CH_3(CH_2)_6CH{=}O$ **c.** $(C_6H_5)_2C{=}O$

d. 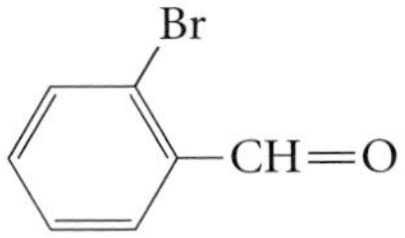   **e.** 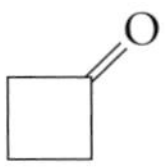   **f.** $(CH_3)_3CCH{=}O$

g. 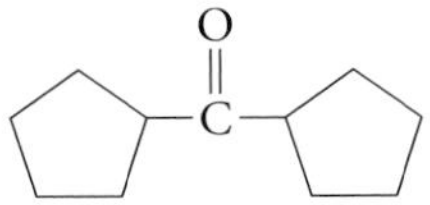   **h.** $CH_3CH{=}CH\overset{\displaystyle O}{\overset{\|}{C}}CH_3$ **i.** $BrCH_2\overset{\displaystyle O}{\overset{\|}{C}}CH_3$

9.31 Write a structural formula for each of the following:

a. 3-octanone   **b.** 4-methylpentanal
c. *m*-chlorobenzaldehyde   **d.** 3-methylcyclohexanone
e. 2-pentenal   **f.** benzyl phenyl ketone
g. cycloheptanone   **h.** *p*-tolualdehyde
i. 2,2-dibromohexanal   **j.** 1-phenyl-2-butanone

9.32 Give an example of each of the following:

a. acetal   **b.** hemiacetal
c. cyanohydrin   **d.** imine
e. oxime   **f.** phenylhydrazone
g. enol   **h.** aldehyde with no α-hydrogen
i. enolate   **j.** hydrazone

9.33 The boiling points of the isomeric carbonyl compounds heptanal, 4-heptanone, and 2,4-dimethyl-3-pentanone are 155°C, 144°C, and 124°C, respectively. Suggest a possible explanation for the observed order. (*Hint:* Recall the effect of chain branching on the boiling points of isomeric alkanes, and how steric effects can influence the association of the molecules.)

= concept connections

Synthesis of Ketones and Aldehydes

9.34 Write an equation for the synthesis of 2-pentanone by
  a. oxidation of an alcohol.   **b.** hydration of an alkyne.

9.35 Write an equation for the synthesis of pentanal from an alcohol.

9.36 Write an equation, using the Friedel–Crafts reaction, for the preparation of

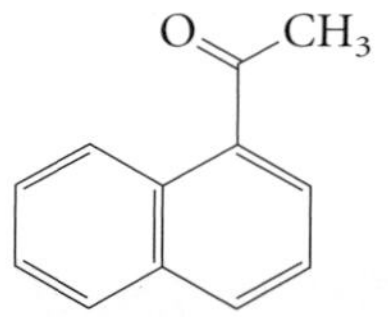

Reactions of Aldehydes and Ketones

9.37 Write an equation for the reaction, if any, of *p*-bromobenzaldehyde with each of the following reagents, and name the organic product.

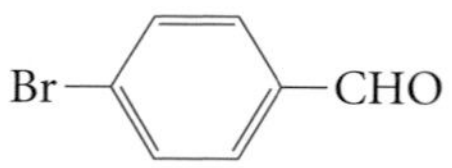

p-bromobenzaldehyde

  a. Tollens' reagent
  c. CrO_3, H^+
  e. methylamine (CH_3NH_2)
  g. cyanide ion
  i. ethylene glycol, H^+

  b. hydroxylamine
  d. ethylmagnesium bromide, then H_3O^+
  f. phenylhydrazine
  h. excess methanol, dry HCl
  j. lithium aluminum hydride

9.38 What simple chemical test can distinguish between the members of the following pairs of compounds? (*Hint:* Think of a reaction that one compound will undergo and the other will not.)
  a. pentanal and 2-pentanone
  b. benzyl alcohol and benzaldehyde
  c. cyclohexanone and 2-cyclohexenone

9.39 Use the structures shown in Figure 9.1 to write equations for the following reactions of natural products:
  a. cinnamaldehyde + Tollens' reagent
  c. carvone + sodium borohydride

  b. vanillin + hydroxylamine
  d. camphor + (1) methylmagnesium bromide and (2) H_3O^+

9.40 Complete each of the following equations:
  a. butanal + excess methanol, H^+ $\longrightarrow$

  b. $CH_3CH(OCH_3)_2$ + H_2O, H^+ $\longrightarrow$

  c. 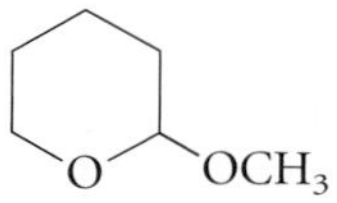 + H_2O, H^+ $\longrightarrow$

  d. 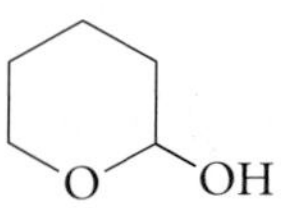 + excess CH_3CH_2OH, H^+ $\longrightarrow$

Reactions with Grignard Reagents and Other Nucleophiles

9.41 Write an equation for the reaction of each of the following with methylmagnesium bromide, followed by hydrolysis with aqueous acid:
  a. acetaldehyde   **b.** acetophenone   **c.** formaldehyde   **d.** cyclopentanone

9.42 Using a Grignard reagent and the appropriate aldehyde or ketone, show how each of the following can be prepared:

a. 1-pentanol   **b.** 3-pentanol   **c.** 2-methyl-2-pentanol
d. 1-cyclopentylcyclopentanol   **e.** 1-phenyl-1-propanol   **f.** 3-butene-2-ol

9.43 Complete the equation for the reaction of

a. cyclohexanone + $Na^{+-}C\equiv CH \longrightarrow \xrightarrow[H^+]{H_2O}$

b. cyclopentanone + HCN $\xrightarrow{KOH}$

c. 2-butanone + $NH_2OH \xrightarrow{H^+}$

d. benzaldehyde + benzylamine $\longrightarrow$

e. propanal + phenylhydrazine $\longrightarrow$

Oxidations and Reductions

9.44 Give the structure of each product.

a. 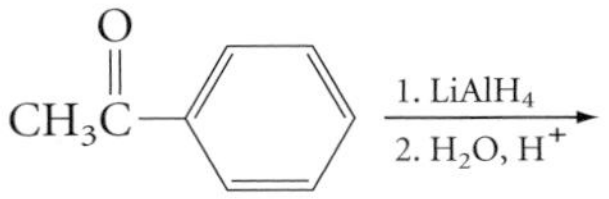 CH$_3$C(=O)—⟨phenyl⟩ $\xrightarrow[\text{2. H}_2\text{O, H}^+]{\text{1. LiAlH}_4}$

b. ⟨phenyl⟩—CH=CH—CH=O $\xrightarrow[\text{Ni, heat}]{\text{excess H}_2}$

c. ⟨cyclohexenone with CH$_3$⟩ $\xrightarrow[\text{2. H}_2\text{O, H}^+]{\text{1. NaBH}_4}$

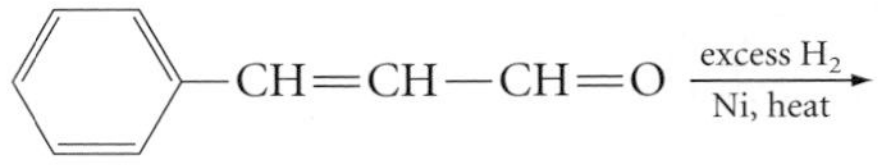

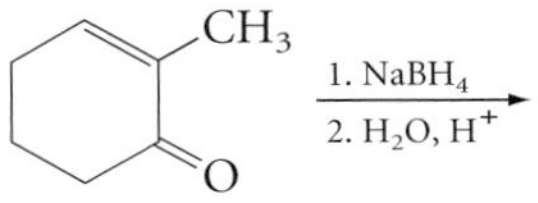

d. 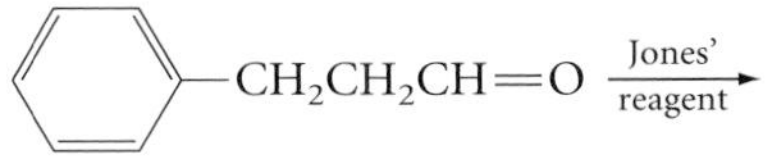 ⟨phenyl⟩—CH$_2$CH$_2$CH=O $\xrightarrow[\text{reagent}]{\text{Jones'}}$

e. CH$_3$CH=CHCHO $\xrightarrow{Ag_2O}$

Enols, Enolates, and the Aldol Reaction

9.45 Write the structural formulas for all possible enols of

a. $CH_3CH_2—C(=O)—CH_3$ **b.** 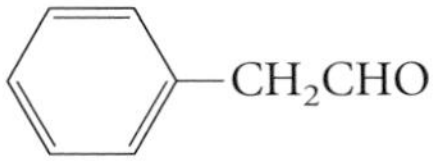 ⟨phenyl⟩—CH$_2$CHO   **c.** 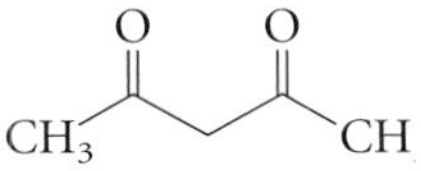 CH$_3$—C(=O)—CH$_2$—C(=O)—CH

9.46 Complete the reaction shown below by drawing the structure of the product.

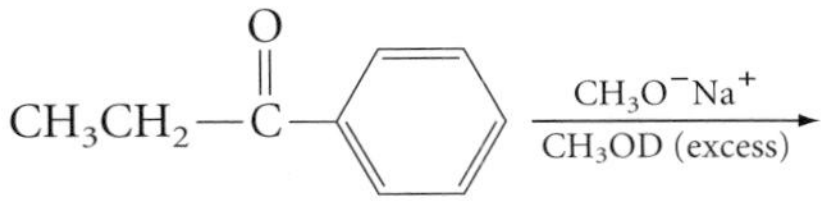

$$CH_3CH_2—C(=O)—\text{⟨phenyl⟩} \xrightarrow[\text{CH}_3\text{OD (excess)}]{\text{CH}_3\text{O}^-\text{Na}^+}$$

9.47 How many hydrogens are replaced by deuterium when each of the following compounds is treated with NaOD in D_2O?

a. 3-methylcyclopentanone   **b.** 3-methylbutanal

9.48 Write out the steps in the mechanism for the aldol condensation of butanal (the first step in the synthesis of the mosquito repellent "6-12," eq. 9.57).

9.49 Lily aldehyde, used in perfumes, can be made starting with a mixed aldol condensation between two different aldehydes. Provide their structures.

lily aldehyde

Puzzles

9.50 Excess benzaldehyde reacts with acetone and base to give a yellow crystalline product, $C_{17}H_{14}O$. Deduce its structure and explain how it is formed.

9.51 Vitamin B_6 (an aldehyde) reacts with an enzyme (partial structure shown below) to form a *coenzyme* that catalyzes the conversion of α-amino acids (Chapter 17) to α-keto acids.

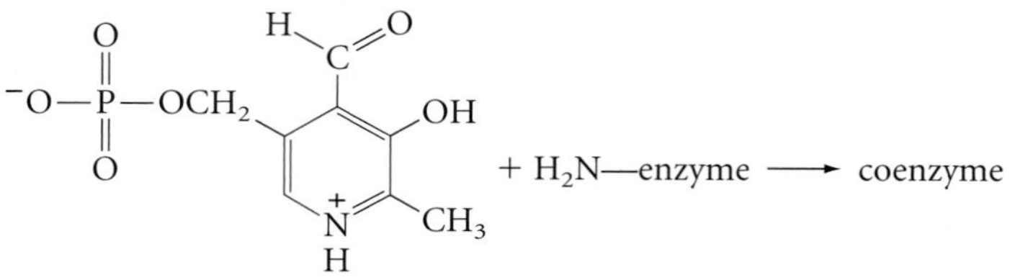

vitamin B_6

$+\ H_2N$—enzyme $\longrightarrow$ coenzyme

a. Draw the structure of the coenzyme.

b. α-Amino acids have the general structure shown below. Draw the structure of an α-keto acid.

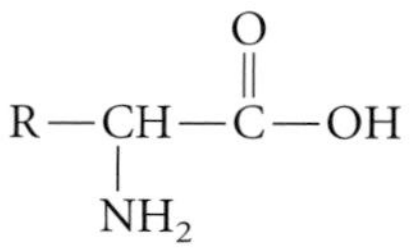

$$R-\underset{\underset{NH_2}{|}}{CH}-\overset{\overset{O}{\|}}{C}-OH \quad \text{α-amino acid}$$

salicylic acid

10

CARBOXYLIC ACIDS AND THEIR DERIVATIVES

The taste of vinegar, the sting of an ant, the rancid smell of butter, the relief derived from aspirin or ibuprofen—all of these are due to compounds that belong to the most important family of organic acids, the **carboxylic acids.** The resilience of polyester and nylon fabrics, the remarkable properties of Velcro, the softness of silk, the strength of bacterial cell walls and our own cell membranes—all of these are due to properties of derivatives of carboxylic acids. The functional group common to all carboxylic acids is the **carboxyl group.** The name is a contraction of the parts: the *carb*onyl and hyd*roxyl* groups. The general formula for a carboxylic acid can be written in expanded or abbreviated forms.

carboxyl group   three ways to represent a carboxylic acid   acid derivatives

In this chapter we will describe the structures, properties, preparation, and reactions of carboxylic acids and will also discuss some common **carboxylic acid derivatives,** in which the hydroxyl group of an acid is replaced by other functional groups.

10.1 Nomenclature of Acids

Because of their abundance in nature, carboxylic acids were among the earliest classes of compounds studied by organic chemists. It is not surprising, then, that many of them have common names. These names usually come from some Latin or Greek word that indicates the original source of the acid. Table 10.1 lists the first ten unbranched carboxylic acids, with their common and IUPAC names. To obtain the IUPAC name of a carboxylic acid, we replace the final *e* in the name of the corresponding alkane with the suffix *-oic* and add the word *acid.* Substituted

▲ The bark of the white willow tree (*Salix alba*) is a source of salicylic acid, from which aspirin (acetylsalicylic acid) is made.

Carboxylic acids are organic acids that contain the **carboxyl group.** In **acid derivatives,** the —OH group is replaced by other groups.

acids are named in two ways. In the IUPAC system, the chain is numbered beginning with the carboxyl carbon atom, and substituents are located in the usual way. If the common name of the acid is used, substituents are located with Greek letters, beginning with the α-carbon atom. IUPAC and common naming systems should not be mixed.

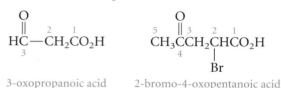

2-bromopropanoic acid	propenoic acid	3-hydroxybutanoic acid
(α-bromopropionic acid)	(acrylic acid)	(β-hydroxybutyric acid)

The carboxyl group has priority over alcohol, aldehyde, or ketone functionality in naming. In the latter cases, the prefix *oxo-* is used to locate the carbonyl group of the aldehyde or ketone, as in these examples:

$$\overset{O}{\underset{3}{\overset{\|}{HC}}}-\overset{2}{CH_2}\overset{1}{CO_2H} \qquad \overset{5}{CH_3}\overset{O}{\overset{\|}{\underset{4}{C}}}\overset{3}{CH_2}\overset{2}{\underset{Br}{CH}}\overset{1}{CO_2H}$$

3-oxopropanoic acid   2-bromo-4-oxopentanoic acid

Stinging ants, source of formic acid, HCOOH

PROBLEM 10.1   Write the structure for

a. 3-bromobutanoic acid.   b. 2-hydroxy-2-methylpropanoic acid.

c. 2-butynoic acid.   d. 5-methyl-6-oxohexanoic acid.

PROBLEM 10.2   Give an IUPAC name for

a. ⬡—CH_2CO_2H.   b. Cl_2CHCO_2H.

c. $CH_3CH=CHCO_2H$.   d. $(CH_3)_3CCO_2H$.

Table 10.1	Aliphatic carboxylic acids			
Carbon atoms	**Formula**	**Source**	**Common name**	**IUPAC name**
1	HCOOH	ants (Latin, *formica*)	formic acid	methanoic acid
2	CH_3COOH	vinegar (Latin, *acetum*)	acetic acid	ethanoic acid
3	CH_3CH_2COOH	milk (Greek, *protos pion*, first fat)	propionic acid	propanoic acid
4	$CH_3(CH_2)_2COOH$	butter (Latin, *butyrum*)	butyric acid	butanoic acid
5	$CH_3(CH_2)_3COOH$	valerian root (Latin, *valere,* to be strong)	valeric acid	pentanoic acid
6	$CH_3(CH_2)_4COOH$	goats (Latin, *caper*)	caproic acid	hexanoic acid
7	$CH_3(CH_2)_5COOH$	vine blossom (Greek, *oenanthe*)	enanthic acid	heptanoic acid
8	$CH_3(CH_2)_6COOH$	goats (Latin, *caper*)	caprylic acid	octanoic acid
9	$CH_3(CH_2)_7COOH$	pelargonium (an herb with stork-shaped seed capsules; Greek, *pelargos,* stork)	pelargonic acid	nonanoic acid
10	$CH_3(CH_2)_8COOH$	goats (Latin, *caper*)	capric acid	decanoic acid

When the carboxyl group is attached to a ring, the ending *-carboxylic acid* is added to the name of the parent cycloalkane.

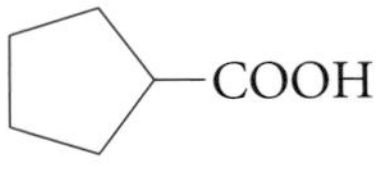

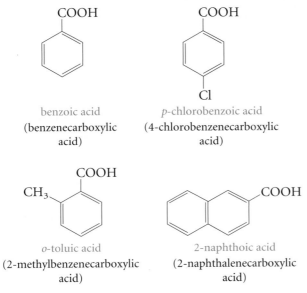

cyclopentanecarboxylic acid 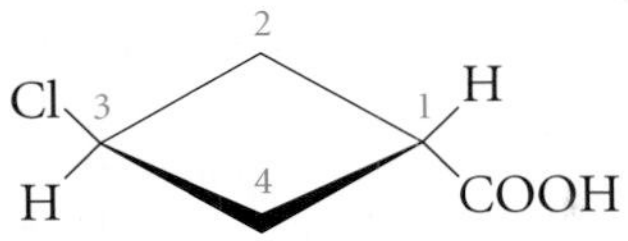   *trans*-3-chlorocyclobutanecarboxylic acid

Aromatic acids are named by attaching the suffix *-oic acid* or *-ic acid* to an appropriate prefix derived from the aromatic hydrocarbon.

benzoic acid
(benzenecarboxylic acid)

p-chlorobenzoic acid
(4-chlorobenzenecarboxylic acid)

o-toluic acid
(2-methylbenzenecarboxylic acid)

2-naphthoic acid
(2-naphthalenecarboxylic acid)

The root of Garden Heliotrope is a source of valeric acid, $CH_3(CH_2)_3COOH$.

PROBLEM 10.3   Write the structure for

a. *trans*-4-methylcyclohexanecarboxylic acid.
b. *m*-nitrobenzoic acid.

PROBLEM 10.4   Give the correct name for

a.  ▷—COOH.   b. 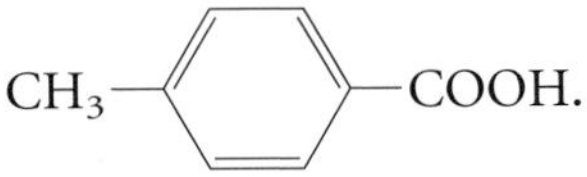 CH_3—⬡—COOH.

Aliphatic dicarboxylic acids are given the suffix *-dioic acid* in the IUPAC system. For example,

$$\overset{1}{HO_2C}—\overset{2}{CH_2}\overset{3}{CH_2}—\overset{4}{CO_2H}\qquad HO_2C—C≡C—CO_2H$$

butanedioic acid   butynedioic acid

Goats, source of caproic, caprylic, and capric acids: $CH_3(CH_2)_nCOOH$, n = 4, 6, 8

Many dicarboxylic acids occur in nature and go by their common names, which are based on their source. Table 10.2 lists some common aliphatic diacids.* The most

---

*The first letter of each word in the sentence "Oh my, such good apple pie" gives, in order, the first letters of the common names of these acids and can help you to remember them.

Table 10.2	Aliphatic dicarboxylic acids		
Formula	**Common name**	**Source**	**IUPAC name**
HOOC—COOH	oxalic acid	plants of the *oxalic* family (for example, sorrel)	ethanedioic acid
HOOC—CH$_2$—COOH	malonic acid	apple (Gk. *malon*)	propanedioic acid
HOOC—(CH$_2$)$_2$—COOH	succinic acid	amber (L. *succinum*)	butanedioic acid
HOOC—(CH$_2$)$_3$—COOH	glutaric acid	gluten	pentanedioic acid
HOOC—(CH$_2$)$_4$—COOH	adipic acid	fat (L. *adeps*)	hexanedioic acid
HOOC—(CH$_2$)$_5$—COOH	pimelic acid	fat (Gk. *pimele*)	heptanedioic acid

Rhubarb, a source of oxalic acid, HOOCCOOH

important commercial compound in this group is adipic acid, used to manufacture nylon.

The two butenedioic acids played a historic role in the discovery of *cis–trans* isomerism and are usually known by their common names **maleic*** and **fumaric** acid.**

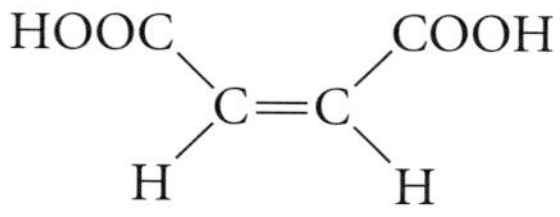

maleic acid
(*cis*-2-butenedioic acid)

and

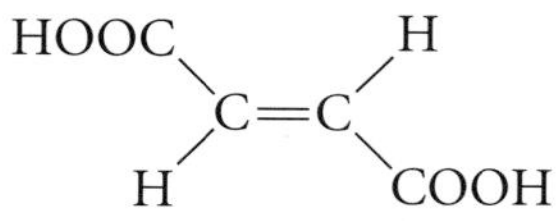

fumaric acid
(*trans*-2-butenedioic acid)

The three benzenedicarboxylic acids are generally known by their common names.

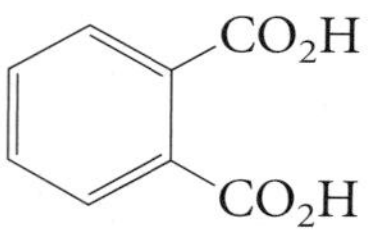

phthalic acid 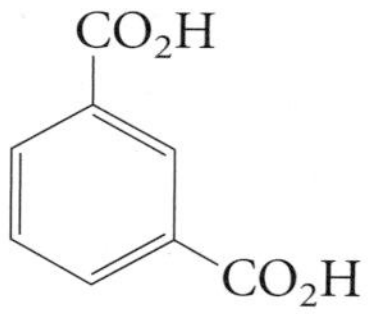   isophthalic acid 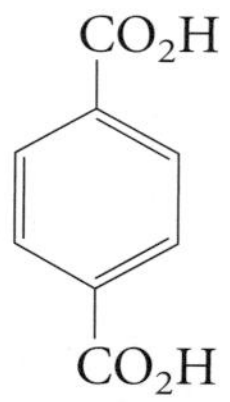   terephthalic acid

All three are important commercial chemicals, used to make polymers and other useful materials.

Finally, it is useful to have a name for an **acyl group.** Particular acyl groups are named from the corresponding acid by changing the *-ic* ending to *-yl*.

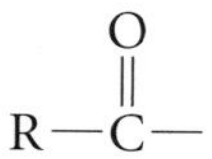

an acyl group 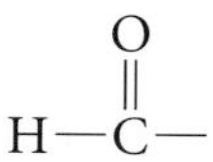   formyl (methanoyl)   acetyl (ethanoyl)   propanoyl 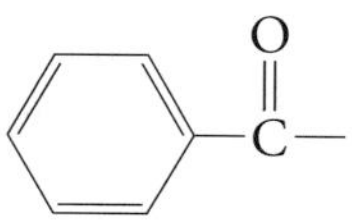   benzoyl

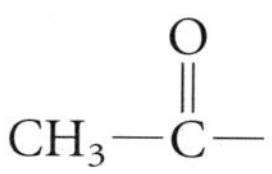

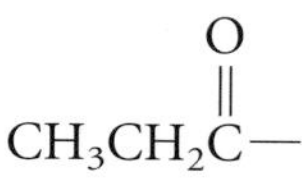

---

*From the Latin *malum* (apple). Malic acid (2-hydroxybutanedioic acid), found in apples, can be dehydrated on heating to give maleic acid.

**Found in fumitory, an herb of the genus *fumaria*.

Name	bp, °C	mp, °C	Solubility, g/100 g H_2O at 25°C
Table 10.3	Physical properties of some carboxylic acids		
formic acid	101	8	
acetic acid	118	17	miscible (∞)
propanoic acid	141	−22	
butanoic acid	164	−8	
hexanoic acid	205	−1.5	1.0
octanoic acid	240	17	0.06
decanoic acid	270	31	0.01
benzoic acid	249	122	0.4 (but 6.8 at 95°C)

PROBLEM 10.5   Write the formula for

    a. 4-acetylbenzoic acid.   b. benzoyl chloride.
    c. butanoyl bromide.   d. formylcyclopentane.

10.2 Physical Properties of Acids

The first members of the carboxylic acid series are colorless liquids with sharp or unpleasant odors. Acetic acid, which constitutes about 4 to 5% of vinegar, gives it its characteristic odor and flavor. Butyric acid gives rancid butter its disagreeable odor, and the goat acids (caproic, caprylic, and capric in Table 10.1) smell like goats. 3-Methyl-2-hexenoic acid, produced by bacteria, is responsible for the offensive odor of human armpits. Table 10.3 lists some physical properties of selected carboxylic acids.

    Carboxylic acids are polar. Like alcohols, they form hydrogen bonds with themselves or with other molecules (Sec. 7.4). They therefore have high boiling points for their molecular weights—higher even than those of comparable alcohols. For example, acetic acid and propyl alcohol, which have the same formula weights (60), boil at 118°C and 97°C, respectively. Carboxylic acids form dimers, with the units neatly held together by *two* hydrogen bonds (see Sec. 7.4).

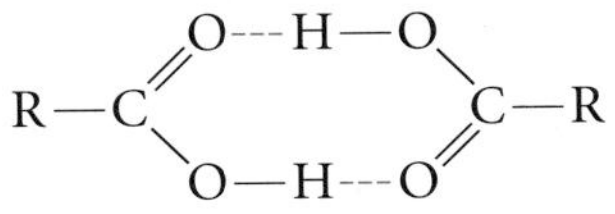

$$R-C\begin{matrix} O \cdots H-O \\ O-H \cdots O \end{matrix}C-R$$

Hydrogen bonding also explains the water solubility of the lower molecular weight carboxylic acids.

10.3 Acidity and Acidity Constants

Carboxylic acids dissociate in water, yielding a **carboxylate anion** and a hydronium ion.

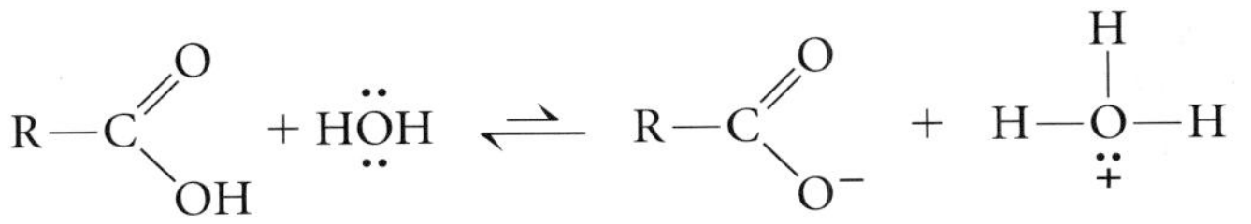

$$R-C\begin{matrix} O \\ OH \end{matrix} + H\ddot{O}H \rightleftharpoons R-C\begin{matrix} O \\ O^- \end{matrix} + H-\overset{H}{\underset{+}{\ddot{O}}}-H \qquad \textbf{(10.1)}$$

                       carboxylate anion   hydronium ion

Their acidity constant K_a is given by the expression

$$K_a = \frac{[RCO_2^-][H_3O^+]}{[RCO_2H]}$$ (10.2)

(Before proceeding further, it would be a good idea for you to review Secs. 7.5 and 7.6.)

Table 10.4 lists the acidity constants for some carboxylic and other acids. In comparing data in this table, remember that the larger the value of K_a or the smaller the value of pK_a, the stronger the acid.

EXAMPLE 10.1

Which is the stronger acid, formic or acetic, and by how much?

Solution   Formic acid is stronger; it has the larger K_a. The ratio of acidities is

$$\frac{2.1 \times 10^{-4}}{1.8 \times 10^{-5}} = 1.16 \times 10^1 = 11.7$$

This means that formic acid is 11.7 times stronger than acetic acid.

PROBLEM 10.6   Using the data given in Table 10.4, determine which is the stronger acid, acetic or chloroacetic, and by how much.

Before we can explain the acidity differences in Table 10.4, we must examine the structural features that make carboxylic acids acidic.

Table 10.4 **The ionization constants of some acids**

Name	Formula	K_a	pK_a
formic acid	HCOOH	2.1×10^{-4}	3.68
acetic acid	CH_3COOH	1.8×10^{-5}	4.74
propanoic acid	CH_3CH_2COOH	1.4×10^{-5}	4.85
butanoic acid	$CH_3CH_2CH_2COOH$	1.6×10^{-5}	4.80
chloroacetic acid	$ClCH_2COOH$	1.5×10^{-3}	2.82
dichloroacetic acid	$Cl_2CHCOOH$	5.0×10^{-2}	1.30
trichloroacetic acid	CCl_3COOH	2.0×10^{-1}	0.70
2-chlorobutanoic acid	$CH_3CH_2CHClCOOH$	1.4×10^{-3}	2.85
3-chlorobutanoic acid	$CH_3CHClCH_2COOH$	8.9×10^{-5}	4.05
benzoic acid	C_6H_5COOH	6.6×10^{-5}	4.18
o-chlorobenzoic acid	$o\text{-}Cl\text{---}C_6H_4COOH$	12.5×10^{-4}	2.90
m-chlorobenzoic acid	$m\text{-}Cl\text{---}C_6H_4COOH$	1.6×10^{-4}	3.80
p-chlorobenzoic acid	$p\text{-}Cl\text{---}C_6H_4COOH$	1.0×10^{-4}	4.00
p-nitrobenzoic acid	$p\text{-}NO_2\text{---}C_6H_4COOH$	4.0×10^{-4}	3.40
phenol	C_6H_5OH	1.0×10^{-10}	10.00
ethanol	CH_3CH_2OH	1.0×10^{-16}	16.00
water	HOH	1.8×10^{-16}	15.74

10.4 What Makes Carboxylic Acids Acidic?

You might wonder why carboxylic acids are so much more acidic than alcohols, since each class ionizes by losing H^+ from a hydroxyl group. There are two reasons, which can best be illustrated with a specific example.

From Table 10.4 we see that acetic acid is approximately 10^{11}, or one hundred thousand million, times stronger an acid than ethanol.

$$CH_3CH_2\ddot{O}H \rightleftharpoons CH_3CH_2\ddot{O}:^- + H^+ \qquad K_a = 10^{-16} \qquad \textbf{(10.3)}$$
$$\text{ethoxide ion}$$

$$CH_3\overset{\displaystyle :O:}{\overset{\|}{C}}-\ddot{O}H \rightleftharpoons CH_3\overset{\displaystyle :O:}{\overset{\|}{C}}-\ddot{O}:^- + H^+ \qquad K_a = 10^{-5} \qquad \textbf{(10.4)}$$
$$\text{acetate ion}$$

The only difference in their structures is the replacement of a CH_2 group (in ethanol) by a carbonyl group. But we saw (Sec. 9.5) that a carbonyl carbon atom carries a substantial *positive* charge ($\delta+$). This charge makes it much easier to place a *negative* charge on the adjacent oxygen atom, which is exactly what happens when we ionize a proton from the hydroxyl group.

In ethoxide ion, *the negative charge is localized on a single oxygen atom*. In acetate ion, on the other hand, *the negative charge can be delocalized through resonance.*

resonance in a carboxylate ion (acetate ion)

The negative charge is spread *equally* over the two oxygens, so that each oxygen in the carboxylate ion carries only half the negative charge. The acetate ion is stabilized by resonance compared to the ethoxide ion, and this stabilization helps to drive the equilibrium more to the right in eq. 10.4 than in eq. 10.3. Consequently, more H^+ is formed from acetic acid than from ethanol.

For both these reasons, the positive charge on the carbonyl carbon and delocalization of the carboxylate ion, carboxylic acids are much more acidic than alcohols.

EXAMPLE 10.2

Phenoxide ions are also stabilized by resonance (Sec. 7.6). Why aren't phenols as strong acids as carboxylic acids?

Solution First, the carbon atom to which the hydroxyl group is attached in a phenol is not as positive as a carbonyl carbon. Second, charge delocalization is not as great in phenoxide ions as in carboxylate ions because the contributors to the resonance hybrid are not equivalent. Some of them put the negative charge on carbon instead of on oxygen and disrupt aromaticity.

PROBLEM 10.7 Write two resonance structures for the benzoate ion ($C_6H_5CO_2^-$) that show how the negative charge is delocalized over the two oxygens. Can the negative charge in the benzoate ion be delocalized into the aromatic ring?

Physical data support the importance of resonance in carboxylate ions. In formic acid molecules, the two carbon–oxygen bonds have different lengths. But in sodium formate, both carbon–oxygen bonds of the formate ion are identical, and their length is between those of normal double and single carbon–oxygen bonds.

formic acid   sodium formate

10.5 Effect of Structure on Acidity; the Inductive Effect Revisited

The data in Table 10.4 show that even among carboxylic acids (where the ionizing functional group is kept constant), acidities can vary depending on what other groups are attached to the molecule. Compare, for example, the K_a of acetic acid with those of mono-, di-, and trichloroacetic acids, and note that the acidity varies by a factor of 10,000.

The most important factor operating here is the inductive effect of the groups close to the carboxyl group. This effect relays charge through bonds, by displacing bonding electrons toward electronegative atoms, or away from electropositive atoms. Recall that *electron-withdrawing groups enhance acidity, and electron-releasing groups reduce acidity* (see Sec. 7.6).

Let us examine the carboxylate ions formed when acetic acid and its chloro derivatives ionize:

acetate   chloroacetate   dichloroacetate   trichloroacetate

Because chlorine is more electronegative than carbon, the C—Cl bond is polarized with the chlorine partially negative and the carbon partially positive. Thus, electrons are pulled away from the carboxylate end of the ion toward the chlorine. The effect tends to spread the negative charge over more atoms than in acetate ion itself and thus stabilizes the ion. The more chlorines, the greater the effect and the greater the strength of the acid.

EXAMPLE 10.3

Explain the acidity order in Table 10.4 for butanoic acid and its 2- and 3-chloro derivatives.

Solution   The 2-chloro substituent increases the acidity of butanoic acid substantially, due to its inductive effect. In fact, the effect is about the same as for chloroacetic and acetic acids. The 3-chloro substituent exerts a similar *but much smaller* effect, because the C—Cl bond is now farther away from the carboxylate group. *Inductive effects fall off rapidly with distance.*

PROBLEM 10.8   Account for the relative acidities of benzoic acid and its *ortho, meta,* and *para* chloro derivatives (Table 10.4).

Phthalic acid, used for making plasticizers, resins, and dyestuffs, is manufactured by similar oxidations, starting with *o*-xylene.

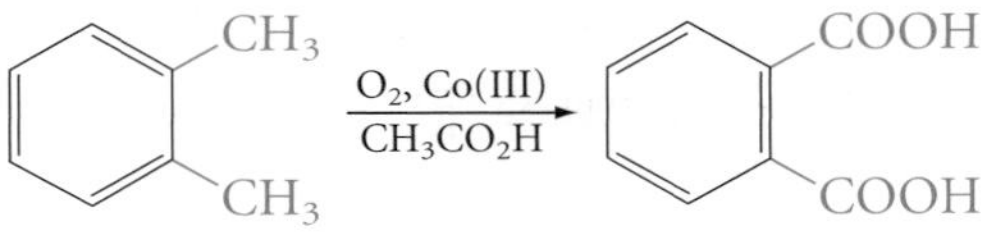

(10.11)

o-xylene   phthalic acid

10.7.c Reaction of Grignard Reagents with Carbon Dioxide

As we saw previously Grignard reagents add to the carbonyl groups of aldehydes or ketones to give alcohols. In a similar way, they add irreversibly to the carbonyl group of carbon dioxide to give acids, after protonation of the intermediate carboxylate salt with a mineral acid like aqueous HCl.

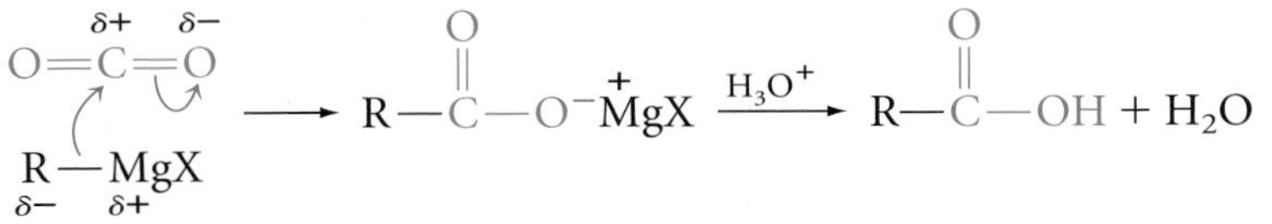

(10.12)

This reaction gives good yields and is an excellent laboratory method for preparing both aliphatic and aromatic acids. Note that the acid obtained has one more carbon atom than the alkyl or aryl halide from which the Grignard reagent is prepared, so the reaction provides a way to increase the length of a carbon chain.

EXAMPLE 10.5

Show how $(CH_3)_3CBr$ can be converted to $(CH_3)_3CCO_2H$.

Solution $(CH_3)_3CBr \xrightarrow[\text{ether}]{Mg} (CH_3)_3CMgBr \xrightarrow[\text{2. H}_3O^+]{\text{1. CO}_2} (CH_3)_3CCO_2H$

PROBLEM 10.10   Show how cyclohexyl chloride can be converted to cyclohexane-carboxylic acid.

PROBLEM 10.11   Devise a synthesis of butanoic acid $(CH_3CH_2CH_2CO_2H)$ from 1-propanol $(CH_3CH_2CH_2OH)$.

10.7.d Hydrolysis of Cyanides (Nitriles)

The carbon–nitrogen triple bond of organic cyanides can be hydrolyzed to a carboxyl group. The reaction requires either acid or base. In acid, the nitrogen atom of the cyanide is converted to an ammonium ion.

(10.13)

a cyanide,   an acid   ammonium
or nitrile   ion

In base, the nitrogen is converted to ammonia and the organic product is the carboxylate salt, which must be neutralized in a separate step to give the acid.

$$R-C\equiv N + 2\,H_2O \xrightarrow{\text{NaOH}} \underset{\text{a carboxylate salt}}{R-\overset{\displaystyle O}{\overset{\|}{C}}-O^-Na^+} + \underset{\text{ammonia}}{NH_3} \qquad (10.14)$$

$$\Big\downarrow H^+$$

$$R-\overset{\displaystyle O}{\overset{\|}{C}}-OH$$

The mechanism of nitrile hydrolysis involves acid or base promoted addition of water across the triple bond. This gives an intermediate imidate that tautomerizes to an amide. The amide is then hydrolyzed to the carboxylic acid. The addition of water to the nitrile resembles the hydration of an alkyne (eq. 3.52). The oxygen of water behaves as a nucleophile and bonds to the electrophilic carbon of the nitrile. Amide hydrolysis will be discussed in Sec. 10.20.

$$\underset{\text{nitrile}}{R-\overset{\delta+}{C}\equiv\overset{\delta-}{N}} \xrightarrow[\text{H}^+\text{ or HO}^-]{H_2O} \underset{\text{imidate}}{R-\overset{\displaystyle OH}{\overset{|}{C}}=NH} \xrightarrow{\text{tautomerization}} \underset{\text{amide}}{R-\overset{\displaystyle O}{\overset{\|}{C}}-NH_2} \xrightarrow[\text{H}^+\text{ or HO}^-]{\text{hydrolysis}} \underset{\text{acid}}{R-\overset{\displaystyle O}{\overset{\|}{C}}-OH} \qquad (10.15)$$

Alkyl cyanides are generally made from the corresponding alkyl halide (usually primary) and sodium cyanide by an S_N2 displacement, as shown in this synthesis of an acid:

$$\underset{\substack{\text{propyl bromide}\\ \text{(1-bromopropane)}}}{CH_3CH_2CH_2Br} \xrightarrow{\text{NaCN}} \underset{\substack{\text{butyronitrile}\\ \text{(butanenitrile)}}}{CH_3CH_2CH_2CN} \xrightarrow[\text{H}^+]{H_2O} \underset{\substack{\text{butyric acid}\\ \text{(butanoic acid)}}}{CH_3CH_2CH_2CO_2H} + NH_4^+ \qquad (10.16)$$

PROBLEM 10.12   Why is it *not* possible to convert bromobenzene to benzoic acid by the nitrile method? How could this conversion be accomplished?

Organic cyanides are commonly named after the corresponding acid, by changing the *-ic* or *-oic* suffix to *-onitrile* (hence butyronitrile in eq. 10.16). In the IUPAC system, the suffix *-nitrile* is added to the name of the hydrocarbon with the same number of carbon atoms (hence butanenitrile in eq. 10.16).

Note that with the hydrolysis of nitriles, as with the Grignard method, the acid obtained has one more carbon atom than the alkyl halide from which the cyanide is prepared. Consequently, both methods provide ways of increasing the length of a carbon chain.

PROBLEM 10.13   Write equations for synthesizing phenylacetic acid ($C_6H_5CH_2CO_2H$) from benzyl bromide ($C_6H_5CH_2Br$) by two routes.

A WORD ABOUT ...

Green Chemistry and Ibuprofen: A Case Study

From our increasing need to respect nature and protect our environment has come a new field of chemistry— "green chemistry"—the design and development of chemistry that is environmentally friendly, chemistry that avoids pollution. This presents many challenges to organic synthesis. One notion of an ideal synthesis is one that provides a useful compound in one step with formation of no disposable by-products by a process that consumes little energy. Such a synthesis would certainly be environmentally friendly! This goal is seldom met, but general principles can be applied to try to approach this ideal.

First let's consider some reactions we have learned. Addition reactions (catalytic hydrogenation and Diels–Alder reactions, for example) do not create any by-products. The same can be said for isomerization reactions. Such reactions are said to be "atom economical"—all of the atoms in the reactants appear in the product.* On the other hand, elimination reactions and substitution reactions necessarily produce by-products. This does not mean that they are bad, but if a synthesis can be devised that focuses on addition and isomerization reactions, less attention will have to be devoted to disposing of, or developing uses for, by-products.

Some other general strategies for the development of green chemistry are to use catalysts to accomplish reactions (rather than stoichiometric reagents), to minimize the use of heavy metals as stoichiometric oxidants (for example,

developed and several of these have been commercialized. The synthesis shown below begins with the reaction of isobutylbenzene with acetic anhydride using HF as the solvent. This is a variation of the Friedel–Crafts acylation in which the anhydride serves as the source of an acylium ion (see Sec. 4.9.d). Through clever engineering processes, the reaction solvent (HF) serves as both the acid catalyst and solvent (recyclable) for the reaction, and water is the only major reaction by-product. The second step is an addition reaction, catalytic hydrogenation of a ketone to an alcohol (see eq. 9.57). The final step is a reaction we have not discussed that involves palladium-catalyzed "insertion" of carbon monoxide into a benzylic C—O bond to give the carboxylic acid (ibuprofen). This reaction is clearly atom economical. Finally, the chemical yields of all the reactions are very high and very little chemical waste is produced.** Although this synthesis is an excellent example of green chemistry in action, there is room for improvement. For example, this produces a racemic mixture of ibuprofen whereas only the (S)-enantiomer is biologically active (see A Word About Enantiomers and Biological Activity in Chapter 5).

It is clear that green chemistry will play an important role in the 21st century. This has been recognized by the Presidential Green Chemistry Challenge, initiated by President Clinton in 1995 to reward the development of environmentally benign chemistry. Although even

isobutylbenzene   ibuprofen

chromium), to focus on the use of molecular oxygen and hydrogen peroxide as oxidants, and to minimize the use of solvents in reactions.

Let's examine a synthesis that does a good job of meeting the goals of green chemistry. Ibuprofen is a very important anti-inflammatory drug. It is the active ingredient of many over-the-counter drugs used to relieve pain from headaches and arthritis. Approximately 25 million pounds of this simple carboxylic acid were produced by synthesis in 2000! A large number of ibuprofen syntheses have been

biological processes seldom meet the notion of an ideal synthesis, to strive for this ideal can only lead to new and better chemistry.

*For more on "atom economy," see B. M. Trost, *Angew. Chem. Int. Ed. Engl.* **1995,** *34,* 259. For more on green chemistry, see W. Leiner, *Science* **1999,** *284,* 1780, and visit the following web site: http://www.epa.gov/greenchemistry.

**For an overview of other syntheses of ibuprofen, consult B. G. Reuben and H. A. Wittcoff, *Pharmaceutical Chemicals in Perspective,* John Wiley and Sons, New York, 1989.

A CLOSER LOOK AT ...

Green Chemistry

Log on to http://www.college.hmco.com and follow the links to the student text web site. Use the Web Link section for Chapter 10 to answer the following questions and find more information on the topics below.

Article Concerning Green Chemistry

1. What is *green chemistry* according to this article?
2. This article reports on an international meeting of scientists called the CHEMRAWN conference. What does CHEMRAWN stand for? What international body sponsors this program?
3. Search the site to find the Twelve Principles of Green Chemistry. What applications of these principles can you find in the article?
4. Return to the top of the article and click the History icon to find a brief description of the Foundations of Green Chemistry. What is the Pollution Prevention

Act of 1990? What are some of the organizations involved in promoting green chemistry? What incentives are these organizations using to promote green chemistry?

Hangers Cleaners and Dry Wash

After looking at the links provided for both companies above, what do you think are the advantages of using CO_2 over traditional dry cleaning solvents? Explain why you think so.

Ionic Liquids and Their Uses

1. What are some of the advantages of using ionic liquids as solvents in chemical reactions?
2. What environmental problems posed by traditional solvents are avoided by using ionic liquids?

For more information on green chemistry, click the additional links provided on the next web site.

10.8 Carboxylic Acid Derivatives

Carboxylic acid derivatives are compounds in which the hydroxyl part of the carboxyl group is replaced by various other groups. All acid derivatives can be hydrolyzed to the corresponding acid. In the remainder of this chapter, we will consider the preparation and reactions of the more important of these acid derivatives. Their general formulas are as follows:

$$\underset{\text{ester}}{R-\overset{\overset{\textstyle O}{\|}}{C}-OR'} \qquad \underset{\text{acyl halide}}{R-\overset{\overset{\textstyle O}{\|}}{C}-X} \quad \binom{X \text{ is usually}}{Cl \text{ or } Br} \qquad \underset{\text{acid anhydride}}{R-\overset{\overset{\textstyle O}{\|}}{C}-O-\overset{\overset{\textstyle O}{\|}}{C}-R} \qquad \underset{\text{primary amide}}{R-\overset{\overset{\textstyle O}{\|}}{C}-NH_2}$$

Esters and amides occur widely in nature. Anhydrides, however, are uncommon in nature, and acyl halides are strictly creatures of the laboratory.

10.9 Esters

An **ester** is a carboxylic acid derivative in which the —OH group is replaced by an —OR group.

Esters are derived from acids by replacing the —OH group by an —OR group. They are named in a manner analogous to carboxylic acid salts. The R part of the —OR group is named first, followed by the name of the acid, with the *-ic* ending changed to *-ate*.

$$\underset{\substack{\text{methyl acetate} \\ \text{(methyl ethanoate)} \\ \text{bp 57°C}}}{CH_3\overset{\overset{\textstyle O}{\|}}{C}-OCH_3} \qquad \underset{\substack{\text{ethyl acetate} \\ \text{(ethyl ethanoate)} \\ \text{bp 77°C}}}{CH_3\overset{\overset{\textstyle O}{\|}}{C}-OCH_2CH_3} \qquad \underset{\substack{\text{methyl butanoate} \\ \text{bp 102.3°C}}}{CH_3CH_2CH_2\overset{\overset{\textstyle O}{\|}}{C}-OCH_3}$$

Notice the different names of the following pair of isomeric esters, where the R and R′ groups are interchanged.

$$CH_3\overset{\displaystyle O}{\overset{\displaystyle \|}{C}}-O-\text{(phenyl)}$$ $$\text{(phenyl)}-\overset{\displaystyle O}{\overset{\displaystyle \|}{C}}-OCH_3$$

phenyl acetate
bp 195.7°C

methyl benzoate
bp 196.6°C

Esters are named with two words that are *not* run together.

EXAMPLE 10.6

Name $CH_3CH_2CO_2CH(CH_3)_2$.

Solution   The related acid is $CH_3CH_2CO_2H$, so the last part of the name is *propanoate* (change the *-ic* of propanoic to *-ate*). The alkyl group that replaces the hydrogen is *isopropyl*, or 2-*propyl*, so the correct name is *isopropyl propanoate*, or 2-*propyl propanoate*.

PROBLEM 10.14   Write the IUPAC name for

a. $H-\overset{\displaystyle O}{\overset{\displaystyle \|}{C}}-OCH_3$.   b. $CH_3CH_2\overset{\displaystyle O}{\overset{\displaystyle \|}{C}}-O-\triangle$.

PROBLEM 10.15   Write the structure of

a.   3-pentyl ethanoate.   b.   ethyl 2-methylpropanoate.

Many esters are rather pleasant-smelling substances and are responsible for the flavor and fragrance of many fruits and flowers. Among the more common are pentyl acetate (bananas), octyl acetate (oranges), ethyl butanoate (pineapples), and pentyl butanoate (apricots). Natural flavors can be exceedingly complex. For example, no fewer than 53 esters have been identified among the volatile constituents of Bartlett pears! Mixtures of esters are used as perfumes and artificial flavors. Low-molecular-weight esters are also used by insects and animals to transmit signals. Female elephants release (Z)-7-dodecen-1-yl acetate to signal their readiness to mate. Many moths release the same ester to attract mates.

Female elephants release the ester (Z)-7-dodecen-1-yl acetate to attract mates.

10.10   Preparation of Esters; Fischer Esterification

When a carboxylic acid and an alcohol are heated in the presence of an acid catalyst (usually HCl or H_2SO_4), an equilibrium is established with the ester and water.

$$R-\overset{\displaystyle O}{\overset{\displaystyle \|}{C}}-OH + HO-R' \underset{}{\overset{H^+}{\rightleftharpoons}} R-\overset{\displaystyle O}{\overset{\displaystyle \|}{C}}-OR' + H_2O \qquad \textbf{(10.17)}$$

acid   alcohol   ester

The process is called **Fischer esterification,** after Emil Fischer (page 163), who developed the method. Although the reaction is an equilibrium, it can be shifted to the right in several ways. If either the alcohol or the acid is inexpensive, a large excess can

Fischer esterification is the acid-catalyzed condensation of a carboxylic acid and an alcohol.

be used. Alternatively, the ester and/or water may be removed as formed (by distillation, for example), thus driving the reaction forward.

PROBLEM 10.16 Following eq. 10.17, write an equation for the preparation of propyl butanoate from the correct acid and alcohol.

10.11 The Mechanism of Acid-Catalyzed Esterification; Nucleophilic Acyl Substitution

We can ask the following simple mechanistic question about Fischer esterification: Is the water molecule formed from the hydroxyl group of the acid and the hydrogen of the alcohol (as shown in color in eq. 10.17), or from the hydrogen of the acid and the hydroxyl group of the alcohol? This question may seem rather trivial, but the answer provides a key to understanding much of the chemistry of acids, esters, and their derivatives.

This question was resolved using isotopic labeling. For example, Fischer esterification of benzoic acid with methanol that had been enriched with the ^{18}O isotope of oxygen gave labeled methyl benzoate.*

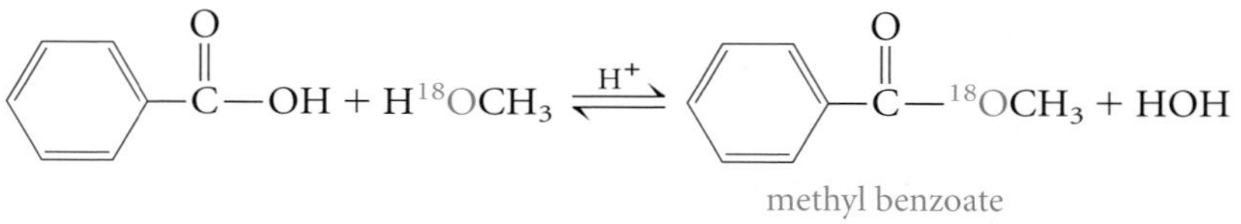

$$ C_6H_5-\overset{\overset{O}{\|}}{C}-OH + H^{18}OCH_3 \rightleftharpoons[H^+] C_6H_5-\overset{\overset{O}{\|}}{C}-{}^{18}OCH_3 + HOH \qquad \textbf{(10.18)} $$

methyl benzoate

None of the ^{18}O appeared in the water. Thus it is clear that *the water was formed using the hydroxyl group of the acid and the hydrogen of the alcohol.* In other words, in Fischer esterification, the —OR group of the alcohol replaces the —OH group of the acid.

How can we explain this experimental fact? A mechanism consistent with this result is as follows (the oxygen atom of the alcohol is shown in color so that its path can be traced):

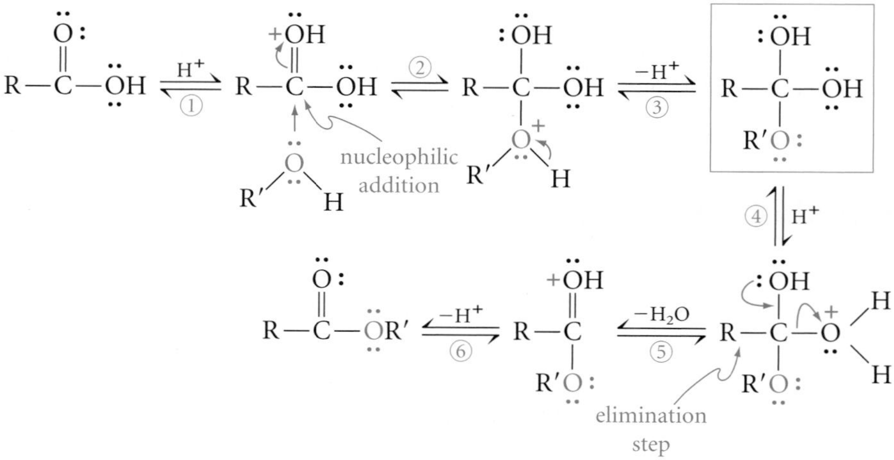

Ch. 10, Ref. 1. View animation of nucleophilic acyl substitution.

(10.19)

*^{18}O is oxygen with two additional neutrons in its nucleus. It is two mass units heavier than ^{16}O. ^{18}O can be distinguished from ^{16}O by mass spectrometry (see Chapter 12).

Let us go through this mechanism, which looks more complicated than it really is, one step at a time.

Step 1. The carbonyl group of the acid is reversibly protonated. This step explains how the acid catalyst works. Protonation increases the positive charge on the carboxyl carbon and enhances its reactivity toward nucleophiles (recall the similar effect of acid catalysts with aldehydes and ketones, eq. 9.9).

Step 2. *This is the crucial step.* The alcohol, as a nucleophile, attacks the carbonyl carbon of the protonated acid. This is the step in which the new C—O bond (the ester bond) is formed.

Steps 3 and 4. These steps are equilibria in which oxygens lose or gain a proton. Such acid–base equilibria are reversible and rapid and go on constantly in any acidic solution of an oxygen-containing compound. In step 4, it doesn't matter which —OH group is protonated since these groups are equivalent.

Step 5. This is the step in which water, one product of the overall reaction, is formed. For this step to occur, an —OH group must be protonated to improve its leaving-group capacity. (This step is similar to the reverse of step 2.)

Step 6. This deprotonation step gives the ester and regenerates the acid catalyst. (This step is similar to the reverse of step 1.)

Some other features of the mechanism in eq. 10.19 are worth examining. The reaction begins with a carboxylic acid, in which the carboxyl carbon is trigonal and sp^2-hybridized. The end product is an ester; the ester carbon is also trigonal and sp^2-hybridized. However, the reaction proceeds through a neutral **tetrahedral intermediate** (shown in a box in eq. 10.19 and in color in eq. 10.20), in which the carbon atom has four groups attached to it and is thus sp^3-hybridized. If we omit all of the proton-transfer steps in eq. 10.19, we can focus on this feature of the reaction:

> A **tetrahedral intermediate** has an sp^3-hybridized carbon atom.

$$\underset{sp^2}{\overset{\displaystyle \text{HO}}{\underset{\displaystyle R}{\diagdown}}\text{C}{=}\text{O}} + \text{R}'\text{OH} \;\rightleftharpoons\; \underset{\substack{\text{tetrahedral}\\\text{intermediate}}}{\overset{\displaystyle \text{R}'\text{O}}{\underset{\underset{sp^3}{\displaystyle \text{HO}\;\; R}}{\diagdown}}\text{C}{-}\text{OH}} \;\rightleftharpoons\; \underset{sp^2}{\overset{\displaystyle \text{R}'\text{O}}{\underset{\displaystyle R}{\diagdown}}\text{C}{=}\text{O}} + \text{H}_2\text{O} \qquad (10.20)$$

The net result of this process is substitution of the —OR′ group of the alcohol for the —OH group of the acid. Hence the reaction is referred to as **nucleophilic acyl substitution.** But the reaction is not a direct substitution. Instead, it occurs in two steps: (1) nucleophilic addition, followed by (2) elimination. We will see in the next and subsequent sections of this chapter that this is a general mechanism for nucleophilic substitutions at the carbonyl carbon atoms of acid derivatives.

> **Nucleophilic acyl substitution** is substitution of another group for the —OH group of a carboxylic acid.

PROBLEM 10.17 Following eq. 10.19, write out the steps in the mechanism for the acid-catalyzed preparation of ethyl acetate from ethanol and acetic acid. In the United States, this method is used commercially to produce more than 100 million pounds of ethyl acetate annually, mainly for use as a solvent in the paint industry, but also as a solvent for nail polish and various glues.

10.12 Lactones

Hydroxy acids contain a hydroxyl group and a carboxyl group.

Lactones are **cyclic esters.**

Hydroxy acids contain both functional groups required for ester formation. If these groups can come in contact through bending of the chain, they may react with one another to form **cyclic esters** called **lactones.** For example,

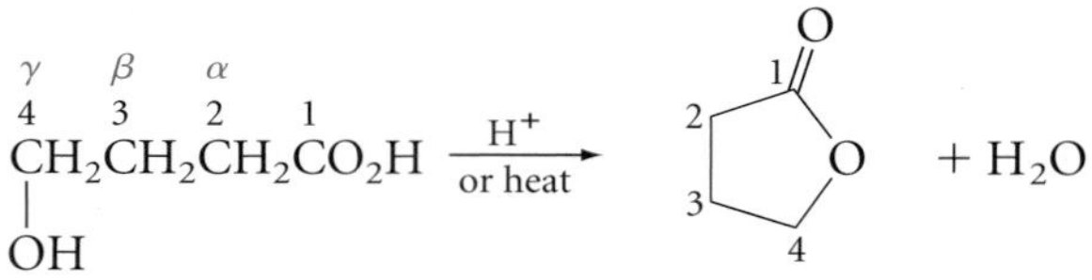

$$\underset{\substack{\text{OH} \\ }}{\overset{\substack{\gamma \quad \beta \quad \alpha \\ 4 \quad 3 \quad 2 \quad 1}}{CH_2CH_2CH_2CO_2H}} \xrightarrow[\text{or heat}]{H^+} \text{[γ-butyrolactone]} + H_2O \qquad (10.21)$$

γ-butyrolactone

Most common lactones have five- or six-membered rings, although lactones with smaller or larger rings are known. Two examples of six-membered lactones from nature are **coumarin,** which is responsible for the pleasant odor of newly mown hay, and **nepatalactone,** the compound in catnip that excites cats. **Erythromycin,** widely used as an antibiotic, is an example of a macrocyclic lactone.*

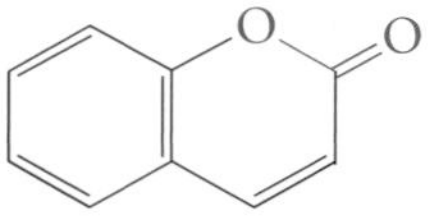

coumarin 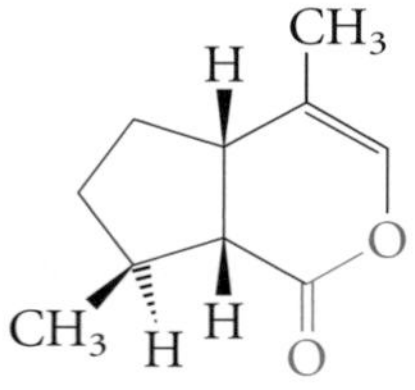   nepatalactone 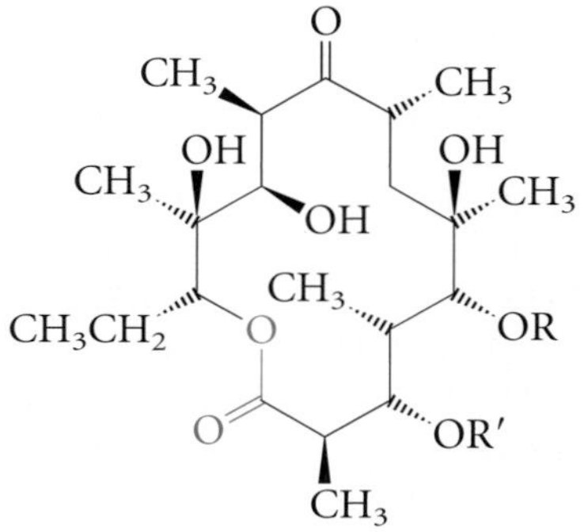   erythromycin

> **PROBLEM 10.18**   Write the steps in the mechanism for the acid-catalyzed reaction in eq. 10.21.

10.13 Saponification of Esters

Saponification is the hydrolysis of an ester with a base.

Esters are commonly hydrolyzed with base. The reaction is called **saponification** (from the Latin *sapon,* soap) because this type of reaction is used to make soaps from fats (Chapter 15). The general reaction is as follows:

$$\underset{\substack{\text{ester}}}{R-\overset{\overset{\displaystyle O}{\|}}{C}-OR'} + \underset{\substack{\text{nucleophile}}}{Na^+HO^-} \xrightarrow[H_2O]{\text{heat}} \underset{\substack{\text{salt of an acid}}}{R-\overset{\overset{\displaystyle O}{\|}}{C}-O^-Na^+} + \underset{\substack{\text{alcohol}}}{R'OH} \qquad (10.22)$$

---

*The R and R′ groups in erythromycin are carbohydrate units (see Chapter 16).

The mechanism is another example of a nucleophilic acyl substitution. It involves nucleophilic attack by hydroxide ion, a strong nucleophile, on the carbonyl carbon of the ester.

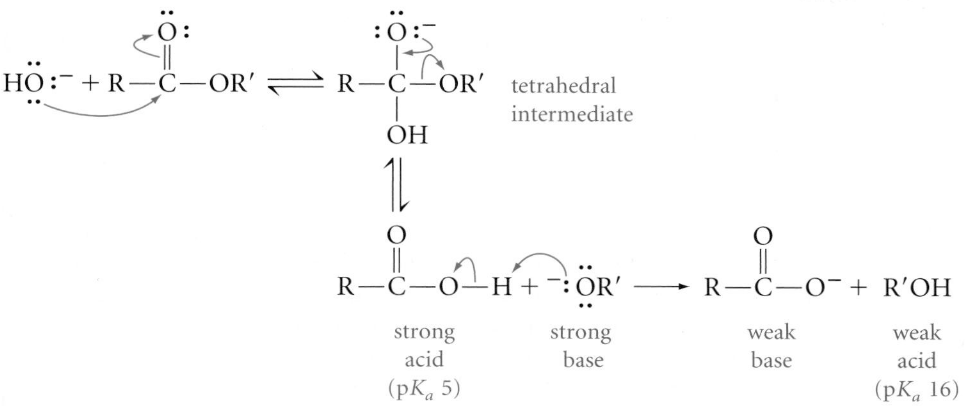

$$
HO:^- + R-\overset{\overset{\displaystyle O:}{\|}}{C}-OR' \rightleftharpoons R-\overset{\overset{\displaystyle :O:^-}{|}}{\underset{\underset{\displaystyle OH}{|}}{C}}-OR' \quad \text{tetrahedral}\atop \text{intermediate}
$$

$$\Updownarrow$$ (10.23)

$$
R-\overset{\overset{\displaystyle O}{\|}}{C}-O-H + {}^-:OR' \longrightarrow R-\overset{\overset{\displaystyle O}{\|}}{C}-O^- + R'OH
$$

strong   strong   weak   weak
acid   base   base   acid
(pK_a 5)   (pK_a 16)

The key step is nucleophilic addition to the carbonyl group. The reaction proceeds via a tetrahedral intermediate, but the reactant and the product are trigonal. *Saponification is not reversible;* in the final step, the strongly basic alkoxide ion removes a proton from the acid to form a carboxylate ion and an alcohol molecule—a step that proceeds completely in the forward direction.

Saponification is especially useful for breaking down an unknown ester, perhaps isolated from a natural source, into its component acid and alcohol for structure determination.

PROBLEM 10.19 Following eq. 10.22, write an equation for the saponification of methyl benzoate (see eq. 10.25 for structure).

10.14 Ammonolysis of Esters

Ammonia converts esters to amides.

$$
R-\overset{\overset{\displaystyle O}{\diagup\!\!\|}}{C}\underset{\displaystyle OR'}{\diagdown} + \overset{..}{N}H_3 \longrightarrow R-\overset{\overset{\displaystyle O}{\diagup\!\!\|}}{C}\underset{\displaystyle NH_2}{\diagdown} + R'OH \qquad (10.24)
$$

ester   amide

For example,

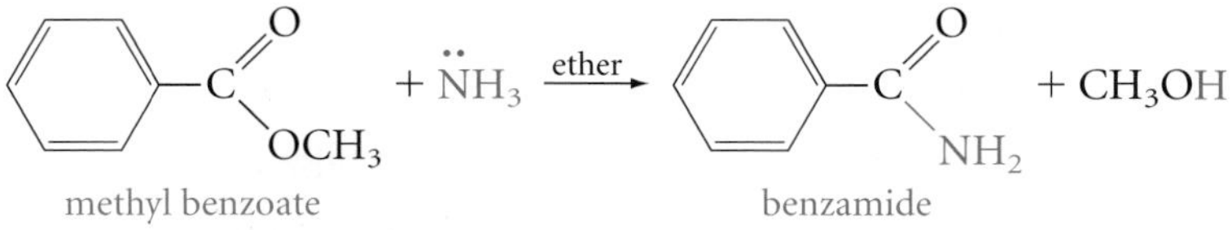

methyl benzoate   benzamide

$$+ \overset{..}{N}H_3 \xrightarrow{\text{ether}} \qquad + CH_3OH \qquad (10.25)$$

The reaction mechanism is very much like that of saponification. The unshared electron pair on the ammonia nitrogen initiates nucleophilic attack on the ester carbonyl group.

$$\underset{R}{\overset{R'O}{\diagdown}}C{=}O + NH_3 \rightleftharpoons \underset{HO}{\overset{R'O}{\diagdown}}\underset{R}{\overset{|}{C}}{-}NH_2 \longrightarrow \underset{R}{\overset{H_2N}{\diagdown}}C{=}O + R'OH \qquad \textbf{(10.26)}$$

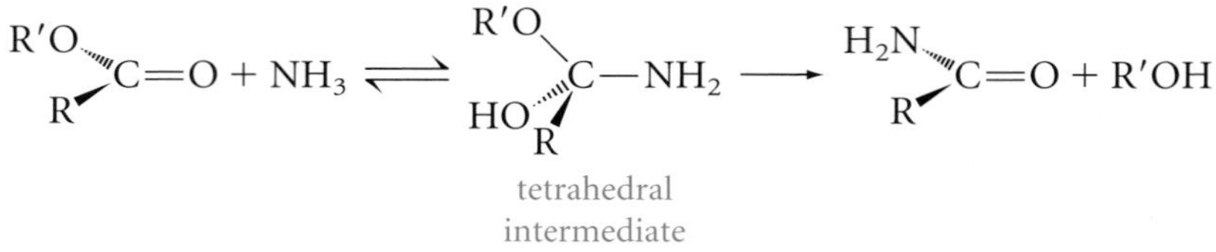

tetrahedral
intermediate

PROBLEM 10.20   The first step in eq. 10.26 really involves two reactions, *addition* of ammonia to the carbonyl carbon to form an ammonium alkoxide followed by a *proton transfer* from the nitrogen to the alkoxide oxygen. Illustrate this process with equations using the arrow-pushing formalism. The second step in eq. 10.26 also involves two steps, *elimination* of an alkoxide ($R'O^-$) followed by deprotonation of the hydroxyl group. Write a detailed mechanism for these steps.

10.15 Reaction of Esters with Grignard Reagents

Esters react with two equivalents of a Grignard reagent to give tertiary alcohols. The reaction proceeds by nucleophilic attack of the Grignard reagent on the ester carbonyl group. The initial product, a ketone, reacts further in the usual way to give the tertiary alcohol.

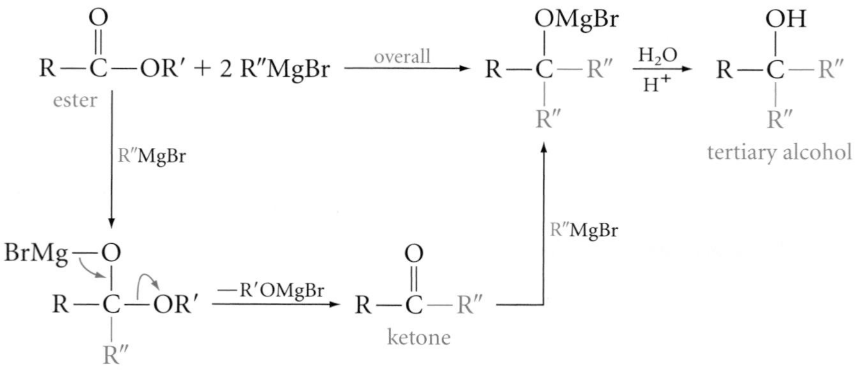

$$\underset{\text{ester}}{R{-}\overset{\overset{O}{\|}}{C}{-}OR'} + 2\ R''MgBr \xrightarrow{\text{overall}} R{-}\underset{\underset{R''}{|}}{\overset{\overset{OMgBr}{|}}{C}}{-}R'' \xrightarrow[H^+]{H_2O} R{-}\underset{\underset{R''}{|}}{\overset{\overset{OH}{|}}{C}}{-}R'' \qquad \textbf{(10.27)}$$

tertiary alcohol

$$R''MgBr \downarrow$$

$$\underset{\underset{R''}{|}}{\overset{BrMg{-}O}{\overset{|}{R{-}C{-}OR'}}} \xrightarrow{-R'OMgBr} \underset{\text{ketone}}{R{-}\overset{\overset{O}{\|}}{C}{-}R''} \xrightarrow{R''MgBr}$$

This method is useful for making tertiary alcohols in which at least two of three alkyl groups attached to the hydroxyl-bearing carbon atom are identical.

PROBLEM 10.21   Using eq. 10.27 as a guide, write the structure of the tertiary alcohol that is obtained from

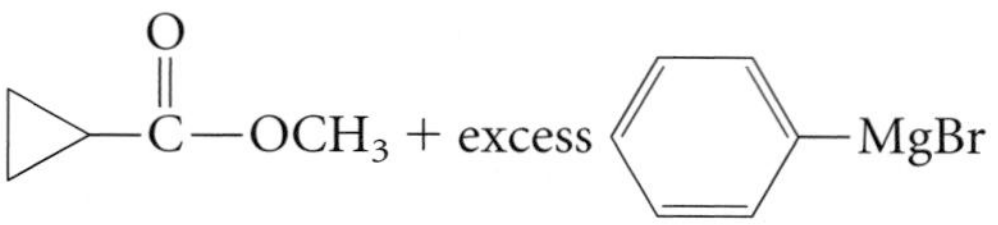

$$\triangleright\!\!-\overset{\overset{O}{\|}}{C}{-}OCH_3 + \text{excess} \quad \bigcirc\!\!-MgBr$$

10.16 Reduction of Esters

Esters can be reduced to primary alcohols by lithium aluminum hydride.

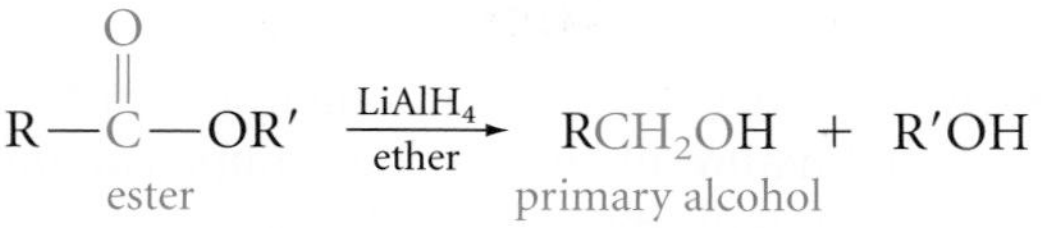

$$
\underset{\text{ester}}{R-\overset{\overset{O}{\|}}{C}-OR'} \xrightarrow[\text{ether}]{\text{LiAlH}_4} \underset{\text{primary alcohol}}{RCH_2OH \;+\; R'OH} \tag{10.28}
$$

The mechanism is similar to the hydride reduction of aldehydes and ketones (eq. 9.33).

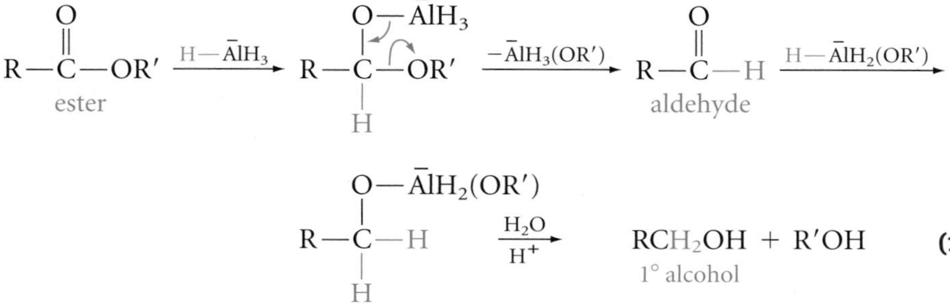

$$
\underset{\text{ester}}{R-\overset{\overset{O}{\|}}{C}-OR'} \xrightarrow{H-\bar{A}lH_3} R-\underset{\underset{H}{|}}{\overset{\overset{O-\bar{A}lH_3}{|}}{C}}-OR' \xrightarrow{-\bar{A}lH_3(OR')} \underset{\text{aldehyde}}{R-\overset{\overset{O}{\|}}{C}-H} \xrightarrow{H-\bar{A}lH_2(OR')}
$$

$$
R-\underset{\underset{H}{|}}{\overset{\overset{O-\bar{A}lH_2(OR')}{|}}{C}}-H \xrightarrow[H^+]{H_2O} \underset{1°\text{ alcohol}}{RCH_2OH \;+\; R'OH} \tag{10.29}
$$

The intermediate aldehyde is not usually isolable and reacts rapidly with additional hydride to produce the alcohol.

It is possible to reduce an ester carbonyl group without reducing a C=C bond in the same molecule. For example,

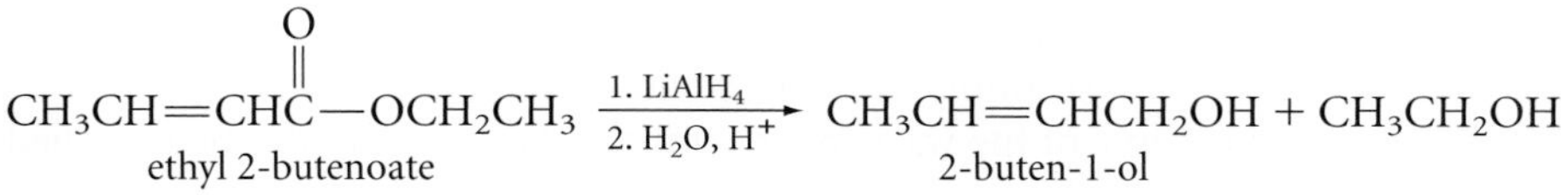

$$
\underset{\text{ethyl 2-butenoate}}{CH_3CH=CH\overset{\overset{O}{\|}}{C}-OCH_2CH_3} \xrightarrow[\text{2. H}_2\text{O, H}^+]{\text{1. LiAlH}_4} \underset{\text{2-buten-1-ol}}{CH_3CH=CHCH_2OH + CH_3CH_2OH} \tag{10.30}
$$

10.17 The Need for Activated Acyl Compounds

As we have seen, most reactions of carboxylic acids, esters, and related compounds involve, as the first step, nucleophilic attack on the carbonyl carbon atom. Examples are Fischer esterification, saponification and ammonolysis of esters, and the first stage of the reaction of esters with Grignard reagents or lithium aluminum hydride. All of these reactions can be summarized by a single mechanistic equation:

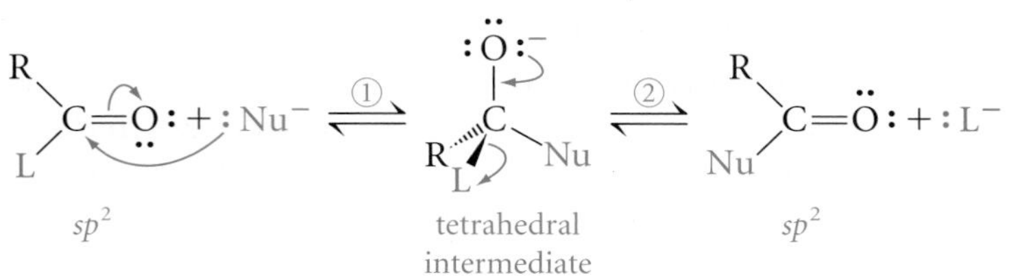

$$
\underset{sp^2}{\underset{L}{\overset{R}{\diagdown}}C=\ddot{O}:\; +\; :Nu^-} \overset{①}{\rightleftharpoons} \underset{\substack{\text{tetrahedral}\\\text{intermediate}}}{R\overset{:\ddot{O}:^-}{\underset{L}{\overset{|}{C}}}Nu} \overset{②}{\rightleftharpoons} \underset{sp^2}{\underset{Nu}{\overset{R}{\diagdown}}C=\ddot{O}:\; +\; :L^-} \tag{10.31}
$$

The carbonyl carbon, initially trigonal, is attacked by a nucleophile $Nu:^-$ to form a tetrahedral intermediate (step 1). Loss of a leaving group $:L^-$ (step 2) then regenerates the carbonyl group with its trigonal carbon atom. The net result is the replacement of L by Nu.

Biochemists look at eq. 10.31 in a slightly different way. They refer to the overall reaction as **acyl transfer.** The acyl group is transferred from L in the starting material to Nu in the product.

An **acyl transfer** is the transfer of an acyl group from a leaving group to a nucleophile.

Regardless of how we consider the reaction, one important feature that can affect the rate of both steps is the nature of the leaving group. *The rates of both steps in a nucleophilic acyl substitution reaction are enhanced by increasing the electron-withdrawing properties of the leaving group.* Step 1 is favored because the more electronegative L is, the more positive the carbonyl carbon becomes, and therefore the more susceptible it is to nucleophilic attack. Step 2 is also facilitated because the more electronegative L is, the better leaving group it becomes.

In general, esters are *less* reactive toward nucleophiles than are aldehydes or ketones because the positive charge on the carbonyl carbon in esters can be delocalized to the oxygen atom.

resonance in aldehydes and ketones

resonance in esters

Consequently, the carbonyl carbon is less positive in esters than it is in aldehydes or ketones and therefore less susceptible to nucleophilic attack.

Now let us examine some of the ways in which the carboxyl group can be modified to *increase* its reactivity toward nucleophiles.

10.18 Acyl Halides

An **acyl halide** is a carboxylic acid derivative in which the —OH group is replaced by a halogen atom.

Acyl halides are among the most reactive of carboxylic acid derivatives. *Acyl chlorides* are more common and less expensive than bromides or iodides. They can be prepared from acids by reaction with thionyl chloride.

$$R-\overset{\overset{\displaystyle O}{\|}}{C}-OH + SOCl_2 \longrightarrow R-\overset{\overset{\displaystyle O}{\|}}{C}-Cl + HCl + SO_2 \qquad \textbf{(10.32)}$$

The mechanism is similar to that for the formation of chlorides from alcohols and thionyl chloride. The hydroxyl group is converted to a good leaving group by the thionyl chloride, followed by a nucleophilic acyl substitution in which chloride is the nucleophile (compare with Sec. 7.10). Phosphorus pentachloride and other reagents can also be used to prepare acyl chlorides from carboxylic acids.

$$R-\overset{\overset{\displaystyle O}{\|}}{C}-OH + PCl_5 \longrightarrow R-\overset{\overset{\displaystyle O}{\|}}{C}-Cl + HCl + POCl_3 \qquad \textbf{(10.33)}$$

Acyl halides react rapidly with most nucleophiles. For example, they are rapidly hydrolyzed by water.

$$
\underset{\text{acetyl chloride}}{CH_3-\overset{\overset{\textstyle O}{\|}}{C}-Cl} + HOH \xrightarrow{\text{rapid}} \underset{\text{acetic acid}}{CH_3-\overset{\overset{\textstyle O}{\|}}{C}-OH} + \underset{\text{(fumes)}}{HCl} \qquad \textbf{(10.34)}
$$

For this reason, acyl halides have irritating odors. Benzoyl chloride, for example, is a lachrymator (tear gas).

EXAMPLE 10.7

Write a mechanism for the reaction shown in eq. 10.34.

Solution Nucleophilic addition of water to the carbonyl group, followed by proton transfer and elimination of HCl from the tetrahedral intermediate, gives the observed products.

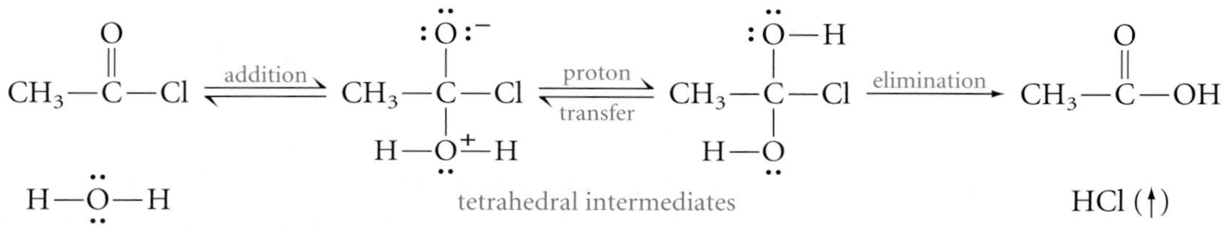

tetrahedral intermediates   HCl (↑)

Acyl halides react rapidly with alcohols to form esters.

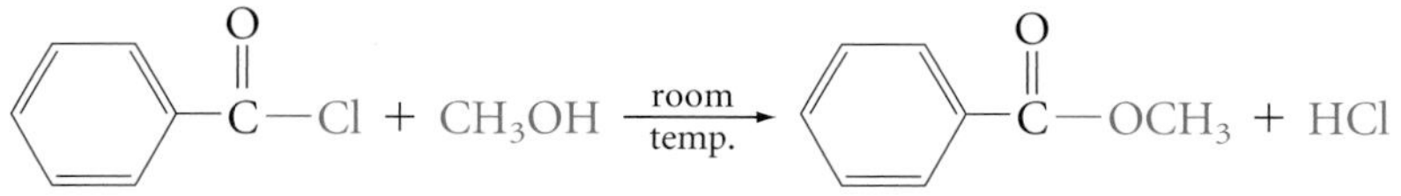

$$
\underset{\text{benzoyl chloride}}{C_6H_5-\overset{\overset{\textstyle O}{\|}}{C}-Cl} + CH_3OH \xrightarrow[\text{temp.}]{\text{room}} \underset{\text{methyl benzoate}}{C_6H_5-\overset{\overset{\textstyle O}{\|}}{C}-OCH_3} + HCl \qquad \textbf{(10.35)}
$$

Indeed, the most common way to prepare an ester *in the laboratory* is to convert an acid to its acid chloride, then react the latter with an alcohol. Even though two steps are necessary (compared with one step for Fischer esterification), the method may be preferable, especially if either the acid or the alcohol is expensive. (Recall that Fischer esterification is an equilibrium reaction and must often be carried out with a large excess of one of the reactants.)

PROBLEM 10.22 Rewrite eq. 10.32 to show the preparation of benzoyl chloride (see eq. 10.35).

PROBLEM 10.23 Explain why acyl halides may be irritating to the nose.

PROBLEM 10.24 Write a mechanism for the reaction shown in eq. 10.35.

Acyl halides react rapidly with ammonia to form amides.

$$
\underset{\text{acetyl chloride}}{CH_3\overset{\overset{\textstyle O}{\|}}{C}-Cl} + 2\,NH_3 \longrightarrow \underset{\text{acetamide}}{CH_3\overset{\overset{\textstyle O}{\|}}{C}-NH_2} + NH_4{}^+Cl^- \qquad \textbf{(10.36)}
$$

The reaction is much more rapid than the ammonolysis of esters. Two equivalents of ammonia are required, however—one to form the amide and one to neutralize the hydrogen chloride.

Acyl halides are used to synthesize aromatic ketones, through Friedel–Crafts acylation of aromatic rings (review Sec. 4.9.d).

PROBLEM 10.25 Devise a synthesis of 4-methylphenyl propyl ketone from toluene and butanoic acid as starting materials.

10.19 Acid Anhydrides

Acid anhydrides are carboxylic acid derivatives formed by condensing two carboxylic acid molecules.

Acid anhydrides are derived from acids by removing water from two carboxyl groups and connecting the fragments.

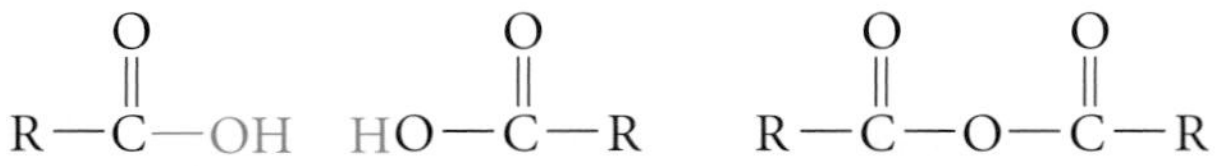

$$\underset{\text{two acid molecules}}{R-\overset{\overset{\textstyle O}{\|}}{C}-OH \quad HO-\overset{\overset{\textstyle O}{\|}}{C}-R} \qquad \underset{\text{an acid anhydride}}{R-\overset{\overset{\textstyle O}{\|}}{C}-O-\overset{\overset{\textstyle O}{\|}}{C}-R}$$

The most important commercial aliphatic anhydride is **acetic anhydride** ($R = CH_3$). About 1 million tons are manufactured annually, mainly to react with alcohols to form acetates. The two most common uses are in making cellulose acetate (rayon) and aspirin.

The name of an anhydride is obtained by naming the acid from which it is derived and replacing the word *acid* with *anhydride*.

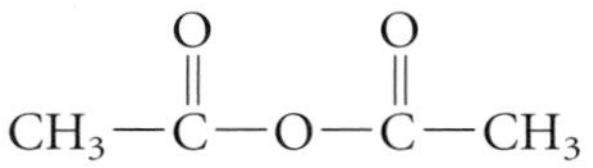

$$CH_3-\overset{\overset{\textstyle O}{\|}}{C}-O-\overset{\overset{\textstyle O}{\|}}{C}-CH_3$$

ethanoic anhydride or acetic anhydride

PROBLEM 10.26 Write the structural formula for

a. propanoic anhydride.
b. benzoic anhydride.

Anhydrides are prepared by dehydration of acids. Dicarboxylic acids with appropriately spaced carboxyl groups lose water on heating to form cyclic anhydrides with five- and six-membered rings. For example,

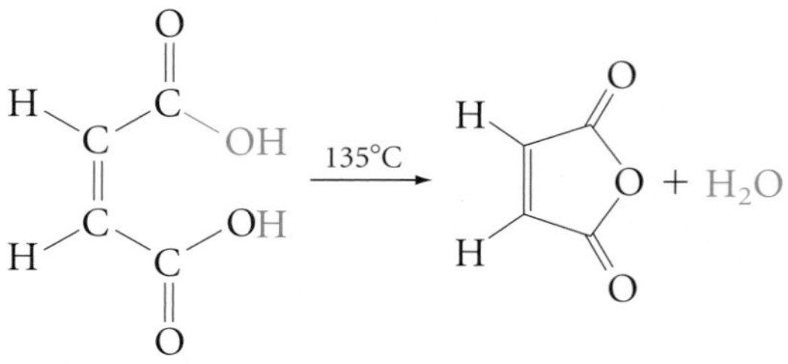

maleic acid   maleic anhydride

$$\xrightarrow{135°C} \quad + \; H_2O \tag{10.37}$$

PROBLEM 10.27　Predict and name the product of the following reaction:

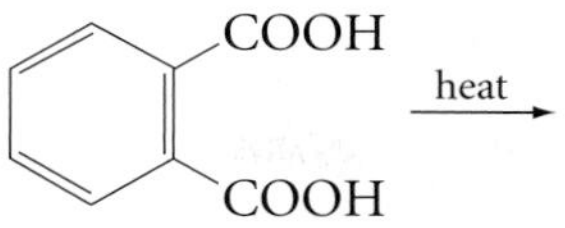

$$\text{(benzene ring with two adjacent COOH groups)} \xrightarrow{\text{heat}}$$

PROBLEM 10.28　Do you expect fumaric acid (page 286) to form a cyclic anhydride on heating? Explain.

Anhydrides can also be prepared from acid chlorides and carboxylate salts in a reaction that occurs by a nucleophilic acyl substitution mechanism. This is a good method for preparing anhydrides derived from two different carboxylic acids, called **mixed anhydrides.**

Mixed anhydrides are prepared from two different carboxylic acids.

$$CH_3CH_2CH_2-\overset{\overset{\displaystyle O}{\|}}{C}-Cl + Na^{+\,-}O-\overset{\overset{\displaystyle O}{\|}}{C}-CH_3 \longrightarrow$$

$$CH_3CH_2CH_2-\overset{\overset{\displaystyle O}{\|}}{C}-O-\overset{\overset{\displaystyle O}{\|}}{C}-CH_3 + NaCl \qquad \textbf{(10.38)}$$
butanoic ethanoic anhydride

Anhydrides undergo nucleophilic acyl substitution reactions. They are more reactive than esters, but less reactive than acyl halides, toward nucleophiles. Some typical reactions of acetic anhydride follow:

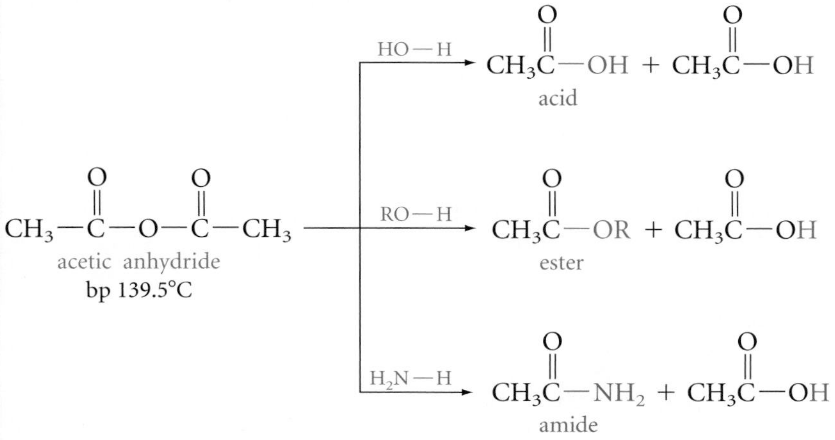

$$CH_3-\overset{\overset{\displaystyle O}{\|}}{C}-O-\overset{\overset{\displaystyle O}{\|}}{C}-CH_3$$
acetic anhydride
bp 139.5°C

$$\xrightarrow{\text{HO}-\text{H}} CH_3\overset{\overset{\displaystyle O}{\|}}{C}-OH + CH_3\overset{\overset{\displaystyle O}{\|}}{C}-OH$$
acid

$$\xrightarrow{\text{RO}-\text{H}} CH_3\overset{\overset{\displaystyle O}{\|}}{C}-OR + CH_3\overset{\overset{\displaystyle O}{\|}}{C}-OH \qquad \textbf{(10.39)}$$
ester

$$\xrightarrow{\text{H}_2\text{N}-\text{H}} CH_3\overset{\overset{\displaystyle O}{\|}}{C}-NH_2 + CH_3\overset{\overset{\displaystyle O}{\|}}{C}-OH$$
amide

Water hydrolyzes an anhydride to the corresponding acid. Alcohols give esters, and ammonia gives amides. In each case, one equivalent of acid is also produced.

PROBLEM 10.29　Write an equation for the reaction of acetic anhydride with 1-butanol ($CH_3CH_2CH_2CH_2OH$).

PROBLEM 10.30　Write equations for the reactions of maleic anhydride (see eq. 10.37) with

a. water.　b. 1-butanol.　c. ammonia.

A WORD ABOUT ...

Thioesters, Nature's Acyl-Activating Groups

Acyl transfer plays an important role in many biochemical processes. However, acyl halides and anhydrides are far too corrosive to be cell constituents—they are hydrolyzed quite rapidly by water and are therefore incompatible with cellular fluid. Most ordinary esters, on the other hand, react too slowly with nucleophiles for acyl transfer to be carried out efficiently at body temperatures. Consequently, other groups have evolved to activate acyl groups in the cell. The most important of these is **coenzyme A** (the A stands for acetylation, one of the functions of this enzyme). Coenzyme A is a complex *thiol* (Figure 10.1). It is usually abbreviated by the symbol **CoA—SH**. Though its structure is made up of three parts—adenosine diphosphate (ADP), pantothenic acid (a vitamin), and 2-aminoethanethiol—it is the thiol group that gives coenzyme A its most important functions.

Coenzyme A can be converted to **thioesters,** the active acyl-transfer agents in the cell. Of the thioesters that coenzyme A forms, the acetyl ester, called **acetyl-coenzyme A** and abbreviated as

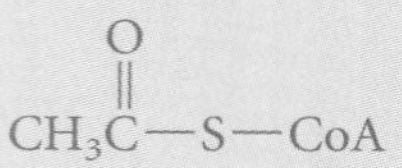

$$CH_3\overset{\displaystyle O}{\overset{\displaystyle \|}{C}}-S-CoA$$

is the most important. Acetyl-CoA reacts with many nucleophiles to transfer the acetyl group.

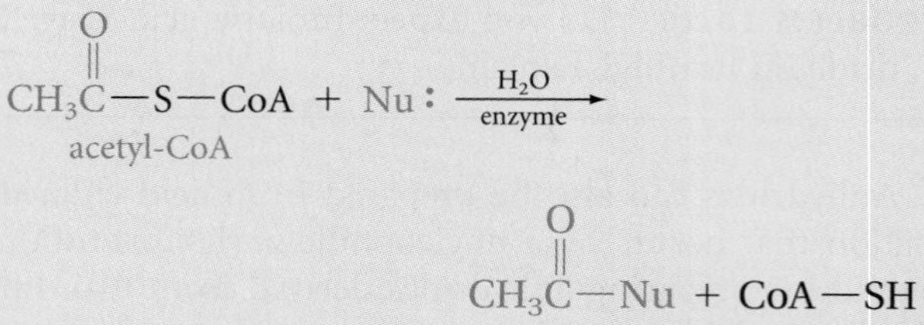

$$CH_3\overset{\displaystyle O}{\overset{\displaystyle \|}{C}}-S-CoA + Nu: \xrightarrow[\text{enzyme}]{H_2O}$$

acetyl-CoA

$$CH_3\overset{\displaystyle O}{\overset{\displaystyle \|}{C}}-Nu + CoA-SH$$

The reactions are usually enzyme-mediated and occur rapidly at ordinary cell temperatures.

Why are thioesters superior to ordinary esters as acyl-transfer agents? Part of the answer lies in the acidity difference between alcohols and thiols (Sec. 7.16). Since thiols are much stronger acids than are alcohols, their conjugate bases, ¯SR, are much weaker bases than ¯OR. Thus the —SR group of thioesters is a much better leaving group, in nucleophilic substitutions, than is the —OR group of ordinary esters. Thioesters are not so reactive that they hydrolyze in cellular fluid, but they are appreciably more reactive than simple esters. Nature makes use of this fact.

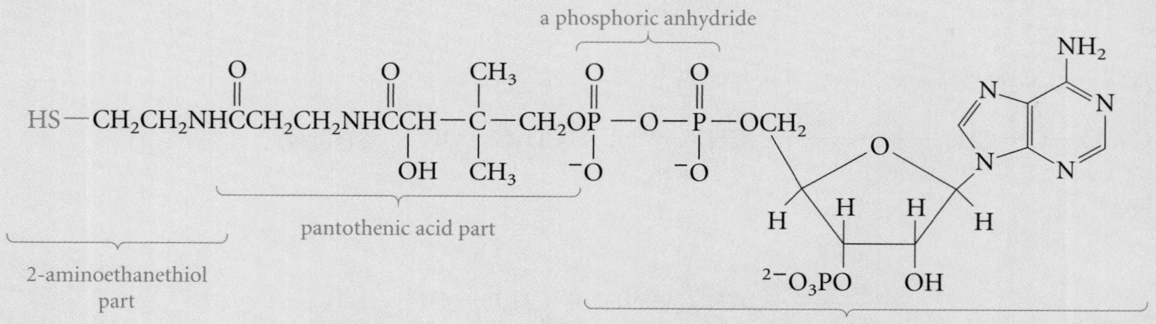

Figure 10.1
Coenzyme A.

The reaction of acetic anhydride with **salicylic acid** (*o*-hydroxybenzoic acid) is used to synthesize **aspirin**. In this reaction, the phenolic hydroxyl group is **acetylated** (converted to its acetate ester).

Annual aspirin production in the United States is more than 24 million pounds, enough to produce over 30 billion standard 5-grain (325 mg) tablets. Aspirin is widely used, either by itself or mixed with other drugs, as an analgesic and antipyretic. It is not without dangers, however. Repeated use may cause gastrointestinal bleeding, and a large single dose (10 to 20 g) can cause death.

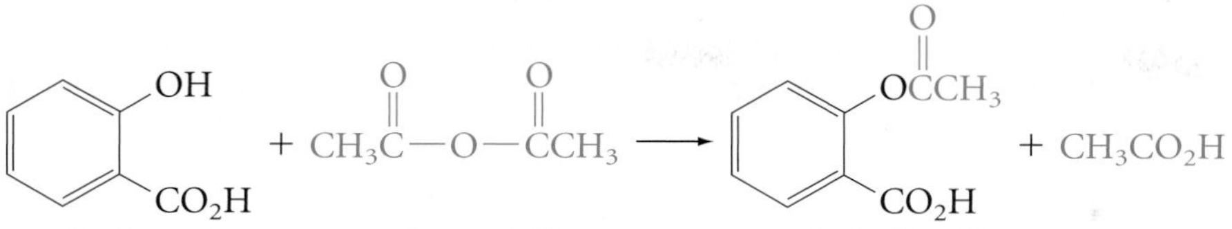

$$\text{salicylic acid} + \text{acetic anhydride} \longrightarrow \text{acetylsalicylic acid (aspirin)} + CH_3CO_2H \qquad \textbf{(10.40)}$$

10.20 Amides

Amides are the least reactive of the common carboxylic acid derivatives. They occur widely in nature. The most important amides are the proteins, whose chemistry we will discuss in Chapter 17. Here we will concentrate on just a few properties of simple amides.

Primary amides have the general formula $RCONH_2$. They can be prepared by the reaction of ammonia with esters (eq. 10.24), with acyl halides (eq. 10.36), or with acid anhydrides (eq. 10.39). Amides can also be prepared by heating the ammonium salts of acids.

Amides are carboxylic acid derivatives in which the —OH group is replaced by —NH_2, —NHR, or —NR_2.

$$R-\overset{O}{\underset{\|}{C}}-OH + NH_3 \longrightarrow R-\overset{O}{\underset{\|}{C}}-O^-NH_4^+ \xrightarrow{\text{heat}} R-\overset{O}{\underset{\|}{C}}-NH_2 + H_2O \qquad \textbf{(10.41)}$$

ammonium salt   amide

Amides are named by replacing the *-ic* or *-oic* ending of the acid name, either the common or the IUPAC name, with the *-amide* ending.

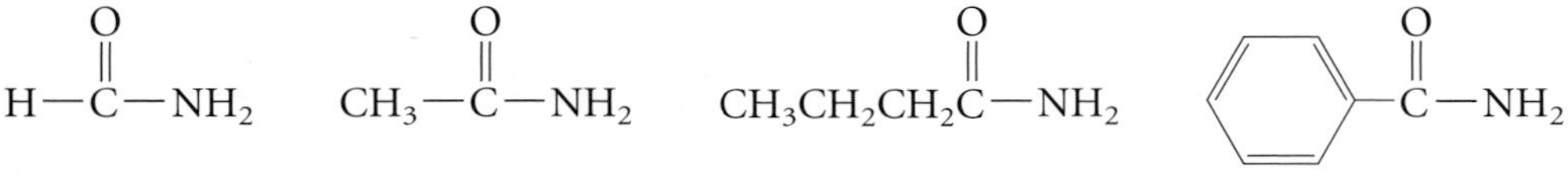

formamide (methanamide)   acetamide (ethanamide)   butanamide   benzamide (benzenecarboxamide)

PROBLEM 10.31

a. Name $(CH_3)_2CHCONH_2$.
b. Write the structure of 1-methylcyclobutanecarboxamide.

The above examples are all primary amides. Secondary and tertiary amides, in which one or both of the hydrogens on the nitrogen atom are replaced by organic groups, are described in the next chapter.

Amides have a planar geometry. Even though the carbon–nitrogen bond is normally written as a single bond, rotation around that bond is restricted, because of resonance.

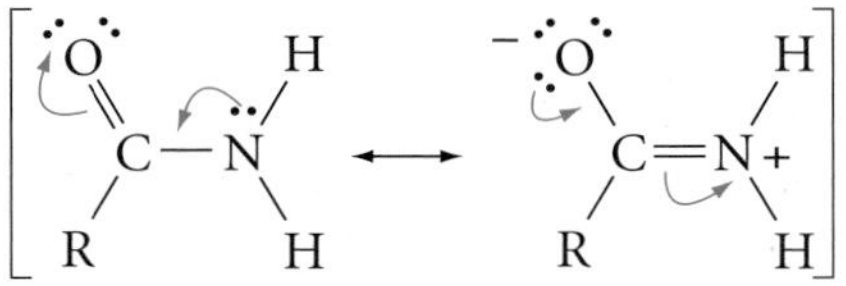

amide resonance

The dipolar contributor is so important that the carbon–nitrogen bond behaves much like a double bond. Consequently, the nitrogen and the carbonyl carbon, and the two atoms attached to each of them, lie in the same plane, and rotation at the C—N bond is restricted. Indeed, the C—N bond in amides is only 1.32 Å long—much shorter than the usual carbon–nitrogen single bond length (which is about 1.47Å).

As the dipolar resonance contributor suggests, amides are highly polar and form strong hydrogen bonds.

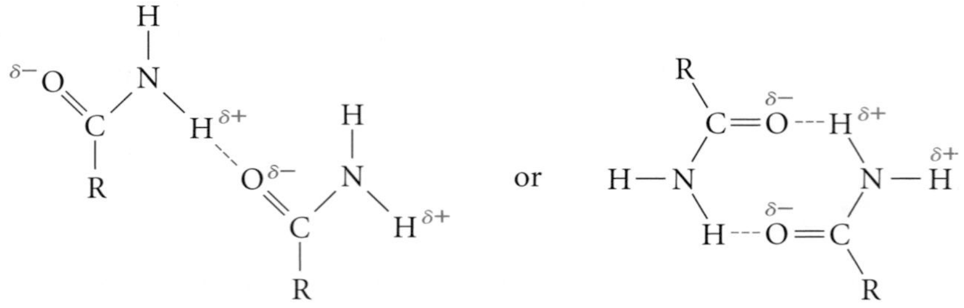

Amides have exceptionally high boiling points for their molecular weights, although alkyl substitution on the nitrogen lowers the boiling and melting points by decreasing the hydrogen-bonding possibilities, as shown in the following two pairs of compounds:

$\underset{\text{formamide}}{\text{H}-\overset{\overset{\textstyle O}{\|}}{\text{C}}-\text{NH}_2}$	$\underset{\textit{N,N}\text{-dimethylformamide}}{\text{H}-\overset{\overset{\textstyle O}{\|}}{\text{C}}-\text{N(CH}_3)_2}$	$\underset{\text{acetamide}}{\text{CH}_3\overset{\overset{\textstyle O}{\|}}{\text{C}}-\text{NH}_2}$	$\underset{\textit{N,N}\text{-dimethylacetamide}}{\text{CH}_3\overset{\overset{\textstyle O}{\|}}{\text{C}}-\text{N(CH}_3)_2}$
bp 210°C	153°C	222°C	165°C
mp 2.5°C	−60.5°C	81°C	−20°C

PROBLEM 10.32 Show that hydrogen bonding is possible for acetamide, but not for *N, N*-dimethylacetamide.

Like other acid derivatives, amides react with nucleophiles. For example, they can be hydrolyzed by water.

$$\underset{\text{amide}}{\text{R}-\overset{\overset{\textstyle O}{\|}}{\text{C}}-\text{NH}_2} + \text{H}-\text{OH} \xrightarrow[\text{HO}^-]{\text{H}^+ \text{ or}} \underset{\text{acid}}{\text{R}-\overset{\overset{\textstyle O}{\|}}{\text{C}}-\text{OH}} + \text{NH}_3 \qquad \textbf{(10.42)}$$

The reactions are slow, and prolonged heating or acid or base catalysis is usually necessary.

PROBLEM 10.33 Using eq. 10.42 as a model, write an equation for the hydrolysis of acetamide.

Amides can be reduced by lithium aluminum hydride to give amines.

$$\underset{\text{amide}}{\text{R}-\overset{\overset{\textstyle O}{\|}}{\text{C}}-\text{NH}_2} \xrightarrow[\text{ether}]{\text{LiAlH}_4} \underset{\text{amine}}{\text{RCH}_2\text{NH}_2} \qquad \textbf{(10.43)}$$

This is an excellent way to make primary amines, whose chemistry is discussed in the next chapter.

PROBLEM 10.34   Using eq. 10.43 as a model, write an equation for the reduction of acetamide with $LiAlH_4$.

Urea is a special amide, a diamide of carbonic acid. A colorless, water-soluble, crystalline solid, urea is the normal end product of protein metabolism. An average adult excretes approximately 30 g of urea in the urine daily. Urea is produced commercially from carbon dioxide and ammonia, mainly for use as a fertilizer.

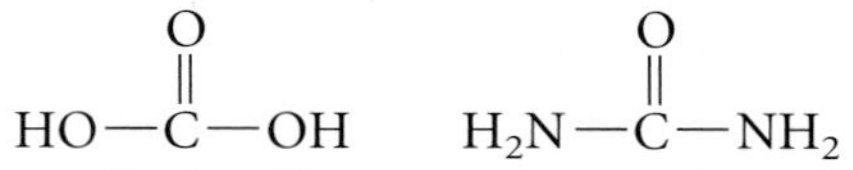

$$
\underset{\text{carbonic acid}}{HO-\overset{\overset{\displaystyle O}{\|}}{C}-OH} \qquad \underset{\substack{\text{urea}\\ \text{mp } 133°C}}{H_2N-\overset{\overset{\displaystyle O}{\|}}{C}-NH_2}
$$

10.21   A Summary of Carboxylic Acid Derivatives

We have studied a rather large number of reactions in this chapter. However, most of them can be summarized in a single chart, shown in Table 10.5.

The four types of acid derivatives are listed at the left of the chart in order of decreasing reactivity toward nucleophiles. Three common nucleophiles are listed across the top. Note that the main organic product in each column is the same, regardless of

---

Table 10.5   Reactions of acid derivatives with certain nucleophiles

Acid derivative	Nucleophile		
	HOH (hydrolysis)	**R'OH (alcoholysis)**	**NH_3 (ammonolysis)**
$R-\overset{\overset{O}{\|}}{C}-Cl$ acyl halide	$R-\overset{\overset{O}{\|}}{C}-OH + HCl$	$R-\overset{\overset{O}{\|}}{C}-OR' + HCl$	$R-\overset{\overset{O}{\|}}{C}-NH_2 + NH_4{}^+Cl^-$
$R-\overset{\overset{O}{\|}}{C}-O-\overset{\overset{O}{\|}}{C}-R$ acid anhydride	$2\,R-\overset{\overset{O}{\|}}{C}-OH$	$R-\overset{\overset{O}{\|}}{C}-OR' + RCO_2H$	$R-\overset{\overset{O}{\|}}{C}-NH_2 + RCO_2H$
$R-\overset{\overset{O}{\|}}{C}-O-R''$ ester	$R-\overset{\overset{O}{\|}}{C}-OH + R''OH$	$R-\overset{\overset{O}{\|}}{C}-OR' + R''OH$ (ester interchange)	$R-\overset{\overset{O}{\|}}{C}-NH_2 + R''OH$
$R-\overset{\overset{O}{\|}}{C}-NH_2$ amide	$R-\overset{\overset{O}{\|}}{C}-OH + NH_3$	————	————
Main organic product	acid	ester	amide

decreasing reactivity (left margin, vertical)

which type of acid derivative we start with. For example, **hydrolysis** gives the corresponding organic acid, whether we start with an acyl halide, acid anhydride, ester, or amide. Similarly, **alcoholysis** gives an ester, and **ammonolysis** gives an amide. Note also that the *other* reaction product is generally the same from a given acid derivative (horizontally across the table), regardless of the nucleophile. For example, starting with an ester, RCO_2R'', we obtain as the second product the alcohol $R''OH$, regardless of whether the reaction type is hydrolysis, alcoholysis, or ammonolysis.

All of the reactions in Table 10.5 take place via attack of the nucleophile on the carbonyl carbon of the acid derivative, as described in eq. 10.31. Indeed, most of the reactions from Sections 10.10 through 10.19 occur by that same mechanism. We can sometimes use this idea to predict new reactions.

For example, the reaction of esters with Grignard reagents (eq. 10.27) involves nucleophilic attack of the Grignard reagent on the ester carbonyl group. Keeping in mind that all acid derivatives are susceptible to nucleophilic attack, it is understandable that acyl halides also react with Grignard reagents to give tertiary alcohols. The first steps involve ketone formation as follows:

$$R-\overset{O}{\overset{\|}{C}}-Cl + R'MgX \longrightarrow R-\overset{\overset{+}{O^-MgX}}{\underset{R'}{\overset{|}{C}}}-Cl \longrightarrow R-\overset{O}{\overset{\|}{C}}-R' + MgXCl \qquad \textbf{(10.44)}$$

The ketone can sometimes be isolated, but usually it reacts with a second mole of Grignard reagent to give a tertiary alcohol.

$$R-\overset{O}{\overset{\|}{C}}-R' + R'MgX \longrightarrow R-\overset{\overset{+}{O^-MgX}}{\underset{R'}{\overset{|}{C}}}-R' \xrightarrow{H_3O^+} R-\overset{OH}{\underset{R'}{\overset{|}{C}}}-R' \qquad \textbf{(10.45)}$$

> **PROBLEM 10.35** Predict the product from the reaction of phenylmagnesium bromide (C_6H_5MgBr) with benzoyl chloride (C_6H_5COCl).

10.22 **The α-Hydrogen of Esters; the Claisen Condensation**

In this final section, we describe an important reaction of esters that resembles the aldol condensation of aldehydes and ketones (Sec. 9.17). It makes use of the α-hydrogen (see page 270) of an ester.

Being adjacent to a carbonyl group, the α-hydrogens of an ester are weakly acidic ($pK_a \sim 23$) and can be removed by a *strong base*. The product is an **ester enolate**.

An **ester enolate** is the anion formed by removing the α-hydrogen of an ester.

$$-\overset{H}{\underset{|}{\overset{|}{C}}}_\alpha-\overset{O}{\overset{\|}{C}}\diagdown_{OR} \underset{\longleftarrow}{\overset{base}{\longrightarrow}} \left[-\overset{..}{C}-\overset{O:}{\overset{\|}{C}}\diagdown_{OR} \longleftrightarrow -C=C\diagup^{\overset{..}{O:}^-}_{\diagdown OR} \right] \qquad \textbf{(10.46)}$$

resonance contributors to an ester enolate

Common bases used for this purpose are sodium alkoxides or sodium hydride. The ester enolate, once formed, can act as a carbon nucleophile and add to the carbonyl group of another ester molecule. This reaction is called the **Claisen condensation.** It is a way of making **β-keto esters.** We will use ethyl acetate as an example to see how the reaction works.

Treatment of ethyl acetate with sodium ethoxide in ethanol produces the β-keto ester **ethyl acetoacetate:**

$$
\underset{\text{ethyl acetate}}{CH_3\overset{\overset{\displaystyle O}{\|}}{C}-OCH_2CH_3} + \underset{\text{ethyl acetate}}{H-\overset{\alpha}{CH_2}-\overset{\overset{\displaystyle O}{\|}}{C}-OCH_2CH_3} \xrightarrow[\substack{\text{in ethanol} \\ 2.\ H_3O^+}]{1.\ NaOCH_2CH_3}
$$

$$
\underset{\substack{\text{ethyl acetoacetate} \\ \text{(ethyl 3-oxobutanoate)}}}{CH_3\overset{\overset{\displaystyle O}{\|}}{C}-CH_2-\overset{\overset{\displaystyle O}{\|}}{C}-OCH_2CH_3} + CH_3CH_2OH \qquad \textbf{(10.47)}
$$

The Claisen condensation takes place in three steps.

Step 1. $\quad CH_3\overset{\overset{\displaystyle O}{\|}}{C}-OCH_2CH_3 + \underset{\text{sodium ethoxide}}{Na^{+\,-}OCH_2CH_3} \rightleftharpoons$

$$
\underset{\text{ester enolate}}{Na^{+\,-}CH_2\overset{\overset{\displaystyle O}{\|}}{C}OCH_2CH_3} + CH_3CH_2OH \qquad \textbf{(10.48)}
$$

Step 2. $\quad CH_3\overset{\overset{\displaystyle O}{\|}}{C}-OCH_2CH_3 + {}^-CH_2\overset{\overset{\displaystyle O}{\|}}{C}OCH_2CH_3 \rightleftharpoons$

$$
CH_3\overset{\overset{\displaystyle {}^-O}{|}}{\underset{\underset{\overset{\displaystyle \|}{O}}{CH_2C-OCH_2CH_3}}{C}}OCH_2CH_3 \rightleftharpoons CH_3\overset{\overset{\displaystyle O}{\|}}{C}CH_2\overset{\overset{\displaystyle O}{\|}}{C}OCH_2CH_3 + {}^-OCH_2CH_3 \qquad \textbf{(10.49)}
$$

Step 3. $\quad CH_3\overset{\overset{\displaystyle O}{\|}}{C}CH_2\overset{\overset{\displaystyle O}{\|}}{C}OCH_2CH_3 + {}^-OCH_2CH_3 \longrightarrow$

$$
\underset{\text{enolate ion of a β-keto ester}}{CH_3\overset{\overset{\displaystyle O}{\|}}{C}-\overset{-}{CH}-\overset{\overset{\displaystyle O}{\|}}{C}OCH_2CH_3} + CH_3CH_2OH \qquad \textbf{(10.50)}
$$

In step 1, the base (sodium ethoxide) removes an α-hydrogen from the ester to form an ester enolate. In step 2, this ester enolate, acting as a nucleophile, adds to the carbonyl group of a second ester molecule, displacing ethoxide ion. This step follows the mechanism in eq. 10.31 and proceeds through a tetrahedral intermediate. These first two steps of the reaction are completely reversible.

Step 3 drives the equilibrium forward. In this step, the β-keto ester is converted to *its* enolate anion. The methylene (CH_2) hydrogens in ethyl acetoacetate are α *to two carbonyl groups* and hence are appreciably more acidic than ordinary α-hydrogens.

They have a pK_a of 12 and are easily removed by the base (ethoxide ion) to form a resonance-stabilized β-keto enolate ion, *with the negative charge delocalized to both carbonyl oxygen atoms.*

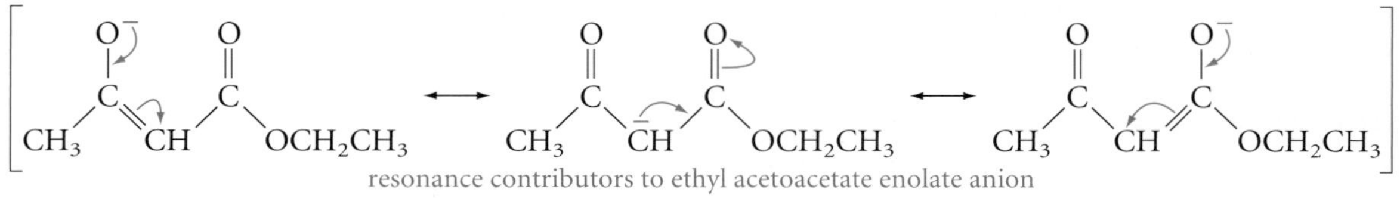

resonance contributors to ethyl acetoacetate enolate anion

To complete the Claisen condensation, the solution is acidified, to regenerate the β-keto ester from its enolate anion.

EXAMPLE 10.8

Identify the product of the Claisen condensation of ethyl propanoate:

$$CH_3CH_2\overset{\displaystyle O}{\overset{\|}{C}}-OCH_2CH_3$$

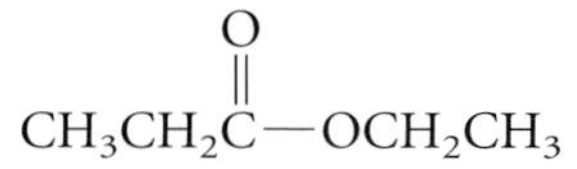

Solution   The product is

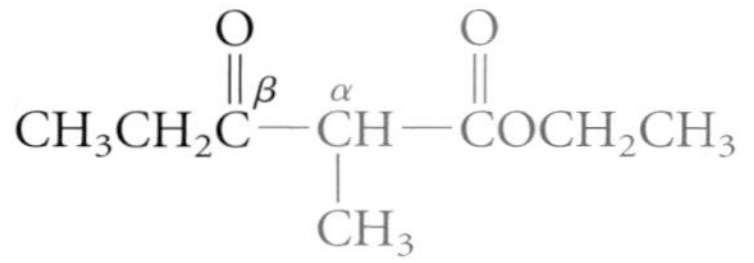

$$CH_3CH_2\overset{\displaystyle O}{\overset{\|}{\underset{}{C}}}\overset{\beta}{}-\overset{\alpha}{\underset{\underset{CH_3}{|}}{CH}}-\overset{\displaystyle O}{\overset{\|}{C}}OCH_2CH_3$$

The α-carbon of one ester molecule displaces the —OR group and becomes joined to the carbonyl carbon of the other ester. The product is always a β-keto ester.

PROBLEM 10.36   Using eqs. 10.48 through 10.50 as a model, write out the steps in the mechanism for the Claisen condensation of ethyl propanoate.

The Claisen condensation, like the aldol condensation, is useful for making new carbon–carbon bonds. The resulting β-keto esters can be converted to a variety of useful products. For example, ethyl acetate can be converted to ethyl butanoate by the following sequence.

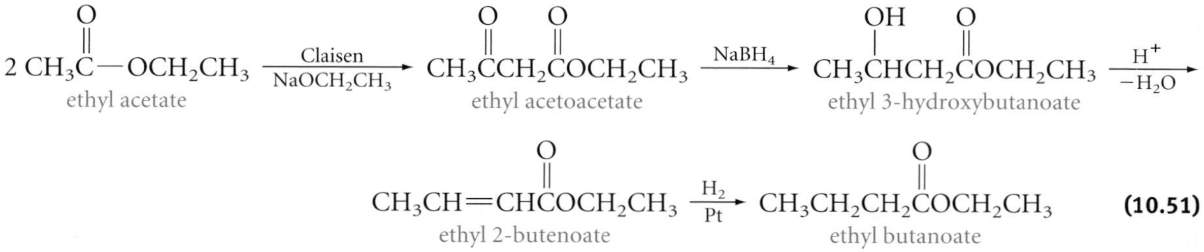

In this way, the acetate chain is lengthened by two carbon atoms. Nature makes use of a similar process, catalyzed by various enzymes, to construct the long-chain carboxylic acids that are components of fats and oils (Chapter 15).

REACTION SUMMARY

1. Preparation of Acids

a. **From Alcohols or Aldehydes (Sec. 10.7)**

$$RCH_2OH \xrightarrow{\text{CrO}_3,\ \text{H}_2\text{SO}_4,\ \text{H}_2\text{O}} RCO_2H \xleftarrow[\text{or O}_2\ \text{or Ag}^+]{\text{CrO}_3,\ \text{H}_2\text{SO}_4,\ \text{H}_2\text{O}} RCH{=}O$$

b. **From Alkylbenzenes (Sec. 10.7)**

$$ArCH_3 \xrightarrow[\text{or O}_2,\ \text{Co}^{+3}]{\text{KMnO}_4} ArCO_2H$$

c. **From Grignard Reagents (Sec. 10.7)**

$$RMgX + CO_2 \longrightarrow RCO_2MgX \xrightarrow{\text{H}_3\text{O}^+} RCO_2H$$

d. **From Nitriles (Sec. 10.7)**

$$RC{\equiv}N + 2\,H_2O \xrightarrow{\text{H}^+\ \text{or HO}^-} RCO_2H + NH_3$$

2. Reactions of Acids

a. **Acid–Base (Secs. 10.4 and 10.6)**

$$RCO_2H \rightleftharpoons RCO_2^- + H^+ \quad \text{(ionization)}$$
$$RCO_2H + NaOH \longrightarrow RCO_2^-Na^+ + H_2O \quad \text{(salt formation)}$$

b. **Preparation of Esters (Secs. 10.10 and 10.12)**

$$RCO_2H + R'OH \xrightarrow{\text{H}^+} RCO_2R' + H_2O$$

c. **Preparation of Acid Chlorides (Sec. 10.18)**

$$RCO_2H + SOCl_2 \longrightarrow RCOCl + HCl + SO_2$$
$$RCO_2H + PCl_5 \longrightarrow RCOCl + HCl + POCl_3$$

d. **Preparation of Anhydrides (Sec. 10.19)**

$$R-\overset{\overset{\displaystyle O}{\|}}{C}-Cl + Na^+\ {}^-O-\overset{\overset{\displaystyle O}{\|}}{C}-R' \longrightarrow R-\overset{\overset{\displaystyle O}{\|}}{C}-O-\overset{\overset{\displaystyle O}{\|}}{C}-R' + NaCl$$

e. **Preparation of Amides (Sec. 10.20)**

$$RCO_2^-NH_4^+ \xrightarrow{\text{heat}} RCONH_2 + H_2O$$

Also see reactions of esters, acid chlorides, and anhydrides in Section 10.21.

3. Reactions of Carboxylic Acid Derivatives

a. **Saponification of Esters (Sec. 10.13)**

$$RCO_2R' + NaOH \longrightarrow RCO_2^-Na^+ + R'OH$$

b. **Ammonolysis of Esters (Sec. 10.14)**

$$RCO_2R' + NH_3 \longrightarrow RCONH_2 + R'OH$$

c. **Esters with Grignard Reagents (Sec. 10.15)**

$$RCO_2R' \xrightarrow{2\ R''MgX} R-\overset{\overset{\displaystyle R''}{|}}{\underset{\underset{\displaystyle R''}{|}}{C}}-OMgX \xrightarrow{\text{H}_3\text{O}^+} R-\overset{\overset{\displaystyle R''}{|}}{\underset{\underset{\displaystyle R''}{|}}{C}}-OH$$
$$+ R'OH$$

d. Reduction of Esters (Sec. 10.16)

$$RCO_2R' + LiAlH_4 \longrightarrow RCH_2OH + R'OH$$

e. Nucleophilic Acyl Substitution Reactions of Acid Chlorides and Anhydrides (Sec. 10.18 and 10.19)

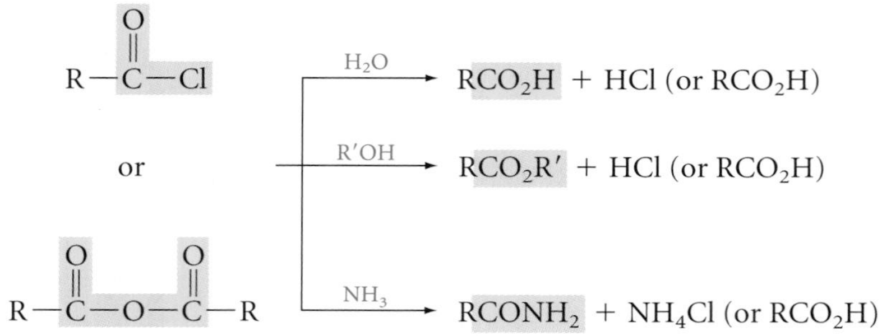

f. Hydrolysis of Amides (Sec. 10.20)

$$RCONH_2 + H_2O \xrightarrow{H^+ \text{ or } HO^-} RCO_2H + NH_3$$

g. Reduction of Amides (Sec. 10.20)

$$RCONH_2 \xrightarrow{LiAlH_4} RCH_2NH_2$$

h. Claisen Condensation (Sec. 10.22)

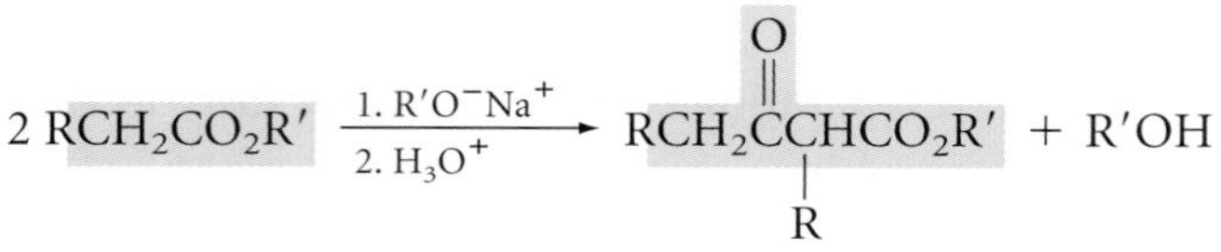

MECHANISM SUMMARY

Nucleophilic Acyl Substitution (Secs. 10.11 and 10.17)

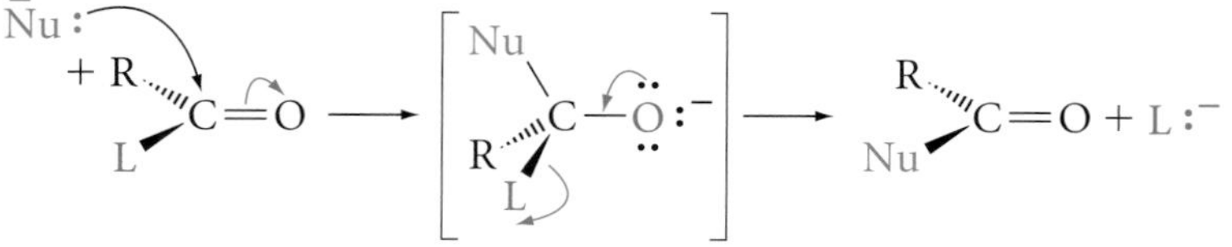

ADDITIONAL PROBLEMS

Nomenclature and Structure of Carboxylic Acids

10.37 Write a structural formula for each of the following acids:

   a. 4-methylpentanoic acid
   b. 2,2-dichlorobutanoic acid
   c. 3-hydroxyhexanoic acid
   d. *p*-toluic acid
   e. cyclobutanecarboxylic acid
   f. 2-propanoylbenzoic acid
   g. phenylacetic acid
   h. 1-naphthoic acid
   i. 2,3-dimethyl-3-butenoic acid
   j. 3-oxobutanoic acid
   k. 2,2-dimethylbutanedioic acid

10.38 Name each of the following acids:

    a. $(CH_3)_2CHCH_2CH_2COOH$

    b. $CH_3CHClCH(CH_3)COOH$

    c.

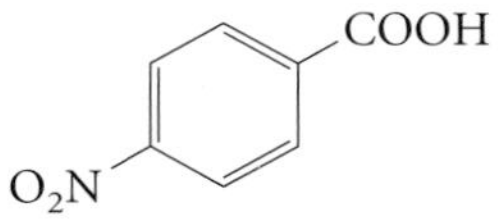

    d.

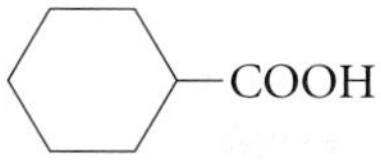

    e. $CH_2{=}CHCOOH$

    g. CH_3CF_2COOH

    f. $CH_3CH(C_6H_5)COOH$

    h. $HC{\equiv}CCH_2CO_2H$

Synthesis and Properties of Carboxylic Acids

10.39 Which will have the higher boiling point? Explain your reasoning.

    a. CH_3CH_2COOH or $CH_3CH_2CH_2CH_2OH$

    b. $CH_3CH_2CH_2CH_2COOH$ or $(CH_3)_3CCOOH$

10.40 In each of the following pairs of acids, which would be expected to be the stronger, and why?

    a. $ClCH_2CO_2H$ and $BrCH_2CO_2H$

    c. CCl_3CO_2H and CF_3CO_2H

    e. $ClCH_2CH_2CO_2H$ and $CH_3CHClCO_2H$

    b. $o\text{-}BrC_6H_4CO_2H$ and $m\text{-}BrC_6H_4CO_2H$

    d. $C_6H_5CO_2H$ and $p\text{-}CH_3OC_6H_4CO_2H$

10.41 Write a balanced equation for the reaction of

    a. $ClCH_2CO_2H$ with KOH.

    b. $CH_3(CH_2)_8CO_2H$ with $Ca(OH)_2$.

10.42 Give equations for the synthesis of

    a. $CH_3CH_2CH_2CO_2H$ from $CH_3CH_2CH_2CH_2OH$

    b. $CH_3CH_2CH_2CO_2H$ from $CH_3CH_2CH_2OH$ (two ways)

    c.

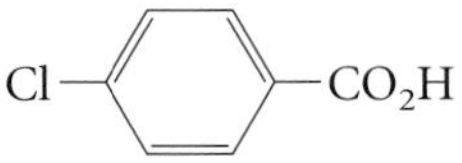

from

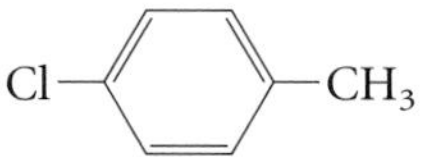

    d.

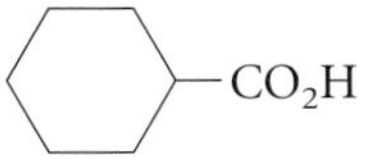

from

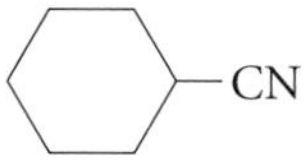

    e. $CH_3OCH_2CO_2H$ from

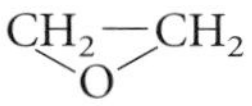

(two steps)

    f.

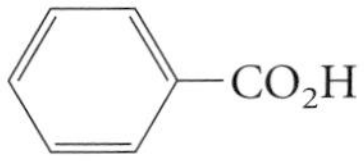

from

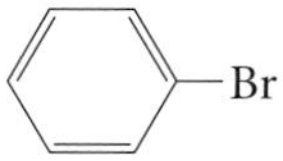

10.43 The Grignard route for the synthesis of $(CH_3)_3CCO_2H$ from $(CH_3)_3CBr$ (Example 10.5) is far superior to the nitrile route. Explain why.

Nomenclature and Structure of Carboxylic Acid Derivatives

10.44 Write a structure for each of the following compounds:

    a. sodium 2-chlorobutanoate

    c. isopropyl acetate

    e. phenyl benzoate

    g. propanoic anhydride

    i. 2-chlorobutanoyl chloride

    k. α-methyl-γ-butyrolactone

    b. calcium acetate

    d. ethyl formate

    f. benzonitrile

    h. o-toluamide

    j. 3-formylcyclopentanecarboxylic acid

10.45 Name each of the following compounds:

a. 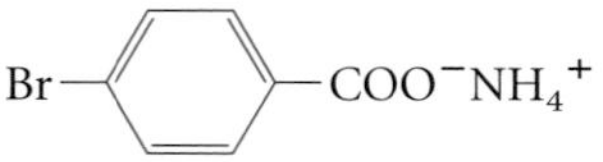 Br—⟨benzene ring⟩—$COO^-NH_4^+$

b. $[CH_3(CH_2)_2CO_2^-]_2Ca^{2+}$

c. $(CH_3)_2CHCOOC_6H_5$

d. $CF_3CO_2CH_3$

e. $HCONH_2$

f. $CH_3(CH_2)_2\overset{O}{\overset{\|}{-C}}-O-\overset{O}{\overset{\|}{C}}-(CH_2)_2CH_3$

10.46 Draw the structure of the mating pheromone of the female elephant, (*Z*)-7-dodecen-1-yl acetate (see page 297).

Synthesis and Reactions of Esters

10.47 Write out each step in the Fischer esterification of benzoic acid with methanol. (You may wish to use eq. 10.19 as a model.)

10.48 Write an equation for the Fischer esterification of pentanoic acid ($CH_3CH_2CH_2CH_2CO_2H$) with ethanol.

10.49 Write an equation for the reaction of ethyl benzoate with
  a. hot aqueous sodium hydroxide.
  b. ammonia (heat).
  c. methylmagnesium iodide (two equivalents), then H_3O^+.
  d. lithium aluminum hydride (two equivalents), then H_3O^+.

⟨benzene ring⟩—CO_2Et

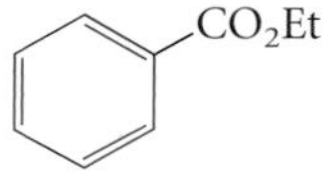

ethyl benzoate

10.50 Write out all the steps in the mechanism for
  a. saponification of $CH_3CH_2CO_2CH_3$.
  b. ammonolysis of $CH_3CH_2CO_2CH_3$.

10.51 Identify the Grignard reagent and the ester that would be used to prepare

a. $CH_3CH_2-\overset{OH}{\overset{|}{\underset{|}{C}}}-CH_2CH_3$
            C_6H_5

b. $CH_3CH_2CH_2C(C_6H_5)_2OH$

Reactions of Carboxylic Acid Derivatives

10.52 Explain each difference in reactivity toward nucleophiles.
  a. Esters are less reactive than ketones.
  b. Benzoyl chloride is less reactive than cyclohexanecarbonyl chloride.

10.53 Write an equation for
  a. hydrolysis of propanoyl chloride.
  b. reaction of benzoyl chloride with methanol.
  c. esterification of 1-pentanol with acetic anhydride.
  d. ammonolysis of butanoyl bromide.
  e. 2-methylpropanoyl chloride + ethylbenzene + $AlCl_3$.
  f. succinic acid + heat (235°C).
  g. phthalic anhydride + methanol (1 equiv.) + H^+.
  h. phthalic anhydride + methanol (excess) + H^+.
  i. adipoyl chloride + ammonia (excess).

10.54 Complete the equation for each of the following reactions:

a. $CH_3CH_2CH_2CO_2H + PCl_5 \longrightarrow$ b. $CH_3(CH_2)_6CO_2H + SOCl_2 \longrightarrow$

c. 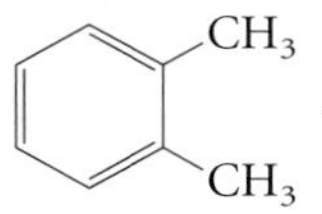 $+ KMnO_4 \longrightarrow$ d.  $-CO_2^-NH_4^+ + heat \longrightarrow$

e. $CH_3(CH_2)_5CONH_2 + LiAlH_4 \longrightarrow$ f. 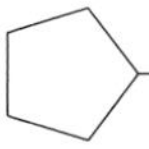 $-CO_2CH_2CH_3 + LiAlH_4 \longrightarrow$

10.55 Considering the relative reactivities of ketones and esters toward nucleophiles, which of the following products seems the more likely?

$$CH_3\overset{O}{\overset{\|}{C}}CH_2CH_2CO_2CH_3 \xrightarrow{NaBH_4} CH_3\overset{O}{\overset{\|}{C}}CH_2CH_2CH_2OH \quad or \quad CH_3\overset{OH}{\overset{|}{C}}HCH_2CH_2CO_2CH_3$$

10.56 Mandelic acid, which has the formula $C_6H_5CH(OH)COOH$, can be isolated from bitter almonds (called *mandel* in German). It is sometimes used in medicine to treat urinary infections. Devise a two-step synthesis of mandelic acid from benzaldehyde, using the latter's cyanohydrin (see Sec. 9.10) as an intermediate.

The Claisen Condensation

10.57 Write the structure of the Claisen condensation product of ethyl phenylacetate ($C_6H_5CH_2CO_2Et$), and show the steps in its formation.

10.58 Diethyl adipate, when heated with sodium ethoxide, gives the product shown, by an *intra*molecular Claisen condensation:

$$CH_3CH_2O\overset{O}{\overset{\|}{C}}-(CH_2)_4-\overset{O}{\overset{\|}{C}}OCH_2CH_3 \xrightarrow[2.\ H_3O^+]{1.\ NaOCH_2CH_3}$$

diethyl adipate   ethyl 2-oxocyclopentanecarboxylate

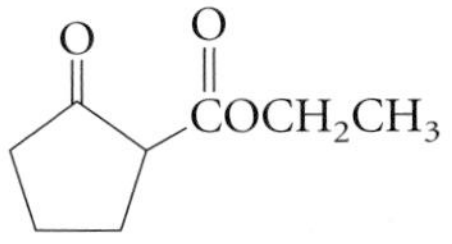

Write out the steps in a plausible mechanism for the reaction.

10.59 Analogous to the mixed aldol condensation (Sec. 9.18), mixed Claisen condensations are possible. Predict the structure of the product obtained when a mixture of ethyl benzoate and ethyl acetate is heated with sodium ethoxide in ethanol.

Miscellaneous Problems

10.60 Write the important resonance contributors to the structure of propanamide and tell which atoms lie in a single plane.

10.61 Consider the structure of the catnip ingredient nepatalactone (p. 300).

a. Show with dotted lines that the structure is composed of two isoprene units.
b. Circle the stereogenic centers and determine their configurations (*R* or *S*).

─────────

⊃ = concept connections

10.62 The lactone shown below, known as "wine lactone," is a sweet and coconut-like smelling odorant isolated recently from white wines such as Gewürztraminer.

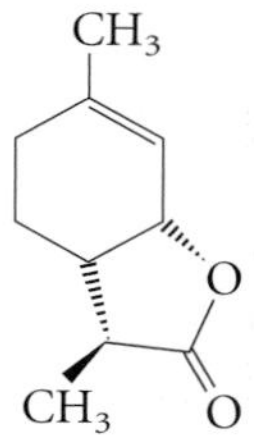

How many stereocenters are present, and what is the configuration (R or S) at each?

10.63 (5R,6S)-6-Acetoxy-5-hexadecanolide is a pheromone that attracts certain disease-carrying mosquitos to sites where they like to lay their eggs. Such compounds might be used to lure these insects away from populated areas to locations where they can be destroyed. The last two steps in a recent synthesis of this compound are shown below. Provide reagents that would accomplish these transformations.

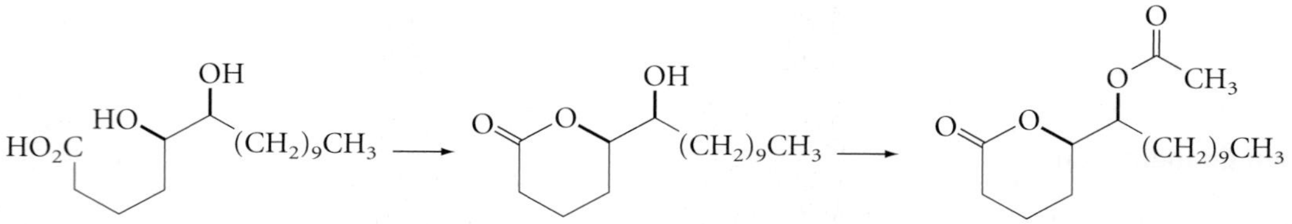

(5R, 6S)-6-Acetoxy-5-hexadecanolide

morphine

11

AMINES AND RELATED NITROGEN COMPOUNDS

In this chapter we will discuss the last of the major families of simple organic compounds—the amines—relatives of ammonia that abound in nature and play an important role in many modern technologies. Examples of important amines include the painkiller morphine, found in poppy seeds, and putrescine, one of several polyamines responsible for the unpleasant odor of decaying flesh. A diamine that is largely the creation of man is 1,6-diaminohexane, used in the synthesis of nylon. Amine derivatives, known as quaternary ammonium salts, also touch our daily lives in the form of synthetic detergents. Several neurotoxins also belong to this family of compounds. They are toxic because they interfere with the key role that acetylcholine, also a quaternary ammonium salt, plays in the transmission of nerve impulses.

$$H_2\ddot{N}(CH_2)_4\ddot{N}H_2 \qquad H_2\ddot{N}(CH_2)_6\ddot{N}H_2 \qquad CH_3-\overset{\overset{\displaystyle CH_3}{|}}{\underset{\underset{\displaystyle CH_3}{|}}{\overset{+}{N}}}-CH_2CH_2-O-\overset{\overset{\displaystyle O}{\|}}{C}-CH_3 \ ^-OH$$

putrescine   1,6-diaminohexane   acetylcholine

In this chapter we will first describe the structure, preparation, chemical properties, and uses of some simple amines. Later in the chapter, we will discuss a few natural and synthetic amines with important biological properties.

11.1 Classification and Structure of Amines

The relation between ammonia and amines is illustrated by the following structures:

H—N̈—H   R—N̈—H   R—N̈—R   R—N̈—R
| | | |
H   H   H   R
ammonia   primary amine   secondary amine   tertiary amine

▲ The painkiller morphine is obtained from opium, the dried sap of the unripe seed of the poppy *Papaver somniferum*.

⊚ **Ch. 11, Ref. 1.** *Examine and manipulate 3D model of trimethylamine.*

Figure 11.1
(a) An orbital view of the pyramidal bonding in trimethylamine. (b) Top view of a space-filling model of trimethylamine. The center ball represents the orbital with the unshared electron pair.

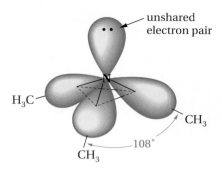

(a)

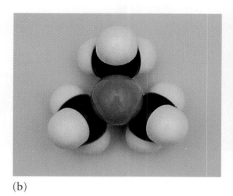

(b)

Amines are organic bases derived from ammonia.

Primary amines have one organic group attached to nitrogen, **secondary** amines have two, and **tertiary** amines have three.

For convenience, amines are classified as **primary, secondary,** or **tertiary,** depending on whether one, two, or three organic groups are attached to the nitrogen. The *R* groups in these structures may be alkyl or aryl, and when two or more *R* groups are present, they may be identical to or different from one another. In some secondary and tertiary amines, the nitrogen may be part of a ring.

PROBLEM 11.1   Classify each of the following amines as primary, secondary, or tertiary:

a. $(CH_3)_3CNH_2$ b. [pyrrolidine ring] $\underset{H}{N}$

c. CH_3—[benzene ring]—NH_2 d. $(CH_3)_2N$—[benzene ring]

The nitrogen atom in amines is trivalent. In addition, the nitrogen carries an unshared electron pair. The nitrogen orbitals are therefore sp^3-hybridized, and the overall geometry is pyramidal (nearly tetrahedral), as shown for trimethylamine in Figure 11.1. From this geometry, one might think that an amine with three different groups attached to the nitrogen would be chiral, with the unshared electron pair acting as the fourth group. This is true in principle, but in practice the two enantiomers usually interconvert rapidly through inversion, via an "umbrella-in-the-wind" type of process, and are not resolvable.

$$R^1 \underset{R^2}{\overset{N}{\diagdown}} R^3 \;\rightleftharpoons\; \left[R^1 - N \underset{R^2}{\overset{R^3}{\diagup}} \right] \;\rightleftharpoons\; R^1 \underset{N}{\diagdown} \overset{R^3 \, R^2}{\diagup} \qquad \textbf{(11.1)}$$

planar transition state

11.2 Nomenclature of Amines

Amines can be named in several different ways. Commonly, simple amines are named by specifying the alkyl groups attached to the nitrogen and adding the suffix -*amine*.

$$CH_3CH_2NH_2 \qquad (CH_3CH_2)_2NH \qquad (CH_3CH_2)_3N$$
ethylamine   diethylamine   triethylamine
(primary)   (secondary)   (tertiary)

In the IUPAC system, the **amino group, —NH$_2$,** is named as a substituent, as in the following examples:

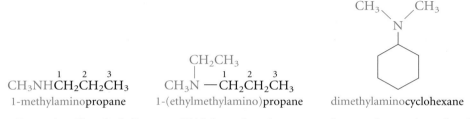

aminoethane   2-aminopentane   *cis*-1,3-diaminocyclobutane

In this system, secondary or tertiary amines are named by using a prefix that includes all but the longest carbon chain, as in

$$\overset{1}{CH_3}NH\overset{2}{CH_2}\overset{3}{CH_2}CH_3$$
1-methylamino**propane**

$$CH_3N - \overset{1}{CH_2}\overset{2}{CH_2}\overset{3}{CH_3} \\ |_{CH_2CH_3}$$
1-(ethylmethylamino)**propane**

dimethylamino**cyclohexane**

Recently, *Chemical Abstracts* (CA) introduced a system for naming amines that is rational and easy to use. In this system, amines are named as **alkanamines.** For example,

$$CH_3CH_2CH_2NH_2 \qquad CH_3CHCH_3 \qquad CH_3CHCH_2CH_2CH_3$$
propanamine   2-propanamine   N-methyl-2-pentanamine

*Amines are named as **alkanamines** in the Chemical Abstracts system.*

EXAMPLE 11.1

Name ⬡—N(CH$_3$)$_2$ by the *Chemical Abstracts* system.

Solution The largest alkyl group attached to nitrogen is used as the root of the name. The compound is *N,N*-dimethylcyclohexanamine.

PROBLEM 11.2 Name $CH_3CH_2CHCH_2CH_3$ by the CA system.
                                      $|$
                                  N(CH$_3$)$_2$

When other functional groups are present, the amino group is named as a substituent:

$$\overset{4}{CH_3}\overset{3}{CH}\overset{2}{CH_2}\overset{1}{CO_2H} \\ \quad |_{NH_2}$$
3-aminobutanoic acid

$$\overset{1}{H_2N}\overset{2}{CH_2}\overset{3}{CH_2}\overset{4}{C}\overset{5}{CH_2CH_3} \\ \qquad\qquad ||_O$$
1-amino-3-pentanone

$$CH_3NH\overset{2}{CH_2}\overset{1}{CH_2}OH$$
2-methylaminoethanol

Aromatic amines are named as derivatives of aniline. In the CA system, aniline is called benzenamine; these CA names are shown in parentheses.

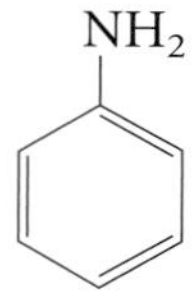

aniline
(benzenamine)

p-bromoaniline
(4-bromobenzenamine)

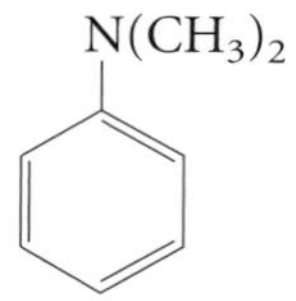

N,N-dimethylaniline
(N,N-dimethylbenzenamine)

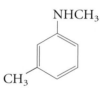

m-methyl-N-methylaniline, or
N-methyl-m-toluidine
(N-methyl-3-methylbenzenamine)

EXAMPLE 11.2

Give an acceptable name for the following compounds:

a. $(CH_3)_2CHCH_2NH_2$ b. $CH_3NHCH_2CH_3$

c.

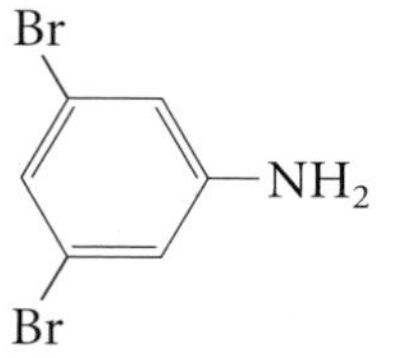

d.

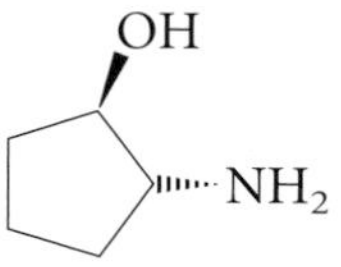

Solution

a. isobutylamine (common); 1-amino-2-methylpropane (IUPAC); 2-methyl-propanamine (CA).
b. ethylmethylamine (common); methylaminoethane (IUPAC); N-methylethanamine (CA).
c. 3,5-dibromoaniline (common, IUPAC); 3,5-dibromobenzenamine (CA).
d. *trans*-2-aminocyclopentanol (only name).

PROBLEM 11.3   Give an acceptable name for the following compounds:

a. $(CH_3)_3CNH_2$ b. $H_2NCH_2CH_2OH$ c. 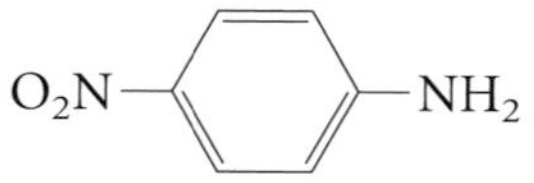 O_2N——NH_2

PROBLEM 11.4   Write the structure for

a. dipropylamine.
c. 2,4,6-trimethylaniline.
b. 3-aminohexane.
d. N,N-diethyl-2-pentanamine.

11.3   Physical Properties and Intermolecular Interactions of Amines

Table 11.1 lists the boiling points of some common amines. Methylamine and ethyl-amine are gases, but primary amines with three or more carbons are liquids. Primary amines boil well above alkanes with comparable molecular weights, but below comparable alcohols, as shown in Table 11.2. Intermolecular N—H··N hydrogen bonds are important and raise the boiling points of primary and secondary amines

Table 11.1	The boiling points of some simple amines	
Name	**Formula**	**bp, °C**
ammonia	NH_3	−33.4
methylamine	CH_3NH_2	−6.3
dimethylamine	$(CH_3)_2NH$	7.4
trimethylamine	$(CH_3)_3N$	2.9
ethylamine	$CH_3CH_2NH_2$	16.6
propylamine	$CH_3CH_2CH_2NH_2$	48.7
butylamine	$CH_3CH_2CH_2CH_2NH_2$	77.8
aniline	$C_6H_5NH_2$	184.0

Table 11.2	A comparison of alkane, amine, and alcohol boiling points*	
alkane	CH_3CH_3 (30) bp −88.6°C	$CH_3CH_2CH_3$ (44) bp −42.1°C
amine	CH_3NH_2 (31) bp −6.3°C	$CH_3CH_2NH_2$ (45) bp +16.6°C
alcohol	CH_3OH (32) bp +65.0°C	CH_3CH_2OH (46) bp +78.5°C

*Molecular weights are given in parentheses.

but are not as strong as the O—H$\cdots$O bonds of alcohols (see Sec. 7.4). The reason for this is that nitrogen is not as electronegative as oxygen.

PROBLEM 11.5 Explain why the tertiary amine $(CH_3)_3N$ boils so much lower than its primary isomer $CH_3CH_2CH_2NH_2$.

All three classes of amines can form hydrogen bonds with the —OH group of water (that is, O—H$\cdots$N). Primary and secondary amines can also form hydrogen bonds with the oxygen atom in water: N—H$\cdots$O. Thus, most simple amines with up to five or six carbon atoms are either completely or appreciably soluble in water.

Now we will describe some ways in which amines can be prepared.

11.4 Preparation of Amines; Alkylation of Ammonia and Amines

Ammonia reacts with alkyl halides to give amines via a two-step process. The first step is a nucleophilic substitution reaction (S_N2).

$$H_3\overset{..}{N}: \ + R \overset{\frown}{—} X \longrightarrow R \overset{+}{—} NH_3 \ X^- \tag{11.2}$$

ammonia   alkylammonium halide

The free amine can then be obtained from its salt by treatment with a strong base.

$$R \overset{+}{—} NH_3 \ X^- + NaOH \longrightarrow RNH_2 + H_2O + Na^+X^- \tag{11.3}$$

primary amine

Primary, secondary, and tertiary amines can be similarly alkylated.

$$\overset{..}{R}NH_2 + R—X \longrightarrow R_2\overset{+}{N}H_2 X^- \xrightarrow{\text{NaOH}} R_2NH \qquad \textbf{(11.4)}$$
primary amine   secondary amine

$$R_2\overset{..}{N}H + R—X \longrightarrow R_3\overset{+}{N}H\ X^- \xrightarrow{\text{NaOH}} R_3N \qquad \textbf{(11.5)}$$
secondary amine   tertiary amine

$$R_3\overset{..}{N} + R—X \longrightarrow R_4N^+\ X^- \qquad \textbf{(11.6)}$$
tertiary amine   quaternary ammonium salt

Ch. 11, Ref. 2. *Examine and manipulate 3D model of tetramethylammonium iodide.*

Unfortunately, mixtures of products are often obtained in these reactions because the starting ammonia or amine and the alkylammonium ion formed in the S_N2 step can equilibrate, as in the following equation:

$$NH_3 + R\overset{+}{N}H_3\ X^- \rightleftharpoons NH_4{}^+X^- + RNH_2 \qquad \textbf{(11.7)}$$

So, in the reaction of ammonia with an alkyl halide (eq. 11.2), some primary amine is formed (eq. 11.7), and it may be further alkylated (eq. 11.4) to give a secondary amine, and so on. By adjusting the ratio of the reactants, however, a good yield of one desired amine may be obtained. For example, with a large excess of ammonia, the primary amine is the major product.

Aromatic amines can often be alkylated selectively.

$$\underset{\text{aniline}}{\boxed{NH_2}} \xrightarrow{CH_3I} \underset{N\text{-methylaniline}}{\boxed{NHCH_3}} \xrightarrow{CH_3I} \underset{N,N\text{-dimethylaniline}}{\boxed{N(CH_3)_2}} \qquad \textbf{(11.8)}$$

The alkylation can be intramolecular, as in the following final step in a laboratory synthesis of nicotine:

$$\xrightarrow[-\text{HBr}]{\text{Intramolecular } S_N2} \qquad \text{nicotine} \qquad \textbf{(11.9)}$$

Tobacco plant.

EXAMPLE 11.3

Write an equation for the synthesis of benzylamine, ⬡—CH_2NH_2.

Solution

$$⬡—CH_2X + 2\overset{..}{N}H_3 \longrightarrow ⬡—CH_2\overset{..}{N}H_2 + NH_4{}^+X^-$$

(X = Cl, Br, or I)

Use of excess ammonia helps prevent further substitution.

PROBLEM 11.6 Complete equations for the following reactions:

  a. $CH_3CH_2CH_2CH_2Br + 2\ NH_3 \longrightarrow$

  b. $CH_3CH_2I + 2(CH_3CH_2)_2NH \longrightarrow$

  c. $(CH_3)_3N + CH_3I \longrightarrow$

  d. $CH_3CH_2CH_2NH_2 + $ $-CH_2Br \longrightarrow$

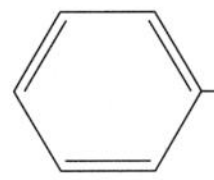

PROBLEM 11.7 Give a synthesis of $-NHCH_2CH_3$ from aniline.

11.5 Preparation of Amines; Reduction of Nitrogen Compounds

All bonds to the nitrogen atom in amines are either N—H or N—C bonds. Nitrogen in ammonia or amines is therefore in a reduced form. It is not surprising, then, that organic compounds in which a nitrogen atom is present in a more oxidized form can be reduced to amines by appropriate reducing agents. Several examples of this useful synthetic approach to amines are described here.

The best route to *aromatic primary amines* is by *reduction of the corresponding nitro compounds,* which are in turn prepared by electrophilic aromatic nitration. The nitro group is easily reduced, either catalytically with hydrogen or by chemical reducing agents.

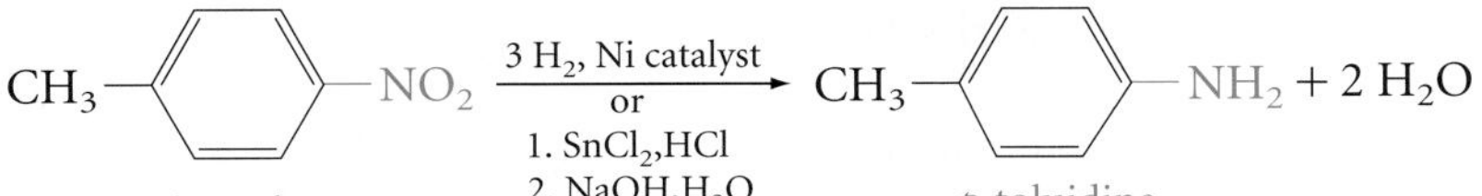

$$CH_3-\underset{\text{p-nitrotoluene}}{\underset{|}{\bigcirc}}-NO_2 \xrightarrow[\substack{\text{or}\\ \text{1. SnCl}_2\text{,HCl}\\ \text{2. NaOH,H}_2\text{O}}]{3\ H_2,\ \text{Ni catalyst}} CH_3-\underset{\text{p-toluidine}}{\underset{|}{\bigcirc}}-NH_2 + 2\ H_2O \qquad \textbf{(11.10)}$$

EXAMPLE 11.4

Devise a synthesis of *p*-chloroaniline, $Cl-\bigcirc-NH_2$, from chlorobenzene.

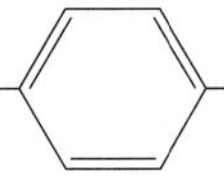

Solution Chlorobenzene is first nitrated; —Cl is an *o,p*-directing group, so the major product is *p*-chloronitrobenzene. This product is then reduced.

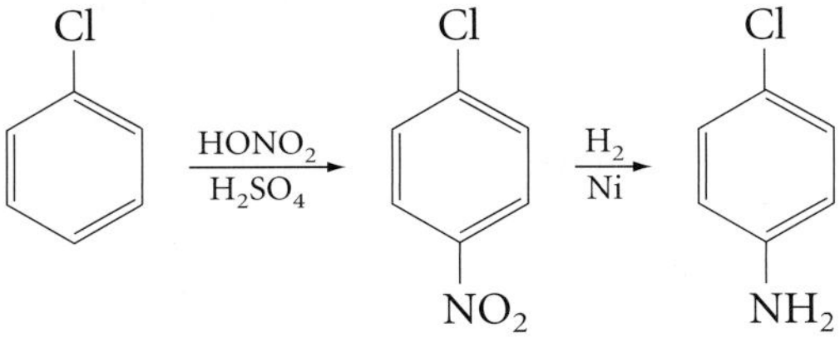

PROBLEM 11.8 Give a synthesis for $H_2N-\bigcirc-CH_3$ from toluene.

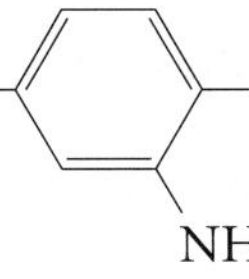

As described in the previous chapter (eq. 10.43), *amides* can be *reduced to amines* with lithium aluminum hydride.

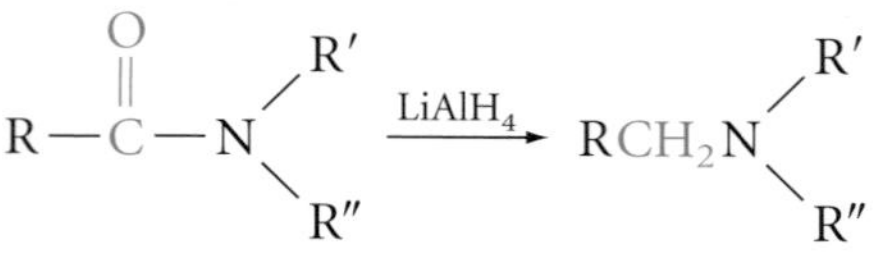

$$\underset{\substack{\| \\ O}}{R-C-N}\overset{R'}{\underset{R''}{<}} \xrightarrow{\text{LiAlH}_4} RCH_2N\overset{R'}{\underset{R''}{<}} \qquad \text{(R' and R'' may be H or organic groups.)} \qquad \textbf{(11.11)}$$

Depending on the structures of R' and R'', we can obtain primary, secondary, or tertiary amines in this way.

EXAMPLE 11.5

Complete the equation $CH_3\overset{O}{\underset{\|}{C}}NHCH_2CH_3 \xrightarrow{\text{LiAlH}_4}$.

Solution   The C=O group is reduced to CH$_2$. The product is the secondary amine $CH_3CH_2NHCH_2CH_3$.

PROBLEM 11.9   Show how $CH_3CH_2N(CH_3)_2$ can be synthesized from an amide.

Reduction of nitriles (cyanides) gives *primary amines*.

$$R-C\equiv N \xrightarrow[\text{or H}_2,\text{ Ni}]{\text{LiAlH}_4} RCH_2NH_2 \qquad \textbf{(11.12)}$$

EXAMPLE 11.6

Complete the equation $NCCH_2CH_2CH_2CH_2CN \xrightarrow[\text{Ni catalyst}]{\text{excess H}_2}$.

Solution   Both CN groups are reduced. The product H_2N—$(CH_2)_6$—NH_2, or 1,6-diaminohexane, is one of two raw materials for the manufacture of nylon (page 402).

PROBLEM 11.10   Devise a synthesis of

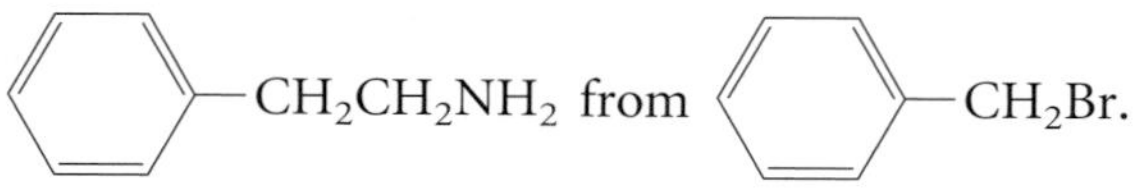

[benzene ring]—$CH_2CH_2NH_2$ from [benzene ring]—CH_2Br.

Aldehydes and ketones undergo **reductive amination** when treated with ammonia, primary, or secondary amines, to give primary, secondary, or tertiary amines, respectively. The most commonly used laboratory reducing agent for this purpose is the metal hydride sodium cyanoborohydride, NaBH$_3$CN.

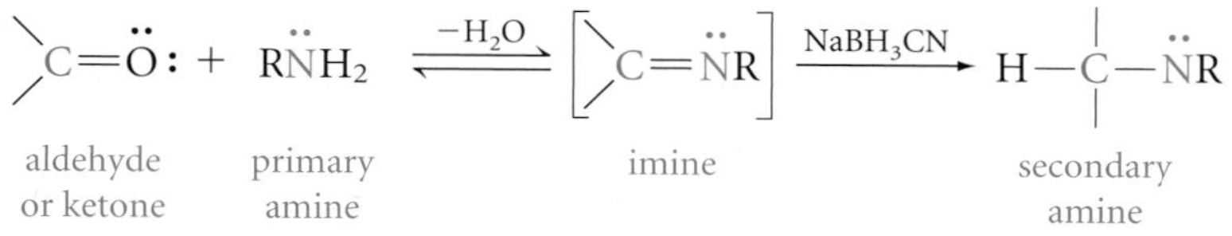

Aldehydes and ketones undergo **reductive amination** *when treated with amines in the presence of NaBH$_3$CN.*

$$\underset{\substack{\text{aldehyde} \\ \text{or ketone}}}{>\!\!C\!\!=\!\!\overset{..}{\underset{..}{O}}:} + \underset{\substack{\text{primary} \\ \text{amine}}}{\overset{..}{R}NH_2} \xrightarrow{-H_2O} \left[\underset{\text{imine}}{>\!\!C\!\!=\!\!\overset{..}{N}R} \right] \xrightarrow{\text{NaBH}_3\text{CN}} \underset{\substack{\text{secondary} \\ \text{amine}}}{H-\underset{|}{\overset{|}{C}}-\overset{..}{N}R} \qquad \textbf{(11.13)}$$

The reaction involves nucleophilic attack on the carbonyl group, leading to an imine (in the case of ammonia or primary amines; compare with eq. 9.31) or an iminium ion with secondary amines. The reducing agent then reduces the $C{=}N$ bond.

PROBLEM 11.11 Using eq. 11.13 as a guide, devise a synthesis of 2-aminopentane (page 323) from 2-pentanone.

Now that we know several ways to make amines, let us examine some of their properties.

11.6 The Basicity of Amines

The unshared pair of electrons on the nitrogen atom dominates the chemistry of amines. Because of this electron pair, amines are both basic and nucleophilic.

Aqueous solutions of amines are basic because of the following equilibrium:

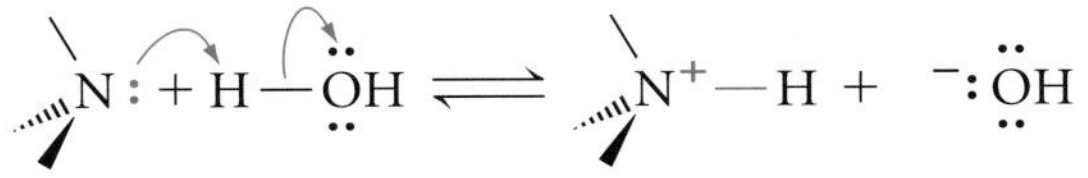

$$\overset{\backslash}{\underset{/}{{\sim}\mathrm{N}{:}}} + \mathrm{H}{-}\ddot{\mathrm{O}}\mathrm{H} \;\rightleftharpoons\; \overset{\backslash}{\underset{/}{{\sim}\mathrm{N}^{+}}}{-}\mathrm{H} \;+\; {^{-}{:}\ddot{\mathrm{O}}\mathrm{H}} \qquad (11.14)$$

amine   ammonium   hydroxide
ion   ion

EXAMPLE 11.7

Write an equation that shows why aqueous solutions of ethylamine are basic.

Solution $CH_3CH_2\ddot{N}H_2 + H_2O \rightleftharpoons CH_3CH_2\overset{+}{N}H_3 + HO^-$

ethylamine   ethylammonium
ion

Amines are more basic than water. They accept a proton from water, producing hydroxide ion, so their solutions are basic.

PROBLEM 11.12 Write an equation representing the equilibrium in an aqueous solution of trimethylamine, $(CH_3)_3N$.

An *amine* and its *ammonium ion* (eq. 11.14) are related as a *base* and its *conjugate acid*. For example, RNH_3^+ is the conjugate acid of the primary amine RNH_2. It is convenient, when comparing basicities of different amines, to compare instead the acidity constants (pK_a's) of their conjugate acids. Equation 11.15 expresses this acidity for a primary alkylammonium ion.

$$\overset{+}{R}NH_3 + H_2O \rightleftharpoons RNH_2 + H_3O^+ \qquad (11.15)$$

conjugate   base
acid

$$K_a = \frac{[RNH_2][H_3O^+]}{[RNH_3^+]}$$

The larger the K_a (or the smaller the pK_a) the stronger $R\overset{+}{N}H_3$ is as an acid, or the weaker RNH_2 is as a base.

EXAMPLE 11.8

The pK_a's of NH_4^+ and $CH_3\overset{+}{N}H_3$ are 9.30 and 10.64, respectively. Which is the stronger base, NH_3 or CH_3NH_2?

Solution NH_4^+ is the stronger acid (lower pK_a). Therefore, NH_3 is the *weaker* base, and CH_3NH_2 is the *stronger*.

Table 11.3 lists some amine basicities. Alkylamines are approximately 10 times as basic as ammonia. Recall that alkyl groups are electron-donating relative to hydrogen. This electron-donating effect stabilizes the ammonium ion (positive charge) relative to the free amine (eq. 11.14). Hence it decreases the acidity of the ammonium ion, or increases the basicity of the amine. In general, *electron-donating groups increase the basicity of amines, and electron-withdrawing groups decrease their basicity.*

PROBLEM 11.13   Do you expect $ClCH_2CH_2NH_2$ to be a stronger or weaker base than $CH_3CH_2NH_2$? Explain.

Table 11.3	**Basicities of some common amines, expressed as pK_a of the corresponding ammonium ions**		
	Formula		
Name	**Amine**	**Ammonium ion**	**pK_a of the ammonium ion**
ammonia	$\overset{..}{N}H_3$	$\overset{+}{N}H_4$	9.30
methylamine	$CH_3\overset{..}{N}H_2$	$CH_3\overset{+}{N}H_3$	10.64
dimethylamine	$(CH_3)_2\overset{..}{N}H$	$(CH_3)_2\overset{+}{N}H_2$	10.71
trimethylamine	$(CH_3)_3\overset{..}{N}$	$(CH_3)_3\overset{+}{N}H$	9.77
ethylamine	$CH_3CH_2\overset{..}{N}H_2$	$CH_3CH_2\overset{+}{N}H_3$	10.67
propylamine	$CH_3CH_2CH_2\overset{..}{N}H_2$	$CH_3CH_2CH_2\overset{+}{N}H_3$	10.58
aniline	$C_6H_5\overset{..}{N}H_2$	$C_6H_5\overset{+}{N}H_3$	4.62
N-methylaniline	$C_6H_5\overset{..}{N}HCH_3$	$C_6H_5\overset{+}{N}H_2(CH_3)$	4.85
N,N-dimethylaniline	$C_6H_5\overset{..}{N}(CH_3)_2$	$C_6H_5\overset{+}{N}H(CH_3)_2$	5.04
p-chloroaniline	*p*-$ClC_6H_4\overset{..}{N}H_2$	*p*-$ClC_6H_4\overset{+}{N}H_3$	3.98

Aromatic amines are much weaker bases than aliphatic amines or ammonia. For example, aniline is less basic than cyclohexylamine by nearly a million times.

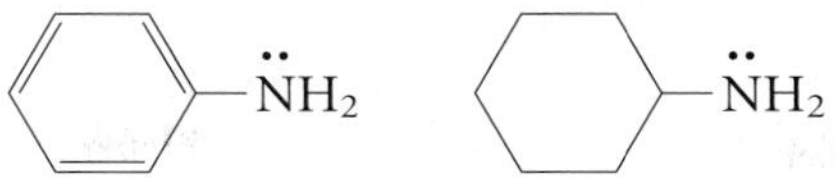

	aniline	cyclohexylamine
pK_a of ammonium ion	4.62	9.8

The reason for this huge difference is the resonance delocalization of the unshared electron pair that is possible in aniline but not in cyclohexylamine.

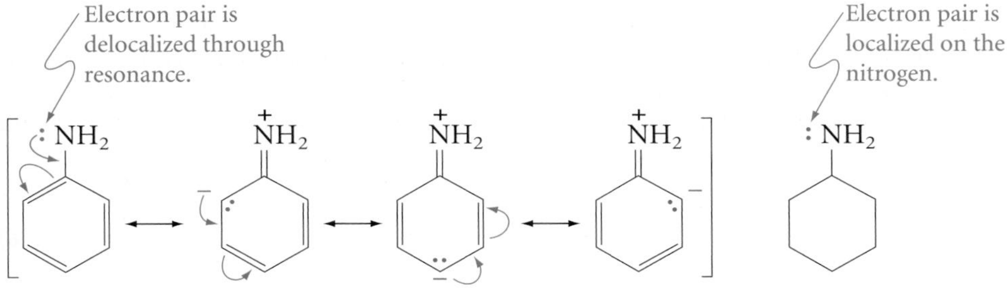

Electron pair is delocalized through resonance.

Electron pair is localized on the nitrogen.

resonance structures of aniline   cyclohexylamine

Resonance stabilizes the unprotonated form of aniline. This shifts the equilibrium in eq. 11.15 to the right, increasing the acidity of the anilinium ion or decreasing the basicity of aniline. Another way to describe the situation is to say that the unshared electron pair in aniline is delocalized and therefore less available for donation to a proton than is the electron pair in cyclohexylamine.

PROBLEM 11.14   Compare the basicities of the last four amines in Table 11.3, and explain the reasons for the observed basicity order.

PROBLEM 11.15   Place aniline, *p*-toluidine, and *p*-nitroaniline in order of increasing basicity.

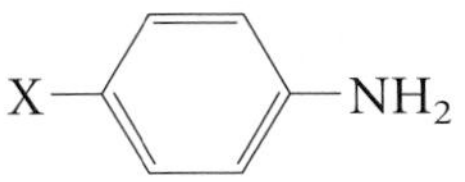

X = H aniline
X = CH$_3$ *p*-toluidine
X = NO$_2$ *p*-nitroaniline

11.7 ## Comparison of the Basicity and Acidity of Amines and Amides

Amines and amides each have nitrogens with an unshared electron pair. There is a huge difference, however, in their basicities. Aqueous solutions of *amines* are *basic*; aqueous solutions of *amides* are essentially *neutral*. Why this striking difference?

The answer lies in their structures, as illustrated in the following comparison of a primary amine with a primary amide:

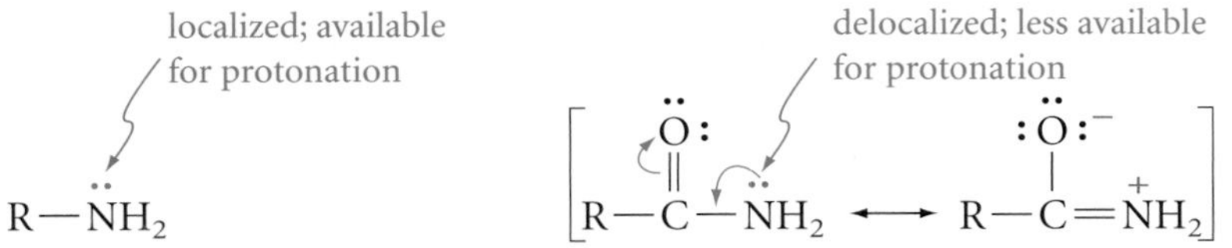

localized; available for protonation

delocalized; less available for protonation

R—NH$_2$
amine

amide

In the amine, the electron pair is mainly localized on the nitrogen. In the amide, the electron pair is delocalized to the carbonyl oxygen. The effect of this delocalization is seen in the low pK_a values for the conjugate acids of amides, compared with those for the conjugate acids of amines, for example:

$$\begin{array}{ccc}
 & & \overset{+}{\ddot{O}}-H \\
 & \overset{+}{CH_3CH_2NH_3} & \overset{\parallel\,..}{CH_3CNH_2} \\
\text{conjugate acid of:} & \text{ethylamine} & \text{acetamide} \\
pK_a: & 10.67 & -0.6
\end{array}$$

Notice that amides are not protonated on nitrogen but on the carbonyl oxygen. This is because protonation on oxygen gives a resonance-stabilized cation, while protonation on nitrogen does not.

Primary and secondary amines and amides have N—H bonds, and one might expect that they could on occasion behave as acids (proton donors).

$$R—\overset{..}{N}H_2 \rightleftharpoons R—\overset{..}{\underset{..}{N}}H^- + H^+ \qquad K_a \cong 10^{-40} \qquad \textbf{(11.16)}$$

Primary amines are exceedingly weak acids, much weaker than alcohols. Their pK_a is about 40, compared with about 16 for alcohols. The main reason for the difference is that nitrogen is much less electronegative than oxygen and thus cannot stabilize a negative charge nearly as well.

Amides, on the other hand, are *much stronger acids than amines;* in fact, their pK_a (about 15) is comparable to that of alcohols:

$$R—\overset{O}{\overset{\parallel}{C}}—\overset{..}{N}H_2 \rightleftharpoons \left[R—\overset{\overset{..}{\ddot{O}}:}{\overset{\parallel}{C}}—\overset{..}{\underset{..}{N}}H \longleftrightarrow R—\overset{:\ddot{O}:^-}{\underset{\mid}{C}}=NH \right] + H^+ \qquad K_a \cong 10^{-15} \qquad \textbf{(11.17)}$$

amidate anion

The **amidate anion** is formed by removal of a proton from the amide nitrogen.

One reason is that the negative charge of the **amidate anion** can be delocalized through resonance. Another reason is that the nitrogen in an amide carries a partial positive charge (see page 331), making it easy to lose the attached proton, which is also positive.

It is important to understand these differences between amines and amides, not only because they involve important chemical principles, but also because they help us understand the chemistry of certain natural products, such as peptides and proteins.

PROBLEM 11.16 Place the following compounds (a) in order of increasing basicity and (b) in order of increasing acidity.

acetanilide   cyclohexylamine   aniline

11.8 Reaction of Amines with Strong Acids; Amine Salts

Amines react with strong acids to form **alkylammonium salts.** An example of this reaction for a primary amine and HCl is as follows:

Alkylamines react with strong acids to form **alkylammonium salts.**

$$R-\overset{..}{N}H_2 \ + HCl \longrightarrow \overset{+}{R N H_3} \ \ Cl^- \qquad (11.18)$$

primary amine   an alkylammonium
                                    chloride

EXAMPLE 11.9

Complete the following acid–base reaction, and name the product.

$$CH_3CH_2NH_2 + HI \longrightarrow$$

Solution

$$CH_3CH_2\overset{..}{N}H_2 \ + HI \longrightarrow CH_3CH_2\overset{\displaystyle H}{\underset{\displaystyle H}{N^+}}-H \ \ I^-$$

ethylamine   ethylammonium iodide

PROBLEM 11.17   Complete the following equation, and name the product:

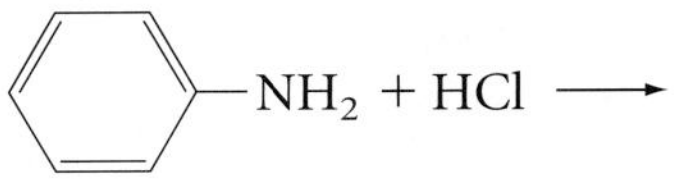

This type of reaction is used to separate or extract amines from neutral or acidic water-insoluble substances. Consider, for example, a mixture of *p*-toluidine and *p*-nitrotoluene, which might arise from a preparation of the amine that for some reason does not go to completion (eq. 11.10). The amine can be separated from the unreduced nitro compound by the following scheme:

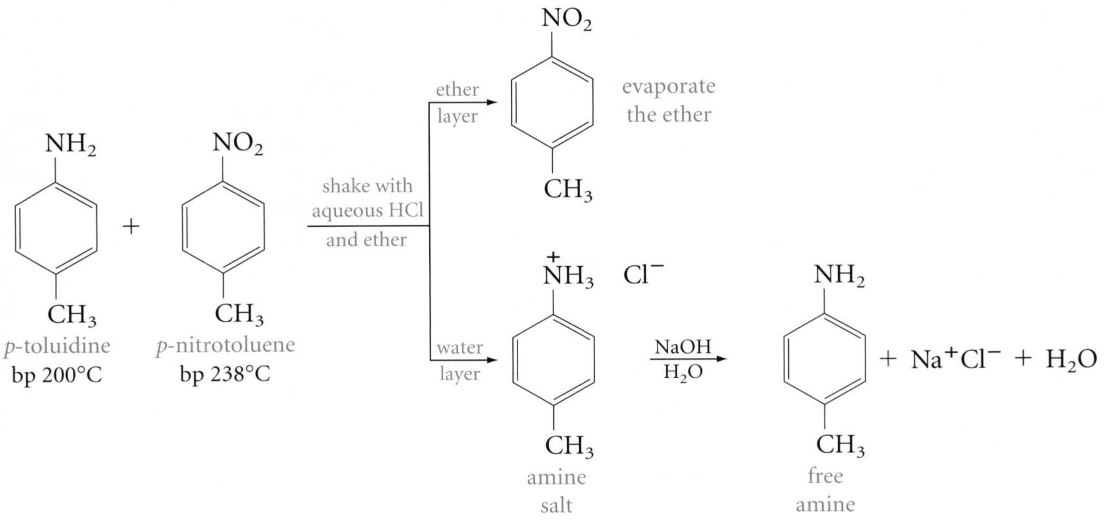

$$(11.19)$$

A WORD ABOUT ...

Alkaloids and the Dart-Poison Frogs

Alkaloids are basic, nitrogen-containing compounds of plant or animal origin, often having complex structures and significant pharmacological properties. Nicotine (from tobacco) and the amine obtained by deprotonation of squalamine (from sharks) are examples of alkaloids already mentioned in this chapter. Several more will appear in Chapter 13.

One prolific producer of alkaloids is the *dart-poison frogs,* a family of colorful and petite amphibians native to Costa Rica, Panama, Ecuador, and Colombia. The intense coloration of these frogs serves as a warning to potential predators who would have an unpleasant experience if they tried to make a meal of one of these tiny creatures. This is because the frogs secrete toxic alkaloids from glands on the surface of their skin. These secretions are so toxic that they have been used by locals to poison blowgun darts used in hunting, hence the name *dart-poison frogs.*

Scientists from the National Institutes of Health have isolated and determined the structures of many of the compounds present in these skin secretions using the techniques of mass spectroscopy and NMR spectroscopy. So far the structures of close to 200 different alkaloids have been determined. The most potent of these toxins is batrachotoxin (from *Phyllobates terribilis),* an example of a steroidal alkaloid. Other toxins include histrionicotoxin

A dart-poison frog.

histrionicotoxin

pumiliotoxin B

(from *Dendrobates histrionicus*) and pumiliotoxin B (from *Dendrobates pumilio*).

Because many of these compounds are produced in only microgram quantities by the frogs, laboratory syntheses have often been developed to confirm structures and also to provide a supply of material for pharmacological studies. It turns out that most of these toxins act on the nervous system by affecting the manner in which ions are transported across cell membranes. As a result of this property, several of these alkaloids are now used as research tools in the field of neuroscience.

In 1992 an alkaloid with painkilling properties far greater than that of morphine was isolated from the Ecuadorian frog *Epipedobates tricolor.* Named epibatidine, it was hoped that this simple alkaloid would be a lead compound for the development of a new family of painkillers. Whether or not epibatidine or related synthetic materials will eventually result in new painkillers remains to be determined, but epibatidine is already being marketed in the form of its tartrate salt for use in biomedical research. Research in this interesting area of natural product chemistry once again confirms that big things can come in small packages.

batrachotoxin

epibatidine

The mixture, neither component of which is water soluble, is dissolved in an inert, low-boiling solvent such as ether and is shaken with aqueous hydrochloric acid. The amine reacts to form a salt, which is ionic and dissolves in the water layer. The nitro compound does not react and remains in the ether layer. The two layers are then separated. The nitro compound can be recovered by evaporating the ether. The amine can be recovered from its salt by making the aqueous layer alkaline with a strong base such as NaOH.

There are many natural and synthetic amine salts of biological interest. Two examples are squalamine, an antimicrobial steroid recently isolated from the dogfish shark (see Chapter 15 for more about steroids), and (+)-methamphetamine hydrochloride, the addictive and toxic stimulant commonly known as "ice" or "meth."

Dogfish shark.

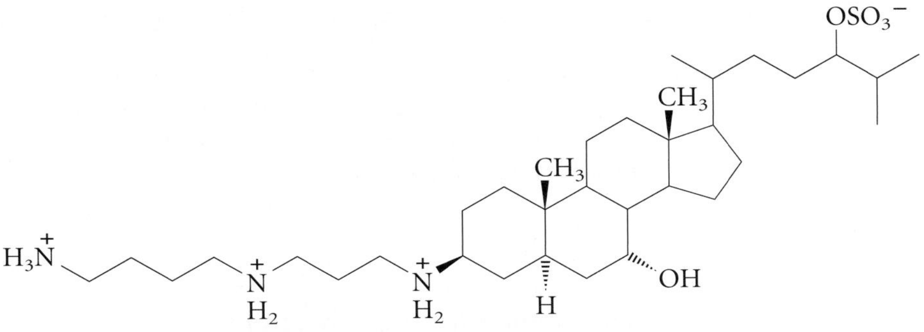

squalamine

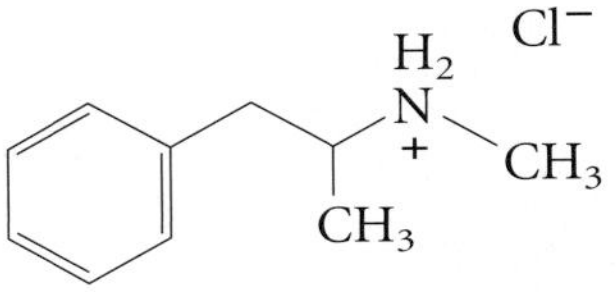

methamphetamine hydrochloride

11.9 Chiral Amines as Resolving Agents

Amines also form salts with organic acids. This reaction is used to resolve enantiomeric acids (Sec. 5.12). For example, (*R*)- and (*S*)-lactic acids can be resolved by reaction with a chiral amine such as (*S*)-1-phenylethylamine:

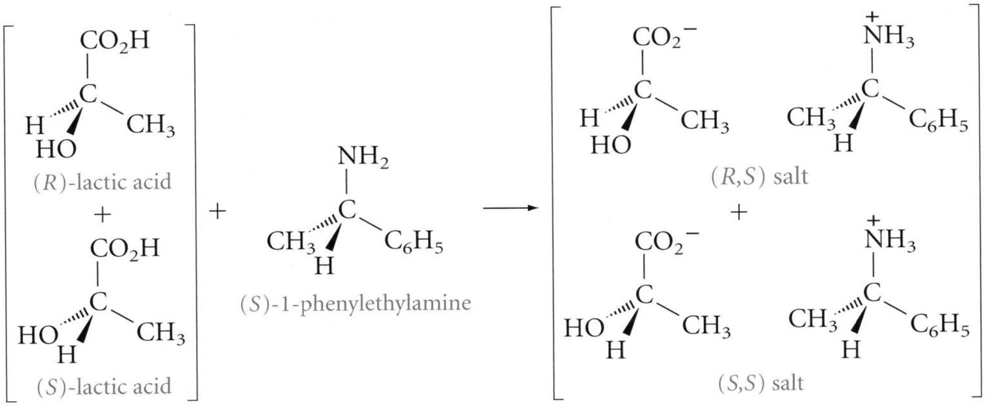

(*R*)-lactic acid

+

(*S*)-lactic acid

+

(*S*)-1-phenylethylamine

→

(*R*,*S*) salt

+

(*S*,*S*) salt

(11.20)

The salts are diastereomers, not enantiomers, and can be separated by ordinary methods, such as fractional crystallization. Once separated, each salt can be treated with a strong acid, such as HCl, to liberate one enantiomer of lactic acid. For example,

$$(R,S) \text{ salt } + \text{ HCl} \longrightarrow \underset{\substack{(R)\text{-lactic acid}}}{\overset{\substack{CO_2H}}{\underset{\substack{H\diagup \quad \diagdown CH_3 \\ HO}}{|\!\!\!\raisebox{-1ex}{C}}}} + \underset{\substack{(S)\text{-1-phenylethyl-} \\ \text{ammonium chloride}}}{\overset{\substack{\overset{+}{N}H_3 \quad Cl^-}}{\underset{\substack{CH_3\diagup \quad \diagdown C_6H_5 \\ H}}{|\!\!\!\raisebox{-1ex}{C}}}} \qquad (11.21)$$

The chiral amine can be recovered for reuse by treating its salt with sodium hydroxide (as in the last step of eq. 11.19).

Numerous chiral amines are available from natural products and can be used to resolve acids. Conversely, some chiral acids are available to resolve amine enantiomers.

So far we have considered reactions in which amines act as bases. Now we will examine some reactions in which they act as nucleophiles.

11.10 Acylation of Amines with Acid Derivatives

Amines are nitrogen nucleophiles. They react with the carbonyl group of acid derivatives (acyl halides, anhydrides, and esters) by nucleophilic acyl substitution (Sec. 10.11).

Looked at from the viewpoint of the amine, we can say that the N—H bond in primary and secondary amines can be *acylated* by acid derivatives. For example, primary and secondary amines react with acyl halides to form amides (compare with eq. 10.36).

$$\underset{\substack{\text{acyl halide}}}{\overset{\substack{O \\ \|}}{R-C-Cl}} + \underset{\substack{\text{primary amine}}}{H_2\overset{\cdot\cdot}{N}-R'} \longrightarrow \underset{\substack{\text{secondary amide}}}{\overset{\substack{O \\ \|}}{R-C-NHR'}} + \text{ HCl} \qquad (11.22)$$

$$\underset{\substack{\text{acyl halide}}}{\overset{\substack{O \\ \|}}{R-C-Cl}} + \underset{\substack{\text{secondary} \\ \text{amine}}}{H\overset{\cdot\cdot}{N}\overset{\diagup R'}{\diagdown R''}} \longrightarrow \underset{\substack{\text{tertiary amide}}}{\overset{\substack{O \\ \|}}{R-C-N}\overset{\diagup R'}{\diagdown R''}} + \text{ HCl} \qquad (11.23)$$

If the amine is inexpensive, two equivalents are used—one to form the amide and the second to neutralize the HCl. Alternatively, an inexpensive base may be added for the latter purpose. This can be sodium hydroxide (especially if R is *aromatic*), or a tertiary amine; having no N—H bonds, tertiary amines cannot be acylated, but they can neutralize the HCl.

EXAMPLE 11.10

Using eq. 10.29 as a guide, write out the steps in the mechanism for eq. 11.22.

Solution

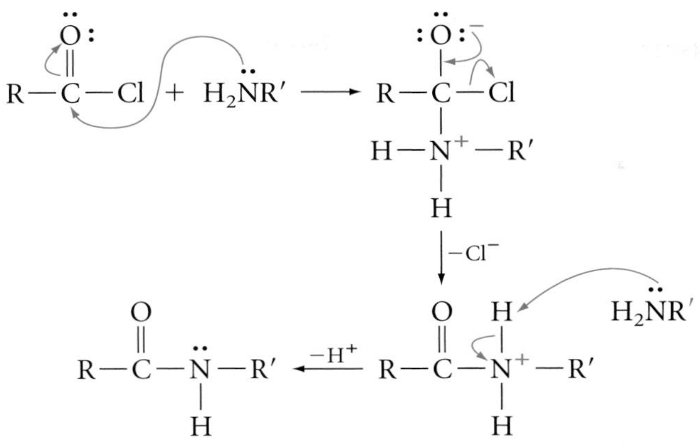

R—C—Cl + H2NR′
R—C—Cl
H—N+—R′
H
−Cl−
H2NR′
R—C—N+—R′
−H+
R—C—N—R′
amide

The first step involves nucleophilic addition to the carbonyl group. Elimination of HCl completes the substitution reaction.

Acylation of amines is put to practical use. For example, the insect repellent Off is the amide formed in the reaction of *m*-toluyl chloride and diethylamine.

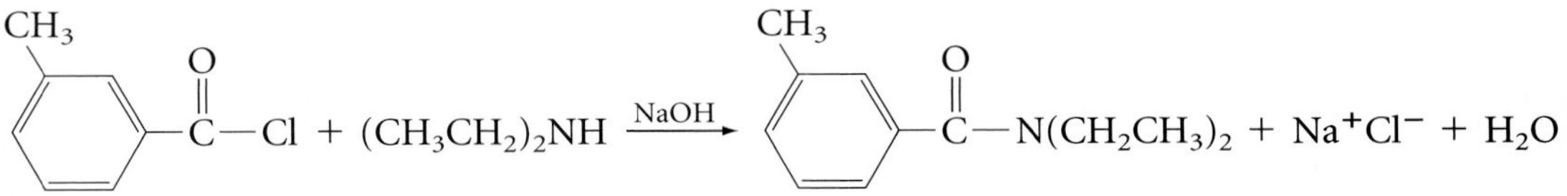

CH3
C—Cl + (CH3CH2)2NH
NaOH
C—N(CH2CH3)2 + Na+Cl− + H2O

(11.24)

m-toluyl chloride — diethylamine — *N*,*N*-diethyl-*m*-toluamide (the insect repellent Off)

PROBLEM 11.18 Write out the steps in the mechanism for the synthesis of Off (eq. 11.24).

The antipyretic (fever-reducing substance) acetanilide is an amide made from aniline and acetic anhydride.

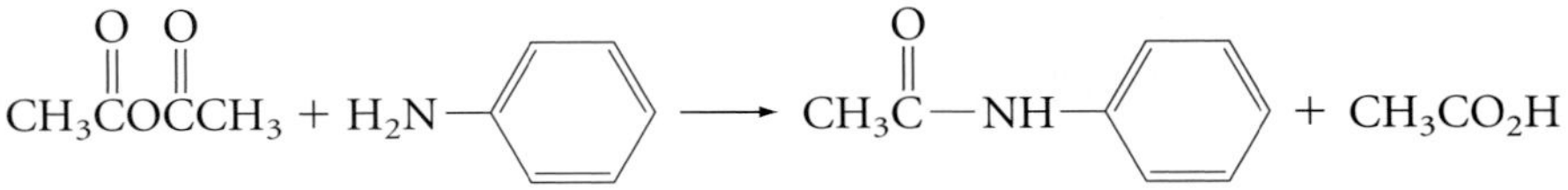

CH3COCCH3 + H2N
CH3C—NH
+ CH3CO2H

(11.25)

acetic anhydride — aniline — acetanilide

PROBLEM 11.19 Provide the structures of the amides obtained from reaction of acetic anhydride with

a. $CH_3CH_2NH_2$.
b. $(CH_3CH_2)_2NH$.

11.11   Quaternary Ammonium Compounds

Tertiary amines react with primary or secondary alkyl halides by an S_N2 mechanism (eq. 11.6). The products are **quaternary ammonium salts,** in which all four hydrogens of ammonium ion are replaced by organic groups. For example,

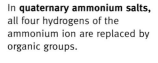

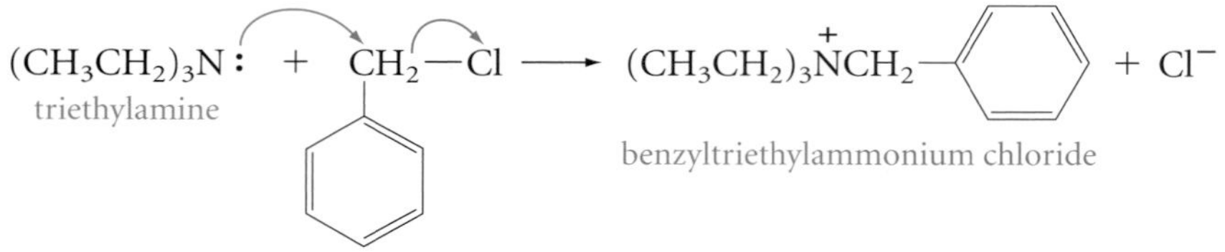

$$(CH_3CH_2)_3N\text{:} + CH_2\text{—}Cl \longrightarrow (CH_3CH_2)_3\overset{+}{N}CH_2\text{—}\langle\text{—}\rangle + Cl^- \qquad (11.26)$$

triethylamine

benzyl chloride

benzyltriethylammonium chloride

Quaternary ammonium compounds are important in biological processes. One of the most common natural quaternary ammonium ions is **choline,** which is present in phospholipids (Sec. 15.6).

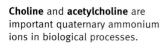

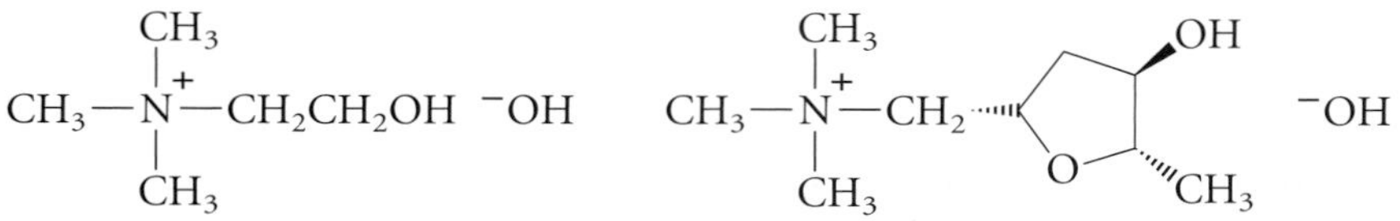

choline   muscarine

Choline is not only involved in various metabolic processes, but is also the precursor of **acetylcholine** (page 321), a compound that plays a key role in the transmission of nerve impulses. The mushroom *Amanita muscaria* contains the deadly neurotoxin muscarine, which structurally resembles acetylcholine and probably interferes with the function of this neurotransmitter.

Amanita muscaria, a source of muscarine.

11.12   Aromatic Diazonium Compounds

Primary aromatic amines react with nitrous acid at 0°C to yield **aryldiazonium ions.** The process is called **diazotization.**

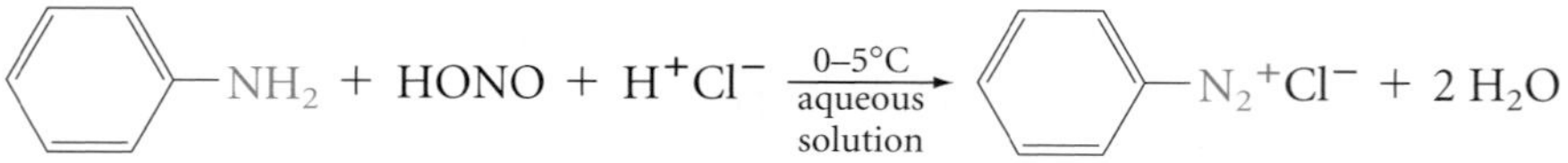

$$\langle\text{—}\rangle\text{—}NH_2 + HONO + H^+Cl^- \xrightarrow[\substack{\text{aqueous}\\ \text{solution}}]{0-5°C} \langle\text{—}\rangle\text{—}N_2^+Cl^- + 2\,H_2O \qquad (11.27)$$

aniline   nitrous acid   benzenediazonium chloride

Diazonium compounds are extremely useful synthetic intermediates. Before we describe their chemistry, let us try to understand the steps in eq. 11.27. First we need to examine the structure of nitrous acid.

    Nitrous acid decomposes rather rapidly at room temperature. It is therefore prepared as needed by treating an aqueous solution of sodium nitrite with a strong acid at ice temperature. At that temperature, nitrous acid solutions are reasonably stable.

$$Na^+NO_2^- + H^+Cl^- \xrightarrow{\text{0-5°C}} H-\overset{..}{\underset{..}{O}}-\overset{..}{N}=\overset{..}{O}: + Na^+Cl^- \qquad \textbf{(11.28)}$$

sodium nitrite   nitrous acid

The reactive species in reactions of nitrous acid is the **nitrosonium ion** NO^+. It is formed by protonation of the nitrous acid, followed by loss of water (compare with eq. 4.18):

$$H\overset{..}{\underset{..}{O}}-\overset{..}{N}=\overset{..}{O}: + H^+ \rightleftharpoons HO\overset{+}{\underset{\underset{H}{|}}{\frown}}\overset{..}{N}=\overset{..}{O}: \rightleftharpoons H_2O + :\overset{+}{N}=\overset{..}{O}: \qquad \textbf{(11.29)}$$

nitrosonium ion

How do the two nitrogens, one from the amine and one from the nitrous acid, become bonded to one another, as they appear in diazonium ions? This happens in the first step of diazotization (eq. 11.30), which involves nucleophilic attack of the primary amine on the nitrosonium ion, followed by proton loss.

$$\overset{..}{Ar}NH_2 + :\overset{+}{N}=\overset{..}{O}: \longrightarrow Ar\overset{+}{\underset{\underset{H}{|}}{\overset{\overset{H}{|}}{N}}}-\overset{..}{N}=\overset{..}{O}: \rightleftharpoons Ar\overset{\overset{H}{|}}{N}-\overset{..}{N}=\overset{..}{O}: + H^+ \qquad \textbf{(11.30)}$$

a primary nitrosamine

Protonation of the oxygen in the resulting nitrosamine, followed by elimination of water then gives the aromatic diazonium ion.

$$Ar\overset{\overset{H}{|}}{N}-\overset{..}{N}=\overset{..}{O}: + H^+ \longrightarrow Ar\overset{\overset{H}{|}}{N}=\overset{+}{N}-\overset{..}{O}H \xrightarrow{-H_2O} Ar\overset{+}{N}\equiv N: \qquad \textbf{(11.31)}$$

aryldiazonium
ion

Notice that in the final product there are no N—H bonds; both hydrogens of the amino group are lost, the first in eq. 11.30 and the second in eq. 11.31. Therefore, *only primary amines can be diazotized.* (Secondary and tertiary amines do react with nitrous acid, but their reactions are different and less important in synthesis.)

Solutions of aryldiazonium ions are moderately stable and can be kept at 0°C for several hours. They are useful in synthesis because the **diazonio group** ($—N_2^+$) can be replaced by nucleophiles; the other product is nitrogen gas.

$$Ar-\overset{+}{N}\equiv N: + Nu:^- \longrightarrow Ar-Nu + N_2 \qquad \textbf{(11.32)}$$

Specific useful examples are shown in eq. 11.33. The nucleophile always takes the position on the benzene ring that was occupied by the diazonio group.

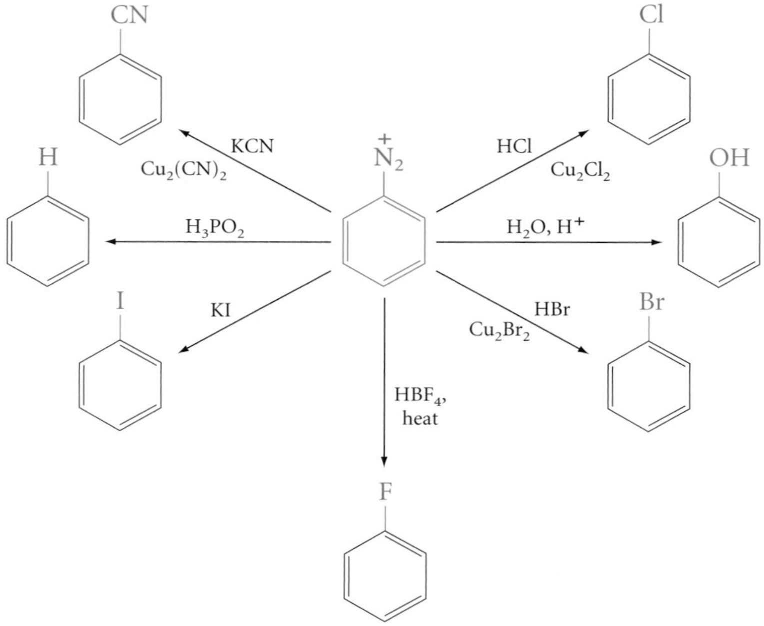

(11.33)

Conversion of diazonium compounds to aryl chlorides, bromides, or cyanides is usually accomplished using cuprous salts, and is known as the **Sandmeyer reaction.** Since a CN group is easily converted to a CO_2H group (eq. 10.13), this provides another route to aromatic carboxylic acids. The reaction with KI gives aryl iodides, usually not easily accessible by direct electrophilic iodination. Similarly, direct aromatic fluorination is difficult, but aromatic fluorides can be prepared from diazonium compounds and tetrafluoroboric acid, HBF_4.

In the **Sandmeyer reaction,** diazonium ions react with cuprous salts to form aryl chlorides, bromides, or cyanides.

Phenols can be prepared by adding diazonium compounds to hot aqueous acid. This reaction is important because there are not many ways to introduce an —OH group directly on an aromatic ring.

Finally, we sometimes use the orienting effect of a nitro or amino group and afterwards remove this substituent from the aromatic ring. This can be done by diazotization followed by reduction. A common reducing agent for this purpose is **hypophosphorous acid,** H_3PO_2.

Hypophosphorous acid is used to reduce the diazonio group to H.

Here are some examples of ways that diazonium compounds can be used in synthesis:

EXAMPLE 11.11

How can *m*-dibromobenzene be prepared?

Solution It *cannot* be prepared by direct electrophilic bromination of bromobenzene, because the Br group is *o,p*-directing (Sec. 4.11). But we can take advantage of the *m*-directing effect of a nitro group and then convert the nitro group to a bromine atom, as follows:

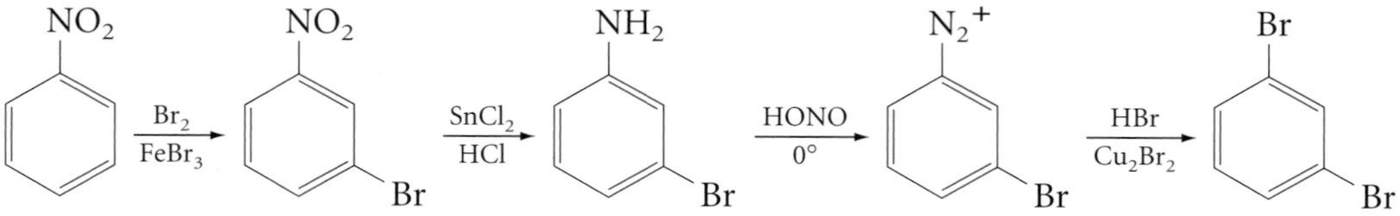

EXAMPLE 11.12

How can *o*-toluic (*o*-methylbenzoic) acid be prepared from *o*-toluidine (*o*-methyl-aniline)?

Solution

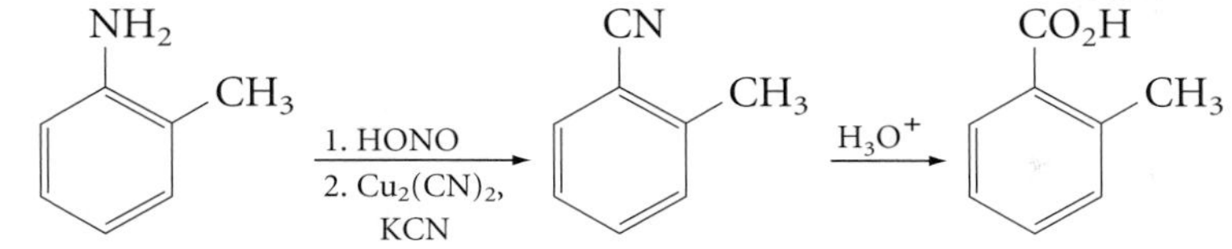

o-toluidine   *o*-toluic acid

EXAMPLE 11.13

Design a route to 1,3,5-tribromobenzene from aniline.

Solution   First brominate; the amino group is *o,p*-directing and ring-activating. Then remove the amino group by diazotization and reduction.

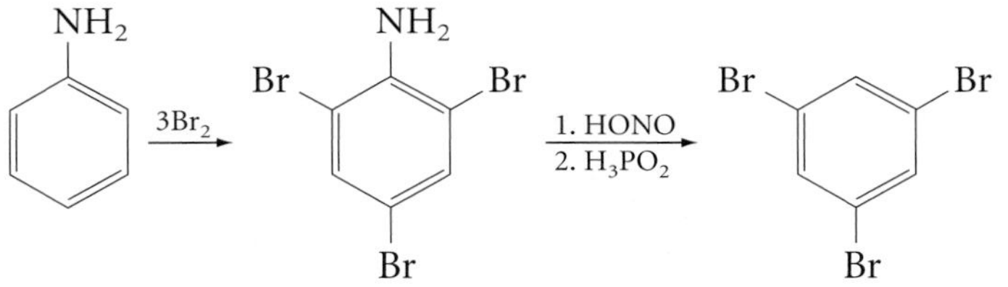

PROBLEM 11.20   Design a synthesis of each of the following compounds, using a diazonium ion intermediate.

  a. *m*-bromochlorobenzene from benzene
  b. *m*-nitrophenol from *m*-nitroaniline
  c. 2,4-difluorotoluene from toluene
  d. 3,5-dibromotoluene from *p*-toluidine

11.13 Diazo Coupling; Azo Dyes

Being positively charged, aryldiazonium ions are electrophiles. They are *weak* electrophiles, however, because the positive charge can be delocalized through resonance.

EXAMPLE 11.14

Write the resonance contributors for the benzenediazonium ion that show how the nitrogen farthest from the benzene ring can become electrophilic.

Solution

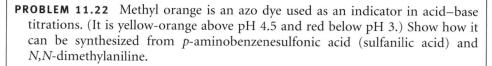

In the second contributor, the nitrogen at the right has only six electrons; it can react as an electrophile.

PROBLEM 11.21   Draw resonance contributors which show that the positive charge in benzenediazonium ion can also be delocalized to the *ortho* and *para* carbons of the benzene ring. (CAREFUL! These contributors have two positive charges and one negative charge.)

Azo compounds contain the azo group, —N=N—.

Diazo coupling is an electrophilic aromatic substitution reaction in which phenols and aromatic amines react with aryldiazonium electrophiles to give azo compounds.

Aryldiazonium ions react with strongly activated aromatic rings (phenols and aromatic amines) to give **azo compounds.** For example,

benzenediazonium ion + phenol $\xrightarrow{HO^-}$ *p*-hydroxyazobenzene yellow leaflets, mp 155–157°C + H_2O **(11.34)**

The nitrogen atoms are retained in the product. This electrophilic aromatic substitution reaction is called **diazo coupling,** because in the product, two aromatic rings are coupled by the azo, or —N=N—, group. *Para* coupling is preferred, as in eq. 11.34, but if the *para* position is blocked by another substituent, *ortho* coupling can occur. *All azo compounds are colored,* and many are used commercially as dyes for cloth and in color photography.*

PROBLEM 11.22   Methyl orange is an azo dye used as an indicator in acid–base titrations. (It is yellow-orange above pH 4.5 and red below pH 3.) Show how it can be synthesized from *p*-aminobenzenesulfonic acid (sulfanilic acid) and *N,N*-dimethylaniline.

$(CH_3)_2N$—N=N—$SO_3^-Na^+$

methyl orange

At this point, we have completed a survey of the main functional groups in organic chemistry. By now, all of the structures in the table inside the front cover of this book should seem familiar to you. In the next chapter, we will describe some modern techniques that help us to assign a structure to a particular molecule. After that, we will conclude with a series of chapters on important commercial and biological applications of organic chemistry.

Methyl orange indicator in basic solution (top) and in acidic solution (bottom).

---

*For an interesting discussion of the diazo copying process, see the article by B. Osterby, *J. Chem. Ed.* **1989,** *66,* 1206–1208.

REACTION SUMMARY

1. Alkylation of Ammonia and Amines to Form Amines (Sec. 11.4)

$$R{-}X + 2\,NH_3 \longrightarrow R{-}NH_2 + NH_4{}^+X^-$$

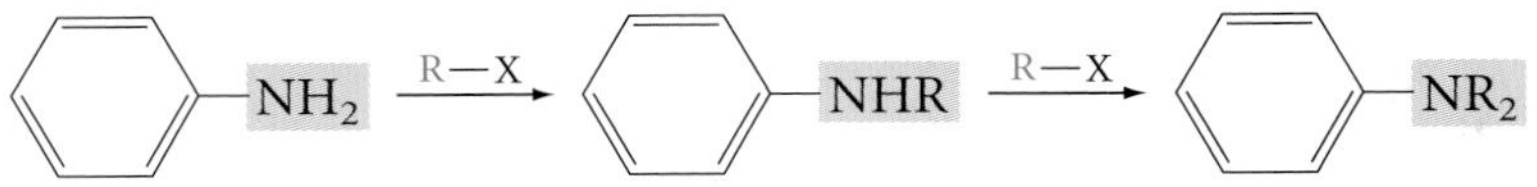

2. Reduction Routes to Amines (Sec. 11.5)

a. Catalytic or Chemical Reduction of the Nitro Group

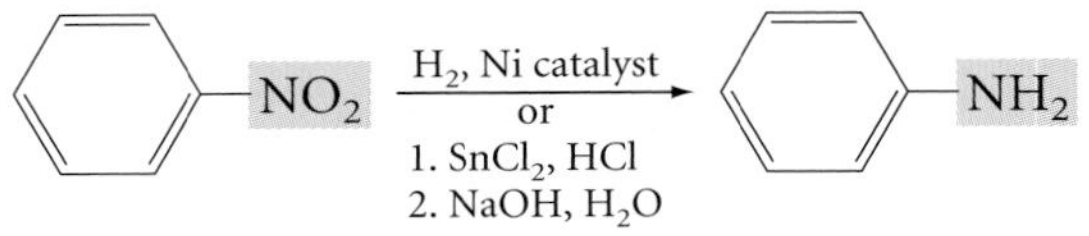

b. Hydride Reduction of Amides and Nitriles

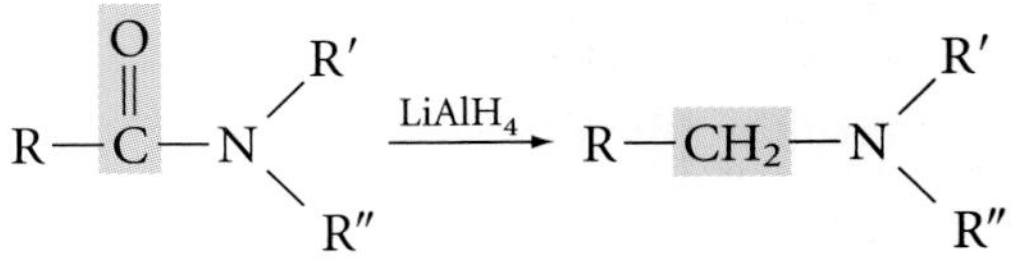

c. Reductive Amination of Aldehydes and Ketones

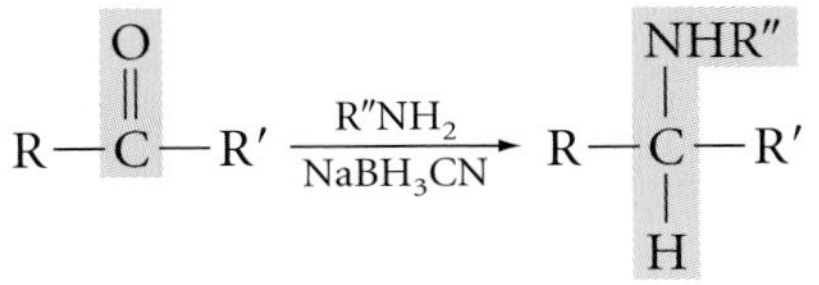

3. Amines as Bases (Secs. 11.6 and 11.8)

$$R{-}NH_2 + H{-}OH \longrightarrow R{-}\overset{+}{N}H_3 + {}^-OH \qquad R{-}NH_2 + H{-}Cl \longrightarrow R{-}\overset{+}{N}H_3 + Cl^-$$

4. Amines as Nucleophiles

a. Acylation of Amines (Sec. 11.10)

Secondary and Tertiary Amides from Primary and Secondary Amines

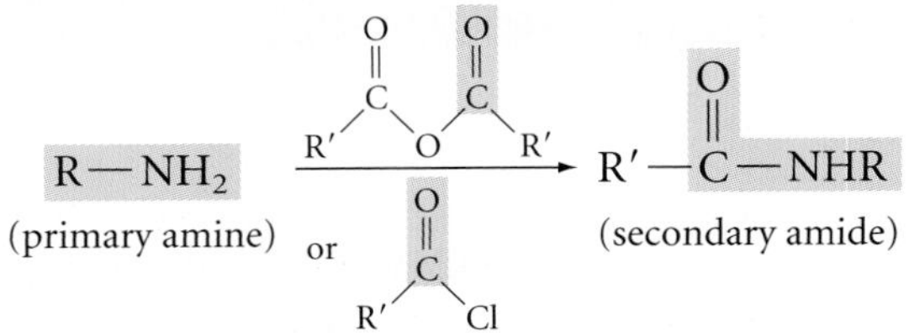

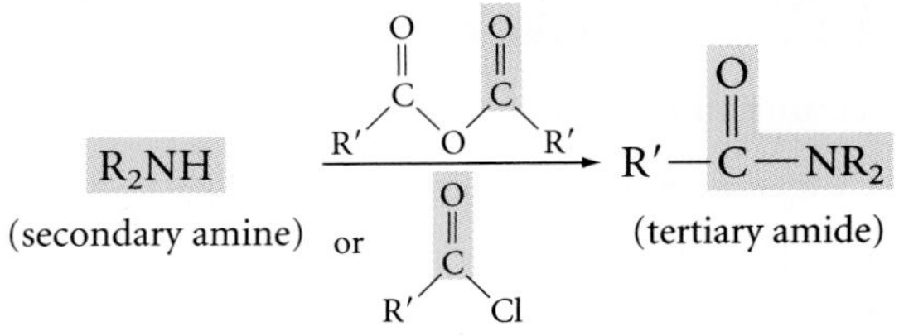

b. Alkylation of Amines: Quaternary Ammonium Salts (Sec. 11.11)

$$R_3N + R'X \longrightarrow R_3\overset{+}{N}{-}R'X^-$$

5. Aryldiazonium Salts: Formation and Reactions (Secs. 11.12 and 11.13)

a. Formation from Aniline and Nitrous Acid (Sec. 11.12)

$$ArNH_2 + HONO \xrightarrow{HX} ArN_2^+X^-$$

(aryldiazonium salt)

b. Reactions to Form Substituted Benzenes (Sec. 11.12)

$$ArN_2^+ + H_2O \xrightarrow{heat} ArOH + N_2 + H^+$$

(phenols)

$$ArN_2^+ + HX \xrightarrow{Cu_2X_2} ArX \quad (X=Cl, Br)$$

$$ArN_2^+ + KI \longrightarrow ArI$$

$$ArN_2^+ + KCN \xrightarrow{Cu_2(CN)_2} ArCN$$

$$ArN_2^+ + HBF_4 \longrightarrow ArF$$

$$ArN_2^+ + H_3PO_2 \longrightarrow ArH$$

c. Diazo Coupling (Sec. 11.13)

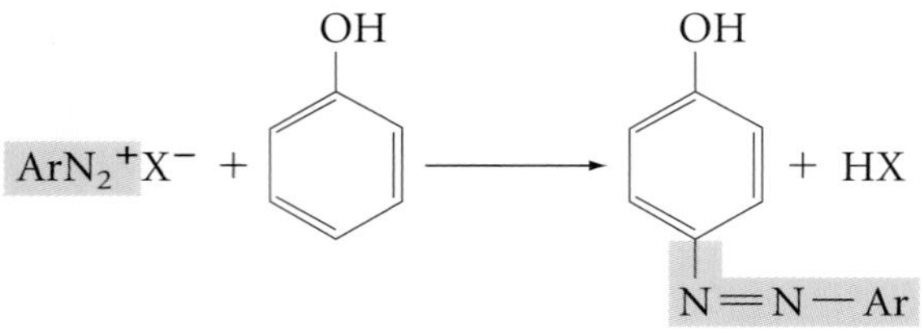

(azo compound)

MECHANISM SUMMARY

Diazotization (Sec. 11.12)

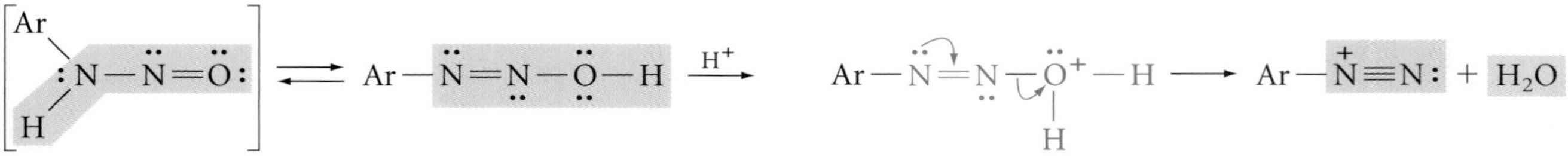

ADDITIONAL PROBLEMS

Nomenclature and Structure of Amines

11.23 Give an example of each of the following:

a. a primary amine	**b.** a cyclic tertiary amine	**c.** a secondary aromatic amine
d. a quaternary ammonium salt	**e.** an aryldiazonium salt	**f.** an azo compound
g. a primary amide		

11.24 Write a structural formula for each of the following compounds:

a. *m*-bromoaniline	**b.** *sec*-butylamine	**c.** 2-aminopentane
d. dimethylpropylamine	**e.** *N*-methylbenzylamine	**f.** 1,3-diaminopropane
g. *N,N*-diethylaminocyclohexane	**h.** tetraethylammonium bromide	**i.** triphenylamine
j. *o*-toluidine	**k.** 3-methyl-2-butanamine	**l.** *N,N*-dimethyl-3-hexanamine

11.25 Write a correct name for each of the following compounds:

a. Cl—⟨benzene⟩—NH$_2$

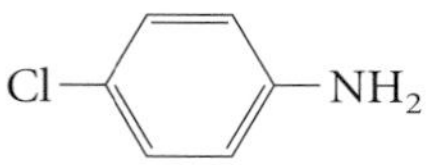

b. CH$_3$NHCH$_2$CH$_2$CH$_3$

c. (CH$_3$CH$_2$)$_2$NCH$_3$

d. (CH$_3$)$_4$N$^+$Cl$^-$

e. CH$_3$CH(OH)CH(NH$_2$)CH$_3$

f. ⟨cyclohexanone with NH$_2$⟩

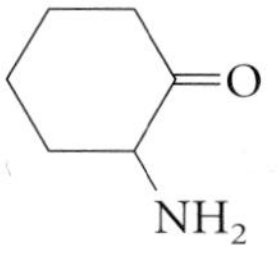

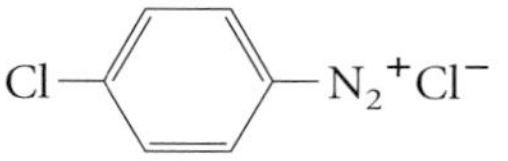

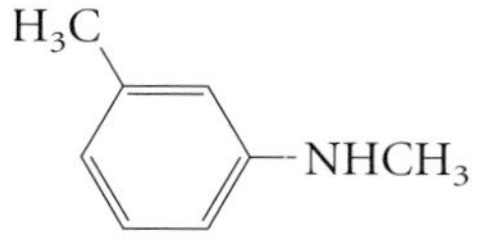

g. Cl—⟨benzene⟩—N$_2$$^+Cl^-$

h. H$_3$C⟨benzene⟩—NHCH$_3$

i. ⟨cyclopentane⟩—NH$_2$

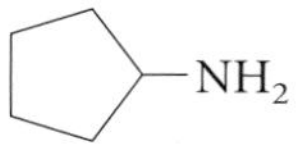

j. H$_2$N(CH$_2$)$_6$NH$_2$

11.26 Draw the structures for, name, and classify as primary, secondary, or tertiary the eight isomeric amines with the molecular formula C$_4$H$_{11}$N.

Properties of Amines and Quaternary Ammonium Salts

11.27 Place the following substances, which have nearly identical formula weights, in order of increasing boiling point: CH$_3$CH$_2$CH$_2$CH$_2$NH$_2$, CH$_3$CH$_2$CH$_2$CH$_2$OH, CH$_3$CH$_2$CH$_2$OCH$_3$, CH$_3$CH$_2$CH$_2$CH$_2$CH$_3$.

11.28 Tell which is the stronger base and why.
  a. aniline or *p*-cyanoaniline   **b.** aniline or diphenylamine

11.29 Write out a scheme similar to eq. 11.19 to show how you could separate a mixture of *p*-toluidine, *p*-methylphenol, and *p*-xylene.

CH$_3$—⟨benzene⟩—NH$_2$ CH$_3$—⟨benzene⟩—OH   CH$_3$—⟨benzene⟩—CH$_3$

p-toluidine   *p*-methylphenol   *p*-xylene

11.30 Draw the important contributors to the resonance hybrid structure of *p*-nitroaniline (page 327).

11.31 Explain why compound A can be separated into its *R*- and *S*-enantiomers, but compound B cannot.

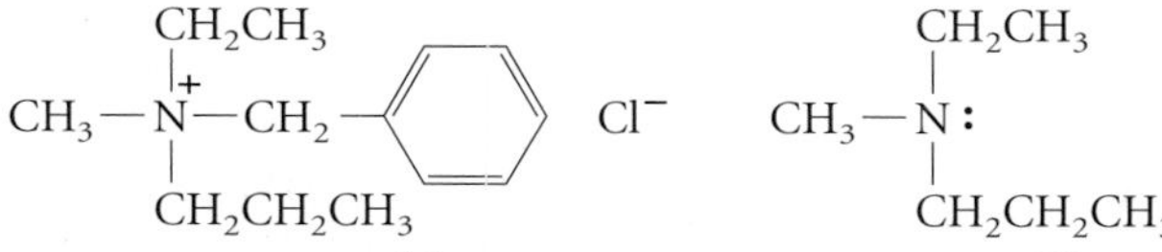

CH$_3$—N$^+$(CH$_2$CH$_3$)(CH$_2$CH$_2$CH$_3$)—CH$_2$—⟨benzene⟩   Cl$^-$ CH$_3$—N:(CH$_2$CH$_3$)(CH$_2$CH$_2$CH$_3$)

compound A   compound B

11.32 Give the priority order of groups in compound A (Problem 11.31), and draw a dash-wedge formula for its *R*-isomer.

Preparation and Reactions of Amines

11.33 Give equations for the preparation of the following amines from the indicated precursor:
  a. *N,N*-diethylaniline from aniline   **b.** *m*-bromoaniline from benzene
  c. *p*-bromoaniline from benzene   **d.** 1-aminohexane from 1-bromopentane

11.34 Complete the following equations:

a. 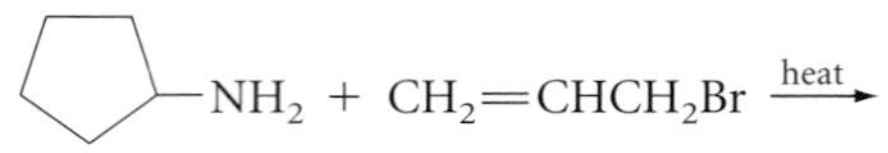 (cyclopentyl)—NH$_2$ + CH$_2$=CHCH$_2$Br $\xrightarrow{\text{heat}}$

b. $CH_3\overset{\displaystyle O}{\overset{\|}{C}}Cl$ + H$_2$NCH$_2$CH$_2$CH(CH$_3$)$_2$ $\longrightarrow$ A $\xrightarrow{\text{LiAlH}_4}$ B

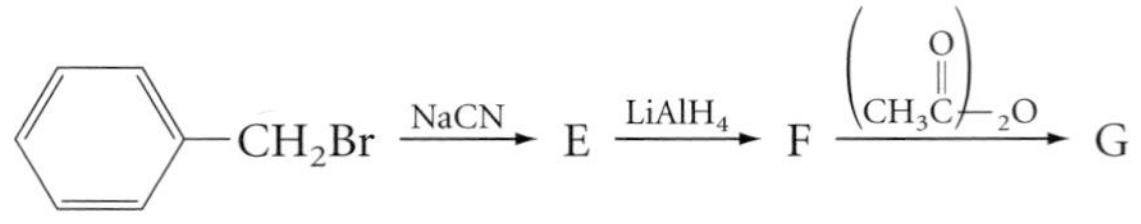

c. 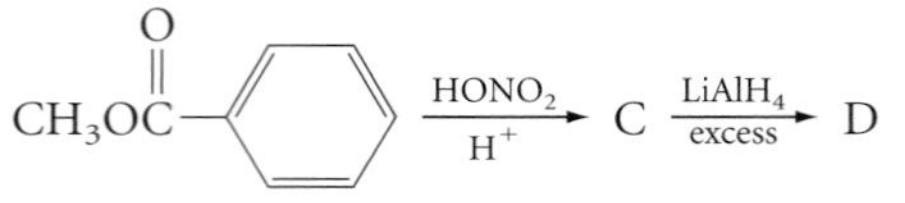 $CH_3O\overset{\displaystyle O}{\overset{\|}{C}}$—(phenyl) $\xrightarrow[\text{H}^+]{\text{HONO}_2}$ C $\xrightarrow[\text{excess}]{\text{LiAlH}_4}$ D

d. (phenyl)—CH$_2$Br $\xrightarrow{\text{NaCN}}$ E $\xrightarrow{\text{LiAlH}_4}$ F $\xrightarrow{(CH_3\overset{\displaystyle O}{\overset{\|}{C}})_2O}$ G

e. (cyclohexyl)=O + (CH$_3$)$_2$CHNH$_2$ $\xrightarrow{\text{NaBH}_3\text{CN}}$ H

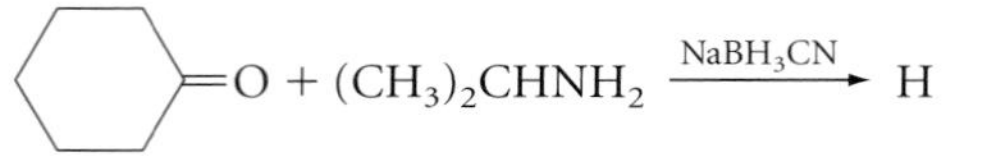

11.35 Write an equation for the reaction of
a. *p*-toluidine (page 333) with hydrochloric acid.
b. triethylamine, (CH$_3$CH$_2$)$_3$N, with sulfuric acid.
c. dimethylammonium chloride, (CH$_3$)$_2\overset{+}{N}$H$_2$Cl$^-$, with sodium hydroxide.
d. *N,N*-dimethylaniline (page 324) with methyl iodide.
e. cyclohexylamine (page 332) with acetic anhydride.

11.36 Write out the steps in the mechanism for the following reaction:

$$CH_3CH_2NH_2 + CH_3\overset{\displaystyle O}{\overset{\|}{C}}O\overset{\displaystyle O}{\overset{\|}{C}}CH_3 \longrightarrow CH_3CH_2NH\overset{\displaystyle O}{\overset{\|}{C}}CH_3 + CH_3COOH.$$

Explain why only one of the hydrogens of the amine is replaced by an acetyl group, even if a large excess of acetic anhydride is used.

Synthetic and Biological Applications of Amine Chemistry

11.37

a. Outline a synthesis of methamphetamine hydrochloride (page 335) that uses 1-phenyl-2-propanone (phenylacetone) and methylamine as carbon sources.
b. Will the product of your proposed synthesis be optically active or racemic? Explain why.

11.38 Choline (Sec. 11.11) can be prepared by the reaction of trimethylamine with ethylene oxide. Write an equation for the reaction, and show its mechanism.

11.39 Acetylcholine (Sec. 11.11) is synthesized in the body's neurons. The enzyme choline acetyltransferase catalyzes its synthesis from acetyl-CoA (page 308) and choline. Write an equation for the reaction, using the formula $CH_3\overset{\displaystyle O}{\underset{\displaystyle O}{\overset{\|}{C}}}$—S—CoA for

acetyl-CoA.

11.40 Decamethonium bromide is used in surgery as a muscle relaxant. It acts by preventing the enzyme acetylcholine esterase from destroying acetylcholine (see page 321), a necessary step in the transmission of nerve impulses. Show how decamethonium bromide can be synthesized from a diamine and an alkyl halide.

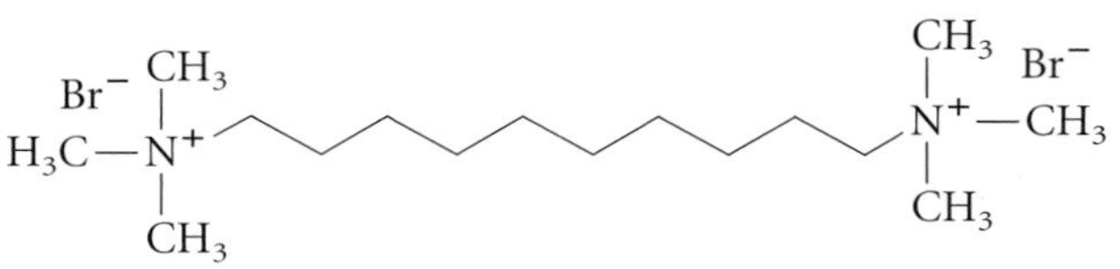

$$Br^- \quad H_3C-\overset{\displaystyle CH_3}{\underset{\displaystyle CH_3}{\overset{|}{\underset{|}{N^+}}}}\text{~~~~~~~~~~}\overset{\displaystyle CH_3}{\underset{\displaystyle CH_3}{\overset{|}{\underset{|}{N^+}}}}-CH_3 \quad Br^-$$

decamethonium bromide

Formation and Reactions of Aryldiazonium Ions

11.41 Primary aliphatic amines (RNH_2) react with nitrous acid in the same way that primary arylamines ($ArNH_2$) do, to form diazonium ions. But alkyldiazonium ions RN_2^+ are much less stable than aryldiazonium ions ArN_2^+ and readily lose nitrogen even at 0°C. Explain the difference.

11.42 Write an equation for the reaction of $CH_3 \!-\!\! \langle \bigcirc \rangle \!-\! N_2^+HSO_4^-$ with

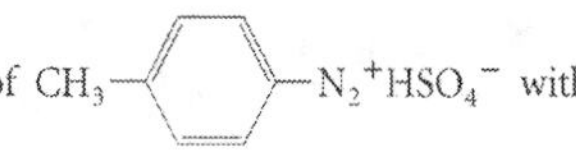

   a. KCN and cuprous cyanide.   **b.** aqueous acid, heat.   **c.** HCl and cuprous chloride.
   d. potassium iodide.   **e.** *p*-methylphenol and HO^-.   **f.** *N,N*-dimethylaniline and base.
   g. hypophosphorous acid.   **h.** fluoroboric acid, then heat.

11.43 Show how diazonium ions could be used to synthesize

   a. *p*-chlorobenzoic acid from *p*-chloroaniline.
   b. *m*-iodochlorobenzene from benzene.
   c. *m*-iodoacetophenone from benzene.
   d. 3-cyano-4-methylbenzenesulfonic acid from toluene.

11.44 Congo red is used as a direct dye for cotton. Write equations to show how it can be synthesized from benzidine and 1-aminonaphthalene-4-sulfonic acid.

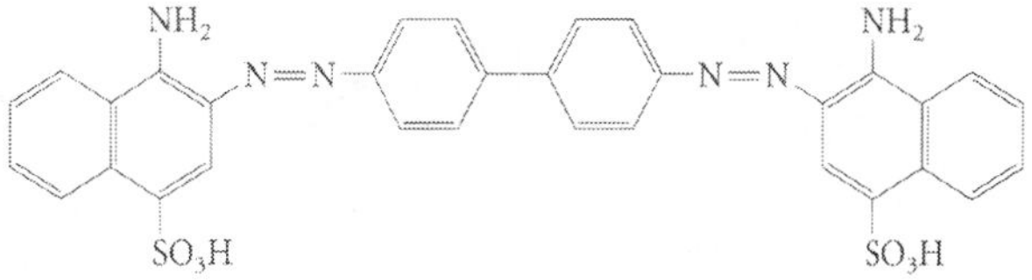

Congo red

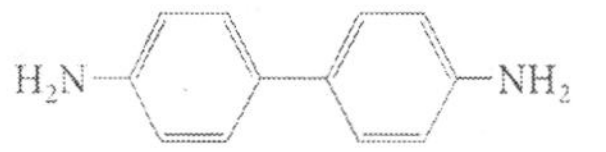

benzidine

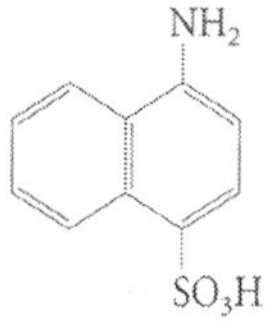

1-aminonaphthalene-4-sulfonic acid

11.45 Sunset yellow is a food dye that can be used to color Easter eggs. Write an equation for an azo coupling reaction that will give this dye.

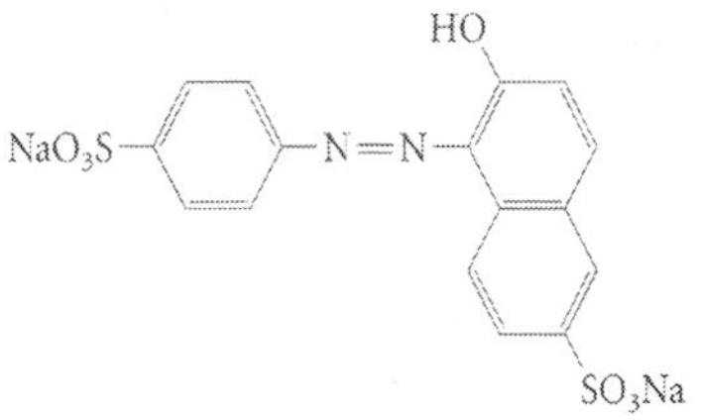

sunset yellow

16

HOCH₂ O OH
HO CH₂OH
OH

β-D-fructofuranose

CARBOHYDRATES

Carbohydrates occur in all plants and animals and are essential to life. Through photosynthesis, plants convert atmospheric carbon dioxide to carbohydrates, mainly cellulose, starch, and sugars. Cellulose is the building block of rigid cell walls and woody tissues in plants, whereas starch is the chief storage form of carbohydrates for later use as a food or energy source. Some plants (cane and sugar beets) produce sucrose, ordinary table sugar. Another sugar, glucose, is an essential component of blood. Two other sugars, ribose and 2-deoxyribose, are components of the genetic materials RNA and DNA. Other carbohydrates are important components of coenzymes, antibiotics, cartilage, the shells of crustaceans, bacterial cell walls, and mammalian cell membranes.

In this chapter, we will describe the structures and a few reactions of the more important carbohydrates.

16.1 Definitions and Classification

The word *carbohydrate* arose because molecular formulas of these compounds can be expressed as *hydrates* of *carbon*. Glucose, for example, has the molecular formula $C_6H_{12}O_6$, which might be written as $C_6(H_2O)_6$. Although this type of formula is useless in studying the chemistry of carbohydrates, the old name persists.

We can now define carbohydrates more precisely in terms of their organic structures. **Carbohydrates** are polyhydroxyaldehydes, polyhydroxyketones, or substances that give such compounds on hydrolysis. The chemistry of carbohy-

▲ Fructose, also known as fruit sugar, is 50% sweeter than sucrose (table sugar) and is the major sugar in honey.

drates is mainly the combined chemistry of two functional groups: the hydroxyl group and the carbonyl group.

Carbohydrates are usually classified according to their structure as **monosaccharides, oligosaccharides,** or **polysaccharides.** The term *saccharide* comes from Latin (*saccharum*, sugar) and refers to the sweet taste of some simple carbohydrates. The three classes of carbohydrates are related to each other through hydrolysis.

Carbohydrates are poly-hydroxyaldehydes, poly-hydroxyketones, or substances that give such compounds on hydrolysis. The hydroxyl group and the carbonyl group are the major functional groups in carbohydrates.

$$\text{Polysaccharide} \xrightarrow[\text{H}^+]{\text{H}_2\text{O}} \text{oligosaccharides} \xrightarrow[\text{H}^+]{\text{H}_2\text{O}} \text{monosaccharides} \qquad \textbf{(16.1)}$$

For example, hydrolysis of starch, a polysaccharide, gives first maltose and then glucose.

$$\underset{\substack{\text{starch}\\\text{(a polysaccharide)}}}{[\text{C}_{12}\text{H}_{20}\text{O}_{10}]_n} \xrightarrow[\text{H}^+]{n\,\text{H}_2\text{O}} \underset{\substack{\text{maltose}\\\text{(a disaccharide)}}}{n\,\text{C}_{12}\text{H}_{22}\text{O}_{11}} \xrightarrow[\text{H}^+]{n\,\text{H}_2\text{O}} \underset{\substack{\text{glucose}\\\text{(a monosaccharide)}}}{2n\,\text{C}_6\text{H}_{12}\text{O}_6} \qquad \textbf{(16.2)}$$

Monosaccharides (or simple sugars, as they are sometimes called) are carbohydrates that cannot be hydrolyzed to simpler compounds. **Polysaccharides** contain many monosaccharide units—sometimes hundreds or even thousands. Usually, but not always, the units are identical. Two of the most important polysaccharides, starch and cellulose, contain linked units of the same monosaccharide, glucose. **Oligosaccharides** (from the Greek *oligos*, few) contain at least two and generally no more than a few linked monosaccharide units. They may be called **disaccharides, trisaccharides,** and so on, depending on the number of units, which may be the same or different. Maltose, for example, is a disaccharide made of two glucose units, but sucrose, another disaccharide, is made of two different monosaccharide units: glucose and fructose.

Monosaccharides (simple sugars) cannot be hydrolyzed to simpler compounds. **Oligosaccharides** contain a few linked monosaccharide units, whereas **polysaccharides** contain many monosaccharide units.

In the next section, we will describe the structures of monosaccharides. Later, we will see how these units are linked to form oligosaccharides and polysaccharides.

16.2 Monosaccharides

Monosaccharides are classified according to the number of carbon atoms present (**triose, tetrose, pentose, hexose,** and so on) and according to whether the carbonyl group is present as an aldehyde (**aldose**) or as a ketone (**ketose**).

There are only two trioses: **glyceraldehyde** and **dihydroxyacetone.** Each has two hydroxyl groups, attached to different carbon atoms, and one carbonyl group.

An **aldose** and a **ketose** contain the aldehyde and ketone functional groups, respectively. A **triose** has three carbon atoms, a **tetrose** has four, and so on.

$$\begin{array}{lll}
\overset{1}{\text{CH}}{=}\text{O} & \overset{1}{\text{CH}_2\text{OH}} & \text{CH}_2\text{OH} \\
| & | & | \\
\overset{2}{\text{CHOH}} & \overset{2}{\text{C}}{=}\text{O} & \text{CHOH} \\
| & | & | \\
\overset{3}{\text{CH}_2\text{OH}} & \overset{3}{\text{CH}_2\text{OH}} & \text{CH}_2\text{OH} \\
\text{glyceraldehyde} & \text{dihydroxyacetone} & \text{glycerol} \\
\text{(an aldose)} & \text{(a ketose)} &
\end{array}$$

Glyceraldehyde is the simplest aldose, and dihydroxyacetone is the simplest ketose. Each is related to glycerol in that each has a carbonyl group in place of one of the hydroxyl groups.

Other aldoses or ketoses can be derived from glyceraldehyde or dihydroxyacetone by adding carbon atoms, each with a hydroxyl group. In aldoses, the chain is numbered from the aldehyde carbon. In most ketoses, the carbonyl group is located at C-2.

$$
\begin{array}{cccccc}
\overset{1}{C}H{=}O & \overset{1}{C}H{=}O & \overset{1}{C}H{=}O & \overset{1}{C}H_2OH & \overset{1}{C}H_2OH & \overset{1}{C}H_2OH \\
\overset{2}{C}HOH & \overset{2}{C}HOH & \overset{2}{C}HOH & \overset{2}{C}{=}O & \overset{2}{C}{=}O & \overset{2}{C}{=}O \\
\overset{3}{C}HOH & \overset{3}{C}HOH & \overset{3}{C}HOH & \overset{3}{C}HOH & \overset{3}{C}HOH & \overset{3}{C}HOH \\
\overset{4}{C}H_2OH & \overset{4}{C}HOH & \overset{4}{C}HOH & \overset{4}{C}H_2OH & \overset{4}{C}HOH & \overset{4}{C}HOH \\
 & \overset{5}{C}H_2OH & \overset{5}{C}HOH & & \overset{5}{C}H_2OH & \overset{5}{C}HOH \\
 & & \overset{6}{C}H_2OH & & & \overset{6}{C}H_2OH \\
\text{tetrose} & \text{pentose} & \text{hexose} & \text{tetrose} & \text{pentose} & \text{hexose} \\
\end{array}
$$

aldoses   ketoses

16.3   Chirality in Monosaccharides; Fischer Projection Formulas and D,L-Sugars

You will notice that glyceraldehyde, the simplest aldose, has one stereogenic carbon atom (C-2) and hence can exist in two enantiomeric forms.

$$
\begin{array}{cc}
CH{=}O & CH{=}O \\
H{-}C{-}OH & HO{-}C{-}H \\
CH_2OH & CH_2OH \\
R\text{-}(+)\text{-glyceraldehyde} & S\text{-}(-)\text{-glyceraldehyde} \\
[\alpha]_D^{25} +8.7(c = 2, H_2O) & [\alpha]_D^{25} -8.7(c = 2, H_2O)
\end{array}
$$

The dextrorotatory form has the R configuration.

It was in connection with his studies on carbohydrate stereochemistry that Emil Fischer invented his system of projection formulas. Because we will be using these formulas here, it might be wise for you to review Sections 5.7 through 5.9. Recall that, in a Fischer projection formula, *horizontal* lines show groups that project *above* the plane of the paper *toward* the viewer; *vertical* lines show groups that project *below* the plane of the paper *away* from the viewer. Thus, R-(+)-glyceraldehyde can be represented as

$$
\begin{array}{ccc}
CH{=}O & & CH{=}O \\
H{-}C{-}OH & \equiv & H{-\!\!\!-}OH \\
CH_2OH & & CH_2OH \\
R\text{-}(+)\text{-glyceraldehyde} & & \text{Fischer projection} \\
 & & \text{formula for} \\
 & & R\text{-}(+)\text{-glyceraldehyde}
\end{array}
$$

with the stereogenic center represented by the intersection of two crossed lines.

Fischer also introduced a stereochemical nomenclature that preceded the *R,S* system and is still in common use for sugars and amino acids. He used a small capital D to represent the configuration of (+)-glyceraldehyde, with the hydroxyl group on the *right*; its enantiomer, with the hydroxyl group on the *left*, was designated L-(−)-glyceraldehyde. The most oxidized carbon (CHO) was placed at the top.

$$
\begin{array}{c}
\text{CHO} \\
\text{H}\!-\!\!\!+\!\!\!-\text{OH} \\
\text{CH}_2\text{OH}
\end{array}
\qquad
\begin{array}{c}
\text{CHO} \\
\text{HO}\!-\!\!\!+\!\!\!-\text{H} \\
\text{CH}_2\text{OH}
\end{array}
$$

D-(+)-glyceraldehyde   L-(−)-glyceraldehyde

Fischer extended his system to other monosaccharides in the following way. If the stereogenic carbon *farthest* from the aldehyde or ketone group had the same configuration as D-glyceraldehyde (hydroxyl on the right), the compound was called a D-sugar. If the configuration at the remote carbon had the same configuration as L-glyceraldehyde (hydroxyl on the left), the compound was an L-sugar.

$$
\begin{array}{c}
\text{CH}\!=\!\text{O} \\
| \\
(\text{CHOH})_n \\
\text{H}\!-\!\!\!+\!\!\!-\text{OH} \\
\text{CH}_2\text{OH}
\end{array}
\quad
\begin{array}{c}
\text{CH}\!=\!\text{O} \\
| \\
(\text{CHOH})_n \\
\text{HO}\!-\!\!\!+\!\!\!-\text{H} \\
\text{CH}_2\text{OH}
\end{array}
\quad
\begin{array}{c}
\text{CH}_2\text{OH} \\
| \\
\text{C}\!=\!\text{O} \\
| \\
(\text{CHOH})_n \\
\text{H}\!-\!\!\!+\!\!\!-\text{OH} \\
\text{CH}_2\text{OH}
\end{array}
\quad
\begin{array}{c}
\text{CH}_2\text{OH} \\
| \\
\text{C}\!=\!\text{O} \\
| \\
(\text{CHOH})_n \\
\text{HO}\!-\!\!\!+\!\!\!-\text{H} \\
\text{CH}_2\text{OH}
\end{array}
$$

a D-aldose   an L-aldose   a D-ketose   an L-ketose

Figure 16.1 shows the Fischer projection formulas for all the D-aldoses through the hexoses. Starting with D-glyceraldehyde, one CHOH unit at a time is inserted in the chain. This carbon, which adds a new stereogenic center to the structure, is shown in black. In each case, the new stereogenic center can have the hydroxyl group at the right or at the left in the Fischer projection formula (*R* or *S* absolute configuration).

EXAMPLE 16.1

Using Figure 16.1, write the Fischer projection formula for L-erythrose.

Solution   L-Erythrose is the enantiomer of D-erythrose. Since both —OH groups are on the right in D-erythrose, they will both be on the left in its mirror image.

$$
\begin{array}{c}
\text{CH}\!=\!\text{O} \\
\text{HO}\!-\!\!\!+\!\!\!-\text{H} \\
\text{HO}\!-\!\!\!+\!\!\!-\text{H} \\
\text{CH}_2\text{OH}
\end{array}
$$

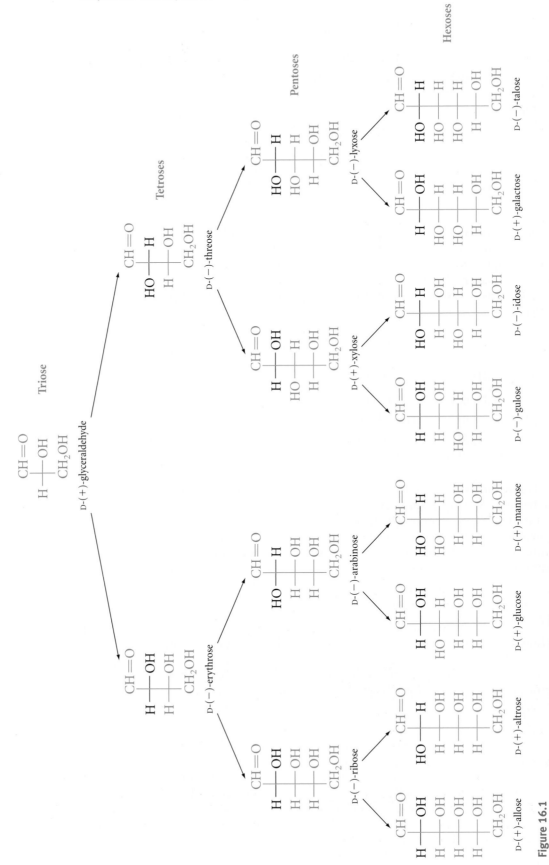

Figure 16.1
Fischer projection formulas and genealogy chart for the D-aldoses with up to six carbon atoms.

EXAMPLE 16.2

Convert the Fischer projection formula for D-erythrose to a three-dimensional structural formula.

Solution

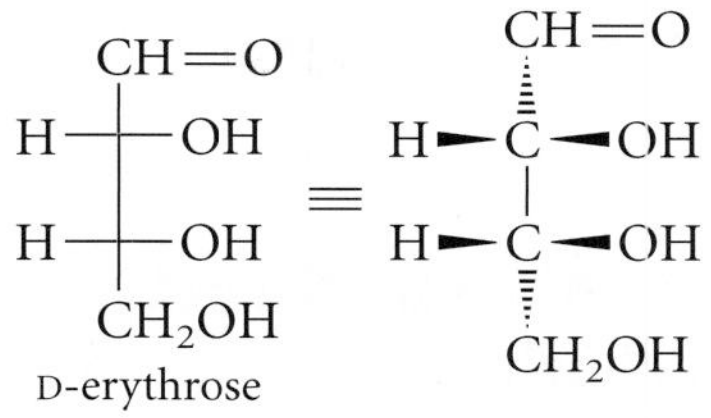

D-erythrose

We can also write the structure as

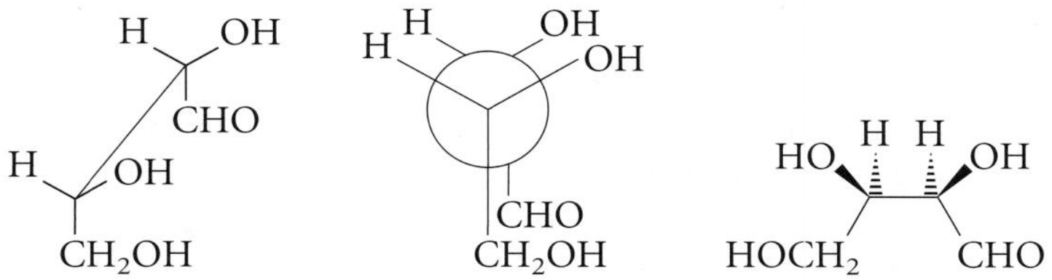

sawhorse   Newman   dash-wedge

and we can then rotate around the central C—C bond to more favorable staggered (instead of eclipsed) conformations, such as

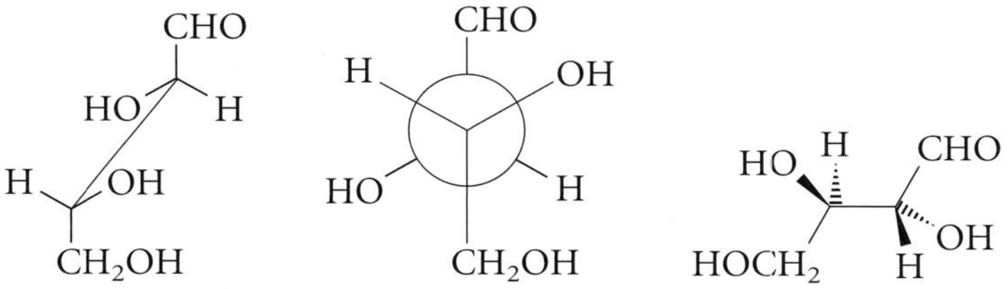

Molecular models may help you to follow these interconversions.

PROBLEM 16.1   Using Figure 16.1, write the Fischer projection formula for

a. L-threose.   b. L-glucose.

PROBLEM 16.2   Convert the Fischer projection formula for D-threose to three-dimensional representations.

PROBLEM 16.3   How many D-aldoheptoses are possible?

How are the sugars with identical numbers of carbon atoms, shown horizontally across Figure 16.1, related to one another? Compare, for example, D-(−)-erythrose and D-(−)-threose. They have the same configuration at C-3 (D, with the OH on the right), but opposite configurations at C-2. These sugars are stereoisomers, but *not* mirror images (*not* enantiomers). In other words, they are *diastereomers* (review Sec. 5.8). Similarly, there are four diastereomeric D-pentoses and eight diastereomeric D-hexoses.

A special name is given to diastereomers that differ in configuration *at only one stereogenic center;* they are called **epimers.** D-(−)-Erythrose and D-(−)-threose are not only diastereomers, they are epimers. Similarly, D-glucose and D-mannose are epimers (at C-2), and D-glucose and D-galactose are epimers (at C-4). Each pair has the same configurations at all stereogenic centers except one.

Epimers are diastereomers that differ in configuration at only one stereogenic center.

> **PROBLEM 16.4**   What pairs of D-pentoses are epimeric at C-3?

Notice that there is no direct relationship between configuration and the sign of optical rotation. Although all the sugars in Figure 16.1 are D-sugars, some are dextrorotatory ($+$) and others are levorotatory ($-$).

16.4 The Cyclic Hemiacetal Structures of Monosaccharides

The monosaccharide structures described so far are consistent with much of the known chemistry of these compounds, but they are oversimplified. We now examine the true structures of these compounds.

We learned earlier that alcohols undergo rapid and reversible addition to the carbonyl group of aldehydes and ketones, to form hemiacetals (review Sec. 9.7). This can happen *intramolecularly* when the hydroxyl and carbonyl groups are properly located in the same molecule (eqs. 9.14 and 9.15), which is the situation in many monosaccharides. *Monosaccharides exist mainly in cyclic, hemiacetal forms* and not in the acyclic aldo- or keto-forms we have depicted so far.

As an example, consider D-glucose. First, let us rewrite its Fischer projection formula in a way that brings the OH group at C-5 within bonding distance of the carbonyl group (as in eq. 9.14). This is shown in Figure 16.2. The Fischer projection is first converted to its three-dimensional (dash-wedge) structure, which is then turned

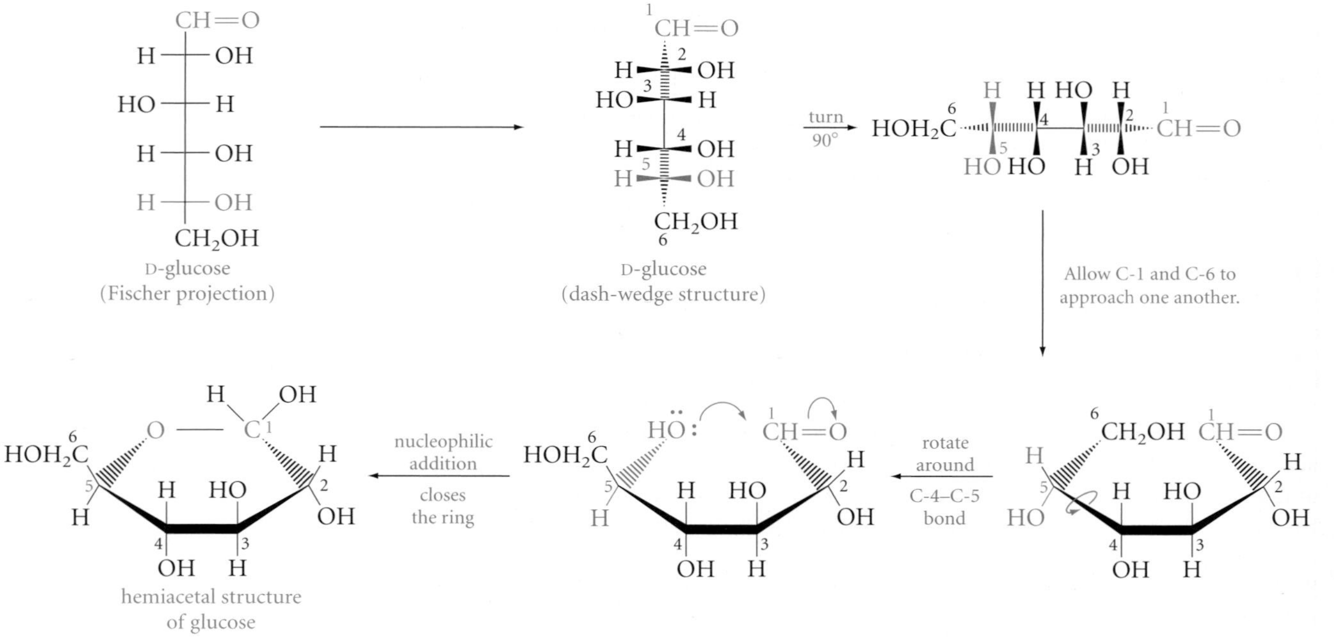

Figure 16.2
Manipulation of the Fischer projection formula of D-glucose to bring the C-5 hydroxyl group in position for cyclization to the hemiacetal form.

on its side and bent around so that C-1 and C-6 are close to one another. Finally, rotation about the C-4—C-5 bond brings the hydroxyl oxygen at C-5 close enough for nucleophilic addition to the carbonyl carbon (C-1). Reaction then leads to the cyclic, hemiacetal structure shown at the bottom left of the figure.

The British carbohydrate chemist W. N. Haworth (Nobel Prize, 1937) introduced a useful way of representing the cyclic forms of sugars. In a **Haworth projection,** the ring is represented as if it were planar and viewed edge on, with the oxygen at the upper right. The carbons are arranged clockwise numerically, with C-1 at the right. Substituents attached to the ring lie above or below the plane. For example, the Haworth formula for D-glucose (Figure 16.2) is written

In **Haworth projections,** the carbohydrate ring is represented as planar and viewed edge on. The carbons are arranged clockwise numerically, with C-1 at the right.

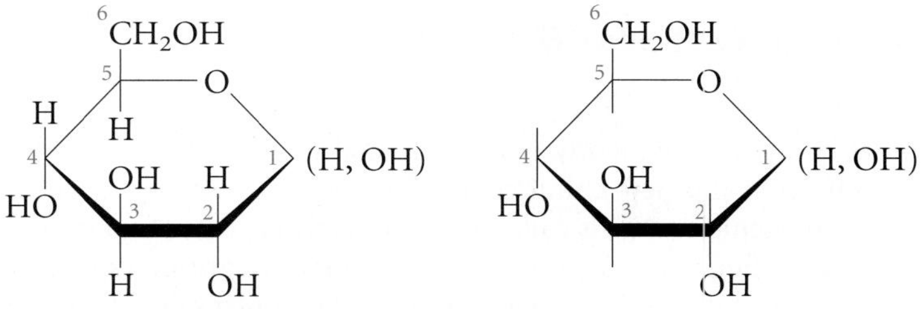

Haworth projection formulas for D-glucose

Sometimes, as in the structure at the right, the ring hydrogens are omitted so that attention can be focused on the hydroxyl groups.

In converting from one type of projection formula to another, notice that hydroxyl groups on the *right* in the Fischer projection are *down* in the Haworth projection (and conversely, hydroxyl groups on the *left* in the Fischer projection are *up* in the Haworth projection). For D-sugars, the terminal —CH₂OH group is *up* in the Haworth projection; for L-sugars, it is down.

EXAMPLE 16.3

Draw the Haworth projection for the six-membered cyclic structure of D-mannose.

Solution   Notice from Figure 16.1 that D-mannose differs from D-glucose *only* in the configuration at C-2. In the Fischer projection formula, the C-2 hydroxyl is on the *left*; therefore, it will be *up* in the Haworth projection. Otherwise, the structure is identical to that of D-glucose.

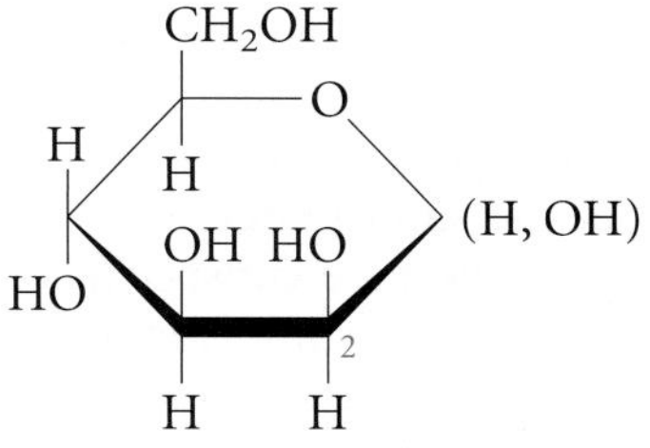

D-mannose

PROBLEM 16.5   Draw the Haworth projection formula for the six-membered cyclic structure of D-galactose.

Now notice three important features of the hemiacetal structure of D-glucose. First, *the ring is heterocyclic,* with five carbons and an oxygen. Carbons 1 through 5 are part of the ring structure, but carbon 6 (the —CH₂OH group) is a substituent on the ring.

Next, C-1 is special. *C-1 is the hemiacetal carbon,* simultaneously an alcohol and an ether carbon (it carries a hydroxyl group, and it is also connected to C-5 by an ether link). In contrast, all the other carbons are monofunctional. C-2, C-3, and C-4 are secondary alcohol carbons; C-6 is a primary alcohol carbon; and C-5 is an ether carbon. These differences show up in the different chemical reactions of D-glucose. Finally, *C-1 in the cyclic, hemiacetal structure is a stereogenic center.* It has four different groups attached to it (H, OH, OC-5, and C-2) and can therefore exist in two configurations, *R* or *S*. Let us consider this last feature in greater detail.

16.5 Anomeric Carbons; Mutarotation

In the acyclic, aldehyde form of glucose, C-1 is achiral, but in the cyclic structures, this carbon becomes chiral. Consequently, *two* hemiacetal structures are possible, depending on the configuration at the new chiral center. The hemiacetal carbon, the carbon that forms the new stereogenic center, is called the **anomeric carbon.** Two monosaccharides that differ only in configuration at the anomeric center are **anomers** (a special kind of epimers). Anomers are called α or β, depending on the position of the hydroxyl group. For monosaccharides in the D-series, the hydroxyl group is "down" in the α anomer and "up" in the β anomer, when the structure is written in the usual way (eq. 16.3).

The hemiacetal carbon in a cyclic monosaccharide is the **anomeric carbon.** Two monosaccharides that differ in configuration only at the anomeric carbon are **anomers.**

(16.3)

α-D-glucose (36%)
(mp 146°C)
$[\alpha]+112°$

D-glucose
(acyclic, aldehyde form)

β-D-glucose (64%)
(mp 150°C)
$[\alpha]+19°$

The α and β forms of D-glucose have identical configurations at every stereogenic center *except at C-1, the anomeric carbon.*

How do we know that monosaccharides exist mainly as cyclic hemiacetals? There is direct physical evidence. For example, if D-glucose is crystallized from methanol, the pure α form is obtained. On the other hand, crystallization from acetic acid gives the β form. The α and β forms of D-glucose are *diastereomers.* Being diastereomers, they have different physical properties, as shown under their structures in eq. 16.3; note that they have different melting points and different specific optical rotations.

The α and β forms of D-glucose interconvert in aqueous solution. For example, if crystalline α-D-glucose is dissolved in water, the specific rotation drops gradually from an initial value of +112° to an equilibrium value of +52°. Starting with the pure crystalline β form results in a gradual rise in specific rotation from an initial +19° to the same equilibrium value of +52°. These changes in optical rotation are called **mutarotation.** They can be explained by the equilibria shown in eq. 16.3. Recall that hemiacetal formation is a *reversible equilibrium process* (Sec. 9.7). Starting with either pure hemiacetal form, the ring can open to the acyclic aldehyde, which can then recyclize to give either the α or the β form. Eventually, an equilibrium mixture is obtained.

Changes in optical rotation due to interconversion of anomers in solution are called **mutarotation.**

At equilibrium, an aqueous solution of D-glucose contains 35.5% of the α form and 64.5% of the β form. There is only about 0.003% of the open-chain aldehyde form present.

EXAMPLE 16.5

Draw the most stable chair conformation of α-D-mannopyranose.

Solution Recall from Example 16.3 that D-mannose differs from D-glucose only at C-2. Using the cyclic structure at the left of eq. 16.6 as a guide, we can write

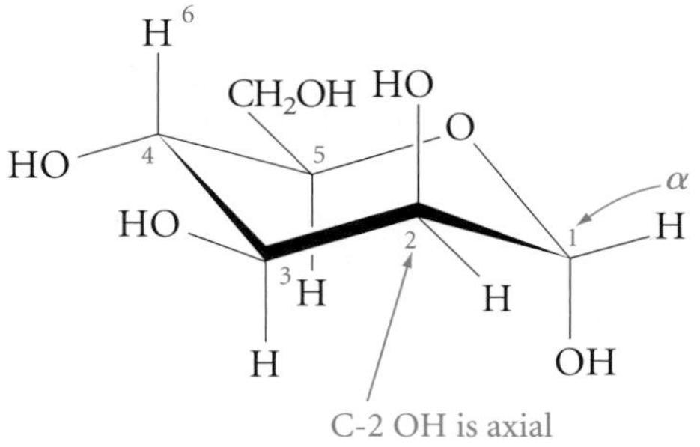

C-2 OH is axial

α-D-mannopyranose

PROBLEM 16.8 D-Galactose differs from D-glucose only in the configuration at C-4. Draw the most stable chair conformation of β-D-galactopyranose.

Now that we have described the structures of monosaccharides, let us examine some of their common reactions.

16.8 Esters and Ethers from Monosaccharides

Monosaccharides contain hydroxyl groups. It is not surprising, then, that they undergo reactions typical of alcohols. For example, they can be converted to esters by reaction with acid halides or anhydrides. The conversion of β-D-glucose to its pentaacetate by reaction with excess acetic anhydride is typical; all five hydroxyl groups, including the hydroxyl at anomeric C-1, are esterified. (To clarify the structure, the ring H's are omitted.)

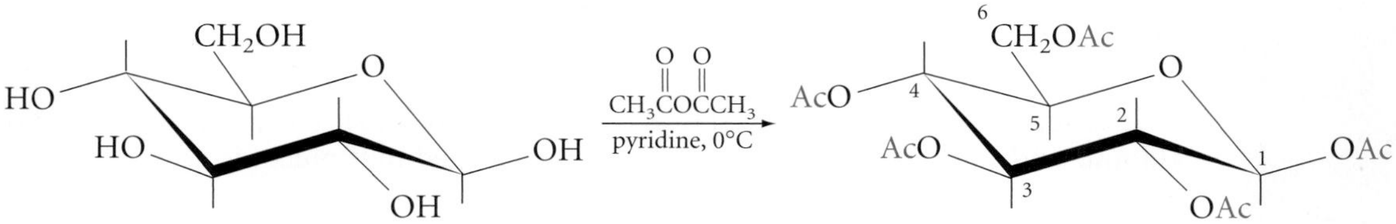

β-D-glucopyranose β-D-glucopyranose pentaacetate   (16.7)

$$Ac = CH_3\overset{\overset{\displaystyle O}{\|}}{C}-$$

The hydroxyl groups can also be converted to ethers by treatment with an alkyl halide and a base (the Williamson synthesis, Sec. 8.5). Because sugars are sensitive to strong bases, the mild base silver oxide is preferred.

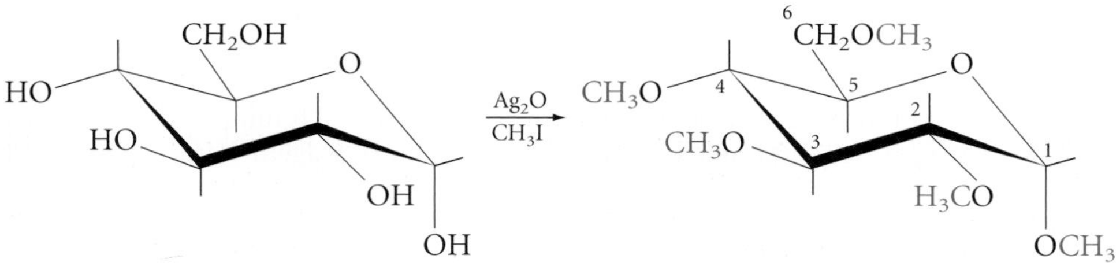

α-D-glucopyranose α-D-glucopyranose pentamethyl ether   (16.8)

Whereas sugars tend to be soluble in water and insoluble in organic solvents, the reverse is true for their esters and ethers. This often facilitates their purification and manipulation with organic reagents.

16.9 Reduction of Monosaccharides

When the carbonyl group of an aldose or ketose is reduced, the product is an alditol, an acyclic polyol.

The carbonyl group of aldoses and ketoses can be reduced by various reagents. The products are **polyols,** called **alditols.** For example, catalytic hydrogenation or reduction with sodium borohydride ($NaBH_4$) converts D-glucose to D-glucitol (also called sorbitol; review Sec. 9.12).

$$
\begin{array}{ccc}
\text{CH}{=}\text{O} & & \text{CH}_2\text{OH} \\
\text{H}\!-\!\text{OH} & & \text{H}\!-\!\text{OH} \\
\text{HO}\!-\!\text{H} & \xrightarrow[\text{or NaBH}_4]{\text{H}_2,\ \text{catalyst}} & \text{HO}\!-\!\text{H} \\
\text{H}\!-\!\text{OH} & & \text{H}\!-\!\text{OH} \\
\text{H}\!-\!\text{OH} & & \text{H}\!-\!\text{OH} \\
\text{CH}_2\text{OH} & & \text{CH}_2\text{OH}
\end{array}
$$

(16.9)

D-glucose (cyclic) ⇌ D-glucose (acyclic) → D-glucitol (sorbitol)

Reaction occurs by reduction of the small amount of aldehyde in equilibrium with the cyclic hemiacetal. As that aldehyde is reduced, the equilibrium shifts to the right, so that eventually all of the sugar is converted. Sorbitol is used commercially as a sweetener and sugar substitute.

PROBLEM 16.9   D-Mannitol, which occurs naturally in olives, onions, and mushrooms, can be made by $NaBH_4$ reduction of D-mannose. Draw its structure.

When the aldehyde group of an aldose is oxidized, the product is an aldonic acid.

16.10 Oxidation of Monosaccharides

Although aldoses exist primarily in cyclic hemiacetal forms, these structures are in equilibrium with a small but finite amount of the open-chain aldehyde. These aldehyde groups can be easily oxidized to acids (review Sec. 9.13). The products are called **aldonic acids.** For example, D-glucose is easily oxidized to D-gluconic acid.

$$
\begin{array}{ccc}
\text{CH}{=}\text{O} & & \text{COOH} \\
\text{H}\!-\!\text{OH} & & \text{H}\!-\!\text{OH} \\
\text{HO}\!-\!\text{H} & \xrightarrow[\text{Ag}^+\ \text{or Cu}^{2+}]{\text{Br}_2,\ \text{H}_2\text{O}} & \text{HO}\!-\!\text{H} \\
\text{H}\!-\!\text{OH} & & \text{H}\!-\!\text{OH} \\
\text{H}\!-\!\text{OH} & & \text{H}\!-\!\text{OH} \\
\text{CH}_2\text{OH} & & \text{CH}_2\text{OH}
\end{array}
$$

(16.10)

D-glucose → D-gluconic acid

Benedict's reagent (blue) reacts with an aldose to give Cu_2O (red precipitate).

The oxidation of aldoses is so easy that they react with such mild oxidizing agents as Tollens' reagent (Ag^+ in aqueous ammonia), Fehling's reagent (Cu^{2+} complexed with tartrate ion), or Benedict's reagent (Cu^{2+} complexed with citrate ion). With Tollens' reagent, they give a silver mirror test (Sec. 9.13), and with the copper reagents, the blue solution gives a red precipitate of cuprous oxide, Cu_2O. A carbohydrate that reacts with Ag^+ or Cu^{2+} is called a **reducing sugar** because reduction of the metal accompanies oxidation of the aldehyde group. These reagents are used in laboratory tests for this property.

An aldose that reduces Ag^+ or Cu^{2+} and is itself oxidized is called a **reducing sugar.**

$$RCH{=}O + 2\,Cu^{2+} + 5\,OH^- \longrightarrow RCO^- + Cu_2O + 3\,H_2O \qquad (16.11)$$

$$\underset{\text{solution}}{\underset{\text{blue}}{}} \qquad\qquad\qquad \underset{\text{precipitate}}{\underset{\text{red}}{}}$$

PROBLEM 16.10 Write an equation for the reaction of D-mannose with Fehling's reagent (Cu^{2+}) to give D-mannonic acid.

Stronger oxidizing agents, such as aqueous nitric acid, oxidize the aldehyde group *and* the primary alcohol group, producing dicarboxylic acids called **aldaric acids.** For example, D-glucose gives D-glucaric acid.

Aldoses are oxidized by aqueous nitric acid to dicarboxylic acids called **aldaric acids.**

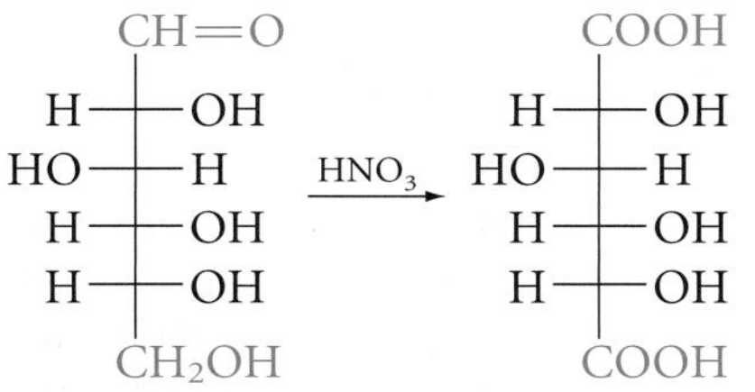

$$(16.12)$$

D-glucose $\xrightarrow{\;HNO_3\;}$ D-glucaric acid

PROBLEM 16.11 Write the structure of D-mannaric acid.

16.11 Formation of Glycosides from Monosaccharides

Because monosaccharides exist as cyclic hemiacetals, they can react with one equivalent of an alcohol to form acetals. An example is the reaction of β-D-glucose with methanol.

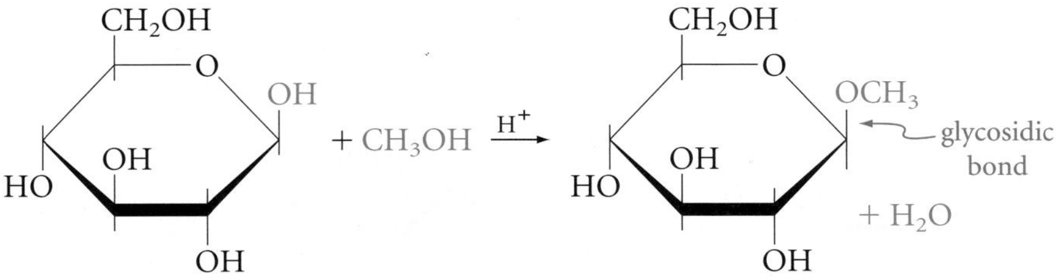

$$(16.13)$$

β-D-glucopyranose $+\ CH_3OH \xrightarrow{\;H^+\;}$ methyl β-D-glucopyranoside (mp 115–116°C) $+\ H_2O$

In a **glycoside,** the anomeric —OH group is replaced by an —OR group. The bond from the anomeric carbon to the —OR group is called the **glycosidic bond.**

Note that *only the —OH on the anomeric carbon is replaced by an —OR group.* Such acetals are called **glycosides,** and the bond from the anomeric carbon to the OR group is called the **glycosidic bond.** Glycosides are named from the corresponding monosaccharide by changing the -*e* ending to -*ide.* Thus, glucose gives glucosides, mannose gives mannosides, and so on.

EXAMPLE 16.6

Write a Haworth formula for ethyl α-D-mannoside.

Solution

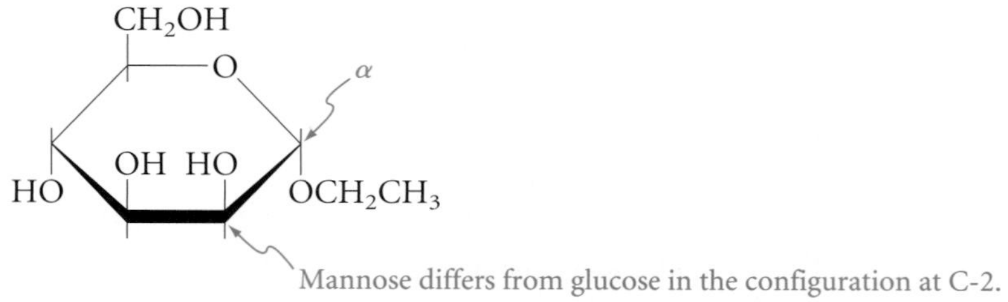

Mannose differs from glucose in the configuration at C-2.

PROBLEM 16.12 Write an equation for the acid-catalyzed reaction of β-D-galactose with methanol.

The mechanism of glycoside formation is the same as that described in eq. 9.13 of Sec. 9.7. The acid catalyst can protonate any of the six oxygen atoms, since each has unshared electron pairs and is basic. However, *only protonation of the hydroxyl oxygen at C-1 leads, after water loss, to a resonance-stabilized carbocation.* In the final step, methanol can attack from either "face" of the six-membered ring, to give either the β-glycoside as shown or the α-glycoside.

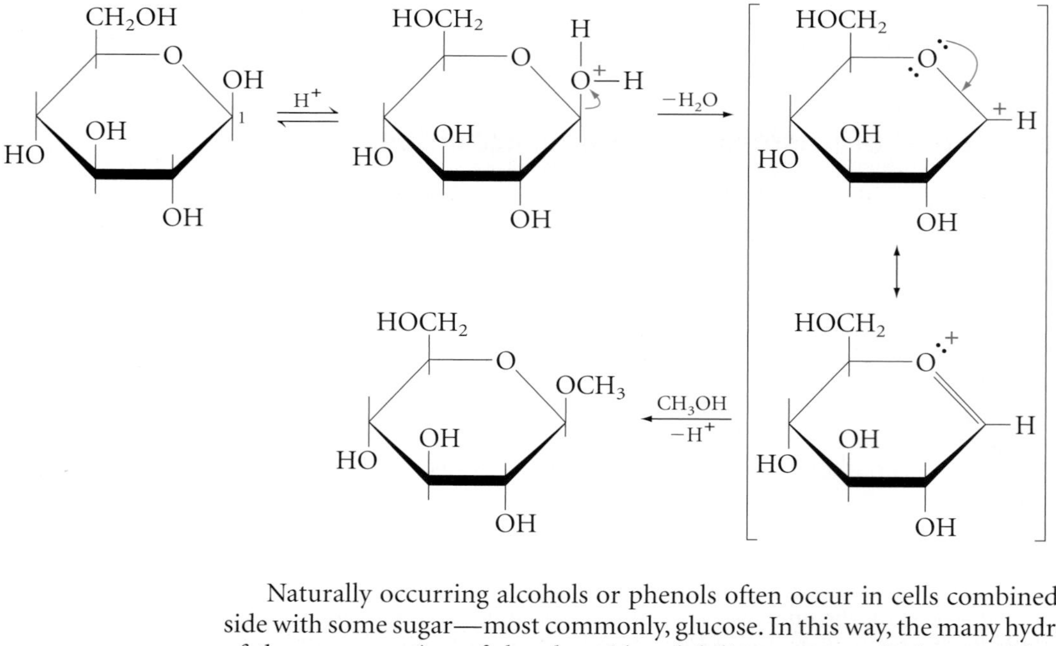

(16.14)

Naturally occurring alcohols or phenols often occur in cells combined as a glycoside with some sugar—most commonly, glucose. In this way, the many hydroxyl groups of the sugar portion of the glycoside solubilize compounds that would otherwise be insoluble in cellular protoplasm. An example is the bitter-tasting glucoside **salicin,** which occurs in willow bark and whose fever-reducing power was known to the ancients.

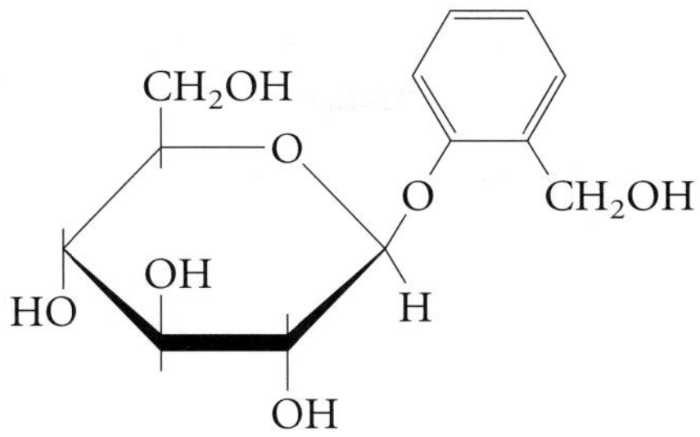

salicin
(the β-D-glucoside of salicyl alcohol)

The glycosidic bond is the key to understanding the structure of oligosaccharides and polysaccharides, as we will see in the following sections.

16.12 Disaccharides

The most common oligosaccharides are **disaccharides.** *In a disaccharide, two monosaccharides are linked by a glycosidic bond between the anomeric carbon of one monosaccharide unit and a hydroxyl group on the other unit.* In this section, we will describe the structure and properties of four important disaccharides.

A **disaccharide** consists of two monosaccharides linked by a glycosidic bond between the anomeric carbon of one unit and a hydroxyl group on the other unit.

16.12.a Maltose

Maltose is the disaccharide obtained by the partial hydrolysis of starch. Further hydrolysis of maltose gives only D-glucose (eq. 16.2). Maltose must, therefore, consist of two linked glucose units. It turns out that the anomeric carbon of the left unit is linked to the C-4 hydroxyl group of the unit at the right as an acetal (glycoside). The configuration at the anomeric carbon of the left unit is α. In the crystalline form, the anomeric carbon of the right unit has the β configuration. Both units are pyranoses, and the right-hand unit fills the same role as the methanol in eq. 16.13.

Some common disaccharides are **maltose, cellobiose, lactose,** and **sucrose.**

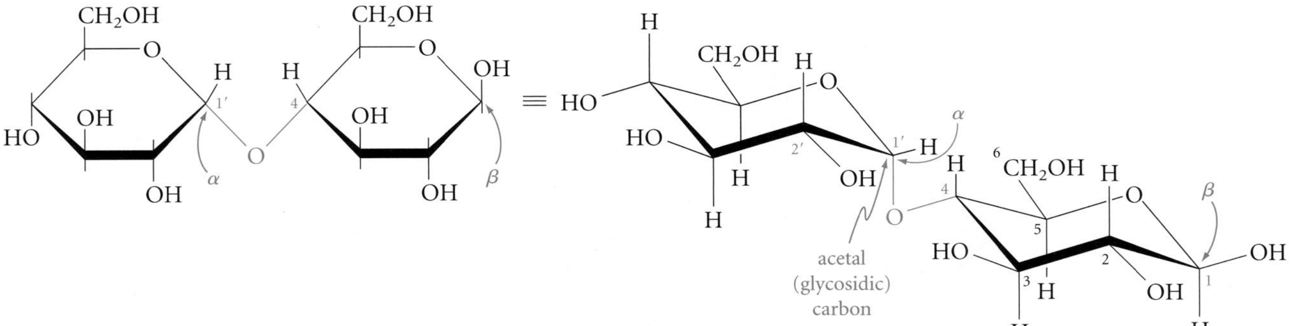

maltose
4-O-(α-D-glucopyranosyl)-β-D-glucopyranose

The systematic name for maltose, shown beneath the common name, describes the structure fully, including the name of each unit (D-glucose), the ring sizes (pyranose), the configuration at each anomeric carbon (α or β), and the location of the hydroxyl group involved in the glycosidic link (4-O).

The anomeric carbon of the right glucose unit in maltose is a hemiacetal. Naturally, when maltose is in solution, this hemiacetal function will be in equilibrium with the open-chain aldehyde form. Maltose therefore gives a positive Tollens' test and other reactions similar to those of the anomeric carbon in glucose.

PROBLEM 16.13   When crystalline maltose is dissolved in water, the initial specific rotation changes and gradually reaches an equilibrium value. Explain.

16.12.b Cellobiose

Cellobiose is the disaccharide obtained by the partial hydrolysis of cellulose. Further hydrolysis of cellobiose gives only D-glucose. Cellobiose must therefore be an isomer of maltose. In fact, *cellobiose differs from maltose* only *in having the β configuration at C-1 of the left glucose unit.* Otherwise, all other structural features are identical, including a link from C-1 of the left unit to the hydroxyl group at C-4 in the right unit.

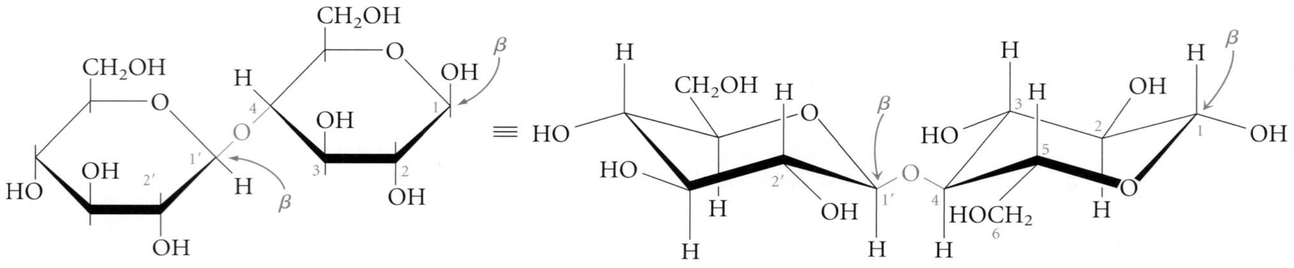

cellobiose
4-O-(β-D-glucopyranosyl)-β-D-glucopyranose

Note that, in the conformational formula for cellobiose, one ring oxygen is drawn to the "rear" and one to the "front" of the molecule. This is the way the rings exist in the cellulose chain.

16.12.c Lactose

Lactose is the major sugar in human and cow's milk (4 to 8% lactose). Hydrolysis of lactose gives equimolar amounts of D-galactose and D-glucose. The anomeric carbon of the galactose unit has the β configuration at C-1 and is linked to the hydroxyl group at C-4 of the glucose unit. The crystalline anomer, with the α configuration at the glucose unit, is made commercially from cheese whey.

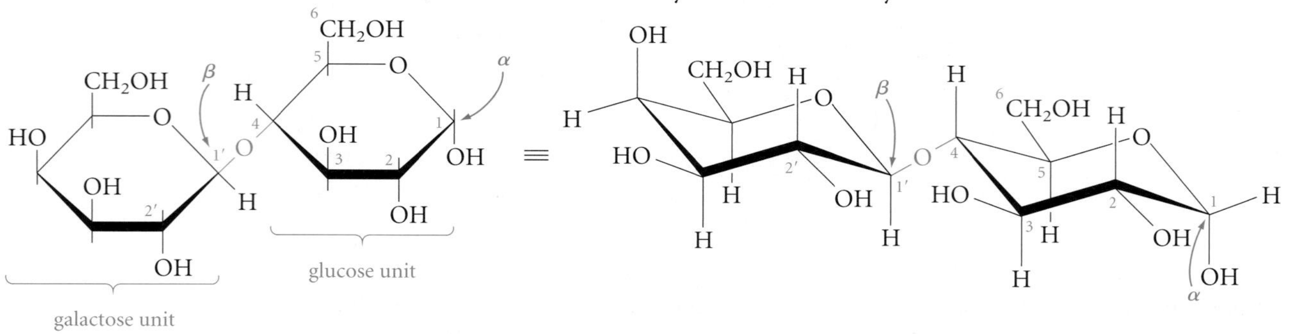

lactose
4-O-(β-D-galactopyranosyl)-α-D-glucopyranose

> **PROBLEM 16.14** Will lactose give a positive Fehling's test? Will it mutarotate?

Some human infants are born with a disease called *galactosemia.* They lack the enzyme that isomerizes galactose to glucose and therefore cannot digest milk. If milk is excluded from such infants' diets, the disease symptoms caused by accumulation of galactose can be avoided.

16.12.d Sucrose

The most important commercial disaccharide is **sucrose,** ordinary table sugar. More than 100 million tons are produced annually worldwide. Sucrose occurs in all photosynthetic plants, where it functions as an energy source. It is obtained commercially from sugar cane and sugar beets, in which it constitutes 14 to 20% of the plant juices.

One of the major engineering advances of the industrial revolution was developed to reduce the cost and labor associated with isolating sucrose from sugar cane and sugar beets. Norbert Rillieux, a free African American living in Louisiana in pre–Civil War days, invented the "triple-effect evaporator" to remove water from the juices of sugar cane and sugar beets in 1844. His invention modernized the sugar industry and versions of his equipment are still used today wherever large amounts of liquid must be quickly evaporated.*

Hydrolysis of sucrose gives equimolar amounts of D-glucose and the ketose D-fructose. *Sucrose differs from the other disaccharides we have discussed in that the anomeric carbons of both units are involved in the glycosidic link.* That is, C-1 of the glucose unit is linked, via oxygen, to C-2 of the fructose unit. A further difference is that the fructose unit is in the furanose form.

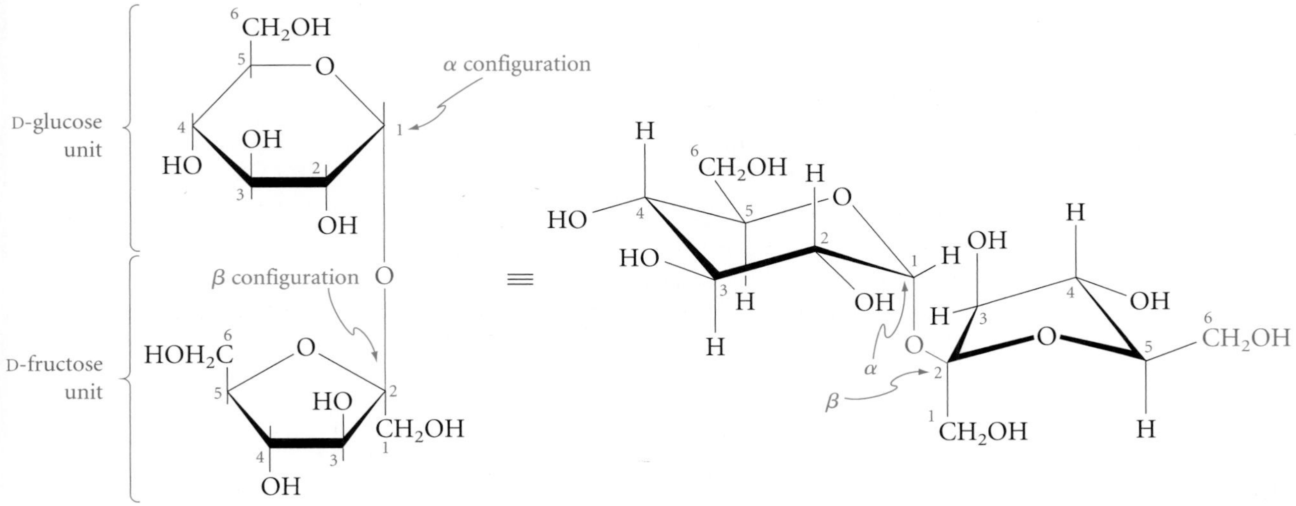

sucrose
α-D-glucopyranosyl-β-D-fructofuranoside
(or β-D-fructofuranosyl-α-D-glucopyranoside)

Since both anomeric carbons are linked in the glycosidic bond, neither monosaccharide unit has a hemiacetal group. Therefore, neither unit is in equilibrium with an acyclic form. Sucrose cannot mutarotate. And, because there is no free or potentially

---

*A marvelous account of this invention can be found in *Prometheans in the Lab* by Sharon B. McGrayne (McGraw-Hill, 2001).

A **WORD ABOUT** ...

Sweetness and Sweeteners

Sweetness is literally a matter of taste. Although individuals vary greatly in their sensory perceptions, it is possible to make some quantitative comparisons of sweetness. For example, we can take some standard sugar solution (say 10% sucrose in water) and compare its sweetness with that of solutions containing other sugars or sweetening agents. If a 1% solution of some compound tastes as sweet as the 10% sucrose solution, we can say that the compound is 10 times sweeter than sucrose.

D-Fructose is the sweetest of the simple sugars—almost twice as sweet as sucrose. D-Glucose is almost as sweet as sucrose. On the other hand, sugars like lactose and galactose have less than 1% of the sweetness of sucrose.

Several products containing saccharin and aspartame.

Many synthetic sweeteners are known, perhaps the most familiar being **saccharin.** It was discovered in 1879 in the laboratory of Professor Ira Remsen at the Johns Hopkins University. Its structure has no relation whatever to that of the saccharides, but saccharin is about 300 times sweeter than sucrose. For most tastes, 0.5 grain (0.03 g) of saccharin is equivalent in sweetness to a heaping teaspoon (10 g) of sucrose. Saccharin is made commercially from toluene, as shown in Figure 16.3.

Saccharin is very sweet yet has virtually no caloric content. It is useful as a sugar substitute for those who must restrict their sugar intake and also for those who wish to control their weight but still have a desire for sweets.

In 1981, **aspartame** became the first new sweetener to be approved by the U.S. Food and Drug Administration (FDA) in nearly 25 years. It is about 160 times sweeter than sucrose. Structurally, aspartame is the methyl ester of a dipeptide of two amino acids that occur naturally in proteins—aspartic acid and phenylalanine—and is sold under the trade name NutraSweet®. (Amino acids and peptides will be discussed in the next chapter.)

$$H_2N-CH-\overset{\overset{\displaystyle O}{\|}}{C}-NHCH-CH_2-\left\langle\begin{array}{c}\\\end{array}\right\rangle$$

$$\underset{CH_2COOH}{}\qquad\overset{CO_2CH_3}{}$$

the methyl ester of
N-L-α-aspartyl-L-phenylalanine
(aspartame)

Aspartame has about the same caloric content as sucrose, but because of its intense sweetness, much less is used, so its energy value becomes insignificant. For a nice review of the synthesis and properties of commercial, synthetic nonnutritive sweeteners, by David J. Ager and coworkers, see *Angew. Chem. Int. Ed.* **1998,** *37,* 1802–1817.

toluene $\xrightarrow{HOSO_2Cl}$ + the para isomer   SO$_2$Cl $\xrightarrow{NH_3}$ o-toluenesulfonamide SO$_2$NH$_2$

$\downarrow$ KMnO$_4$

sodium saccharin $\xleftarrow{NaOH}$ saccharin $\xleftarrow[-H_2O]{heat}$ CO$_2$H / SO$_2$NH$_2$

Figure 16.3
Synthesis of saccharin.

free aldehyde group, sucrose cannot reduce Tollens', Fehling's, or Benedict's reagent. Sucrose is therefore referred to as a *nonreducing sugar,* in contrast with the other disaccharides and monosaccharides we have discussed, all of which are reducing sugars.

PROBLEM 16.15   Although β-D-glucose is a reducing sugar, methyl β-D-glucopyranoside (eq. 16.13) is not. Explain.

Sucrose has an optical rotation of $[\alpha] = +66°$. When sucrose is hydrolyzed to an equimolar mixture of D-glucose and D-fructose, the optical rotation changes value and sign and becomes $[\alpha] = -20°$. This is because the equilibrium mixture of D-glucose anomers (α and β) has a rotation of +52°, but the mixture of fructose anomers has a strong negative rotation, $[\alpha] = -92°$. In the early days of carbohydrate chemistry, glucose was called **dextrose** (because it was dextrorotatory), and fructose was called **levulose** (because it was levorotatory). Because hydrolysis of sucrose inverts the sign of optical rotation (from + to −), enzymes that bring about sucrose hydrolysis are called **invertases,** and the resulting equimolar mixture of glucose and fructose is called **invert sugar.** A number of insects, including the honeybee, possess invertases. Honey is largely a mixture of D-glucose, D-fructose, and some unhydrolyzed sucrose. It also contains flavors from the particular flowers whose nectars are collected.

Hydrolysis of sucrose ($[\alpha] = +66°$), catalyzed by enzymes called **invertases,** produces a mixture of glucose and fructose called **invert sugar** ($[\alpha] = -20°$).

16.13   Polysaccharides

Polysaccharides contain many linked monosaccharides and vary in chain length and molecular weight. Most polysaccharides give a single monosaccharide on complete hydrolysis. The monosaccharide units may be linked linearly, or the chains may be branched. In this section, we will describe a few of the more important polysaccharides.

16.13.a   Starch and Glycogen

Starch is the energy-storing carbohydrate of plants. It is a major component of cereals, potatoes, corn, and rice. It is the form in which glucose is stored by plants for later use.

Starch is made up of glucose units joined mainly by 1,4-α-glycosidic bonds, although the chains may have a number of branches attached through 1,6-α-glycosidic bonds. Partial hydrolysis of starch gives maltose, and complete hydrolysis gives only D-glucose.

Starch can be separated by various techniques into two fractions: amylose and amylopectin. In **amylose,** which constitutes about 20% of starch, the glucose units (50 to 300) are in a continuous chain, with 1,4 linkages (Figure 16.4).

Amylopectin (Figure 16.5) is highly branched. Although each molecule may contain 300 to 5000 glucose units, chains with consecutive 1,4 links average only 25 to 30 units in length. These chains are connected at branch points by 1,6 linkages. Because of this highly branched structure, starch granules swell and eventually form colloidal systems in water.

Glycogen is the energy-storing carbohydrate of animals. Like starch, it is made of 1,4- and 1,6-linked glucose units. Glycogen has a higher molecular weight than starch (perhaps 100,000 glucose units) and its structure is even more branched than that of amylopectin, with a branch every 8 to 12 glucose units. Glycogen is produced from glucose that is absorbed from the intestines into the blood; transported to the liver, muscles, and elsewhere; and then polymerized enzymatically. Glycogen helps maintain the glucose balance in the body, by removing and storing excess glucose from ingested food and later supplying it to the blood when various cells need it for energy.

The polysaccharide **starch** contains glucose units joined by 1,4-α-glycosidic bonds. **Amylose** is an unbranched form of starch, whereas **amylopectin** is highly branched. **Glycogen,** found in animals, is more highly branched than amylopectin and has a higher molecular weight.

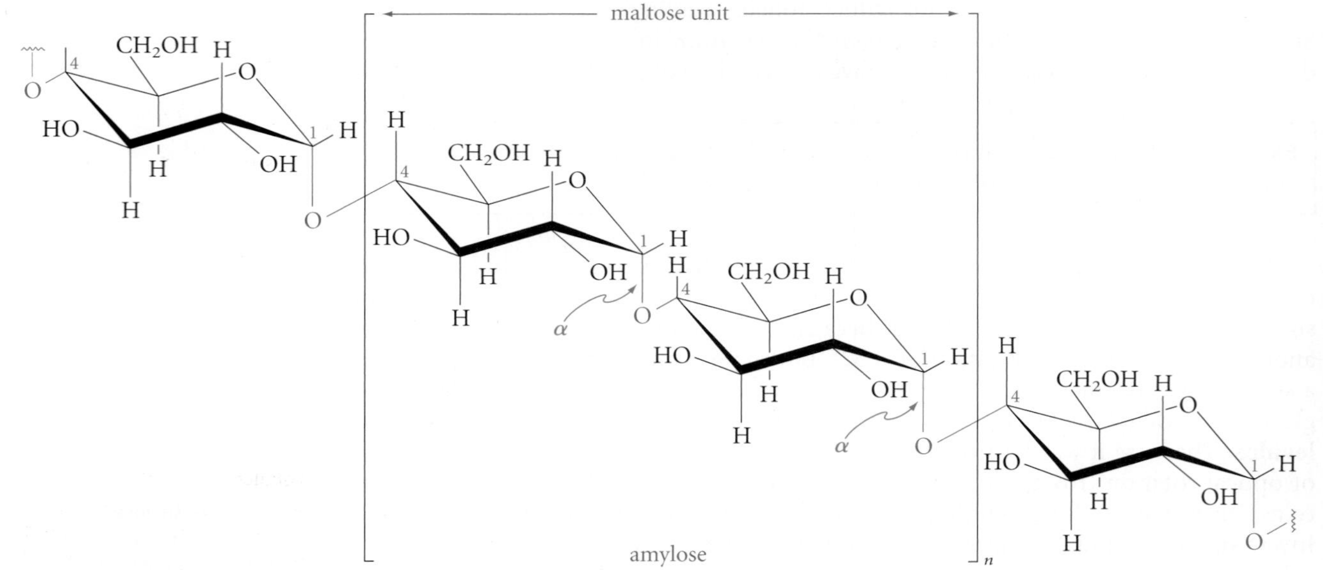

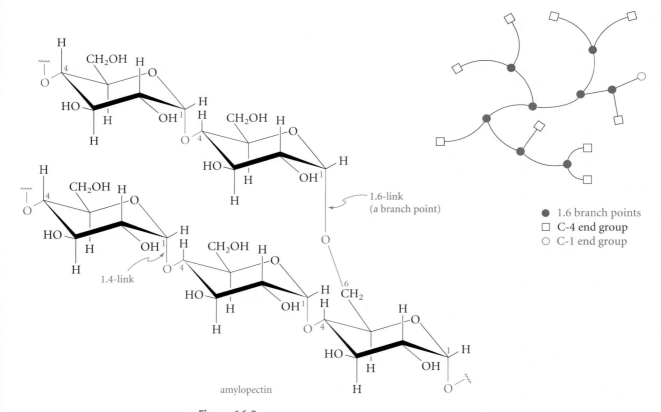

Figure 16.4
Structure of the amylose fraction of starch.

Figure 16.5
Structure of the amylopectin fraction of starch. Adapted from Peter M. Collins and Robert J. Ferrier, *Monosaccharides: Their Chemistry and Their Roles in Natural Products*, p. 491. Copyright © 1995 by John Wiley & Sons, Ltd. Reprinted with permission.

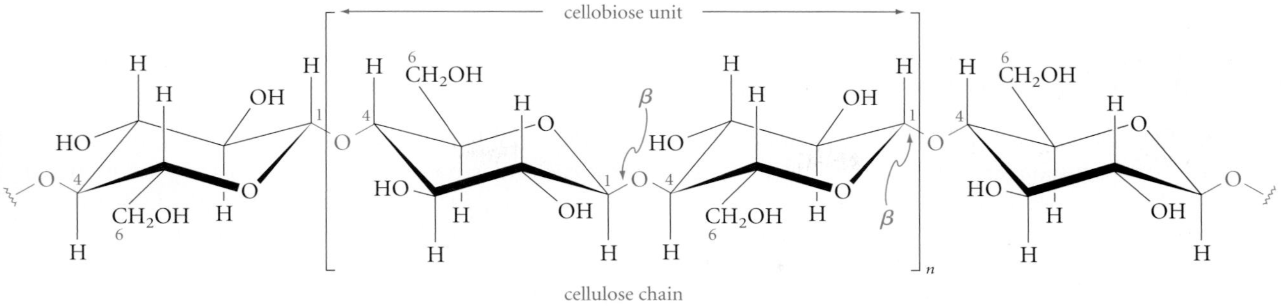

Figure 16.6
Partial structure of a cellulose molecule showing the β linkages of each glucose unit.

16.13.b Cellulose

Cellulose is an *unbranched* polymer of glucose joined by 1,4-β-glycosidic bonds. X-ray examination of cellulose shows that it consists of linear chains of cellobiose units, in which the ring oxygens alternate in "forward" and "backward" positions (Figure 16.6). These linear molecules, containing an average of 5000 glucose units, aggregate to give fibrils bound together by hydrogen bonds between hydroxyls on adjacent chains. Cellulose fibers having considerable physical strength are built up from these fibrils, wound spirally in opposite directions around a central axis. Wood, cotton, hemp, linen, straw, and corncobs are mainly cellulose.

Although humans and other animals can digest starch and glycogen, they cannot digest cellulose. This is a truly striking example of the specificity of biochemical reactions. *The only chemical difference between starch and cellulose is the stereochemistry of the glucosidic link*—more precisely, the stereochemistry at C-1 of each glucose unit. The human digestive system contains enzymes that can catalyze the hydrolysis of α-glucosidic bonds, but it lacks the enzymes necessary to hydrolyze β-glucosidic bonds. Many bacteria, however, do contain β-glucosidases and can hydrolyze cellulose. Termites, for example, have such bacteria in their intestines and thrive on wood (cellulose) as their main food. Ruminants (cud-chewing animals such as cows) can digest grasses and other forms of cellulose because they harbor the necessary microorganisms in their rumen.

Cellulose is the raw material for several commercially important derivatives. *Each glucose unit in cellulose contains three hydroxyl groups.* These hydroxyl groups can be modified by the usual reagents that react with alcohols. For example, cellulose reacts with acetic anhydride to give **cellulose acetate.**

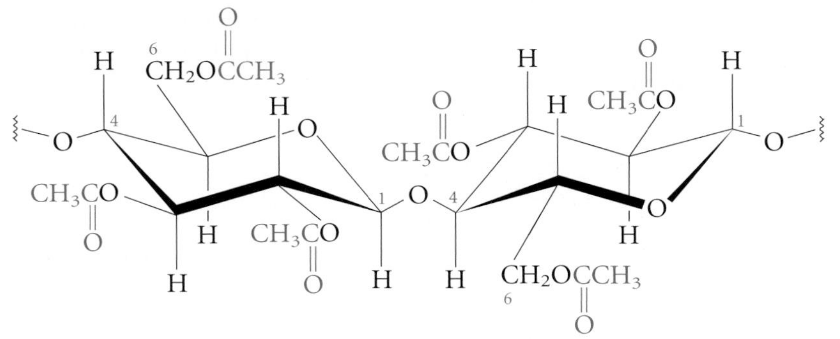

Cellulose is an unbranched polymer of glucose joined by 1,4-β-glycosidic bonds. In **cellulose acetate,** the free hydroxyl groups are acetylated; in **cellulose nitrate,** they are nitrated. **Guncotton,** a highly nitrated cellulose, is an explosive.

segment of a cellulose acetate molecule

Fat Substitutes from Carbohydrates

Like it or not, many of us eat numerous food additives as a consequence of our diets. We are all familiar with artificial sweeteners (see "A Word About Sweetness and Sweeteners" earlier in this chapter), which satisfy our desire for the taste of sugar without providing the caloric intake and weight problems associated with sweets. Fats also taste good and are an even better source of energy than carbohydrates. Every gram of fat we eat provides us with twice the energy (in calories) of a gram of carbohydrate. Therefore it is not surprising that considerable effort has gone into developing artificial fats.

Several products containing Olestra, Oatrim, and Z-Trim.

One family of compounds that have been used as artificial fats is polyesters of sucrose. They are prepared by esterifying sucrose (table sugar) with derivatives of fatty acids obtained from vegetable oils such as cottonseed or soybean oil. Esters of this type are marketed under the name **Olestra.** Olestra has been used to replace fat in a number of snacks and can replace cooking oil in preparing fried foods such as potato chips. Olestra has a structure similar to that of a normal fat, being a fatty acid ester of a polyol. It has the physical and textural properties of fats, but because it is so large it is not hydrolyzed by the esterases that normally begin the process of fat storage and metabolism. It simply passes through the digestive tract and is excreted without being used as an energy source.

ester linkages

$$R = C(CH_2)_nCH_3$$

triglycerides
R = long chain saturated and unsaturated alkyl

Olestra

Fats are triesters of the polyol glycerol. They are good at delivering flavors and provide food with a pleasing texture. When we eat fats, they are metabolized when needed as an energy source, or stored for later use when carbohydrates are no longer available as an energy source. Both metabolism and storage of fats involve hydrolysis of the ester linkages by esterases (see "A Word About Prostaglandins, Aspirin, and Pain" in Chapter 15) in the gut to give glycerol and fatty acids.

Other carbohydrate-derived fat substitutes have been developed from oats. Slightly less than 50% of the weight of oats consists of fiber, cellulose-like polysaccharides not easily digested by humans. Some of this fiber is water soluble and some of it is not. Both kinds of fibers have been processed to give materials (Oatrim and Z-Trim) that mimic the texture of fat but have very little caloric value. These products are being used in the preparation of some low-fat food products such as cheese and baked goods. For an interesting overview of fat substitutes see "Fake Fats in Real Food" by Elisabeth M. Kirschner in *Chemical and Engineering News,* **1997,** April 21, page 19.

Cellulose with about 97% of the hydroxyl groups acetylated is used to make acetate rayon.

Cellulose nitrate is another useful cellulose derivative. Like glycerol, cellulose can be converted with nitric acid to a nitrate ester (compare eq. 7.41). The number of hydroxyl groups nitrated per glucose unit determines the properties of the product.

Guncotton, a highly nitrated cellulose, is an efficient explosive used in smokeless powders.

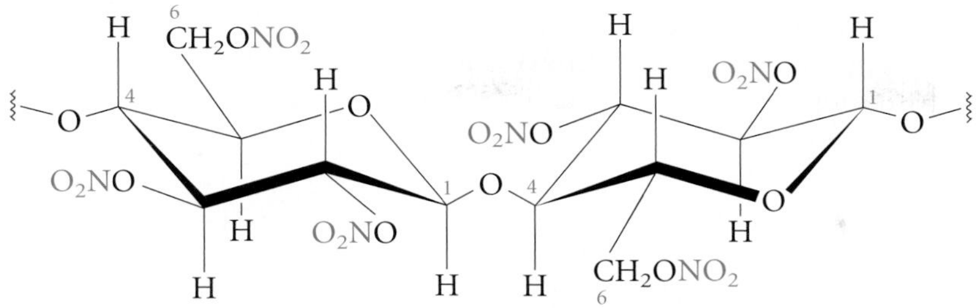

segment of a cellulose nitrate molecule

16.13.c Other Polysaccharides

Chitin is a nitrogen-containing polysaccharide that forms the shells of crustaceans and the exoskeletons of insects. It is similar to cellulose, except that the hydroxyl group at C-2 of each glucose unit is replaced by an acetylamino group, CH_3CONH-.

Chitin is found in crustacean shells and insect exoskeletons; pectins, found in fruits, are used in making jellies.

Pectins, which are obtained from fruits and berries, are polysaccharides used in making jellies. They are linear polymers of D-galacturonic acid, linked with 1,4-α-glycosidic bonds. D-Galacturonic acid has the same structure as D-galactose, except that the C-6 primary alcohol group is replaced by a carboxyl group.

Numerous other polysaccharides are known, such as gum arabic and other gums and mucilages, chondroitin sulfate (found in cartilage), the blood anticoagulant heparin (found in the liver and heart), and the dextrans (used as blood plasma substitutes).

Some saccharides have structures that differ somewhat from the usual polyhydroxyaldehyde or polyhydroxyketone pattern. In the final sections of this chapter, we will describe a few such modified saccharides that are important in nature.

16.14 Sugar Phosphates

Phosphate esters of monosaccharides are found in all living cells, where they are intermediates in carbohydrate metabolism. Some common **sugar phosphates** are the following:

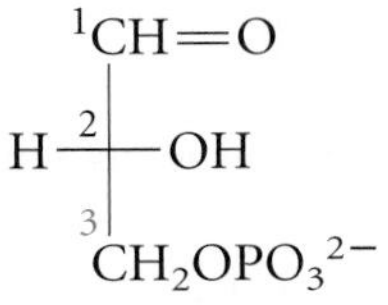

D-glyceraldehyde-3-phosphate 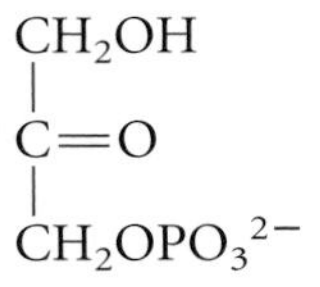   dihydroxyacetone phosphate 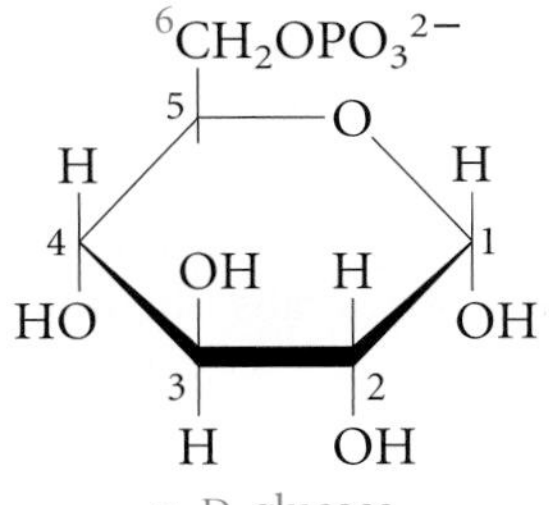 α-D-glucose-6-phosphate 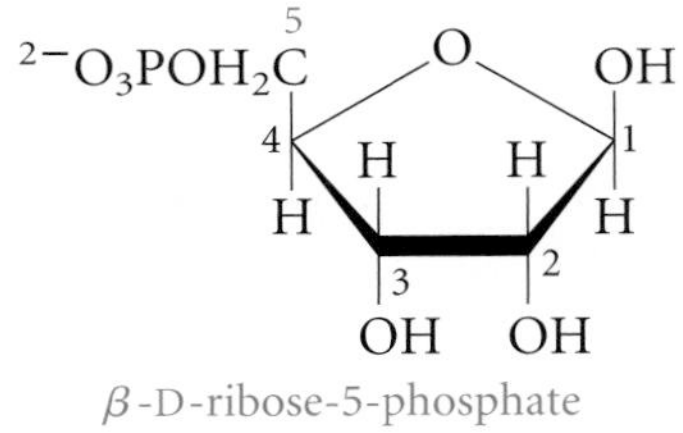 β-D-ribose-5-phosphate

Phosphates of the five-carbon sugar ribose and its 2-deoxy analog are important in nucleic acid structures (DNA, RNA) and in other key biological compounds (Sec. 18.12).

16.15 Deoxy Sugars

In **deoxy sugars,** one or more of the hydroxyl groups is replaced by a hydrogen atom. The most important example is **2-deoxyribose,** the sugar component of DNA. It lacks the hydroxyl group at C-2 and occurs in DNA in the furanose form.

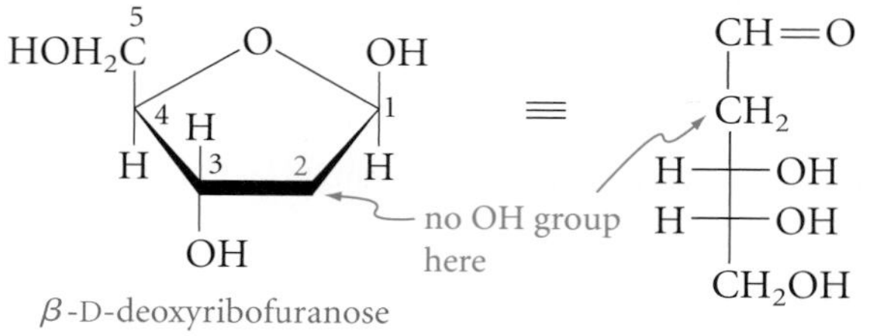

β-D-deoxyribofuranose
(the sugar of DNA)

16.16 Amino Sugars

In **amino sugars,** an —OH group is replaced by an —NH$_2$ group.

In **amino sugars,** one of the sugar hydroxyl groups is replaced by an amino group. Usually the —NH$_2$ group is also acetylated. **D-Glucosamine** is one of the more abundant amino sugars.

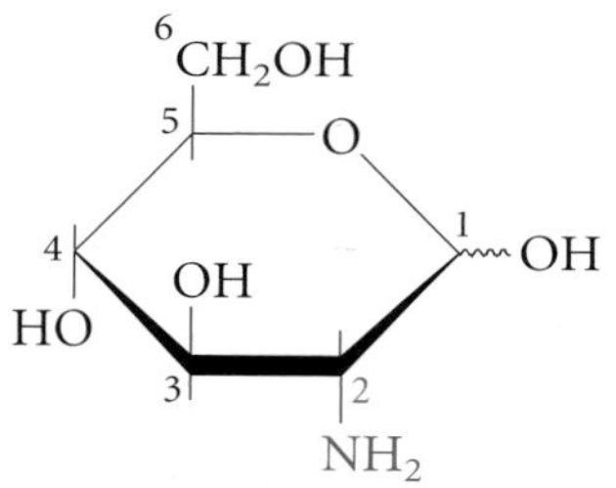

D-glucosamine
α mp 88°C
β [mp 110°C (decomposes)]

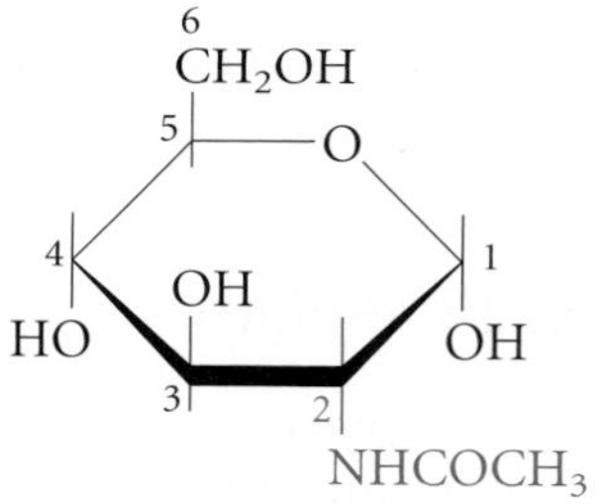

N-acetyl-α-D-glucosamine
[mp 211°C (decomposes)]

In its *N*-acetyl form, β-D-glucosamine is the monosaccharide unit of chitin, which forms the shells of lobsters, crabs, shrimp, and other shellfish.

16.17 Ascorbic Acid (Vitamin C)

L-**Ascorbic acid (vitamin C)** has a five-membered lactone ring containing an **enediol.** The hydroxyl proton at C-3 is acidic.

L-**Ascorbic acid (vitamin C)** resembles a monosaccharide, but its structure has several unusual features. The compound has a five-membered unsaturated lactone ring (review Sec. 10.12) with two hydroxyl groups attached to the doubly bonded carbons. This **enediol** structure is relatively uncommon.

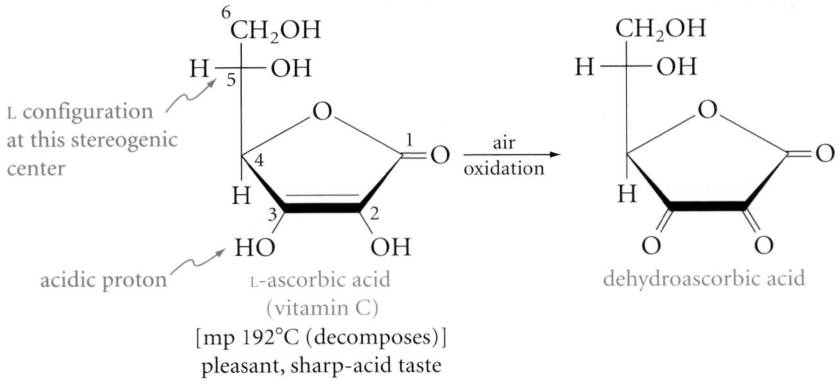

L configuration at this stereogenic center

acidic proton

L-ascorbic acid
(vitamin C)
[mp 192°C (decomposes)]
pleasant, sharp-acid taste

air
oxidation

dehydroascorbic acid

As a consequence of this structural feature, ascorbic acid is easily oxidized to dehydroascorbic acid. Both forms are biologically effective as a vitamin.

There is no carboxyl group in ascorbic acid, but it is nevertheless an acid with a pK_a of 4.17. The proton of the hydroxyl group at C-3 is acidic, because the anion that results from its loss is resonance stabilized and similar to a carboxylate anion.

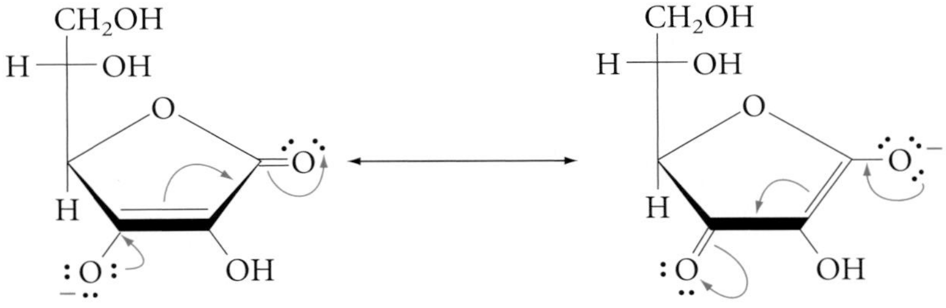

resonance stabilized ascorbate anion

Humans, monkeys, guinea pigs, and a few other vertebrates lack an enzyme that is essential for the biosynthesis of ascorbic acid from D-glucose. Hence ascorbic acid must be included in the diet of humans and these other species. Ascorbic acid is abundant in citrus fruits and tomatoes. Its lack in the diet causes scurvy, a disease that results in weak blood vessels, hemorrhaging, loosening of teeth, lack of ability to heal wounds, and eventual death. Ascorbic acid is needed for collagen synthesis (collagen is the structural protein of skin, connective tissue, tendon, cartilage, and bone). In the eighteenth century, British sailors were required to eat fresh limes (a vitamin C source) to prevent outbreaks of the dreaded scurvy; hence their nickname "limeys."

REACTION SUMMARY

1. Reactions of Monosaccharides

a. Mutarotation (Sec. 16.5)

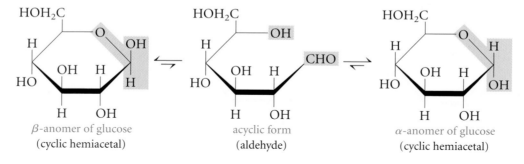

β-anomer of glucose (cyclic hemiacetal) ⇌ acyclic form (aldehyde) ⇌ α-anomer of glucose (cyclic hemiacetal)

b. Esterification (Sec. 16.8)

$\xrightarrow{\text{Ac}_2\text{O}}$ $Ac = CH_3\overset{\overset{O}{\|}}{C}-$

c. Etherification (Sec. 16.8)

$\xrightarrow[\text{or } CH_3I,\ Ag_2O]{\text{NaOH, } (CH_3)_2SO_4}$

d. Reduction (Sec. 16.9)

$$
\begin{array}{c}
CH{=}O \\
| \\
(CHOH)_n \\
| \\
CH_2OH \\
\text{aldose}
\end{array}
\xrightarrow[\substack{\text{or}\\ NaBH_4}]{H_2,\ \text{catalyst}}
\begin{array}{c}
CH_2OH \\
| \\
(CHOH)_n \\
| \\
CH_2OH \\
\text{alditol}
\end{array}
$$

e. Oxidation (Sec. 16.10)

$$
\begin{array}{c}
CO_2H \\
| \\
(CHOH)_n \\
| \\
CO_2H \\
\text{aldaric acid}
\end{array}
\xleftarrow{HNO_3}
\begin{array}{c}
CH{=}O \\
| \\
(CHOH)_n \\
| \\
CH_2OH \\
\text{aldose}
\end{array}
\xrightarrow[\substack{\text{or}\\ Ag^+ \text{ or } Cu^{2+}}]{Br_2,\ H_2O}
\begin{array}{c}
CO_2H \\
| \\
(CHOH)_n \\
| \\
CH_2OH \\
\text{aldonic acid}
\end{array}
$$

f. Preparation of Glycosides (Sec. 16.11)

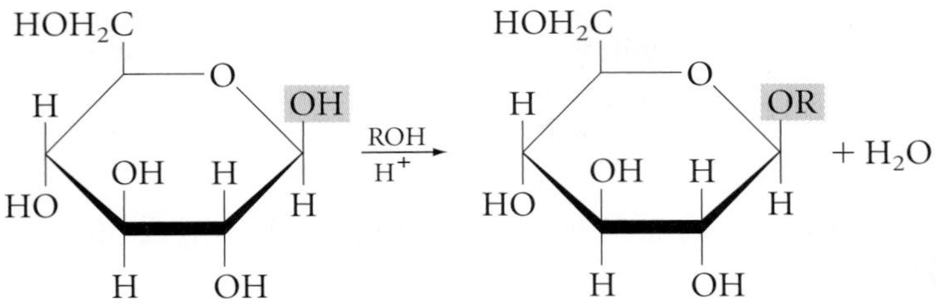

glycoside formation

---

2. Hydrolysis of Polysaccharides (Sec. 16.1)

polysaccharide $\xrightarrow{\text{H}_3\text{O}^+}$ oligosaccharide $\xrightarrow{\text{H}_3\text{O}^+}$ monosaccharide

---

ADDITIONAL PROBLEMS

Nomenclature and Structure of Carbohydrates

16.16 Define each of the following, and give the structural formula of one example:

  a. aldohexose
  d. disaccharide
  g. pyranose

  b. ketopentose
  e. polysaccharide
  h. glycoside

  c. monosaccharide
  f. furanose
  i. anomeric carbon

16.17 Explain, using formulas, the difference between a D-sugar and an L-sugar.

16.18 Three of the four hydroxyl groups at the stereogenic centers in the Fischer projection of D-talose (Figure 16.1) are on the left, yet it is called a D-sugar. Explain.

16.19 What term would you use to describe the stereochemical relationship between D-xylose and D-lyxose?

Monosaccharides: Fischer and Haworth Projections

16.20 Construct an array analogous to Figure 16.1 for the D-ketoses as high as the hexoses. Dihydroxyacetone should be at the head, in place of glyceraldehyde.

16.21 Using Figure 16.1 if necessary, write a Fischer projection formula and a Haworth projection formula for

  a. methyl α-D-glucopyranoside.
  c. β-D-arabinofuranose.

  b. α-D-gulopyranose.
  d. methyl α-L-glucopyranoside.

16.22 Draw the Fischer projection formula for

  a. L-(−)-mannose.

  b. L-(+)-fructose.

16.23 At equilibrium in aqueous solution, D-ribose exists as a mixture containing 20% α-pyranose, 56% β-pyranose, 6% α-furanose, and 18% β-furanose forms. Draw Haworth formulas for each of these forms.

16.24 Write Fischer, Haworth, and conformational structures for β-D-allose.

16.25 D-Threose can exist in a furanose form but *not* in a pyranose form. Explain. Draw the β-furanose structure.

16.26 Draw the Fischer and Newman projection formulas for L-erythrose.

16.27 L-Fucose is a component of bacterial cell walls. It is also called 6-deoxy-L-galactose. Using the description in Sec. 16.15, write its Fischer projection formula.

Anomers and Mutarotation

16.28 The solubilities of α- and β-D-glucose in water at 25°C are 82 and 178 g/100 mL, respectively. Why are their solubilities not identical?

16.29 The specific rotations of pure α- and β-D-fructofuranose are +21° and −133°, respectively. Solutions of each isomer mutarotate to an equilibrium specific rotation of −92°. Assuming that no other forms are present, calculate the equilibrium concentrations of the two forms.

16.30 Starting with β-D-glucose and using acid (H^+) as a catalyst, write out all the steps in the mechanism for the mutarotation process.

16.31 Lactose exists in α and β forms, with specific rotations of +92.6° and +34°, respectively.
  a. Draw their structures.
  b. Solutions of each isomer mutarotate to an equilibrium value of +52°. What is the percentage of each isomer at equilibrium?

Reactions of Monosaccharides

16.32 Oxidation of either D-erythrose or D-threose with nitric acid gives tartaric acid. In one case, the tartaric acid is optically active; in the other, it is optically inactive. How can these facts be used to assign stereochemical structures to erythrose and threose?

16.33 Write a structure for
  a. D-galactonic acid.   **b.** D-galactaric acid.

16.34 Using complete structures, write out the reaction of D-galactose with
  a. bromine water.   **b.** nitric acid.
  c. sodium borohydride.   **d.** acetic anhydride.

16.35 Reduction of D-fructose with $NaBH_4$ gives a mixture of D-glucitol and D-mannitol. What does this result prove about the configurations of D-fructose, D-mannose, and D-glucose?

16.36 Although D-galactose contains five stereogenic centers (in its cyclic form) and is optically active, its oxidation with nitric acid gives an optically inactive dicarboxylic acid (called galactaric or mucic acid). What is the structure of this acid, and why is it optically inactive?

16.37 Write an equation for the acid-catalyzed hydrolysis of salicin (page 463). Notice that one of the products is structurally similar to aspirin (eq. 10.40), perhaps accounting for salicin's fever-reducing property.

16.38 Write a balanced equation for the reaction of D-(+)-glucose (use either an acyclic or a cyclic structure, whichever seems more appropriate) with each of the following:
  a. acetic anhydride (excess)   **b.** bromine water
  c. hydrogen, catalyst   **d.** hydroxylamine (to form an oxime)
  e. methanol, H^+ **f.** hydrogen cyanide (to form a cyanohydrin)
  g. Fehling's reagent

───────────────

 = concept connections

Disaccharides

16.39 Write equations that clearly show the mechanism for the acid-catalyzed hydrolysis of

    a. maltose to glucose.
    b. lactose to galactose and glucose.
    c. sucrose to fructose and glucose.

16.40 What would be the structures of the monosaccharide esters obtained from hydrolysis of Olestra ("A Word About Fat Substitutes from Carbohydrates," page 470) with an aqueous acid?

16.41 Write equations for the reaction of maltose with

    a. methanol and H^+.
    b. Tollens' reagent.
    c. bromine water.
    d. acetic anhydride.

16.42 Trehalose is a disaccharide that is the main carbohydrate component in the blood of insects. Its structure is

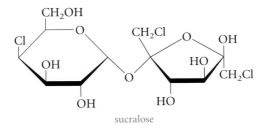

trehalose

    a. What are its hydrolysis products?
    b. Will trehalose give a positive or a negative test with Fehling's reagent? Explain.

16.43 Explain why sucrose is a nonreducing sugar but maltose is a reducing sugar.

16.44 Sucralose is a chlorinated derivative of sucrose that is about 600 times sweeter than its parent. It is hydrolyzed only slowly at the mildly acidic pH of soft drinks. Draw the structure of the hydrolysis products.

sucralose

16.45 An important portion of bacterial cell walls is a disaccharide in which C-1 of N-acetylglucosamine (Sec. 16.16) is linked to the C-4 oxygen of N-acetylmuramic acid by a β-glycosidic bond. Draw the structure of this disaccharide.

N-acetylmuramic acid

Polysaccharides

16.46 Using the descriptions in Sec. 16.13.c, write a formula for

    a. chitin.   **b.** pectin.

16.47 Hemicelluloses are noncellulose materials produced by plants and found in straw, wood, and other fibrous tissues. Xylans are the most abundant hemicelluloses. They consist of 1,4-β-linked D-xylopyranoses. Draw the structure for the repeating unit in xylans.

Other Sugars and Sweeteners

16.48 Write the main contributors to the resonance hybrid anion formed when ascorbic acid acts as an acid (loses the proton from the —OH group on C-3).

16.49 Inositols are hexahydroxycyclohexanes, with one hydroxyl group on each carbon atom of the ring. Although not strictly carbohydrates, they are obviously similar to pyranose sugars and do occur in nature. There are nine possible stereoisomers. Draw Haworth formulas for all possibilities (all are known), and tell which are chiral.

16.50 Aspartame has the S configuration at each of its two stereogenic centers (see "A Word About Sweetness and Sweeteners," page 466). Draw a structure of aspartame that shows its stereochemistry.

$$H-\underset{\underset{NH_2}{|}}{\overset{\overset{H}{|}}{C}}-COOH \qquad H_3C-\underset{\underset{NH_2}{|}}{\overset{\overset{H}{|}}{C}}-COOH$$

glycine   alanine

AMINO ACIDS, PEPTIDES, AND PROTEINS

Proteins are naturally occurring polymers composed of **amino acid** units joined one to another by amide (or peptide) bonds. Spider webs, animal hair and muscle, egg whites, and hemoglobin (the molecule that transports oxygen in the body to where it is needed) are all proteins. **Peptides** are oligomers of amino acids that play important roles in many biological processes. For example, the peptide hormone insulin controls our blood sugar levels, bradykinin controls our blood pressure, and oxytocin regulates uterine contraction and lactation. Thus, proteins, peptides, and amino acids are essential to the structure, function, and reproduction of living matter. In this chapter we will first discuss the structure and properties of amino acids, then the properties of peptides, and finally the structures of proteins.

17.1 Naturally Occurring Amino Acids

The amino acids obtained from protein hydrolysis are **α-amino acids.** That is, the amino group is on the α-carbon atom, the one adjacent to the carboxyl group.

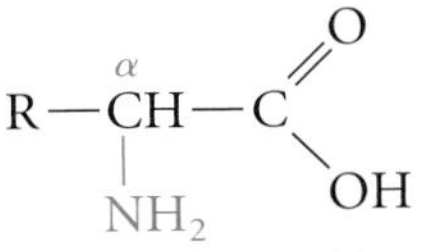

$$R-\underset{\underset{NH_2}{|}}{\overset{\alpha}{C}H}-C\overset{\displaystyle O}{\underset{\displaystyle OH}{\big<}}$$

an α-amino acid

With the exception of glycine, where R = H, α-amino acids have a stereogenic center at the α-carbon. All except glycine are therefore optically active. They have the L configuration relative to glyceraldehyde (Figure 17.1, page 480). Note that the Fischer convention, used with carbohydrates, is also applied to amino acids.

Table 17.1 lists the 20 α-amino acids commonly found in proteins. The amino acids are known by common names. Each also has a three-letter abbreviation

▲ Silk, produced by numerous insects and spiders, is the common name for β-keratin (Figure 17.10), a fibrous protein composed largely of the amino acids glycine and alanine.

Figure 17.1
Naturally occurring α-amino acids have the L configuration.

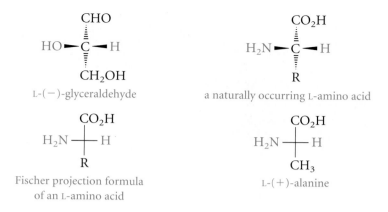

L-(−)-glyceraldehyde

a naturally occurring L-amino acid

Fischer projection formula
of an L-amino acid

L-(+)-alanine

Table 17.1	Names and formulas of the common amino acids		
Name	**Three-letter abbreviation (isoelectric point) one-letter abbreviation**	**Formula**	**R**
A. One amino group and one carboxyl group			
1. glycine	Gly (6.0) G	$H-CH-CO_2H$ $\ \ \ \ \ \ \ \ \mid$ $\ \ \ \ \ \ \ \ NH_2$	
2. alanine	Ala (6.0) A	$CH_3-CH-CO_2H$ $\ \ \ \ \ \ \ \ \ \mid$ $\ \ \ \ \ \ \ \ \ NH_2$	
3. valine	Val (6.0) V	$CH_3CH-CH-CO_2H$ $\ \ \ \ \ \mid \ \ \ \ \mid$ $\ \ \ \ CH_3 \ \ NH_2$	R is hydrogen or an alkyl group.
4. leucine	Leu (6.0) L	$CH_3CHCH_2-CH-CO_2H$ $\ \ \ \ \ \mid \ \ \ \ \ \ \ \ \mid$ $\ \ \ \ \ CH_3 \ \ \ \ \ NH_2$	
5. isoleucine	Ile (6.0) I	$CH_3CH_2CH-CH-CO_2H$ $\ \ \ \ \ \ \ \ \mid \ \ \ \mid$ $\ \ \ \ \ \ \ CH_3 \ NH_2$	
6. serine	Ser (5.7) S	$CH_2-CH-CO_2H$ $\ \mid \ \ \ \ \mid$ $OH \ \ NH_2$	
7. threonine	Thr (5.6) T	$CH_3CH-CH-CO_2H$ $\ \ \ \ \ \mid \ \ \ \mid$ $\ \ \ \ OH \ \ NH_2$	R contains an alcohol function.
8. cysteine	Cys (5.0) C	$CH_2-CH-CO_2H$ $\ \mid \ \ \ \ \mid$ $SH \ \ NH_2$	
9. methionine	Met (5.7) M	$CH_3S-CH_2CH_2-CH-CO_2H$ $\ \ \ \ \ \ \ \ \ \ \ \ \ \ \ \ \ \mid$ $\ \ \ \ \ \ \ \ \ \ \ \ \ \ \ \ \ NH_2$	R contains sulfur.

Table 17.1 *(continued)*

Name	Three-letter abbreviation (isoelectric point) one-letter abbreviation	Formula	R
10. proline	Pro (6.3) P	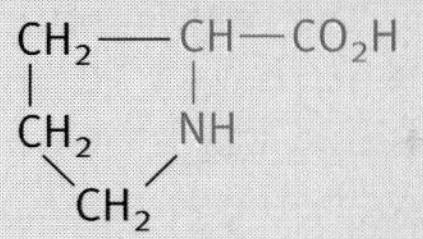	The amino group is secondary and part of a ring.
11. phenylalanine	Phe (5.5) F	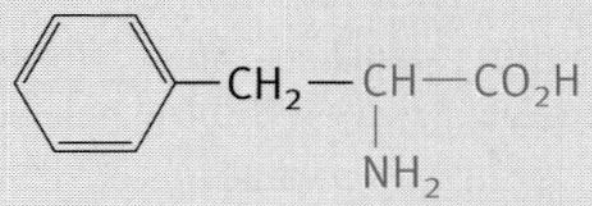	
12. tyrosine	Tyr (5.7) Y	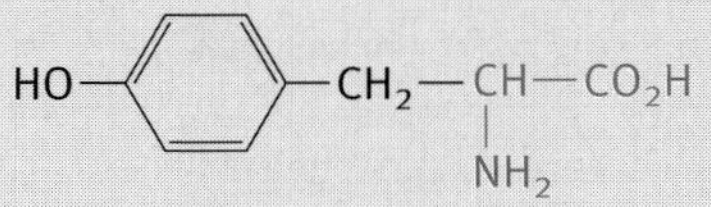	One hydrogen in alanine is replaced by an aromatic or heteroaromatic (indole) ring.
13. tryptophan	Trp (5.9) W	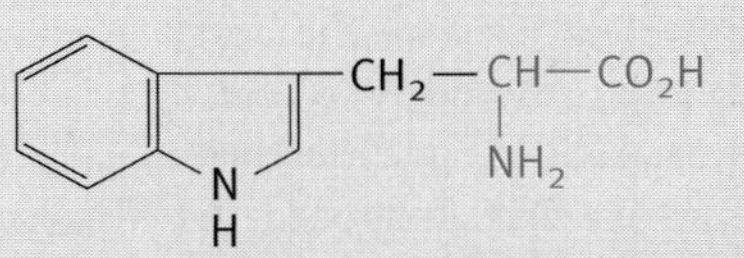	

B. One amino group and two carboxyl groups

14. aspartic acid	Asp (3.0) D	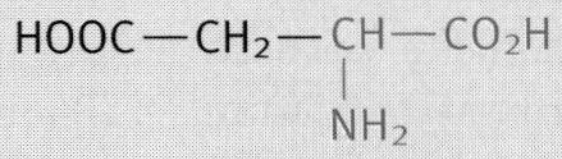 $HOOC-CH_2-CH-CO_2H$, NH_2	
15. glutamic acid	Glu (3.2) E	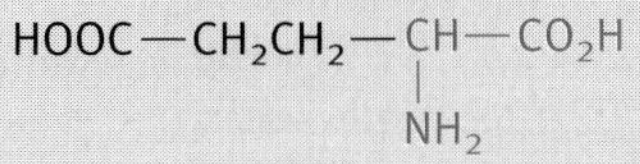 $HOOC-CH_2CH_2-CH-CO_2H$, NH_2	
16. asparagine	Asn (5.4) N	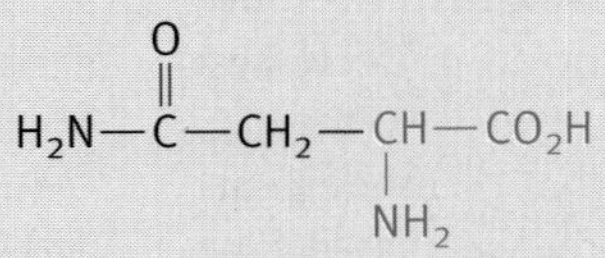	
17. glutamine	Gln (5.7) Q	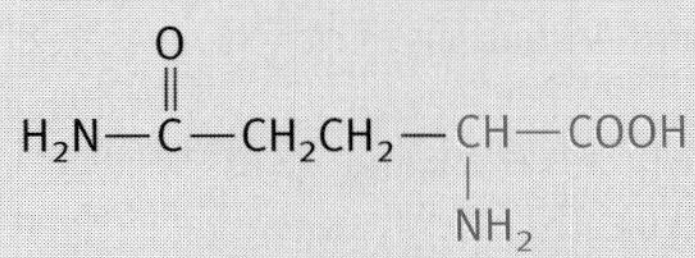	

C. One carboxyl group and two basic groups

18. lysine	Lys (9.7) K	$CH_2CH_2CH_2CH_2-CH-CO_2H$, NH_2 , NH_2	The second basic group is a primary amine, a guanidine, or an imidazole.
19. arginine	Arg (10.8) R		
20. histidine	His (7.6) H	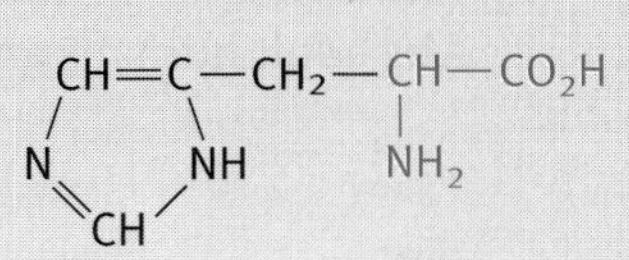	

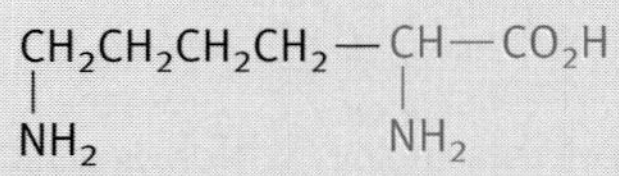

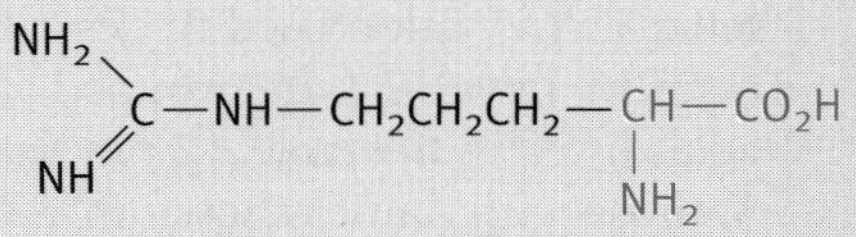

A WORD ABOUT ...

Amino Acid Dating

The question of age is one of the first that archaeologists seek to answer when they find artifacts or skeletons at a "dig." Are the bones or the pot shards ancient or modern? If they are ancient, *how* ancient? Knowing the age helps to answer other questions, such as how the people lived, with what other groups they traded or had contact, whom they followed and who followed them, and so on.

Chemists have helped archaeologists with this problem. One of the best known methods is carbon-14, or radioactive, dating, first proposed in 1947 by Willard F. Libby (Nobel Prize in 1960). The isotope ^{14}C decays with a half-life of 5730 years, which is sufficiently long for a steady-state equilibrium concentration to be established in the biosphere. A tiny but constant fraction (about $1.2 \times 10^{-10}\%$) of the carbon in live plants and animals is ^{14}C. After death, when ^{14}C is no longer taken in from the environment (as food, carbon dioxide, and so on), its concentration decreases. Knowing the decay rate of ^{14}C and comparing the ^{14}C content of the ancient material with that of modern allows the age of the ancient material to be calculated. The practical limit of the method is about 10 half-lives of ^{14}C, or about 50,000 years.

The extent of racemization of chiral amino acids found in fossil bones, shells, and teeth offers another method of dating ancient material. In living systems, amino acids have the L configuration and are optically pure. Once death occurs, however, the biochemical reactions that prevent equilibration of the L and D forms are terminated, and gradual thermal equilibration of the two forms begins. This reaction can be used for dating because the amount of racemization is a function of the material's age.

Amino acid dating is used to determine the age of bones such as these mammoth bones.

$$NH_2 \overset{CO_2H}{\underset{R}{\rule[-0.5em]{0pt}{0pt}|}} H \rightleftharpoons H \overset{CO_2H}{\underset{R}{\rule[-0.5em]{0pt}{0pt}|}} NH_2$$

L form   D form

Racemization rates differ for different amino acids. For example, the half-life at 25°C and pH 7 for aspartic acid is about 3000 years; for alanine, it is about 12,000 years. The racemization rate also depends on temperature. For example, the half-life at 0°C for aspartic acid increases to about 430,000 years. So it is necessary, for accurate dating, to know the temperature at which the material was stored. Fortunately, the temperature below ground at a given depth and climate often remains constant over long periods and can be estimated fairly accurately.

Accuracy can be improved by using a combination of dating methods or by calibrating one method with another. One advantage of amino acid dating is that the sample size can be very much smaller than is needed for carbon-14 dating. Also, a range of time spans can be covered by using different amino acids. Amino acid dating can be extended back in time well beyond the limits of ^{14}C dating, even to ice ages 100 to 400 thousand years ago!

Proteins are composed of **α-amino acids,** carboxylic acids with an amino group on the **α-carbon atom. Peptides** consist of a few linked amino acids.

based on this name, which is used when writing the formulas of peptides, and a one-letter abbreviation used to describe the amino acid sequence in a protein. The amino acids in Table 17.1 are grouped to emphasize structural similarities. Of the 20 amino acids listed in the table, 12 can be synthesized in the body from other foods. The other 8, those with names shown in color and referred to as **essential amino acids,** cannot be synthesized by adult humans and therefore must be included in the diet in the form of proteins.

17.2 The Acid–Base Properties of Amino Acids

The carboxylic acid and amine functional groups are *simultaneously* present in amino acids, and we might ask whether they are mutually compatible, since one group is acidic and the other is basic. Although we have represented the amino acids in Table 17.1 as having amino and carboxyl groups, these structures are oversimplified.

Amino acids with one amino group and one carboxyl group are better represented by a **dipolar ion structure.***

$$R-\underset{\underset{^+NH_3}{|}}{CH}-\overset{\overset{O}{\|}}{C}-O^-$$

dipolar structure of an α-amino acid

The amino group is protonated and present as an ammonium ion, whereas the carboxyl group has lost its proton and is present as a carboxylate anion. This dipolar structure is consistent with the saltlike properties of amino acids, which have rather high melting points (even the simplest, glycine, melts at 233°C) and relatively low solubilities in organic solvents.

Amino acids are *amphoteric* (page 209). They can behave as acids and donate a proton to a strong base, or they can behave as bases and accept a proton from a strong acid. These behaviors are expressed in the following equilibria for an amino acid with one amino and one carboxyl group:

$$\underset{\substack{amino\ acid \\ at\ low\ pH \\ (acid)}}{\underset{^+NH_3}{RCHCO_2H}} \underset{\underset{H^+}{\overset{HO^-}{\rightleftharpoons}}}{} \underset{\substack{dipolar\ ion \\ form \\ (neutral)}}{\underset{^+NH_3}{RCHCO_2^-}} \underset{\underset{H^+}{\overset{HO^-}{\rightleftharpoons}}}{} \underset{\substack{amino\ acid \\ at\ high\ pH \\ (base)}}{\underset{NH_2}{RCHCO_2^-}} \qquad \textbf{(17.1)}$$

Figure 17.2 shows a titration curve for alanine, a typical amino acid of this kind. At low pH (acidic solution), the amino acid is in the form of a substituted ammonium

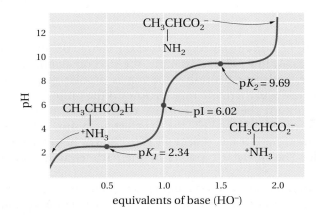

Figure 17.2
Titration curve for alanine, showing how its structure varies with pH.

---

*Such structures are sometimes called *zwitterions* (from a German word for hybrid ions).

ion. At high pH (basic solution), it is present as a substituted carboxylate ion. At some intermediate pH (for alanine, pH 6.02), the amino acid is present as the dipolar ion. A simple rule to remember for any acidic site is that *if the pH of the solution is* less *than the* pK_a, *the proton is on; if the pH of the solution is* greater *than the* pK_a, *the proton is off.*

EXAMPLE 17.1

Starting with alanine hydrochloride (its structure at low pH in hydrochloric acid is shown in the lower left corner of the curve in Figure 17.2), write equations for its reaction with one equivalent of sodium hydroxide and then with a second equivalent of sodium hydroxide.

Solution

$$CH_3CHCO_2H + Na^+ HO^- \longrightarrow CH_3CHCO_2^- + Na^+ Cl^- + H_2O \qquad \textbf{(17.2)}$$

with $\overset{|}{+}NH_3 \ Cl^-$ on the left (ammonium salt) and $\overset{|}{+}NH_3$ on the right (dipolar ion)

$$CH_3CHCO_2^- + Na^+ HO^- \longrightarrow CH_3CHCO_2^-Na^+ + H_2O \qquad \textbf{(17.3)}$$

with $\overset{|}{+}NH_3$ on the left (dipolar ion) and $\overset{|}{}NH_2$ on the right (carboxylate salt)

The first equivalent of base removes a proton from the carboxyl group to give the dipolar ion, and the second equivalent of base removes a proton from the ammonium ion to give the sodium carboxylate.

PROBLEM 17.1 Starting with the sodium carboxylate salt of alanine, write equations for its reaction with one equivalent of hydrochloric acid and then with a second equivalent, and explain what each equivalent of acid does.

PROBLEM 17.2 Which group in the ammonium salt form of alanine is more acidic, the $-\overset{+}{N}H_3$ group or the $-CO_2H$ group?

PROBLEM 17.3 Which group in the carboxylate salt form of alanine is more basic, the $-NH_2$ group or the $-CO_2^-$ group?

Note from Figure 17.2 and from eq. 17.1 that the charge on an amino acid changes as the pH changes. At low pH, for example, the sign on alanine is positive, at high pH it is negative, and near neutrality the ion is dipolar. If placed in an electric field, the amino acid will therefore migrate toward the cathode (negative electrode) at low pH and toward the anode (positive electrode) at high pH (Figure 17.3). At some intermediate pH, called the **isoelectric point (pI)**, the amino acid will be dipolar and have a net charge of zero. It will be unable to move toward either electrode. The isoelectric points of the various amino acids are listed in Table 17.1.

> The **isoelectric point (pI)** of an amino acid is the pH at which it is mainly in its dipolar form and has no net charge.

In general, amino acids with one amino group and one carboxyl group, and no other acidic or basic groups in their structure, have two pK_a values: one around 2 to 3 for proton loss from the carboxyl group and the other around 9 to 10 for proton loss from the ammonium ion. The isoelectric point is about halfway between the two pK_a values, near pH 6.

R is neutral

$$\underset{\substack{\overset{|}{+}NH_3}}{RCHCO_2H} \xrightarrow{\;pK_a = 2\text{–}3\;} \underset{\substack{\overset{|}{+}NH_3}}{RCHCO_2^-} \xrightarrow{\;pK_a = 9\text{–}10\;} \underset{\substack{\overset{|}{}NH_2}}{RCHCO_2^-} \qquad \textbf{(17.4)}$$

low pH ————————————————————→ high pH

net charge   +1   0   −1

EXAMPLE 17.2

Write the structure of leucine

a. at the pI.   b. at high pH.   c. at low pH.

Solution

a. $(CH_3)_2CHCH_2\underset{\substack{\overset{|}{+}NH_3}}{CHCO_2^-}$ b. $(CH_3)_2CHCH_2\underset{\substack{\overset{|}{}NH_2}}{CHCO_2^-}$

   dipolar and neutral   negative

c. $(CH_3)_2CHCH_2\underset{\substack{\overset{|}{+}NH_3}}{CHCO_2H}$

   positive

PROBLEM 17.4   Write the structure for the predominant form of each of the following amino acids at the indicated pH. If placed in an electric field, toward which electrode (+ or −) will each amino acid migrate?

a. methionine at its pI   b. serine at low pH   c. phenylalanine at high pH

The situation is more complex with amino acids containing two acidic or two basic groups.

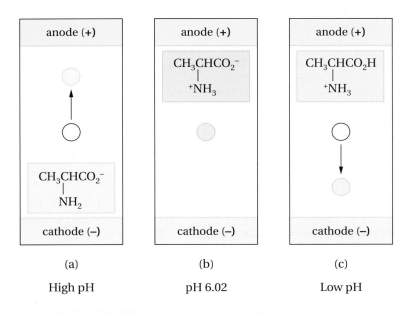

(a) High pH   (b) pH 6.02   (c) Low pH

Figure 17.3
The migration of an amino acid (such as alanine) in an electric field depends on pH: (a) At high pH, the alanine is negatively charged and migrates toward the positive anode. (b) At the isoelectric point (pH 6.02), the alanine is neutral and does not migrate. (c) At low pH, the alanine is positively charged and migrates toward the negatively charged cathode.

17.3 The Acid–Base Properties of Amino Acids with More Than One Acidic or Basic Group

Aspartic and glutamic acids (numbers 14 and 15 in Table 17.1) have two carboxyl groups and one amino group. In strong acid (low pH) all three of these groups are in their acidic form (protonated). As the pH is raised and the solution becomes more basic, each group in succession gives up a proton. The equilibria are shown for **aspartic acid,** with the three pKa values over the equilibrium arrows:

$$\underset{\substack{|\\ {}^+NH_3}}{HO_2CCH_2CHCO_2H} \xrightleftharpoons{\substack{pK_a=\\2.09}} \underset{\substack{|\\ {}^+NH_3}}{HO_2CCH_2CHCO_2^-} \xrightleftharpoons{\substack{pK_a=\\3.86}} \underset{\substack{|\\ {}^+NH_3}}{{}^-O_2CCH_2CHCO_2^-} \xrightleftharpoons{\substack{pK_a=\\9.82}} \underset{\substack{|\\ NH_2}}{{}^-O_2CCH_2CHCO_2^-} \qquad (17.5)$$

low pH ——————————————————————————————→ high pH
net charge $+1$ .... 0 .... -1 .... -2

The isoelectric point for aspartic acid, the pH at which it is mainly in the neutral dipolar form, is 2.87 (in general, the pI is close to the average of the two pK_a's on either side of the neutral, dipolar species).

EXAMPLE 17.3

Which carboxyl group is the stronger acid in the most acidic form of aspartic acid?

Solution   As shown at the extreme left of eq. 17.5, the first proton to be removed from the most acidic form of aspartic acid is the proton on the carboxyl group nearest the ${}^+NH_3$ substituent. The ${}^+NH_3$ group is electron-withdrawing due to its positive charge and enhances the acidity of the carboxyl group closest to it.

PROBLEM 17.5   Use eq. 17.5 to tell which is the least acidic group in aspartic acid and why.

The situation differs for amino acids with two basic groups and only one carboxyl group (numbers 18, 19, and 20 in Table 17.1). With **lysine,** for example, the equilibria are

$$\underset{\substack{|\\ {}^+NH_3}}{\overset{}{CH_2(CH_2)_3}}\underset{\substack{|\\ {}^+NH_3}}{CHCO_2H} \xrightleftharpoons{\substack{pK_a=\\2.18}} \underset{\substack{|\\ {}^+NH_3}}{\overset{}{CH_2(CH_2)_3}}\underset{\substack{|\\ {}^+NH_3}}{CHCO_2^-} \xrightleftharpoons{\substack{pK_a=\\8.95}} \underset{\substack{|\\ {}^+NH_3}}{\overset{}{CH_2(CH_2)_3}}\underset{\substack{|\\ NH_2}}{CHCO_2^-} \xrightleftharpoons{\substack{pK_a=\\10.53}} \underset{\substack{|\\ NH_2}}{\overset{}{CH_2(CH_2)_3}}\underset{\substack{|\\ NH_2}}{CHCO_2^-}$$

low pH ——————————————————————————————→ high pH
net charge $+2$ .... $+1$ .... 0 .... -1

$$(17.6)$$

The pI for lysine comes in the basic region, at 9.74.

The second basic groups in arginine and histidine are not simple amino groups. They are a **guanidine** group and an **imidazole** ring, respectively, shown in color. The most protonated forms of these two amino acids are

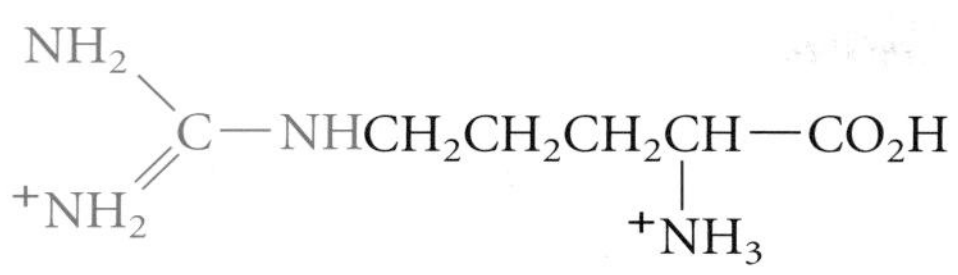

arginine at pH 1 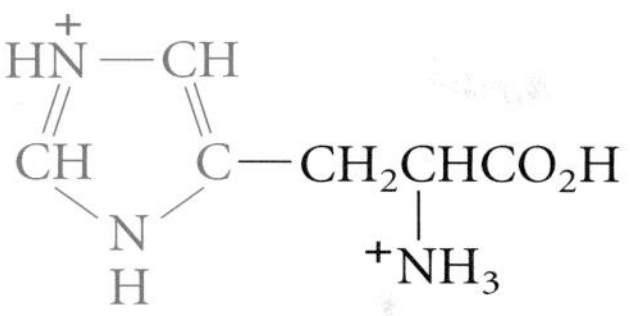   histidine at pH 1

PROBLEM 17.6   Arginine shows three pK_a's: at 2.17 (the —COOH group), at 9.04 (the —$\overset{+}{N}H_3$ group), and at 12.48 (the guanidinium ion). Write equilibria (similar to eq. 17.6) for its dissociation. At approximately what pH will the isoelectric point come, and what is the structure of the dipolar ion?

Table 17.2 summarizes the approximate pK_a values and isoelectric points for the three types of amino acids.

Table 17.2	Approximate acidity constants and isoelectric points (pI) for the three types of amino acids				
			pK_a		
Type		**1**	**2**	**3**	**pI**
1 acidic and 1 basic group		2.3	9.4	—	6.0
2 acidic and 1 basic group		2.2	4.1	9.8	3.0
1 acidic and 2 basic groups	(Lys, Arg)	2.2	9.0	11.5	10.0
	(His)	1.8	6.0	9.2	7.6

17.4   Electrophoresis

As we have seen, the charge on an amino acid depends on the pH of its environment. **Electrophoresis** is an important method for separating amino acids that takes advantage of these charge differences. In a typical electrophoresis experiment, a mixture of amino acids is placed on a solid support, for example, paper, and the support is bathed in an aqueous solution at a controlled pH. An electrical field is then applied across the paper. Amino acids that are positively charged at that pH migrate toward the negatively charged cathode. Amino acids that are negatively charged migrate toward the positively charged anode. Migration ceases when the electrical field is turned off.

An example of an electrophoresis experiment is illustrated in Figure 17.3. In this case, the α-amino acid alanine migrates toward the anode at high pH, toward the cathode at low pH, and does not migrate when the pH is adjusted to the isoelectric point of the amino acid. Because most amino acids do not have exactly the same charge at

Electrophoresis is a method for separating amino acids and proteins, based on their charge differences.

a given pH, mixtures of amino acids can be separated by electrophoresis because the components migrate toward the anode or cathode at different rates. Electrophoresis is also used to separate peptides and proteins because they behave much like their building blocks, the amino acids, at a given pH.

EXAMPLE 17.4

Predict the direction of migration (toward the positive or negative electrode) of alanine in an electrophoresis apparatus at pH 5. Do the same for aspartic acid.

Solution   A pH of 5 is *less* than the pI of alanine (~6). Therefore, the dipolar ions will be protonated (positive) and migrate toward the negative electrode. But pH 5 is *greater* than the pI of aspartic acid (~3). Therefore aspartic acid will exist mainly as the −1 ion (eq. 17.5) and migrate toward the positive electrode. A mixture of the two amino acids could therefore easily be separated in this way.

PROBLEM 17.7   Predict the direction of migration in an electrophoresis apparatus (toward the positive or negative electrode) of each component of the following amino acid mixtures:

   a. glycine and lysine at pH 7   b. phenylalanine, leucine, and proline at pH 6

17.5 Reactions of Amino Acids

In addition to their acidic and basic behavior, amino acids undergo other reactions typical of carboxylic acids or amines. For example, the carboxyl group can be esterified:

$$R-\underset{\underset{+NH_3}{|}}{CH}-CO_2^- + R'OH + H^+ \xrightarrow{heat} R-\underset{\underset{+NH_3}{|}}{CH}-CO_2R' + H_2O \qquad (17.7)$$

The amino group can be acylated to an amide:

$$R-\underset{\underset{+NH_3}{|}}{CH}-CO_2^- + R'-\overset{\overset{O}{||}}{C}-Cl \xrightarrow{2\,HO^-} R-\underset{\underset{R'C-NH}{|}}{CH}-CO_2^- + 2\,H_2O + Cl^- \qquad (17.8)$$

These types of reactions are useful in temporarily modifying or protecting either of the two functional groups, especially during the controlled linking of amino acids to form peptides or proteins.

PROBLEM 17.8   Using eqs. 17.7 and 17.8 as models, write equations for the following reactions:

   a. glutamic acid + CH_3OH + HCl $\longrightarrow$
   b. proline + benzoyl chloride + NaOH $\longrightarrow$
   c. phenylalanine + acetic anhydride $\xrightarrow{heat}$

17.6 The Ninhydrin Reaction

Ninhydrin is a useful reagent for detecting amino acids and determining the concentrations of their solutions. It is the hydrate of a cyclic triketone, and when it reacts with an amino acid, a violet dye is produced. The overall reaction, whose mechanism is complex and need not concern us in detail here, is as follows:

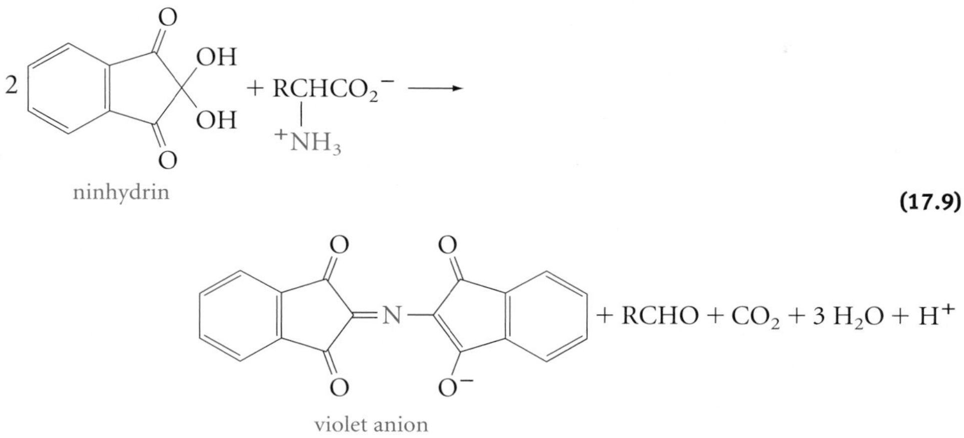

$$2 \text{ ninhydrin} + \text{RCHCO}_2^-\ (^+\text{NH}_3) \longrightarrow$$

ninhydrin

(17.9)

$$\text{violet anion} + \text{RCHO} + \text{CO}_2 + 3\,\text{H}_2\text{O} + \text{H}^+$$

violet anion

Only the nitrogen atom of the violet dye comes from the amino acid; the rest of the amino acid is converted to an aldehyde and carbon dioxide. Therefore, *the same violet dye is produced from all α-amino acids with a primary amino group,* and the intensity of its color is directly proportional to the concentration of the amino acid present. Only proline, which has a secondary amino group, reacts differently to give a yellow dye, but this, too, can be used for analysis.

PROBLEM 17.9 Write an equation for the reaction of alanine with ninhydrin.

17.7 Peptides

Amino acids are linked in peptides and proteins by an amide bond between the carboxyl group of one amino acid and the α-amino group of another amino acid. Emil Fischer, who first proposed this structure, called this amide bond a **peptide bond.** A molecule containing only *two* amino acids (the shorthand aa is used for amino acid) joined in this way is a **dipeptide:**

An amide bond linking two amino acids is called a **peptide bond.** A peptide has an **N-terminal amino acid** with a free $^+\text{NH}_3$ group and a **C-terminal amino acid** with a free CO_2^- group.

$$\underset{\text{N-terminal aa}}{\underset{^+\text{NH}_3}{\text{R}-\text{CH}}}-\overset{\text{O}}{\overset{\|}{\text{C}}}\underset{\text{peptide bond}}{\;}\text{NH}-\underset{\text{C-terminal aa}}{\underset{\text{R}'}{\text{CH}}}-\text{CO}_2^-$$

$$\xleftarrow{\quad\text{aa}_1\quad}\;\xleftarrow{\quad\text{aa}_2\quad}$$

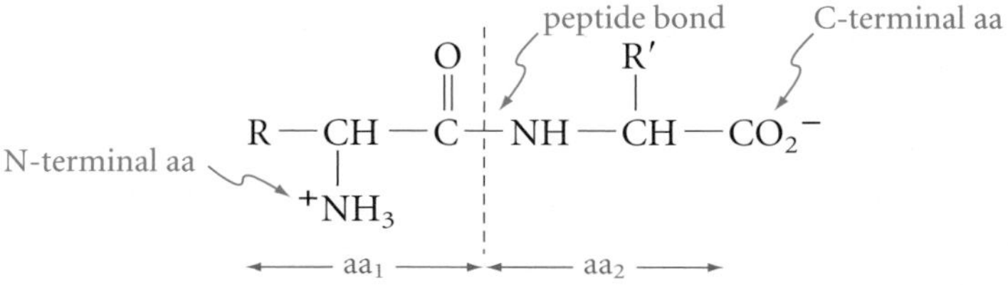

By convention, the peptide bond is written with the amino acid having a free $^+\text{NH}_3$ group at the left and the amino acid with a free CO_2^- group at the right. These amino acids are called, respectively, the **N-terminal amino acid** and the **C-terminal amino acid.**

EXAMPLE 17.5

Write the dipeptide structures that can be made by linking alanine and glycine with a peptide bond.

Solution   There are two possibilities:

$$\underset{\text{glycylalanine}}{H_3\overset{+}{N}-CH_2-\overset{\overset{\displaystyle O}{\|}}{C}-NH-\underset{\underset{\displaystyle CH_3}{|}}{CH}-CO_2^-} \qquad \underset{\text{alanylglycine}}{H_3\overset{+}{N}-\underset{\underset{\displaystyle CH_3}{|}}{CH}-\overset{\overset{\displaystyle O}{\|}}{C}-NH-CH_2-CO_2^-}$$

In glycylalanine, glycine is the N-terminal amino acid, and alanine is the C-terminal amino acid. In alanylglycine, these roles are reversed. The two dipeptides are structural isomers.

We often write the formulas for peptides in a kind of shorthand by simply linking the three-letter abbreviations for each amino acid, *starting with the N-terminal one at the left*. For example, glycylalanine is Gly—Ala, and alanylglycine is Ala—Gly.

PROBLEM 17.10   In Example 17.5 the formulas for Gly—Ala and Ala—Gly are written in their dipolar forms. At what pH do you expect these structures to predominate? Draw the expected structure of Gly—Ala in solution at pH 3; at pH 9.

PROBLEM 17.11   Write the dipolar structural formula for

a. valylalanine.   b. alanylvaline.

EXAMPLE 17.6

Consider the abbreviated formula Gly—Ala—Ser for a tripeptide. Which is the N-terminal amino acid, and which is the C-terminal amino acid?

Solution   Such formulas always read from the N-terminal amino acid at the left to the C-terminal amino acid at the right. Glycine is the N-terminal amino acid, and serine is the C-terminal amino acid. Both the amino group *and* the carboxyl group of the middle amino acid, alanine, are tied up in peptide bonds.

PROBLEM 17.12   Write out the complete structural formula for Gly—Ala—Ser.

PROBLEM 17.13   Write out the *abbreviated* formulas for all possible tripeptide isomers of Gly—Ala—Ser.

The complexity that is possible in peptide and protein structures is truly astounding. For example, Problem 17.13 shows that there are 6 possible arrangements of 3 different amino acids in a tripeptide. For a tetrapeptide this number jumps to 24, and for an octapeptide (constructed from 8 different amino acids) there are 40,320 possible arrangements!

Now we must introduce one small additional complication before we consider the structures of particular peptides and proteins.

A WORD ABOUT ...

Some Naturally Occurring Peptides

Peptides with just a few linked amino acids per molecule have been isolated from living matter, where they often perform important roles in biology. Here are a few examples.

Bradykinin is a nonapeptide present in blood plasma and involved in regulating blood pressure. Several peptides found in the brain act as chemical transmitters of nerve impulses. One of these is the undecapeptide **substance P,** thought to be a transmitter of pain impulses. Notice that the C-terminal amino acid, methionine, is present as the primary amide. This is quite common in peptide chains and is symbolized by placing an NH_2 at the right-hand end of the formula.

$$Arg—Pro—Pro—Gly—Phe—Ser—$$
$$Pro—Phe—Arg$$

bradykinin

$$Arg—Pro—Lys—Pro—Gln—Gln—Phe—Phe—$$
$$Gly—Leu—Met—NH_2$$

substance P

Oxytocin and **vasopressin** are two cyclic nonapeptide hormones produced by the posterior pituitary gland. Oxytocin regulates uterine contraction and lactation and may be administered when it is necessary to induce labor at childbirth. Note that its structure includes two cysteine units joined by a disulfide bond and, once again, the C-terminal amino acid is present as the amide. Vasopressin differs from oxytocin only in the substitution of Phe for Ile and Arg for Leu. Vasopressin regulates the excretion of water by the kidneys and also affects blood pressure. The disease *diabetes insipidus,* in which too much urine is excreted, is a consequence of vasopressin deficiency and can be treated by administering this hormone.

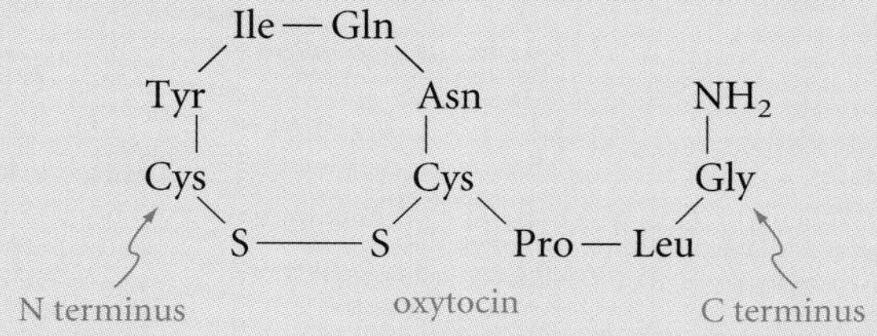

oxytocin

Cyclosporin A is another cyclic peptide that was first isolated from the fungus *Trichoderma polysporum.* Cyclosporin A has immunosuppressive activity and is used to help prevent organ rejection after transplant operations. Notice that cyclosporin A contains several unusual amino acids and that a number of the amide nitrogens are methylated.

One area of current research is modifying the structures of natural, biologically important peptides (by replacing one amino acid with another or by altering side-chain structures or by replacing portions of the peptide chain with nonpeptidic structures) with the intent of developing new and useful drugs.

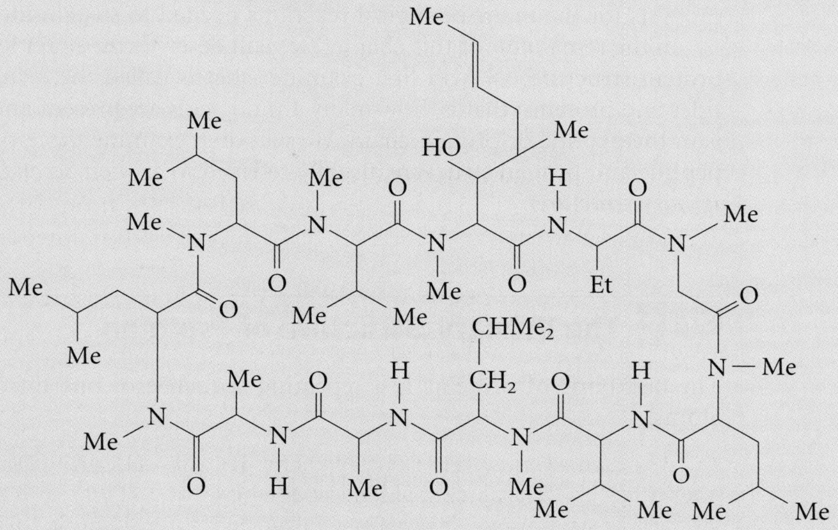

cyclosporin A

17.8 The Disulfide Bond

The **disulfide bond** is an S—S single bond. In proteins, it links two **cysteine** amino acid units.

Aside from the peptide bond, the only other type of covalent bond between amino acids in peptides and proteins is the **disulfide bond.** It links two **cysteine** units. Recall that thiols are easily oxidized to disulfides (eq. 7.49). Two cysteine units can be linked by a disulfide bond.

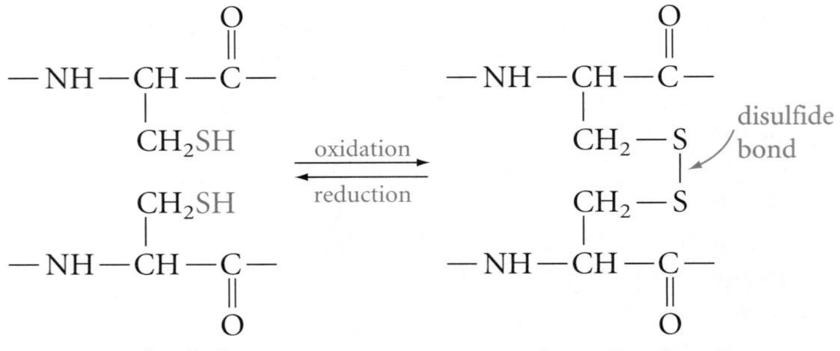

$$(17.10)$$

two cysteine units   —Cys—S—S—Cys—

If the two cysteine units are in different parts of the *same* chain of a peptide or protein, a disulfide bond between them will form a "loop," or large ring. If the two units are on different chains, the disulfide bond will cross-link the two chains. We will see examples of both arrangements. Disulfide bonds can easily be broken by mild reducing agents (see "A Word About Hair, Curly or Straight" on page 225).

17.9 Proteins

Proteins are biopolymers composed of many amino acids connected to one another through amide (peptide) bonds. They play numerous roles in biological systems. Some proteins are major components of structural tissue (muscle, skin, nails, hair). Others transport molecules from one part of a living system to another. Yet others serve as catalysts for the many biological reactions needed to sustain life.

In the remainder of this chapter, we will describe the main features of peptide and protein structure. We will first examine what is called the *primary structure* of peptides and proteins; that is, how many amino acids are present and what their sequence is in the peptide or protein chain. We will then examine three-dimensional aspects of peptide and protein structure, usually referred to as their *secondary, tertiary,* and *quaternary structures.*

17.10 The Primary Structure of Proteins

The backbone of proteins is a repeating sequence of one nitrogen and two carbon atoms.

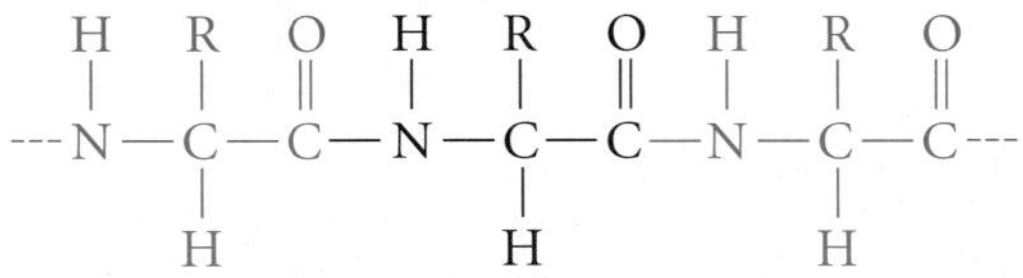

protein chain, showing amino acids linked by amide bonds

Things we must know about a peptide or protein, if we are to write down its structure, are (1) which amino acids are present and how many of each there are and (2) the sequence of the amino acids in the chain. In this section, we will briefly describe ways to obtain this kind of information.

17.10.a Amino Acid Analysis

Since peptides and proteins consist of amino acids held together by amide bonds, they can be hydrolyzed to their amino acid components. This hydrolysis is typically accomplished by heating the peptide or protein with 6 M HCl at 110°C for 24 hours. Analysis of the resulting amino acid mixture requires a procedure that separates the amino acids from one another, identifies each amino acid present, and determines its amount.

An instrument called an **amino acid analyzer** performs these tasks automatically in the following way. The amino acid mixture from the complete hydrolysis of a few milligrams of the peptide or protein is placed at the top of a column packed with material that selectively absorbs amino acids. The packing is an insoluble resin that contains strongly acidic groups. These groups protonate the amino acids. Next, a buffer solution of known pH is pumped through the column. The amino acids pass through the column at different rates, depending on their structure and basicity, and are thus separated.

The column effluent is met by a stream of ninhydrin reagent. Therefore, the effluent is alternately violet or colorless, depending on whether or not an amino acid is being eluted from the column. The intensity of the color is automatically recorded as a function of the volume of effluent. Calibration with known amino acid mixtures allows each amino acid to be identified by the appearance time of its peak. Furthermore, the intensity of each peak gives a quantitative measure of the amount of each amino acid that is present. Figure 17.4 shows a typical plot obtained from an automatic amino acid analyzer.

PROBLEM 17.14   Show the products expected from complete hydrolysis of Gly—Ala—Ser.

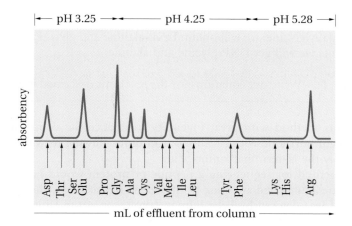

Figure 17.4
Different amino acids in a peptide hydrolyzate are separated on an ion-exchange resin. Buffers with different pH's elute the amino acids from the column. Each amino acid is identified by comparing the peaks with the standard elution profile shown near the bottom on the figure. The amount of each amino acid is proportional to the area under its peak. This sample contains eight amino acids: aspartic acid, glutamic acid, glycine, alanine, cysteine, methionine, phenylalanine, and arginine.

17.10.b Sequence Determination

Frederick Sanger[*] devised a method for sequencing peptides based on the observation that the N-terminal amino acid differs from all others in the chain by having a free amino group. If that amino group were to react with some reagent prior to hydrolysis, then after hydrolysis, that amino acid would be labeled and could be identified. **Sanger's reagent** is 2,4-dinitrofluorobenzene, which reacts with the NH_2 group of amino acids and peptides to give yellow 2,4-dinitrophenyl (DNP) derivatives.

2,4-dinitrofluoro-
benzene

N-terminal
amino acid

(17.11)

DNP-peptide, labeled
at the N terminus

Hydrolysis of a peptide treated this way (eq. 17.11) would give the DNP derivative of the N-terminal amino acid; other amino acids in the chain would be unlabeled. In this way, the N-terminal amino acid could be identified.

EXAMPLE 17.7

How might alanylglycine be distinguished from glycylalanine?

Solution   Each dipeptide will give one equivalent each of alanine and glycine on hydrolysis. Therefore, we cannot distinguish between them without applying a sequencing method.

Treat the dipeptide with 2,4-dinitrofluorobenzene and *then* hydrolyze. If the dipeptide is alanylglycine, we will obtain DNP-alanine and glycine; if the dipeptide is glycylalanine, we will get DNP-glycine and alanine.

PROBLEM 17.15   Write out equations for the reactions described in Example 17.7.

Sanger used his method with great ingenuity to deduce the complete sequence of insulin, a protein hormone with 51 amino acid units. But the method suffers in that it identifies only the N-terminal amino acid.

An ideal method for sequencing a peptide or protein would have a reagent that clips off just one amino acid at a time from the end of the chain, and identifies it. Just such a method was devised by Pehr Edman (professor at the University of Lund in Sweden), and it is now widely used.

Edman's reagent is phenyl isothiocyanate, $C_6H_5N{=}C{=}S$. The steps in selectively labeling and releasing the N-terminal amino acid are shown in Figure 17.5. In the first step, the N-terminal amino acid acts as a nucleophile toward the $C{=}S$ bond

---

[*]Frederick Sanger (Cambridge University, England) received *two* Nobel Prizes, the first in 1958 for his landmark work in amino acid sequencing and the second in 1980 for methodology in the base sequencing of RNA and DNA.

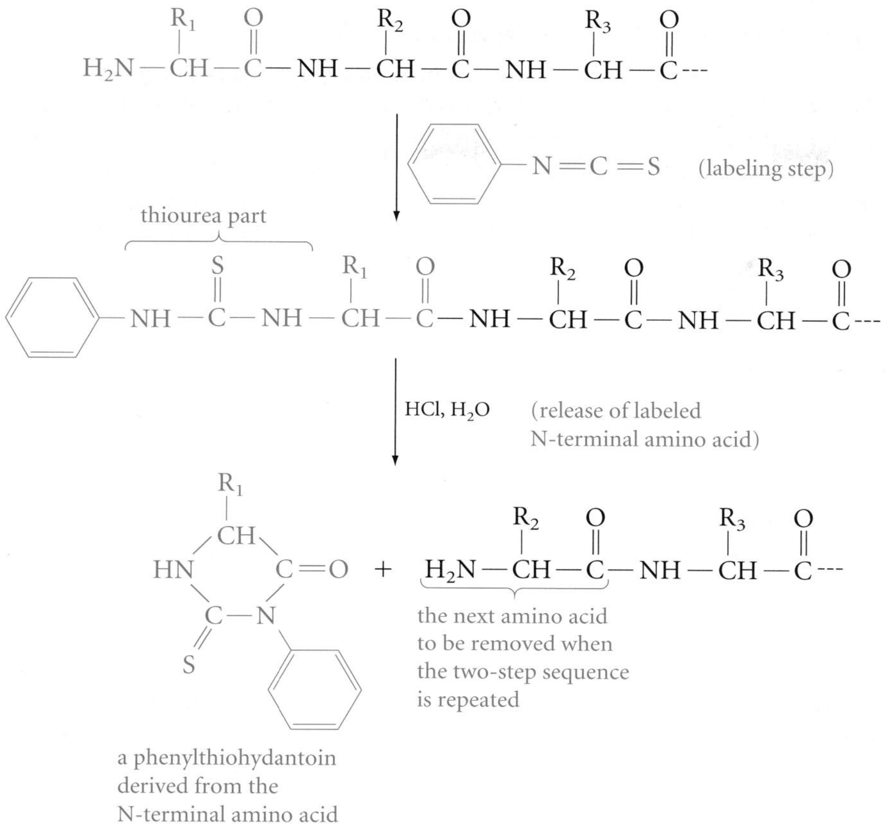

$$H_2N-\underset{\underset{R_1}{|}}{CH}-\underset{\underset{O}{\parallel}}{C}-NH-\underset{\underset{R_2}{|}}{CH}-\underset{\underset{O}{\parallel}}{C}-NH-\underset{\underset{R_3}{|}}{CH}-\underset{\underset{O}{\parallel}}{C}---$$

(labeling step)

thiourea part

HCl, H$_2$O   (release of labeled N-terminal amino acid)

a phenylthiohydantoin derived from the N-terminal amino acid

the next amino acid to be removed when the two-step sequence is repeated

Figure 17.5
The Edman degradation of peptides.

of the reagent to form a thiourea derivative. In the second step, the N-terminal amino acid is removed in the form of a heterocyclic compound, a phenylthiohydantoin. The specific phenylthiohydantoin that is formed can be identified by comparison with reference compounds separately prepared from the known amino acids. Then the two steps are repeated, to identify the next amino acid, and so on. The method has been automated, so currently amino acid "sequenators" can easily determine, in a day, the sequence of the first 50 or so amino acids in a peptide, starting at the N-terminal end. But the Edman method cannot be used indefinitely, due to the gradual buildup of impurities.

> **PROBLEM 17.16**   Write the structure of the phenylthiohydantoin derived from the first cycle of Edman degradation of Phe—Ala—Ser.

17.10.c Cleavage of Selected Peptide Bonds

If a protein contains several hundred amino acid units, it is best to first partially hydrolyze the chain to smaller fragments that can be separated and subsequently sequenced by the Edman method. Certain chemicals or enzymes are used to cleave proteins at *particular* peptide bonds. For example, the enzyme *trypsin* (an intestinal digestive enzyme) specifically hydrolyzes polypeptides only at the carboxy end of arginine and lysine. A few of the many reagents of this type are listed in Table 17.3.

Table 17.3	Reagents for specific cleavage of polypeptides
Reagent	**Cleavage site**
trypsin	carboxyl side of Lys, Arg
chymotrypsin	carboxyl side of Phe, Tyr, Trp
cyanogen bromide (CNBr)	carboxyl side of Met
carboxypeptidase	the C-terminal amino acid

EXAMPLE 17.8

Consider the following peptide:

$$\text{Ala—Gly—Tyr—Trp—Ser—Lys—Gly—Leu—Met—Gly}$$

By referring to Table 17.3, determine what fragments will be obtained when this peptide is hydrolyzed with

a. trypsin.   b. chymotrypsin.   c. cyanogen bromide.

Solution

a. The enzyme trypsin will split the peptide on the carboxyl side of lysine, to give

$$\text{Ala—Gly—Tyr—Trp—Ser—Lys and Gly—Leu—Met—Gly}$$

b. The enzyme chymotrypsin will split the peptide on the carboxyl sides of tyrosine and tryptophan, to give

$$\text{Ala—Gly—Tyr and Trp and Ser—Lys—Gly—Leu—Met—Gly}$$

c. Cyanogen bromide will split the peptide on the carboxyl side of methionine, thus splitting off the C-terminal glycine and leaving the rest of the peptide untouched. (Carboxypeptidase would do the same thing, confirming that the C-terminal amino acid is glycine.)

PROBLEM 17.17 Determine what fragments will be obtained if bradykinin (its abbreviated formula is given on page 491) is hydrolyzed enzymatically with

a. trypsin.   b. chymotrypsin.

During the past 15 years, the methods discussed here have been improved and expanded; the separation and sequencing of peptides and proteins can now be accomplished even if only minute amounts are available.

17.11 The Logic of Sequence Determination

A specific example will illustrate the reasoning that is used to fully determine the amino acid sequence in a particular peptide with 30 amino acid units.

First we hydrolyze the peptide completely, subject it to amino acid analysis, and find that it has the formula

$$\text{Ala}_2\text{ArgAsnCys}_2\text{GlnGlu}_2\text{Gly}_3\text{His}_2\text{Leu}_4\text{LysPhe}_3\text{ProSerThrTyr}_2\text{Val}_3$$

Using the Sanger method, we find that the N-terminal amino acid is Phe.

Since the chain is rather long, we decide to simplify the problem by digesting the peptide with chymotrypsin. (We select chymotrypsin because the peptide contains three Phe's and two Tyr's and will undoubtedly be cleaved by that reagent.) When we carry out this cleavage, we get three fragment peptides. In addition, we get two equivalents of Phe and one of Tyr. After separation, we subject the three fragment peptides to Edman degradation and obtain their structures.

A. Leu—Val—Cys—Gly—Glu—Arg—Gly—Phe

B. Val—Asn—Gln—His—Leu—Cys—Gly—Ser—His—Leu—Val—Glu—
Ala—Leu—Tyr

    27   28   29   30
C. Thr—Pro—Lys—Ala

We still cannot write a unique structure for the intact peptide, but we can say that the C-terminal amino acid must be Ala and that the last four amino acids must be in the sequence shown for fragment C. We deduce this because we know that Ala is *not* cleaved at its carboxyl end by chymotrypsin, yet it appears at the C-terminal end of one of the fragments. (Note that the C-terminal amino acids in fragments A and B are Phe and Tyr, both cleaved at the carboxyl ends by chymotrypsin.) That the C-terminal amino acid is Ala can be confirmed using carboxypeptidase. We can number the amino acids in fragment C as 27 through 30 in the chain.

What do we do next? Cyanogen bromide is no help, because the peptide does not contain Met. But the peptide does contain Lys and Arg, so we go back to the beginning and digest the intact peptide with trypsin, which cleaves peptides on the carboxy side of these amino acid units. We obtain (not surprisingly) some Ala (the C-terminal amino acid) because it comes right after a Lys. We also obtain two peptides. One of them is relatively short, so we determine its sequence by the Edman method and find it to be

    23   24   25   26   27   28   29
D. Gly—Phe—Phe—Tyr—Thr—Pro—Lys

Because the last three amino acids in fragment D *overlap* with 27, 28, and 29 of fragment C, we can number the rest of that chain, back to 23. We now note that amino acids 23 and 24 appear at the end of fragment A, so originally A must have been connected to C. The only place left for fragment B is in front of A. This leaves only one of the Phe's unaccounted for, and it must occupy the N-terminal position (recall the Sanger result). We can now write out the complete sequence!

The blue vertical arrows show the cleavage points with chymotrypsin, and the red ones show the cleavage points with trypsin.

The peptide just used for illustration is the B chain of the protein hormone **insulin,** whose structure was first determined by Sanger and is shown schematically in Figure 17.6. Insulin consists of an A chain with 21 amino acid units and a B chain with 30 amino acid units. The two chains are joined by two disulfide bonds, and the A chain also contains a small disulfide loop.

A WORD ABOUT ...

Protein Sequencing and Evolution

There are many reasons why it is important to determine the sequences of amino acids in proteins. First, we must know the detailed structures of proteins if we are to understand, at a molecular level, the way they function. The amino acid sequence is the link between the genetic message coded in DNA and the three-dimensional protein structure that forms the basis for biological function.

There are medical benefits to knowing amino acid sequences. Certain genetic diseases, such as sickle-cell anemia, can result from the change of a single amino acid unit in a protein. In this case, it is a replacement of glutamic acid at position 6 in the β chain of hemoglobin by a valine unit. Thus, sequence determination is an important part of medical pathology. One future possible application of genetic engineering is devising ways to correct these amino acid sequence errors.

Another important reason for determining amino acid sequences is that they provide a chemical tool for studying our evolutionary history. Proteins resemble one another in amino acid sequence if they have a common evolutionary ancestry. Let's look at a specific example.

Cytochrome c, an enzyme important in the respiration of most plants and animals, is an electron-transport protein with 104 linked amino acids. It is involved in oxidation-reduction processes. In these reactions, cytochrome c must react with and transfer an electron from one enzyme complex (cytochrome reductase) to another (cytochrome oxidase).

Cytochrome c probably evolved more than 1.5 billion years ago, even before the evolutionary divergence of plants and animals. The function of this protein has been preserved all that time! We know this because the cytochrome c isolated from any eukaryotic microorganism (one that contains a cell nucleus) reacts *in vitro* with the cytochrome oxidase of every other species tested so far. For example, wheat germ cytochrome c, a plant-derived enzyme, reacts with human cytochrome c oxidase. Also, the three-dimensional structures of cytochrome c isolated from such diverse species as tuna and photosynthetic bacteria are very similar.

Although *the shape and functions are similar* for cytochrome c samples isolated from different sources, *the amino acid sequence varies somewhat* from one species to another. The amino acid sequence of cytochrome c isolated from humans differs from that of monkeys in *just 1* (out of 104) amino acid residues! On the other hand, cytochrome c from dogs, a more distant evolutionary relative of humans, differs from the human protein by 11 amino acid residues.

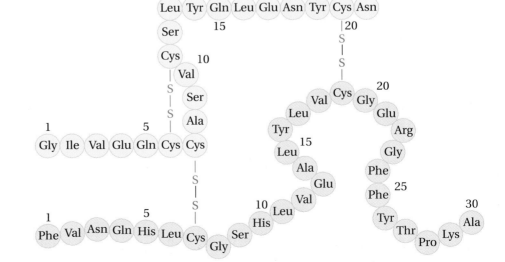

Figure 17.6
Primary structure of beef insulin. The A chain is shown in blue, and the B chain, whose structure determination is described in the text, is shown in pink with black letters.

> **PROBLEM 17.18** How could the A and B chains of insulin be separated chemically? (*Hint:* See eq. 7.49.)

17.12 Peptide Synthesis

Once we know the amino acid sequence in a peptide or protein, we are in a position to synthesize it from its amino acid components. Why would we want to do this? There are several reasons. For example, we might wish to verify a particular peptide structure by comparing the properties of the synthetic and natural substances. Or we might wish to study the effect of substituting one amino acid for another on the biological properties of a peptide or protein. Such modified proteins could be valuable in treating disease or in understanding how the protein functions.

Many methods have been developed to link amino acids in a controlled manner. They require careful strategy. Amino acids are bifunctional. To link the carboxyl group of one amino acid to the amino group of a second amino acid, we must first prepare each compound by protecting the amino group of the first and the carboxyl group of the second.

$$\underset{\text{aa}_1}{\text{H}_2\text{N}-\overset{\overset{\text{R}_1}{|}}{\text{CH}}-\text{CO}_2\text{H}} \xrightarrow[\text{amino group}]{\text{protect the}} \boxed{\text{P}_1}-\text{NH}-\overset{\overset{\text{R}_1}{|}}{\text{CH}}-\text{CO}_2\text{H} \qquad \textbf{(17.12)}$$

$$\underset{\text{aa}_2}{\text{H}_2\text{N}-\overset{\overset{\text{R}_2}{|}}{\text{CH}}-\text{CO}_2\text{H}} \xrightarrow[\text{carboxyl group}]{\text{protect the}} \text{H}_2\text{N}-\overset{\overset{\text{R}_2}{|}}{\text{CH}}-\overset{\overset{\text{O}}{||}}{\text{C}}-\boxed{\text{P}_2} \qquad \textbf{(17.13)}$$

In this way, we can control the linking of the two amino acids so that the carboxyl group of aa_1 combines with the amino group of aa_2.

$$\boxed{\text{P}_1}-\text{NH}\overset{\overset{\text{R}_1}{|}}{\text{CH}}\text{CO}_2\text{H} + \text{H}_2\text{N}-\overset{\overset{\text{R}_2}{|}}{\text{CH}}-\overset{\overset{\text{O}}{||}}{\text{C}}-\boxed{\text{P}_2} \xrightarrow{-\text{H}_2\text{O}}$$

$$\underset{\text{doubly protected dipeptide}}{\boxed{\text{P}_1}-\text{NHCH}-\overset{\overset{\text{O}}{||}}{\text{C}}-\text{NH}-\overset{\overset{\text{R}_2}{|}}{\text{CH}}-\overset{\overset{\text{O}}{||}}{\text{C}}-\boxed{\text{P}_2}} \qquad \textbf{(17.14)}$$

peptide bond

Later we will give specific examples of protecting groups and of a reagent that can be used to form the peptide bond.

EXAMPLE 17.9

What would happen if we tried to combine aa_1 with aa_2 without using protecting groups?

Solution Since each amino acid could react either as an amine or as a carboxylic acid, we could get not only aa_1—aa_2 but also aa_2—aa_1, aa_1—aa_1, and aa_2—aa_2. Furthermore, since the resulting dipeptides would still have a free amino and a free carboxyl group, we could also get trimers, tetramers, and so on. In other words, a mess.

After the peptide bond is formed, we must be able to *remove the protecting groups under conditions that do not hydrolyze the peptide bond.* Or, if more amino acids are to be added to the chain, we must be able to *selectively* remove one of the two protecting groups from the doubly protected dipeptide before joining the next amino acid to it. All of this can be quite a tricky and tedious process. Yet these methods were used by Vincent du Vigneaud* and his colleagues to synthesize oxytocin and vasopressin (page 491), the first naturally occurring polypeptides to be synthesized in the laboratory.

In 1965, R. Bruce Merrifield** developed a technique that revolutionized peptide synthesis. This **solid-phase technique** avoids many of the tedious aspects of previous methods and is now universally used. The principle is to *assemble the peptide chain while one end of it is chemically anchored to an insoluble inert solid.* In this way, excess reagents and by-products can be removed simply by washing and filtering the solid. The growing peptide chain does not need to be purified at any intermediate stage. When the peptide is fully constructed, it is cleaved chemically from the solid support.

Typically, the solid phase is a cross-linked polystyrene in which some (usually 1 to 10%) of the aromatic rings contain chloromethyl ($ClCH_2$—) groups.

Peptides are synthesized by Merrifield's **solid-phase technique,** in which one end of the growing peptide is chemically attached to an insoluble inert solid.

$$\sim\sim CH_2-CH-CH_2-CH-CH_2-CH-CH_2-CH-CH_2-CH-CH_2-CH\sim\sim$$

CH₂Cl   CH₂Cl

The polymer behaves chemically like benzyl chloride, an alkyl halide that is quite reactive in nucleophilic substitution reactions (S_N2).

The steps in a Merrifield synthesis are summarized for a dipeptide in Figure 17.7. In step 1, the polymer is first treated with an N-protected amino acid. The carboxylate ion acts as an oxygen nucleophile and displaces the chloride ion from the polymer, thus forming an ester link. *The first amino acid attached to the polymer will eventually become the C-terminal amino acid of the synthetic peptide.*

Many protecting groups are known, but in solid-phase peptide synthesis, the most frequently used N-protecting group is the ***t*-butoxycarbonyl (Boc)** group. The amino acid is protected by reaction with di-*t*-butyl dicarbonate.

The most frequently used N-protecting group in solid-phase peptide synthesis is the ***t*-butoxycarbonyl (Boc)** group.

$$(CH_3)_3CO-C \begin{matrix} O \\ \| \\ \\ O \end{matrix} \quad R \\ + \; H_3\overset{+}{N}-CH-CO_2^- \quad \xrightarrow{\text{base}} \quad \underbrace{(CH_3)_3CO-\overset{O}{\overset{\|}{C}}}_{\boxed{P}}-NH-CH-CO_2H$$

$$(CH_3)_3CO-C \begin{matrix} \\ \\ \| \\ O \end{matrix}$$

di-*t*-butyl dicarbonate

(17.15)

---

*Vincent du Vigneaud (Cornell University) was awarded the 1955 Nobel Prize in chemistry for this achievement.

R. Bruce Merrifield (Rockefeller University) received the 1984 Nobel Prize in chemistry for his contribution, which not only revolutionized peptide synthesis but also affected many other areas of chemistry through the use of polymer-bound reagents. For an account of the history of this discovery, see the article by Merrifield in *Science* **1986, 232, 341, and for a personal history, see *Chemistry in Britain* **1987**, 816.

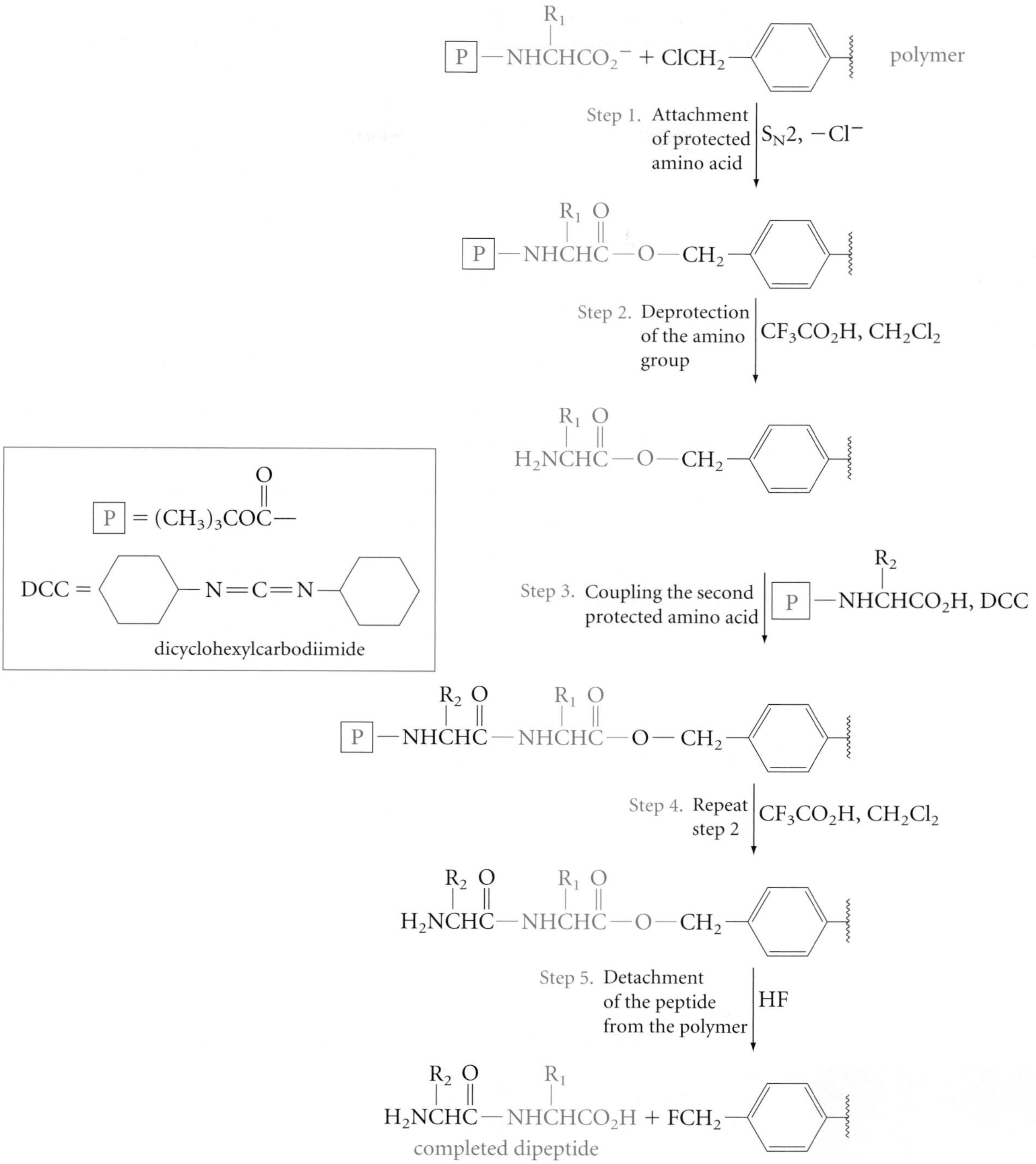

Figure 17.7
Solid-phase synthesis of a dipeptide.

After the amino acid is attached to the polymer, the protecting group is removed. This is accomplished (step 2) by reaction with acid under mild conditions.

$$
\text{CH}_3-\overset{\overset{\displaystyle \text{H}-\text{CH}_2}{|}}{\underset{\underset{\displaystyle \text{CH}_3}{|}}{\text{C}}}-\text{O}-\overset{\overset{\displaystyle \text{O}}{\|}}{\text{C}}-\text{NHCH}-\overset{\overset{\displaystyle \text{R}}{|}}{\underset{\underset{\displaystyle \text{O}}{\|}}{\text{C}}}-\text{O}-\text{CH}_2 \quad \xrightarrow[\text{CH}_2\text{Cl}_2]{\text{CF}_3\text{CO}_2\text{H}}
$$

ester group still intact

$$
\text{CH}_3-\overset{\overset{\displaystyle \text{CH}_2}{\|}}{\underset{\underset{\displaystyle \text{CH}_3}{|}}{\text{C}}} \;+\; \text{CO}_2 \;+\; \text{H}_2\text{N}-\overset{\overset{\displaystyle \text{R}_1}{|}}{\text{CH}}-\overset{\overset{\displaystyle }{\underset{\underset{\displaystyle \text{O}}{\|}}{\text{C}}}}{}-\text{O}-\text{CH}_2
$$

2-methylpropene

deprotected amino acid   **(17.16)**

The by-products of deprotection are gaseous (2-methylpropene and carbon dioxide) and are thus easily removed from the reaction mixture. The ester group that links the first amino acid to the polymer is *not* hydrolyzed under these conditions.

In step 3, the next N-protected amino acid is linked to the first one. This is accomplished with the aid of **dicyclohexylcarbodiimide (DCC)**. DCC is able to link carboxyl and amino groups in a peptide bond; in the process, the DCC is converted to dicyclohexylurea.

Dicyclohexylcarbodiimide (DCC) is a reagent used to link carboxyl and amino groups in a peptide bond.

$$
\overset{\overset{\displaystyle \text{O}}{\|}}{\text{C}}-\text{OH} \;+\; \text{H}_2\text{N} \quad + \quad \bigcirc\!\!-\text{N}=\text{C}=\text{N}-\bigcirc \quad \longrightarrow
$$

dicyclohexylcarbodiimide
(DCC)

     (17.17)

$$
\overset{\overset{\displaystyle \text{O}}{\|}}{\text{C}}-\text{NH} \quad + \quad \bigcirc\!\!-\text{NH}-\overset{\overset{\displaystyle \text{O}}{\|}}{\text{C}}-\text{NH}-\bigcirc
$$

peptide bond

dicyclohexylurea

Steps 2 and 3 may be repeated to add a third amino acid, a fourth, and so on. Finally, when the desired amino acids have been connected in the proper sequence and the N-terminal amino group has been deprotected (step 4 in Figure 17.7), the complete polypeptide chain is detached from the polymer. This can be accomplished by treatment with anhydrous hydrogen fluoride, which cleaves the benzyl ester without hydrolyzing the amide bonds in the polypeptide (step 5 in Figure 17.7).

All operations in solid-phase peptide synthesis have been automated. The reactions occur in a single reaction vessel, with reagents and wash solvents automatically added from reservoirs by mechanical pumps. Working around the clock, the programmer can incorporate eight or more amino acids into a polypeptide in a day. Merrifield synthesized the nonapeptide bradykinin (page 491) in just 27 hours using

Automated polypeptide synthesizer.

this technique. And, in 1969, he used the automated synthesizer to prepare the enzyme ribonuclease (124 amino acid residues), the first enzyme to be prepared synthetically from its amino acid components. The synthesis, which required 369 chemical reactions and 11,391 steps, was completed in only six weeks. Automated computerized peptide synthesis, though still not without occasional problems, is now a fairly routine matter.

PROBLEM 17.19 Write all the equations for the synthesis of Gly—Ala—Phe using Merrifield solid-phase technology.

We have seen how the primary structure of peptides and proteins can be determined and how peptides can be synthesized in the laboratory. Now let us examine some further details of protein structure.

17.13 Secondary Structure of Proteins

Because proteins consist of long chains of amino acids strung together, one might think that their shapes are rather amorphous, or "floppy" and ill-defined. This is incorrect. Many proteins have been isolated in pure crystalline form and are polymers with very well defined shapes. Indeed, even in solution, the shapes seem to be quite regular. Let us examine some of the structural features of peptide chains that are responsible for their definite shapes.

17.13.a Geometry of the Peptide Bond

We pointed out earlier that simple amides have a planar geometry, that the amide C—N bond is shorter than usual, and that rotation around that bond is restricted (Sec. 10.20). Bond planarity and restricted rotation, which are consequences of resonance, are also important in peptide bonds.

   X-ray studies of crystalline peptides by Linus Pauling and his colleagues determined the precise geometry of peptide bonds. The characteristic dimensions, which are common to all peptides and proteins, are shown in Figure 17.8.

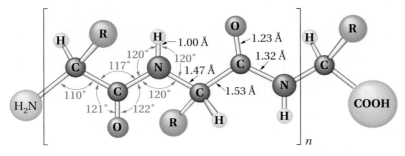

Figure 17.8
The characteristic bond angles and bond lengths in peptide bonds.

Things to notice about peptide geometry are as follows: (1) The amide group is flat; the carbonyl carbon, the nitrogen, and the four atoms connected to them all lie in a single plane. (2) The short amide C—N distance (1.32 Å, compared with 1.47 Å for the other C—N bond) and the 120° bond angles around that nitrogen show that it is essentially sp^2-hybridized and that the bond between it and the carbonyl carbon is like a double bond. (3) Although each amide group is planar, two adjacent amide groups need not be *co*planar because of rotation about the other single bonds in the chain; that is, rotation can occur around the two single bonds to the —CHR— group.

The rather rigid geometry and restricted rotation of the peptide bond help to impart a definite shape to proteins.

17.13.b Hydrogen Bonding

We pointed out earlier that amides readily form *inter*molecular hydrogen bonds between the carbonyl group and the N—H group, bonds of the type C=O···H—N. Such bonds are present and important in peptide chains. The chain may coil in such a way that the N—H of one peptide bond can hydrogen-bond with a carbonyl group of another peptide bond farther down the *same* chain, thus rigidifying the coiled structure. Alternatively, carbonyl groups and N—H groups on *different* peptide chains may hydrogen-bond, linking the two chains. Although a single hydrogen bond is relatively weak (perhaps only 5 kcal/mol of energy), the possibility of forming multiple intrachain or interchain hydrogen bonds makes this a very important factor in protein structure, as we will now see.

17.13.c The α Helix and the Pleated Sheet

X-ray studies of α-keratin, a structural protein present in hair, wool, horns, and nails, showed that some feature of the structure repeats itself every 5.4 Å. Using molecular models with the correct geometry of the peptide bond, Linus Pauling was able to suggest a structure that explains this and other features of the x-ray studies. Pauling proposed that the polypeptide chain coils about itself in a spiral manner to form a helix, held rigid by intrachain hydrogen bonds. The α **helix,** as it is called, is right-handed and has a pitch of 5.4 Å, or 3.6 amino acid units (Figure 17.9).*

The α **helix** and the **pleated sheet** are two common secondary structures of proteins or segments of proteins.

Note several features of the α helix. Proceeding from the N terminus (at the top of the structure as drawn in the figure), each carbonyl group points ahead or down toward the C terminus and is hydrogen-bonded to an N—H bond farther down the chain. The N—H bonds all point back toward the N terminus. All the hydrogen bonds are roughly aligned with the long axis of the helix. The very large number of hydrogen bonds (one for each amino acid unit) strengthens the helical structure. The R groups of the individual amino acid units are directed *outward* and do not disrupt the central core of the helix. It turns out that the α helix is a natural pattern into which many proteins or segments of proteins fold.

---

*Linus Pauling (1901–1994) made many contributions to our knowledge of organic structures. He did fundamental work on the theory of resonance, on the measurement of bond lengths and energies, and on the structure of proteins and the mechanism of antibody action. He received the Nobel Prize in chemistry in 1954 and the Nobel Peace Prize in 1962.

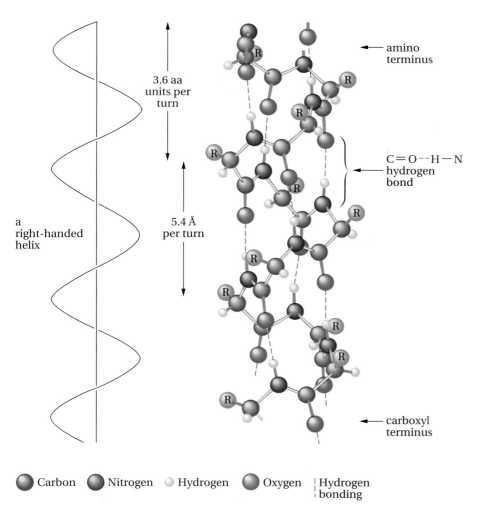

3.6 aa
units per
turn

amino
terminus

C=O--H—N
hydrogen
bond

a
right-handed
helix

5.4 Å
per turn

R

carboxyl
terminus

● Carbon ● Nitrogen ○ Hydrogen ● Oxygen ┊Hydrogen
                                              ┊bonding

Figure 17.9
Segment of an α helix, showing three turns of the helix, with 3.6 amino acid units per turn. Hydrogen bonds are shown as dashed colored lines.

The structural protein β-keratin, obtained from silk fibroin, shows a different repeating pattern (7 Å) in its x-ray structure. To explain the data, Pauling suggested a **pleated-sheet** arrangement of the peptide chain (Figure 17.10). In the pleated sheet, peptide chains lie side by side and are held together by *interchain* hydrogen bonds. Adjacent chains run in opposite directions. The repeating unit in each chain, which is stretched out compared with the α helix, is about 7 Å. In the pleated-sheet structure, the R groups of amino acid units in any one chain alternate above and below the mean plane of the sheet. If the R groups are large, there will be appreciable steric repulsion between them on adjacent chains. For this reason, the pleated-sheet structure is important *only* in proteins that have a high percentage of amino acid units with *small* R groups. In the β-keratin of silk fibroin, for example, 36% of the amino acid units are glycine (R = H) and another 22% are alanine (R = CH$_3$).

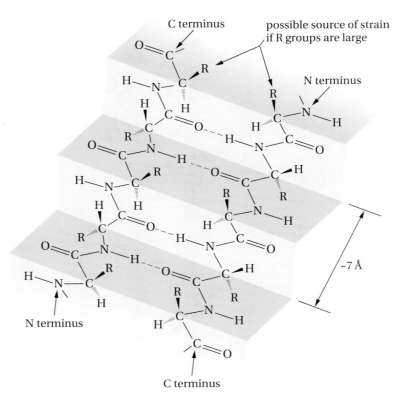

Figure 17.10
A segment of the pleated-sheet structure of β-keratin. Adjacent chains run in opposite directions and are held together by hydrogen bonds (shown in color). R groups project above or below the mean plane of the sheet.

17.14 Tertiary Structure: Fibrous and Globular Proteins

We may well ask how materials as rigid as horses' hoofs, as springy as hair, as soft as silk, as slippery and shapeless as egg white, as inert as cartilage, and as reactive as enzymes can all be made of the same building blocks: amino acids and proteins. The key lies mainly in the amino acid makeup itself. So far we have focused on the protein backbone and its shape. But what about the diverse R groups of the various amino acids? How do they affect protein structure?

Some amino acids have nonpolar R groups, simple alkyl or aromatic groups. Others have highly polar R groups, with carboxylate or ammonium ions and hydroxyl or other polar groups. Still others have flat, rigid aromatic rings that may interact in specific ways. *Different R groups affect the gross properties of a protein.*

PROBLEM 17.20   Which amino acids in Table 17.1 have nonpolar R groups? highly polar groups? relatively flat R groups?

Proteins generally fall into one of two main classes: **fibrous** or **globular. Fibrous proteins** are animal structural materials and hence are water insoluble. They, in turn, fall into three general categories: the **keratins,** which make up protective tissue, such as skin, hair, feathers, claws, and nails; the **collagens,** which form connective tissue, such as cartilage, tendons, and blood vessels; and the **silks,** such as the fibroin of spider webs and cocoons.

Keratins and collagens have helical structures, whereas silks have pleated-sheet structures. A large fraction of the R groups attached to these frameworks are nonpolar, accounting for the insolubility of these proteins in water. In hair, three α helices are braided to form a rope, the helices being held together by disulfide cross-links. The ropes are further packed side by side in bundles that ultimately form the hair fiber. The α-keratin of more rigid structures, such as nails and claws, is similar to that of hair, except that there is a higher percentage of cysteine amino acid units in the polypeptide chain. Therefore, there are more disulfide cross-links, giving a firmer, less flexible overall structure.

To summarize, nonpolar R groups and disulfide cross-links, together with helical or sheetlike backbones, tend to give fibrous proteins their rather rigid, insoluble structures.

Globular proteins are very different from fibrous proteins. They tend to be water soluble and have roughly spherical shapes, as their name suggests. Instead of being structural, globular proteins perform various other biological functions. They may be **enzymes** (biological catalysts), **hormones** (chemical messengers that regulate biological processes), **transport proteins** (carriers of small molecules from one part of the body to another, such as hemoglobin, which transports oxygen in the blood), or **storage proteins** (which act as food stores; ovalbumin of egg white is an example).

Globular proteins have more amino acids with polar or ionic side chains than the water-insoluble fibrous proteins. An enzyme or other globular protein that carries out its function mainly in the aqueous medium of the cell will adopt a structure in which the nonpolar, hydrophobic R groups point in toward the center and the polar or ionic R groups point out toward the water.

Globular proteins are mainly helical, but they have folds that permit the overall shape to be globular. One of the 20 amino acids, proline has a secondary amino group. Wherever a proline unit occurs in the primary peptide structure, there will be no N—H group available for intrachain hydrogen bonding.

> Fibrous proteins (including **keratins, collagens,** and **silks**) have rather rigid shapes and are not water soluble.

> Globular proteins (including **enzymes, hormones, transport proteins,** and **storage proteins**) are spherical in shape and tend to be water soluble.

Proline units tend therefore to disrupt an α helix, and we frequently find them at "turns" in a protein structure.

Myoglobin, the oxygen-transport protein of muscle, is a good example of a globular protein (Figure 17.11). It contains 153 amino acid units, yet is extremely compact, with very little empty space in its interior. Approximately 75% of the amino acid units in myoglobin are part of eight right-handed α-helical sections. There are four

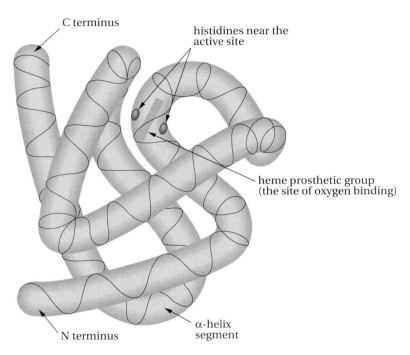

Figure 17.11
Schematic drawing of myoglobin. Each of the tubular sections is a segment of α helix, but the overall shape is globular.

proline units, and each occurs at or near "turns" in the structure. There are also three other "turns" caused by structural features of other R groups. The interior of myoglobin consists almost entirely of nonpolar R groups, such as those of leucine, valine, phenylalanine, and methionine. The only interior polar groups are two histidines. These perform a necessary function at the *active site* of the protein, where the nonprotein portion, a molecule of the porphyrin *heme* (page 392) binds the oxygen. The outer surface of the protein includes many highly polar amino acid residues (lysine, arginine, glutamic acid, and so on).

> The particular amino acid content of a protein influences its overall shape or **tertiary structure.**

> The structure of the aggregate formed by subunits of a high-molecular-weight protein is called its **quaternary structure.**

To summarize this section, we see that the particular amino acid content of a peptide or protein influences its shape. These interactions are mainly a consequence of disulfide bonds and of the polarity or nonpolarity of the R groups, their shape, and their ability to form hydrogen bonds. When we refer to the **tertiary structure** of a protein, we refer to all the contributions of these factors to its three-dimensional structure.

17.15 Quaternary Protein Structure

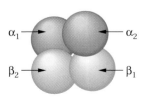

Figure 17.12
Schematic drawing of the four hemoglobin subunits.

Some high-molecular-weight proteins exist as aggregates of several subunits. These aggregates are referred to as the **quaternary structure** of the protein. Aggregation helps to keep nonpolar portions of the protein surface from being exposed to the aqueous cellular environment. **Hemoglobin,** the oxygen-transport protein of red cells, provides an example of such aggregation. It consists of four almost spherical units, two α units with 141 amino acids and two β units with 146 amino acids. The four units come together in a tetrahedral array, as shown in Figure 17.12.

Many other proteins form similar aggregates. Some are active only in their aggregate state, whereas others are active only when the aggregate dissociates into subunits. Aggregation in quaternary structures, then, provides an additional control mechanism over biological activity.

REACTION SUMMARY

1. Reactions of Amino Acids

a. Acid–Base Reactions (Secs. 17.2 and 17.3)

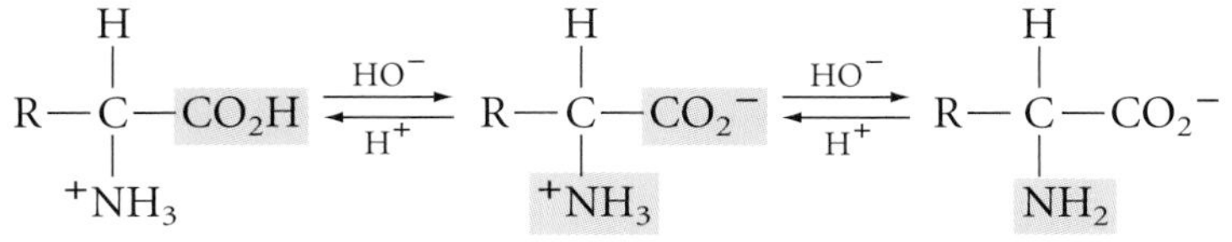

dipolar ion

b. Esterification (Sec. 17.5)

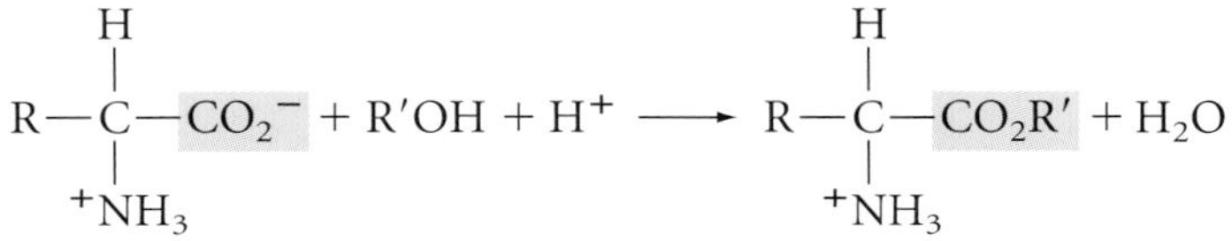

c. Amide Formation (Sec. 17.5)

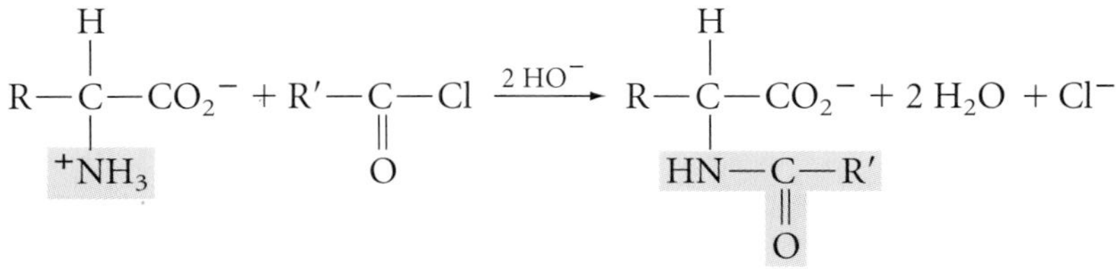

d. Ninhydrin Reaction (Sec. 17.6)

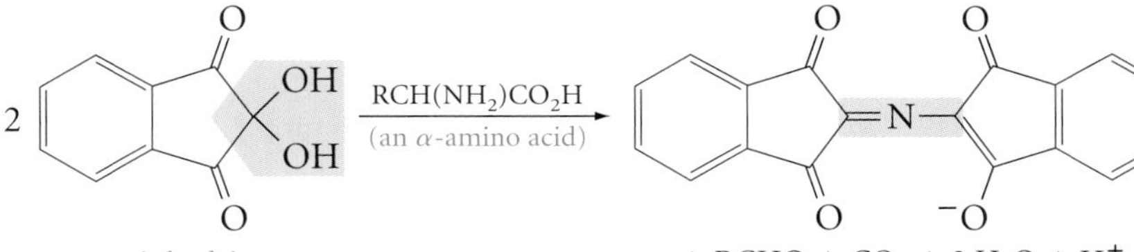

ninhydrin

(purple)

$+ RCHO + CO_2 + 3\ H_2O + H^+$

2. Reactions of Proteins and Peptides

a. Hydrolysis (Sec. 17.10)

$$\text{proteins} \xrightarrow[\substack{H_2O \\ \Delta}]{HCl} \text{peptides} \xrightarrow[\substack{H_2O \\ \Delta}]{HCl} \alpha\text{-amino acids}$$

b. Sanger's Reagent (Sec. 17.10; used to identify the N-terminal amino acid of a peptide or protein)

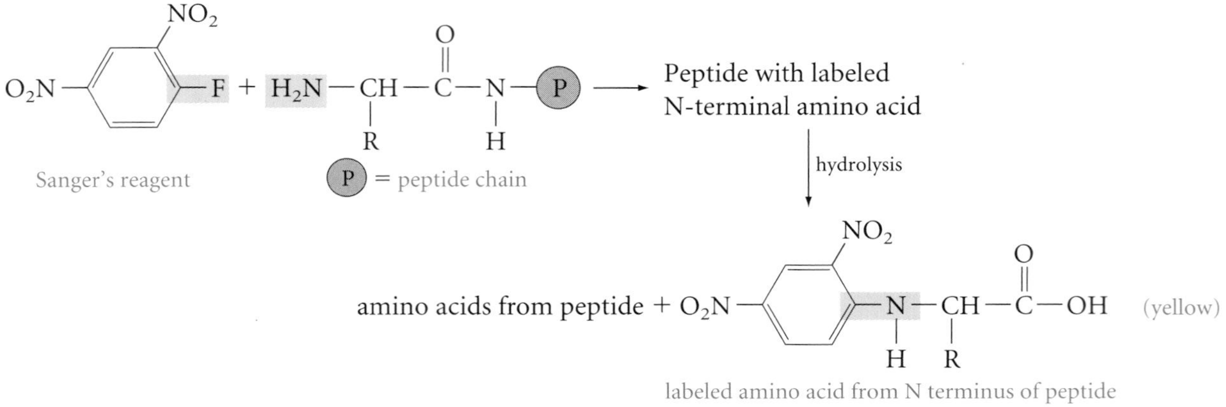

Sanger's reagent   P = peptide chain

Peptide with labeled
N-terminal amino acid

hydrolysis

amino acids from peptide + O_2N—

(yellow)

labeled amino acid from N terminus of peptide

c. Edman degradation (Sec. 17.10; used to determine the amino acid sequence of a peptide)

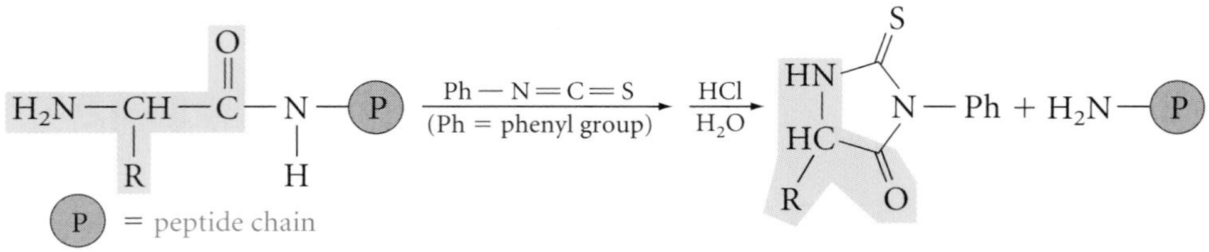

$\dfrac{Ph—N=C=S}{(Ph = phenyl\ group)}$ $\dfrac{HCl}{H_2O}$

P = peptide chain

d. Solid-Phase Peptide Synthesis (Sec. 17.12)

Details are given in Figure 17.7.

ADDITIONAL PROBLEMS (See pages 480–481 for structures of amino acids.)

Amino Acids: Definitions, Formulas, and Properties

17.21 Give a definition or illustration of each of the following terms:

a. peptide bond
b. dipolar ion
c. dipeptide
d. L configuration of amino acids
e. essential amino acid
f. amino acid with a nonpolar R group
g. amino acid with a polar R group
h. amphoteric compound
i. isoelectric point
j. ninhydrin

17.22 Draw a Fischer projection for L-leucine. What is the priority order of the groups attached to the stereogenic center? What is the absolute configuration, R or S?

17.23 Write Fischer projection formulas for

a. L-phenylalanine.
b. L-valine.

17.24 Illustrate the amphoteric nature of amino acids by writing an equation for the reaction of alanine in its dipolar ion form with one equivalent of

a. hydrochloric acid.
b. sodium hydroxide.

17.25 Write the formula for each of the following in its dipolar ion form:

    a. valine   **b.** serine

    c. proline   **d.** tryptophan

17.26 Locate the most acidic proton in each of the following species, and draw the structure of the product formed by reaction with one equivalent of base (HO^-).

    a. $HOOC-CH_2CH_2CHCO_2H$
                              $\overset{+}{N}H_3$

    b. $HOCH_2-CHCO_2^-$
                          $\overset{+}{N}H_3$

    c. $(CH_3)_2CHCHCO_2H$
                        $\overset{+}{N}H_3$

    d. NH_2
                $\searrow$
                   $C-NHCH_2CH_2CH_2CHCO_2^-$
       $H_2\overset{+}{N}$ $/\!/$ NH_2

17.27 What species is obtained by adding a proton to each of the following?

    a. $CH_3CH-CHCO_2^-$
                OH $\overset{+}{N}H_3$

    b. $^-O_2CCH_2CH-CO_2^-$
                         $\overset{+}{N}H_3$

17.28 Protonated alanine, $CH_3CH(\overset{+}{N}H_3)CO_2H$, has a pK_a of 2.34, whereas propanoic acid, $CH_3CH_2CO_2H$, has a pK_a of 4.85. Explain the increase in acidity due to replacing an α-hydrogen with an $-\overset{+}{N}H_3$ substituent.

17.29 The pK_a's of glutamic acid are 2.19 (the α carboxyl group), 4.25 (the other carboxyl group), and 9.67 (the α ammonium ion). Write equations for the sequence of reactions that occurs when base is added to a strongly acidic (pH = 1) solution of glutamic acid.

17.30 The pK_a's of arginine are 2.17 for the carboxyl group, 9.04 for the ammonium ion, and 12.48 for the guanidinium ion. Write equations for the sequence of reactions that occurs when acid is gradually added to a strongly alkaline solution of arginine.

17.31 Draw the structure of histidine at pH 1, and show how the positive charge in the second basic group (the imidazole ring) can be delocalized.

17.32 Predict the direction of migration in an electrophoresis apparatus of each component in a mixture of asparagine, histidine, and aspartic acid at pH 6.

Reactions of Amino Acids

17.33 Write equations for the reaction of valine with

    a. $CH_3CH_2OH + HCl$.   **b.** C_6H_5COCl + base.   **c.** acetic anhydride.

17.34 Write equations for the following reactions:

    a. serine + excess acetic anhydride $\longrightarrow$

    b. threonine + excess benzoyl chloride $\longrightarrow$

    c. glutamic acid + excess methanol + HCl $\longrightarrow$

17.35 Write the equations that describe what occurs when valine is treated with ninhydrin.

  $\circlearrowright$ = concept connections

Peptides

17.36 Write structural formulas for the following peptides:

    a. alanylalanine
    b. valyltryptophan
    c. tryptophanylvaline
    d. glycylalanylglycine

17.37 Write an equation for the hydrolysis of

    a. leucylserine.
    b. serylleucine.
    c. valyltyrosylmethionine.

17.38 Write an equation for the acid-catalyzed hydrolysis of the artificial sweetener aspartame (page 466).

17.39 Write formulas that show how the structure of alanylglycine changes as the pH of the solution changes from 1 to 10. Estimate the pI (isoelectric point) of this dipeptide.

17.40 Use the one-letter abbreviations to write out all possible tetrapeptides containing one unit each of glycine, alanine, valine, and leucine. How many structures are possible?

17.41 Write the structure of the product expected from the reaction of glycylcysteine with a mild oxidizing agent, such as hydrogen peroxide (see Sec. 17.8).

17.42 Write equations for the following reactions of Sanger's reagent:

    a. 2,4-dinitrofluorobenzene + glycine →
    b. excess 2,4-dinitrofluorobenzene + lysine →

17.43 Examine the structure of cyclosporin A (page 491).

    a. By drawing a dashed line at each peptide bond, deduce how many amino acid units are present in cyclosporin A.
    b. Three of the units are identical. What are they?
    c. Two of the units are unusual and are *not* listed in Table 17.1. What are their structures?
    d. There are only three other types of amino acids present, though there may be more than one unit of each. What are they, and how many units are there of each?

17.44 Verify that the one-letter representation of bradykinin (page 491) is RPPGFSPFR. What is the one-letter representation of substance P (page 491)?

Primary Structure of Peptides and Proteins

17.45 A pentapeptide was converted to its 2,4-dinitrophenyl (DNP) derivative, then completely hydrolyzed and analyzed quantitatively. It gave DNP-methionine, 2 moles of methionine, and 1 mole each of serine and glycine. The peptide was then partially hydrolyzed, the fragments were converted to their DNP derivatives, and each of them was hydrolyzed and analyzed quantitatively. Two tripeptides and two dipeptides isolated in this way gave the following products:

Tripeptide A: DNP-methionine and 1 mole each of methionine and glycine
Tripeptide B: DNP-methionine and 1 mole each of methionine and serine
Dipeptide C: DNP-methionine and 1 mole of methionine
Dipeptide D: DNP-serine and 1 mole of methionine

Deduce the structure of the original pentapeptide, and explain your reasoning.

17.46 Write the equations for the removal of one amino acid from the peptide alanylglycylvaline by the Edman method. What is the name of the remaining dipeptide?

17.47 Express the B chain of insulin (Figure 17.6) using one-letter abbreviations for the amino acids, and compare it with the space taken by the three-letter abbreviations on page 498.

17.48 Insulin (Figure 17.6), when subjected to the Edman degradation, gives *two* phenylthiohydantoins. From which amino acids are they derived? Draw their structures.

17.49 The following compounds are isolated as hydrolysis products of a peptide: Ala—Gly, Tyr—Cys—Phe, Phe—Leu—Trp, Cys—Phe—Leu, Val—Tyr—Cys, Gly—Val, and Gly—Val—Tyr. Complete hydrolysis of the peptide shows that it contains one unit of each amino acid. What is the structure of the peptide, and what are its N- and C-terminal amino acids?

17.50 Simple pentapeptides called *enkephalins* are abundant in certain nerve terminals. They have opiate-like activity and are probably involved in organizing sensory information pertaining to pain. An example is *methionine enkephalin*, Tyr— Gly— Gly—Phe—Met. Write out its complete structure, including all the side chains.

17.51 Angiotensin II is an octapeptide with vasoconstrictor activity. Complete hydrolysis gives one equivalent each of Arg, Asp, His, Ile, Phe, Pro, Tyr, and Val. Reaction with Sanger's reagent gives, after hydrolysis,

$$
\text{O}_2\text{N} \overset{\displaystyle \text{NO}_2}{-\!\!\!\!\bigcirc\!\!\!\!-} \text{NHCHCO}_2\text{H}
$$
$$
\underset{\displaystyle \text{CH}_2\text{CO}_2\text{H}}{|}
$$

and seven amino acids. Treatment with carboxypeptidase gives Phe as the first released amino acid. Treatment with trypsin gives a dipeptide and a hexapeptide, whereas with chymotrypsin, two tetrapeptides are formed. One of these tetrapeptides, by Edman degradation, had the sequence Ile—His—Pro—Phe. From these data, deduce the complete sequence of angiotensin II.

17.52 *Endorphins* were isolated from the pituitary gland in 1976. They are potent pain relievers. β-Endorphin is a polypeptide containing 32 amino acid residues. Digestion of β-endorphin with trypsin gave the following fragments:

Lys
Gly—Gln
Asn—Al—His—Lys
Asn—Ala—Ile—Val—Lys
Tyr—Gly—Gly—Phe—Leu—Met—Thr—Ser—Glu—Lys
Ser—Gln—Thr—Pro—Leu—Val—Thr—Leu—Phe—Lys

From these data only, what is the C-terminal amino acid of β-endorphin?

Treatment with cyanogen bromide gave the hexapeptide

Tyr—Gly—Gly—Phe—Leu—Met

and a 26 amino acid fragment. From these data only, what is the N-terminal amino acid of β-endorphin? Digestion of β-endorphin with chymotrypsin gave, among other fragments, a 15-unit fragment identified as

Leu—Met—Thr—Ser—Glu—Lys—Ser—Gln—Thr—Pro—Leu—Val—Thr—Leu—Phe

You should now be able to locate 22 of the 32 amino acid units. Write out as much as you can of the sequence. What further information do you need to complete the sequence?

17.53 *Glucagon* is a polypeptide hormone secreted by the pancreas when the blood sugar level is low. It increases the blood sugar level by stimulating the breakdown of glycogen in the liver. The primary structure of glucagon is

His—Ser—Glu—Gly—Thr—Phe—Thr—Ser—Asp—Tyr—Ser—Lys—Tyr—Leu—Asp—Ser—Arg—Arg—Ala— Gln—Asp—Phe—Val—Gln—Trp—Leu—Met—Asn—Thr

What fragments would you expect to obtain from digestion of glucagon with

a. trypsin? **b.** chymotrypsin?

Peptide Synthesis

17.54 The attachment of the N-protected C-terminal amino acid to the polymer in solid-phase peptide synthesis (Figure 17.7) is an S_N2 displacement reaction. What is the nucleophile? What is the leaving group? Write an equation that clearly shows the reaction mechanism.

17.55 The detachment of the peptide chain from the polymer in solid-phase peptide synthesis (Figure 17.7) occurs by an acid-catalyzed S_N2 mechanism. Write an equation to show this mechanism.

17.56 Write out all the steps in a Merrifield solid-phase synthesis of Leu—Pro.

Protein Structure

17.57 Write the structure for glycylglycine, and show the resonance contributors to the peptide bond. At which bond is rotation restricted?

17.58 In a globular protein, which of the following amino acid's side chains are likely to point toward the center of the structure? Which will point toward the surface when the protein is dissolved in water?

a. arginine   **b.** phenylalanine   **c.** isoleucine
d. glutamic acid   **e.** asparagine   **f.** tyrosine

PHOTO CREDITS

APPENDIX

Table A Bond energies for the dissociation of selected bonds in the reaction $A-X \rightarrow A \cdot + X \cdot$ (in kcal/mol)

I. Single bonds — Bond energies (kcal/mol)

A—X	X = H	F	Cl	Br	I	OH	NH₂	CH₃	CN
CH₃—X	105	108	84	70	57	92	85	90	122
CH₃CH₂—X	100	108	80	68	53	94	84	88	
(CH₃)₂CH—X	96	107	81	68	54	94	84	86	
(CH₃)₃C—X	96		82	68	51	93	82	84	
H—X	104	136	103	88	71	119	107	105	124
X—X	104	38	59	46	36			90	
Ph—X	111	126	96	81	65	111	102	101	
CH₃C(O)—X	86	119	81	67	50	106	96	81	
H₂C=CH—X	106								
HC≡C—X	132								

II. Multiple bonds — Bond energies (kcal/mol)

H₂C=CH₂	163
HC≡CH	230
H₂C=NH	154
HC≡N	224
H₂C=O	175
C≡O	257

Table B Bond lengths of selected bonds (in angstroms, Å)

I. Single bonds		II. Double bonds				III. Triple bonds	
Bond	Length (Å)	Bond	Length (Å)	Bond	Length (Å)	Bond	Length (Å)
H—H	0.74	H—C=	1.08	C=C	1.33	C≡C	1.20
H—F	0.92	H—Ph	1.08	C=O	1.21	C≡N	1.16
H—Cl	1.27	H—C≡	1.06			C≡O	1.13
H—Br	1.41	C—C	1.54				
H—I	1.61	C—N	1.47				
H—OH	0.96	C—O	1.43				
H—NH₂	1.01	C—F	1.38				
H—CH₃	1.09	C—Cl	1.77				
F—F	1.42	C—Br	1.94				
Cl—Cl	1.98	C—I	2.21				
Br—Br	2.29						
I—I	2.66						

Table C **Typical acidities of organic functional groups**

(Left scale: Strong Acid at top → Weak Acid at bottom; right scale: Weak Base at top → Strong Base at bottom)

Name and Example*	pK_a	Conjugate Base
Hydrochloric acid, **HCl**	−7	Cl^-
Sulfuric acid, $\mathbf{H}_2SO_4$	−3	HSO_4^-
Sulfonic acid	0–2	
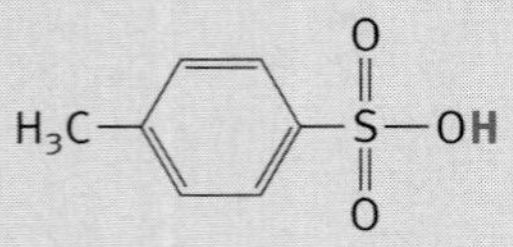	−1	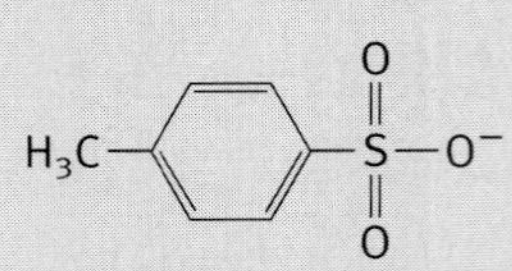
Carboxylic acid	3–5	
CH_3—C(=O)—**OH**	4.74	CH_3—C(=O)—O^-
Arylammonium ion	4–5	
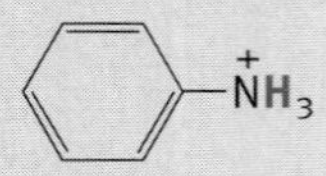	4.6	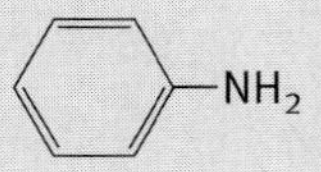
Ammonium ion, $\overset{+}{N}\mathbf{H}_4$	9.3	NH_3
Phenol	9–10	
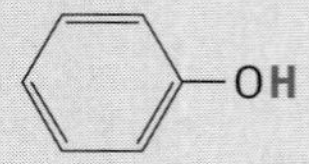	10	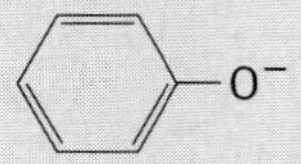
β-diketone	9–10	
CH_3—C(=O)—C**H_2**—C(=O)—CH_3	9	CH_3—C(=O)—$\bar{C}H$—C(=O)—CH_3
Thiol	8–12	
CH_3CH_2S**H**	10.6	$CH_3CH_2S^-$
β-ketoester	10–11	
CH_3—C(=O)—C**H_2**—C(=O)—OCH_2CH_3	10.7	CH_3—C(=O)—$\bar{C}H$—C(=O)—OCH_2CH_3
Alkylammonium ion	10–12	
$CH_3CH_2\overset{+}{N}\mathbf{H}_3$	10.7	$CH_3CH_2NH_2$
Water, $\mathbf{H}_2O$	15.7	HO^-

Table C	Typical acidities of organic functional groups (continued)		

Name and Example*	pK_a		Conjugate Base
Alcohol	15–19		
CH_3CH_2OH	15.9		$CH_3CH_2O^-$
Amide	15–19		
$CH_3-\overset{O}{\overset{\|}{C}}-NH_2$	15		$CH_3-\overset{O}{\overset{\|}{C}}-\bar{N}H$
Aldehyde, ketone	17–20		
$CH_3-\overset{O}{\overset{\|}{C}}-CH_3$	19		$CH_3-\overset{O}{\overset{\|}{C}}-\bar{C}H_2$
Ester	23–25		
$CH_3-\overset{O}{\overset{\|}{C}}-OCH_2CH_3$	24.5		$\bar{C}H_2-\overset{O}{\overset{\|}{C}}-OCH_2CH_3$
Alkyne	23–25		
$H-C\equiv C-H$	24		$H-C\equiv C^-$
Ammonia, NH_3	33		$\bar{N}H_2$
Hydrogen, H_2	35		H^-
Alkylamine	~40		
$\langle\text{cyclohexyl}\rangle-NH_2$	42		$\langle\text{cyclohexyl}\rangle-\bar{N}H$
Alkene	~45		
$H_2C=CH_2$	44		$H_2C=\bar{C}H$
Aromatic hydrocarbon	41–43		
$\langle\text{aryl}\rangle-H$	43		$\langle\text{aryl}\rangle^-$
Alkane	50–60		
CH_4	50		$\bar{C}H_3$

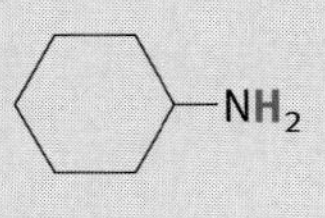

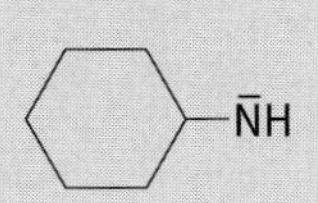

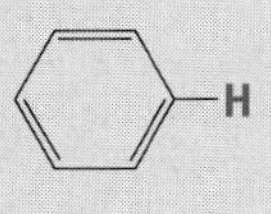

Left margin: Strong Acid ↑   Weak Acid ↓

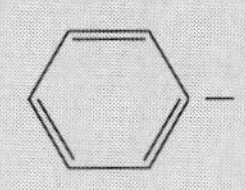

Right margin: Weak Base ↓   Strong Base ↓

*Some inorganic acids are included for comparison.

INDEX/GLOSSARY

Key terms, which appear in **boldface,** *are followed by their definitions.*

aldehydes by, 277
aromatic ketones by, 254
Frogs, alkaloids in, 334
Fructofuranose, 485
Fructose, 448, 458, 465
FT spectrometer. *See Fourier transform (FT-NMR) spectrometer*
Fuel oil, 106 (Table 3.3)
Fuels
alkanes as, 61–62
automotive, 107
Fuller, R. Buckminster, 139
Fullerene derivatives, 139
Fullerenes, 138–139
Fumaric acid, 286
Functional group bands   *Absorptions in IR spectroscopy that indicate the presence of a specific functional group, such as carbonyl or hydroxyl groups,* 364
Functional groups   *Groups of atoms that have characteristic chemical properties regardless of the molecular framework to which they are attached: usually characteristic of a class of organic compounds—for example, C═C for alkenes and C—O—C for ethers,* 33–34, 36–37
Furan, 388–389, 390–391
Furanose   *Monosaccharide with a five-membered ring oxygen heterocycle,* 457
Furfural, 388
2-Furfuraldehyde, 388
Fused polycyclic hydrocarbons, 140
Fused-ring five-membered heterocyclic compounds, 393, 398–399

Galactose, 452
Galactosemia, 465
Gallstones, 443
Garlic, 226
Gas, arrow symbol for, 269
Gasoline, 107, 237
General anesthetics, 240
Genetic code   *The relationship between the base sequence in DNA and the amino acid sequence in a protein,* 531–534
Genetics
genetic code, 531–534
human genome, 531
mutations, 531
Geodesic domes, 139
Geometric isomers, 60, 82
Geometric stereoisomerism, 58
Geraniol, 33, 219
Gibbs free-energy difference, 90
Gilbert, W., 522
Globin, 392
Globular proteins   *Proteins that tend to be spherical in shape and are soluble in water—for example, some enzymes and hormones,* 507–508
Glucitol, 460
Glucofuranose, 457
Glucopyranose, 458, 459, 461

Glucopyranose pentaacetate, 459
Glucopyranose pentamethyl ether, 459
Glucosamine, 472
Glucose, 448, 449
acyclic form, 460, 474
ascorbic acid from, 473
chair conformation of, 58, 460
diastereomers of, 456
Fischer projection formula for, 67, 454
formation of, 449
furanose form, 457
Haworth formula for, 455
hemiacetal structure of, 260, 455–456, 474
from hydrolysis of sucrose, 465
pyranose form, 458
reaction with methanol, 461
rotation in solution, 457
stereochemistry of, 451, 456
Glucosidic link
stereochemistry of, 469
Glutamic acid, 481 (Table 17.1), 486
Glutamine, 481 (Table 17.1)
Glutaric acid, 286 (Table 10.2)
Glyceraldehyde, 251, 449, 450, 452
Glycerol, 220, 427–431, 431, 449
Glyceryl dipalmitoöleate, 430
Glyceryl stearopalmitoöleate, 429
Glyceryl trilaurate, 435
Glyceryl trinitrate, 220
Glyceryl trioleate, 431
Glyceryl tripalmitate, 430, 431
Glyceryl tristearate, 431
Glycine, 480 (Table 17.1)
Glycogen, 467
Glycols   *Compounds with two hydroxyl groups on adjacent carbons,* 101, 220–221
cyclic acetals from, 260–261
from epoxides, 243
Glycosides   *Acetals derived by replacing the anomeric hydroxyl (OH) group of a monosaccharide with an alkoxy (OR) group,* 461–463, 475
Glycosidic bond   *The bond from the anomeric carbon to the alkoxy (OR) group in a glycoside,* 461
Glycylalanine, 490, 494
Goodyear, C., 412
Gore-Tex, 197
Gout, 394
Graft copolymers, 414
Graphite, 138
Grass, color of, 392
Green chemistry, 295
Grignard, V., 234
Grignard reagent(s)   *Alkylmagnesium halides; a kind of organometallic compound,* 244
addition to aldehydes and ketones, 262–264, 278
carboxylic acids from, 293
nomenclature of, 235
preparation of, 234–236, 246
reaction with acyl halides, 312
reaction with carbon dioxide, 293, 315

reaction with esters, 302, 312, 315
structure of, 234–236
Guanidine group, 487
Guanine, 395, 516, 517, 519, 524
Gulose, 452
Gum arabic, 471
Gum benzoin, 116
Guncotton, 471
Gutta-percha, 412

Hair
chemical composition and reactions of, 225
Half-headed arrow, 65
Halides, 187, 190, 194–195
See also Alkyl halides
Halo group, 133 (Table 4.1)
Halogen
functional groups containing, 36 (Table 1.6)
Halogen compounds, organic, 180–200
biological methylations, 192
dehydrohalogenation of, 193–194
nucleophilic substitution reactions of, 180–192, 199
polyhalogenated aliphatic compounds, 196–197
from the sea, 199
Halogen nucleophiles, 181, 183 (Table 6.1)
Halogen substituents
alkanes, 47
cycloalkanes, 54
Halogenation
α-halogenation, 276
of acetone, 276
of alkanes, 63, 65–66, 67, 82
of alkenes, 83, 109
of alkynes, 105, 107–108
of benzene, 117, 125, 127, 141
free-radical chain mechanism of, 65–66
Halomon, 199
Halons   *Polyhalogenated compounds containing bromine, chlorine, and fluorine,* 197
Halothane, 240
Hard water, 436
Hardening   *The process of converting oils to fats by catalytic hydrogenation of double bonds,* 431, 445
Haworth projection, 455
Haworth, W. N., 455
Head-to-tail, 1,4-addition, 412
Heavy water, 236
Heeger, A., 411
Helium, 8
Heme, 392, 508
Hemiacetal(s)   *A compound with one alkoxy group and one hydroxyl group bonded to the same carbon atom,* 258–260, 454
Hemiacetal carbon, 456
Hemoglobin, 372–373, 392, 508
Heparin, 471
Heptane, 33, 43 (Table 2.1), 107
Heptanedioic acid, 286 (Table 10.2)